T0305551

Wastewater Treatment

Wastewater Treatment: Recycling, Management, and Valorization of Industrial Solid Wastes bridges the gap between the theory and applications of wastewater treatments, principles of diffusion, and the mechanism of biological and industrial treatment processes. It presents the practical applications that illustrate the treatment of several types of data, providing an overview of the characterization and treatment of wastewaters, and then examining the different biomaterials and methods for the evaluation of the treatment of biological wastewaters. Further, it considers the various types of industrial wastewater treatment, separation, and characterization of industrial wastewater. The book serves as a valuable resource for practicing engineers and students who are interested in the field of wastewater treatment.

Features:

- Presents the latest technologies in water treatment, including nanomaterials for industrial wastewater
- Covers different treatments for various industrial wastewaters, including chemical and pharmaceutical waste
- Includes forward-thinking analysis including conclusions and recommendations for water reuse programs

Wastewater Treatment

Recycling, Management, and Valorization of Industrial Solid Wastes

Edited by

Irene Samy Fahim and Lobna A. Said

CRC Press is an imprint of the
Taylor & Francis Group, an informa business

Designed cover image: Shutterstock

First edition published 2023
by CRC Press
6000 Broken Sound Parkway NW, Suite 300, Boca Raton, FL 33487-2742

and by CRC Press
4 Park Square, Milton Park, Abingdon, Oxon, OX14 4RN

CRC Press is an imprint of Taylor & Francis Group, LLC

Library of Congress Cataloging-in-Publication Data

Names: Samy Fahim, Irene, editor. | Said, Lobna, editor.
Title: Wastewater treatment : recycling, management, and valorization of industrial solid wastes / edited by Irene Samy Fahim and Lobna Said. Other titles: Wastewater treatment (CRC Press : 2023)
Description: First edition. | Boca Raton : CRC Press, [2023] | Includes bibliographical references and index. | Identifiers: LCCN 2022055157 (print) | LCCN 2022055158 (ebook) | ISBN 9781032404691 (hbk) | ISBN 9781032407227 (pbk) | ISBN 9781003354475 (ebk)
Subjects: LCSH: Sewage--Purification. | Factory and trade waste. | Recycling (Waste, etc.)
Classification: LCC TD430 .W36274 2023 (print) | LCC TD430 (ebook) | DDC 628.3--dc23/eng/20221206
LC record available at https://lccn.loc.gov/2022055157
LC ebook record available at https://lccn.loc.gov/2022055158

ISBN: 978-1-032-40469-1 (hbk)
ISBN: 978-1-032-40722-7 (pbk)
ISBN: 978-1-003-35447-5 (ebk)

DOI: 10.1201/9781003354475

Typeset in Times
by KnowledgeWorks Global Ltd.

Contents

About the Editors

Irene Samy Fahim is an assistant professor in the Industrial and Service Engineering and Management department, Nile University (NU), Cairo, Egypt, and the American University in Cairo. Irene has several scientific publications related to investigation of natural fiber-reinforced polymers, reinforcement of plastic waste with treated natural fibers, and characterization of natural polymeric nanocomposites. She participated in the Fulbright Junior Faculty program for renewable energy, 2016, and participated in Entrepreneurship and Leadership Program, 1000Women, Goldman Sachs, 2016. She received the Newton Mosharfa institutional link award for two years in collaboration with Nottingham University, UK for manufacturing plastic bags from natural materials. Irene is interested in studying sustainability concepts and is a volunteer in the IEEE Smart Village committee. She is also an active member in the organizing committee for IEEE Conferences such as IEEE Power Africa 2017, where Irene plans to extend the IEEE Smart Village committee in Egypt to supply solar electricity in poor villages. She is also one of the members in the Events Committee of the IEEE Humanitarian Activities, 2018. She is also the technical chair for the first IEEE SIGHT Egypt ideation camp 2018.

Lobna A. Said is a full-time assistant professor at the Faculty of Engineering and Applied Science and the Nano-Electronics Integrated System Research Center (NISC), NU. She received B.Sc., M.Sc., and Ph.D. degrees in electronics and communications from Cairo University, Egypt, in 2007, 2011, and 2016, respectively. She has an H-index of 15, and more than 660 citations based on the Scopus database. She has over 85 publications distributed between high-impact journals, conferences, and book chapters. She was involved in many research grants as a senior researcher, or as a Co-PI from different national organizations. Her research interests are interdisciplinary, including system modeling, control techniques, optimization techniques, analog and digital integrated circuits, fractional-order circuits and systems, nonlinear analysis, and chaos theory. She has received the Recognized Reviewer Award from many international journals. She is the vice-chair of research activities at the IEEE Computational Intelligence Egypt Chapter. She has received the Excellence Award from the Center for the Development of Higher Education and Research in 2016. She is the winner of Dr. Hazem Ezzat Prize for the outstanding researcher, NU 2019. She is one of the top ten researchers at NU for the year 2018–2019. Recently, she was selected as a member of the Egyptian Young Academy of Sciences (EYAS) to empower and encourage young Egyptian scientists in science and technology and build knowledge-based societies. In 2020, she was selected to be an affiliate member of the African Academy of Science (AAS). She is in the technical program committee for many international conferences.

About the Editors

Contributors

A. B. Abdel-Aziz
October High Institute for Engineering and Technology
Giza, Egypt

Enas M. Abou-Taleb
National Research Centre
Giza, Egypt

Imran Ahmad
Universiti Teknologi Malaysia (UTM)
Kuala Lumpur, Malaysia

P. B. Irénikatché Akponikpè
University of Parakou
Parakou, Benin

Khloud Ahmed
Nile University
Giza, Egypt

Nour F. Attia
National Institute of Standards
Giza, Egypt

Xiang Chen
Guangdong University of Technology
Guangzhou, PR China

Shyma Mohamed El Saeed
Egyptian Petroleum Research Institute
Cairo, Egypt

Sally E. A. Elashery
Cairo University
Giza, Egypt

Ghadir A. El-Chaghaby
Agricultural Research Center
Giza, Egypt

Eman M. Abd El-Monaem
Alexandria University
Alexandria, Egypt

Reem M. Eltaweel
Nile University
Giza, Egypt

Abdelazeem S. Eltaweil
Alexandria University
Alexandria, Egypt

Irene Samy Fahim
Nile University
Giza, Egypt

Nesreen A. Fatthallah
Egyptian Petroleum Research Institute
Cairo, Egypt

El Sayed Gamal Zaki
Egyptian Petroleum Research Institute
Cairo, Egypt

Zhifeng Hao
Guangdong University of Technology
Guangzhou, PR China

Hassan Hefni Hassan
Egyptian Petroleum Research Institute
Cairo, Egypt

Mohamed S. Hellal
National Research Centre
Giza, Egypt

Shimaa A. Higazy
Egyptian Petroleum Research Institute
Cairo, Egypt

Shima Husien
Nile University
Giza, Egypt

G. Esaie Kpadonou
University of Parakou
Parakou, Benin

Hekmat R. Madian
Egyptian Petroleum Research Institute
Cairo, Egypt

Nermine El Sayed Maysour Refaat
Egyptian Petroleum Research Institute
Cairo, Egypt

Nagwan G. Mostafa
Cairo University
Giza, Egypt

Kalyani Rajashree Naik
Indian Institute of Technology
Kharagpur, India

Ashish Kumar Nayak
Indian Institute of Technology
Kharagpur, India

Anjali Pal
Indian Institute of Technology
Kharagpur, India

Ahmed G. Radwan
Cairo University
and
Nile University
Giza, Egypt

Sayed Rashad
Agricultural Research Center
Giza, Egypt

Pushpa Ruwali
M. B. Govt. P. G. College
Haldwani, India

Lobna A. Said
Nile University
Giza, Egypt

Alyaa I. Salim
Nile University
Giza, Egypt

Krishika Sambyal
Chandigarh University
Gharuan, Punjab, India

Mohamed S. Selim
Egyptian Petroleum Research Institute
Cairo, Egypt
and
Guangdong University of Technology
Guangzhou, PR China

Rahul Vikram Singh
Academy of Scientific and Innovative Research (AcSIR)
Ghaziabad, India

Pierre G. Tovihoudji
University of Parakou
Parakou, Benin

Sissou Zakari
University of Parakou
Parakou, Benin

1 Water Importance and Pollution Sources—Recommended Limits of Pollutants

A. B. Abdel-Aziz, Alyaa I. Salim, and Irene Samy Fahim

1.1 INTRODUCTION

Water is one of the most important resources for human life on Earth. Water quality depends on the geological environment, reproduction, and human activities such as domestic, industrial, and agricultural uses [1]. Providing clean water with specific properties is essential to industry, although that produces large amounts of wastewater contaminated with various toxic materials. The massive industrial development all over the world leads to the production of industrial wastewater with high concentrations of many harmful chemicals, suspended solids (SS), oxygen demand (OD), and pungent odors. To counter this massive pollution in industrial wastewater effluents, hundreds of new economies are employed every year. The main industries that cause this damage are oil, leather, dyes, coal, plastic, rubber, paint, and pharmaceuticals. The most common water pollution factors are illustrated in Figure 1.1. In addition, treated wastewater recycling plays an important role in the global water crisis [2]. Recycled wastewater has many uses in various fields according to the grade of treatment like irrigation and generating energy.

1.2 WASTEWATER CLASSIFICATION

Wastewater is classified into two main categories:

1.2.1 Domestic Wastewater

Domestic wastewater refers to the flow that is discharged from human activities such as food preparation, clothing, and cleaning [3], which are produced in huge volumes and are abundant with carbon, nitrogen, and phosphorous [4].

DOI: 10.1201/9781003354475-1

FIGURE 1.1 Sources of water pollutants.

1.2.2 Industrial Wastewater

Industrial wastewater refers to the flow from manufacturing activities such as foods, cosmetics, and paints [5–7].

There are various chemical and biological pollution species present in water with different concentrations, as shown in Figure 1.2. They are defined as micro-pollutants, which are hazardous to human health [8, 9]. The damages of these pollutants are also summarized in Table 1.1.

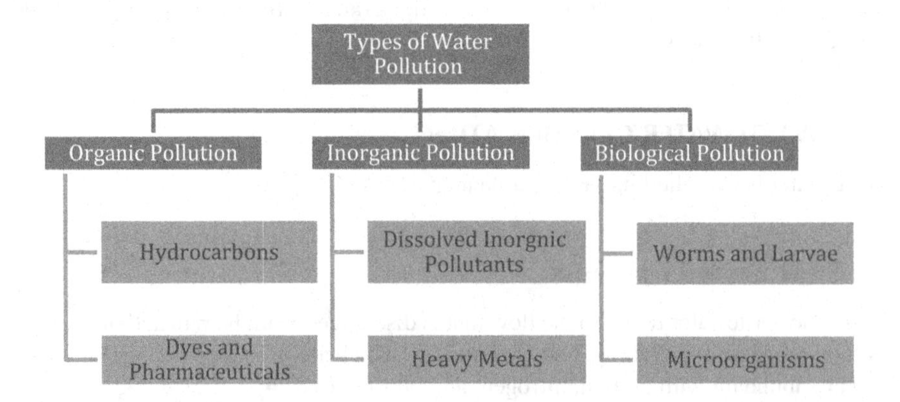

FIGURE 1.2 Types of water pollution.

TABLE 1.1
Water Important Pollutants and Their Issues [10]

Pollutant	Issues
Heavy metals	Highly toxic and environmentally hazardous
Suspended solids	Sludge deposits
Dissolved inorganic constituents (e.g., Ca, Na, and SO_4)	Initially added to domestic water supplies and may have to be removed from wastewater to reuse
Pathogenic organisms	Infection diseases
Refractory organics (e.g., surfactants, phenols)	Resist conventional wastewater treatment

Dyes, paints, and textile wastewater are generated in large amounts. Therefore, small amounts of dye particles in water can cause visual impairment, color shifting, prevent sunlight from entering the water, and cause adverse effects on aquatic plants. It has been proven that the treatment of these pollutants is extremely important to treat the wastewater completely from all species of pollutants [11, 12].

1.3 DYE, PAINT, AND TEXTILE MANUFACTURING

1.3.1 Dyes

Dyes are compounds that change the color of a substrate and are used in fabrics, paper, plastic, paints, pharmaceuticals, foods, cosmetics, leather, and other materials. Dyes give the crystalline form of the colored substances [13, 14]. The pigment is a special type of dye which has different particle size. Additionally, dyes comprise synthetic organic compounds that might be water soluble or organic soluble, while pigments have extremely low solubility and stay in particulate form. Therefore, dyes are not ultraviolet (UV) stable, whereas pigments are usually UV stable [15]. Dyes can adhere to other substrate surfaces by physical adsorption, mechanical forces, covalent bonds, or by making complexes with the surface of the substrate. The color of dyes depends on the amount of absorbed visible light.

1.3.1.1 Dye Classification

Dyes have many classifications, according to their solubility, applications, and chemical structure [16]:

1. Dyes may be soluble in water or organic solvents as shown in Table 1.2 [17].
2. Dyes are widely used to apply color on different materials, but this literature is mainly focused on textiles and paints. Textiles have various types of fibers, and each one of these fibers can be dyed with a different type of dye like cationic dyes, anionic dyes, organic soluble dyes, and pigments [17]. These textile substrates include cellulose (cotton and linen), cellulose ester, protein (wool and silk), polypropylene, polyacrylics, polyesters, polyamides, and polyvinyl chloride. Generally, basic soluble dyes are used in

TABLE 1.2
Classification of Dyes according to Their Solubility

	Water Soluble							Water Insoluble					
	Cationic	Anionic						Soluble in Organic Solvents					Low Soluble in Organic Solvents
Dye Type	Basic	Acidic	Mordant	Direct	Reactive	Solubilized Vat	Sulfur	Disperse	Solvent	Mordant	Vat	Sulfur	Pigment
Azo	√	√	√	√	√	---	---	√	√	√	---	---	√
Di- and triphenyl methane	√	√	---	---	---	---	---	---	√	√	---	---	√
Xanthene	√	√	---	---	---	---	---	---	√	√	---	---	√
Acridine	√	---	---	---	---	---	---	---	---	---	---	---	√
Quinoline	√	√	---	√	---	---	---	---	√	---	---	---	√
Methine	√	---	---	√	---	---	---	√	---	---	---	---	√
Azine	√	√	---	---	---	---	---	---	√	---	---	---	---
Thiazine	√	---	---	---	---	---	---	---	√	√	√	---	---
Sulfur	---	---	---	---	---	---	√	---	---	---	---	√	---
Anthraquinone	√	√	√	---	√	√	---	√	√	√	√	---	√
Indigoid	---	√	---	---	---	√	---	---	---	---	√	---	√
Phthalocyanine	---	√	---	√	√		---	---	√	---	---	---	√

√: Soluble; ---: Insoluble

dyeing inks and modified record systems while acidic soluble dyes are used in foods, drugs, cosmetics, varnishes, inks, plastics, and resins. On the other hand, disperse dyes are used on plastics and pigments, inks, cosmetics, waxes, and carbon paper.

3. Dyes are classified according to their chemical structure. Dyes may include groups like nitro, nitroso, azoic, triphenylmethyl, phthalein, indigoid, and anthraquinone dyes as illustrated in Figure 1.3 [16].

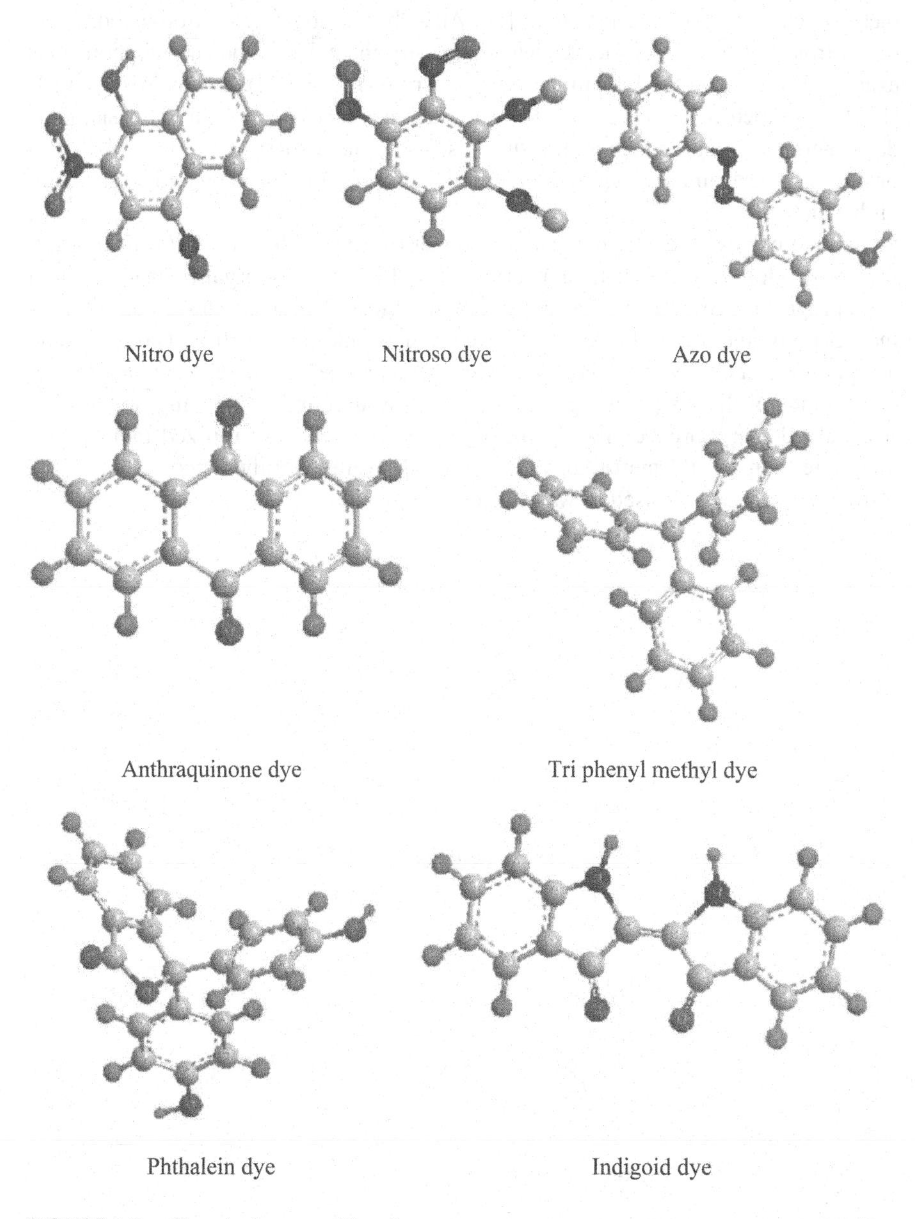

FIGURE 1.3 Chemical composition dyes.

1.3.1.2 Synthetic Dyes

Synthetic dyes consist of two major groups: (1) chromophore and (2) auxochrome. Chromophore is a group of atoms that controls the dye color by absorbing and emitting the visible light. These chromophores contain centers that are based on diverse functional groups of unsaturated bonds which absorb the energy of visible light, excite, and then emit this energy again coming back to their unexcited energy levels appearing the color to eyes, like azo (–N=N–), anthraquinone, methine (–CH=), nitro (–NO_2), aril methane, carbonyl (–C=O), and others [18]. Auxochrome contains electron-withdrawing or electron-donating groups to deepen or intensify the color of the chromophores, for example, sulfonate (–SO_3H), amine (–NH_2), carboxyl (–COOH), and hydroxyl (–OH) [14, 18]. Auxochromes include many types such as reactive, acid, direct, basic, mordant, pigment, vat, anionic, ingrain, sulfur, solvent, and disperse dye. Figure 1.4 demonstrates some examples of chromophores and auxochromes in the chemical structure of dye molecules.

Synthetic dyes have common applications in various fields like paints and textiles. Dyes give a long-wearing color appearance with highly stable organic structures and inert properties. Synthetic dyes are among the most significant classes of environmental pollutants that adversely influence human and aquatic life. The genotoxic and carcinogenic effects of diazo dyes from industrial effluents represent a series of environmental threats polluting the aquatic ecosystem and influencing human life. It is valued that moreover 10^4 distinct dyes and pigments are utilized industrially, and more than 7×10^5 metric tons of dyes are delivered worldwide consistently and 5–10% of them are released in wastewater [19].

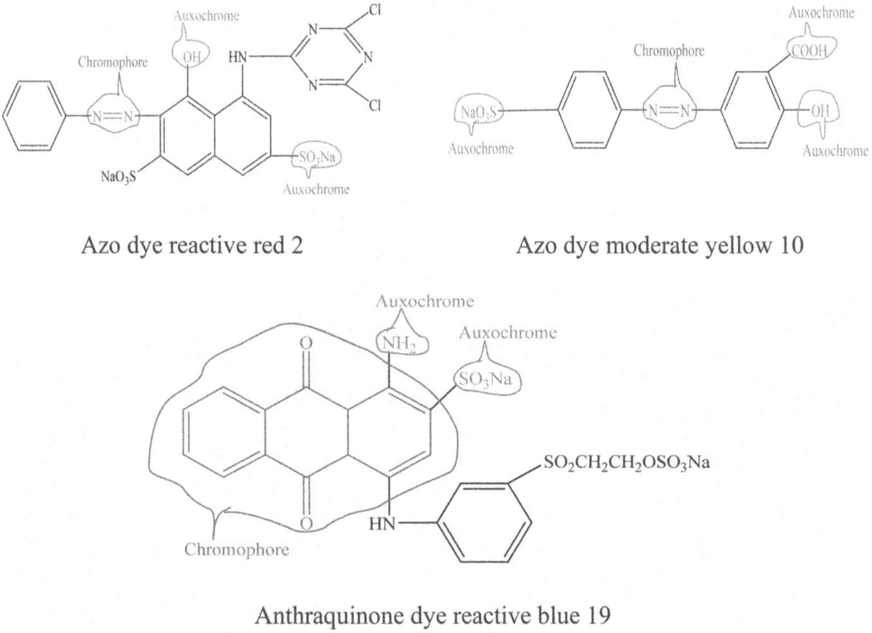

Azo dye reactive red 2

Azo dye moderate yellow 10

Anthraquinone dye reactive blue 19

FIGURE 1.4 Chromophores and auxochromes for azo and anthraquinone dyes.

1.3.1.3 Azo Dyes

Azo dyes account for about 50% of dyestuff effluents of various industries such as textiles, leather, food, and paper. Azo dyes are the largest and the most widely recognized production volume of synthetic dyes, which has (–N=N–) as a chromophore group, as shown in Figure 1.5 [14, 20]. Azo dyes are synthesized from an aromatic primary amine using the diazotization process, followed by attachment with one or more electron-rich groups such as amine and hydroxyl as shown in Eq. (1.1). Several active groups, such as –OH, $-NH_2$, and –SH, make covalent bonds with fibers of textile substrates (wool, cotton, silk, and nylon). They are mainly used for the colors yellow, orange, and red [18]. For certain colors, as a rule, a mixture of red, yellow, and blue dyes is applied in the dye baths. This mixture can contain a wide range of chromophores, of which azo, phthalocyanine, and anthraquinone dyes are among the most important groups. Azo dye molecules may be monoazo, diazo, triazo, or polyazo dye molecules, which indicate the number of azo groups in the same molecule [21, 22].

$$\underset{}{R{-}NH_2} \xrightarrow[\substack{NaNO_2 \\ \text{Diazotization}}]{HCl} \underset{\text{Diazonium compound}}{[R{-}N{=}N]^+ Cl^-} \xrightarrow[\substack{NaOH \\ \text{Coupling}}]{R{-}H} \underset{\text{Azo colorant}}{R{-}N{=}N{-}R} \tag{1.1}$$

The annual production of azo dyes into wastewater effluents is estimated at one million tons, which affect the photosynthesis process of living organisms [23]. Moreover, the partial degradation of the azo dye molecules leads to form a carcinogenic compound, so that the degradation of azo dyes requires an additional step after the reduction of azo groups. Microbial degradation is necessary to treat water effluents from the hazardous aromatic amines produced from the degradation of azo dyes. As well, azo dyes have harmful effects on humans and aquatic life. These have influenced urgent calls to process azo dyestuff to eliminate them or convert them into beneficial and safe products.

1.3.1.4 Anthraquinone Dyes

Anthraquinone dyes are the second most important textile dyes after azo dyes and have a wide range of colors in almost the entire visible spectrum. They are generally used for the colors violet, blue, and green. They have a general formula of anthracene

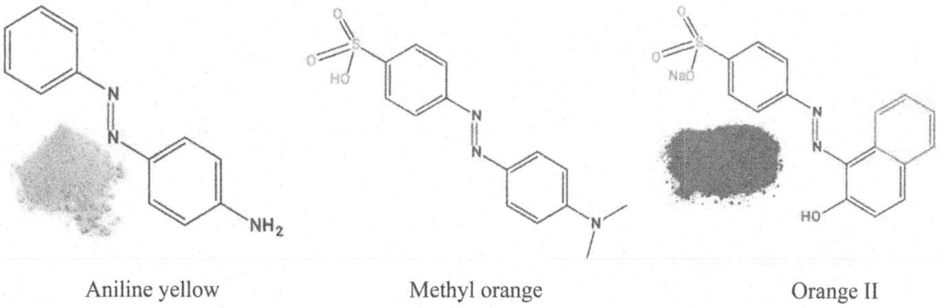

Aniline yellow Methyl orange Orange II

FIGURE 1.5 Azo dyes.

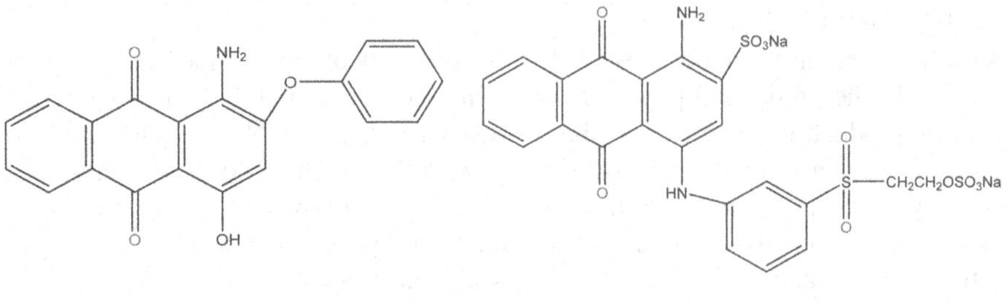

C.I Disperse red 60

C.I Reactive blue 19

FIGURE 1.6 Anthraquinone dyes.

with quinone nucleus as a chromophore group and are attached to another auxochrome group such as $-NH_2$ or $-OH$ as shown in Figure 1.6 [18, 22, 24].

1.3.1.5 Naming Dyes

There are about 8,000 synthetic dyes that are listed by Color Index (CI) number, under 40,000 trade names. The CI number is a five-digit code used to identify these substances worldwide. The first two digits indicate the structural category of the dye. Each dye is categorizing according to a code name, which specifies its color, type, and order number [25]. For example, the Reactive Black 5 is a di azo dye, which was given as a representative example to show the color index classification as shown in Figure 1.7. Reactive Black 5 molecular formula is $C_{26}H_{21}N_5Na_4O_{19}S_6$, molar mass is 991.82 g/mol., and its molecular structure is di azo dye. The CI number is 20505; CAS registry number is 12225-25-1/17095-24-8. Its application class is reactive, and commercial names are Reactive Black 5, Remazol black B, Remazol Black GF 17095-24-8, and Diamira Black B.

1.3.1.6 Dye Wastewater

Dye wastewater is viewed as one of the most destructive carcinogenic substances for humans and aquatic organisms [26]. This is due to the fact that the dye not only causes infection but also causes eutrophication and unattractive pollution. The dye is stable and can stay in the environment for a long time. For example, the half-life of hydrolyzed reactive blue 19 (RB19) is about 46 years at pH 7 and 25°C [27]. Dye wastewater

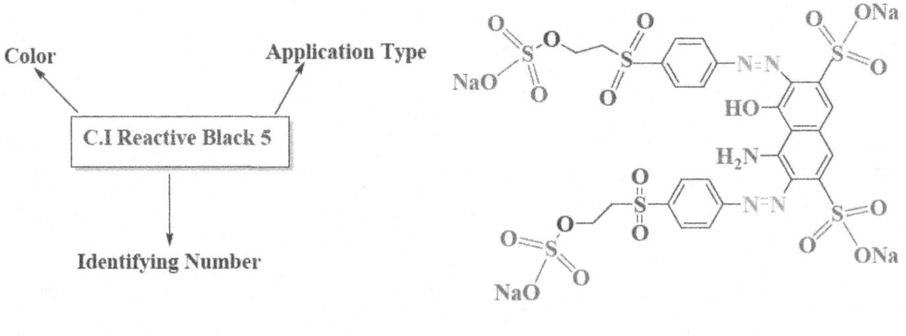

Color

Application Type

C.I Reactive Black 5

Identifying Number

FIGURE 1.7 Color index classification of Reactive Black 5.

is treated using different materials. Nanomaterials and its applications in industrial wastewater treatment have established adequate attention. The unique structural features and flexibility of nanomaterials provide higher removal efficiency and are economically viable for large-scale process industries. The higher ratio of surface area to particle size at nanoscale leads to change in electrical, magnetic, and optical properties [11, 28]. Harmful pollutants such as heavy metals, hydrocarbons, and dyes are treated with nanomaterials. Recent literature on the treatment of wastewater has focused on engineered nanomaterials.

In recent years, graphene oxide has become one of the most used nanomaterials to treat the industrial wastewater. The graphene oxides have favorable properties like a planar shape and an extremely large surface area per layer [29]. The oxygen-containing functional groups in the form of epoxy, hydroxyl, and carboxyl groups give graphene oxide its hydrophilic property. Using graphene oxide as an adsorbent for the purpose of removing dyes has been reported by several authors in different methods of synthesis [29–33]. On the other hand, carbon nanotubes (CNTs) have been proven to be effective adsorbents due to their nanoscale dimensions, huge specific surface area, and multilayer structures [34–38]. Because of their demonstrated biocompatibility and high magnetic characteristics, super paramagnetic iron oxide nanoparticles (SPIONs) have gotten a lot of interest [39]. Nowadays, SPIONs have a wide range of application where they are favorable adsorbents for water pollution. Different literatures have been reported utilizing SPIONs [40].

1.3.2 Textile Industry

The textile industry is one of the most important industries that expand significantly in all developing countries. It is responsible for the production of home and decorative textiles. It is one of the most important industrial wastewater producers and is characterized by high water consumption. It is estimated that 15% of the dyes produced every year are from the textile industry [41, 42]. In the textile industry, up to 200,000 tons of these dyes are lost through wastewater annually, which escape during traditional wastewater treatment processes and remain in the environment. The textile wastewater effluents have high stability against temperature, light, chemicals, detergents, and soap [43]. Antimicrobial agents are used in textile manufacturing to resist biodegradation; these are not biodegradable and are mixed with the wastewater [44]. Environmental legislation obliges textile industries to remove dye from the wastewater effluents before disposing in water bodies.

use of continuous or semi-continuous batches or processes for dying textile materials depends on (1) the type of material such as fiber, yarn, fabric construction, and garment; (2) size of the dye; and (3) the quality requirements within the dye fabric. Therefore, the batch process is the most common method for dyeing textile materials [45].

1.3.2.1 Raw Materials of Textile Manufacturing

The used raw materials for this industry are classified into natural raw materials like cotton and wool, which are the most commonly used natural fibers, and synthetic raw materials like polyester and nylon. Also, the natural fibers can be classified according to their source into plant fibers (cellulosic), such as cotton and linen, and animal fibers (protein), such as wool, hair, and silk [46, 47].

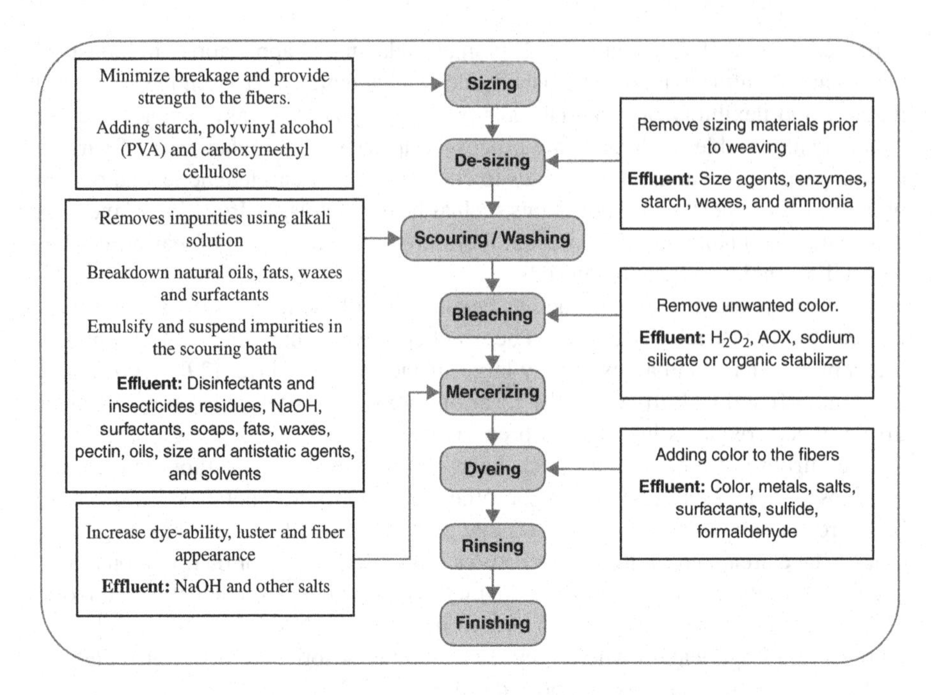

FIGURE 1.8 Textile finishing industry operations; purposes and pollutants in each step.

1.3.2.2 Textile Manufacturing Process

In Egypt, the textile sector contains around 300 companies, ranging from small to very large companies. The most common textile processing facility is de-sizing, scouring, bleaching, mercerizing, and dyeing processes, where each process consumes different chemicals [18]. Textile manufacturing process includes several steps that contain many wet operations, which produce large amounts of wastewater effluents as shown in Figure 1.8. Generally, the textile wastewater effluent is alkaline and is characterized by high levels of BOD5, chemical oxygen demand (COD), and total organic carbon (TOC). The wastewater contains many pollutants such as suspended solids, mineral oils (e.g., fat, anti-foaming agents, lubricants, and non-biodegradable or weakly biodegradable surfactants), dyes, heavy metals (e.g., chromium, copper, zinc, lead, and nickel), and other organic compounds such as phenols [48].

1.3.2.3 Industrial Textile Wastewater Processing

Industrial textile wastewater process contains pre-finishing operation on each industrial step as the following illustration:

1. *Scouring fiber*: In this process, hot water and detergents are utilized to remove dirt, oil, vegetables, and different contaminants. The scouring wastewater is described by strongly alkaline and huge extents of biological oxygen demand (BOD) and COD pollution.

2. *Finishing operations*: These are fundamental procedures of fabric preparation like de-sizing, bleaching, mercerizing, dyeing, printing, and other specific treatments. Fabrics are treated with chemicals and liquor baths, and frequently multiple washing, rinsing, and drying steps are necessitated that produce a lot of wastewater, which is illustrated in the following [48]:
   a. *De-sizing operations*: In this process, chemicals used in this process are enzymes, sulfuric acid, detergents, and alkali. De-sizing process generates large amounts of wastewater with high concentrations of organic matter (e.g., PVA, poly acrylates, and carboxyl methyl cellulose), solids, and significant BOD and COD loads (35–50% of total load).
   b. *Mercerizing process*: It occurs in the presence of caustic soda, which is responsible for alkalinity, is then treated by washing with hot water to remove the excess of the caustic soda.
   c. *Dyeing*: Wastewater from dyeing can contain color pigments, halogens, heavy metals (e.g., chromium, nickel, lead, copper, and zinc), organic compounds (e.g., amines), and other chemicals used in dye formation. Dyeing wastewater is characterized by extremely high BOD and COD values above 5000 mg/L, and high concentrations of salts from reactive dye use with an average range of 2,000–3,000 ppm. During dying process, large amounts of unused dyes, especially reactive dyes, are left in wastewater due to hydrolysis of reactive dye under alkaline conditions where hydrolyzed dye has no affinity for fiber substrate. So, dye hydrolysis increases the loss of dyes [44]. In alkaline conditions, the reactive form of the dye, vinyl sulfonate ($-SO_4-CH=CH_2$), is generated in the dye bath, which forms bonds with fabrics (Figure 1.9). Unfortunately, the reactive group further undergoes hydrolysis, producing hydrolyzed products that don't have affinity to fabric and enter wastewater [17, 50].

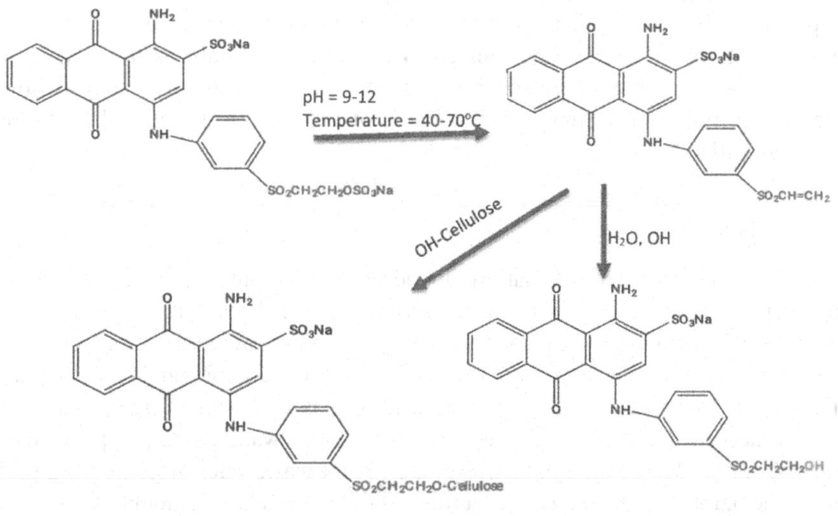

FIGURE 1.9 Reactive dye reaction pathway for the formation of vinyl sulfonate and its reaction with cellulose or hydrolysis.

d. *Printing process*: Use color pigments or dyes, organic solvents, and binder resins to accelerate the printing operation. The wastewater of washing after printing operations has an oily appearance and a significant content of volatile organic compound from the used solvents.
e. *Mothproofing*: Mothproofing agents are based mainly on permethrin, cyfluthrin, and other biocides, characterized by high toxicity, which are transferred with washing to wastewater that is very harmful to aquatic life [46, 51].

1.3.2.4 Textile Wastewater Treatment

Water is the main input to the textile industry where the production operation includes many wet processes (like sizing, scouring, de-sizing, bleaching, dyeing, and finishing). These processes consume a huge amount of water that is determined to be around 200 L/kg of product. With increasing of textile products and the usage of synthetic dyes, the textile wastewater increases. Textile factories are considered as the most polluting of all industrial sectors, and their produced amount of wastewater is one of the main causes of serious pollution problems [14, 19]. The traditional physiochemical and biological treatment methods are ineffective in treating textile wastewater. Therefore, researchers in the fields of analytics, the environment, and technology have tried to develop methods for removing accumulated pollutants [52].

Wastewater treatment in the complicated textile industry can be carried out in various steps to remove all pollutants. Pretreatment steps for wastewater include (i) high load flows of COD that contain non-biodegradable compounds using chemical oxidation, (ii) reduction of heavy metals using chemical precipitation, coagulation, and flocculation, etc., and (iii) treatment of colored or high total dissolved solids (TDS) currents using reverse osmosis [53]. The wastewater treatment steps include fat separators, oil-water separators for separating floating solids, filtration for separating filterable solids, compensation, sedimentation for reducing suspended solids with clarifiers, biological treatment, typically aerobic treatment for reducing soluble organic substances (BOD), biological nutrient removal for the reduction of nitrogen and phosphorus, chlorination of the wastewater if disinfection is required, and drainage and disposal of residues in designated hazardous waste landfills [46–48, 51].

1.3.3 Paints

Paints are pigment solutions that are applied to a substrate in order to improve its color. They can be made into any color and stored as liquids, but most dry out into solids. Paint manufacturing has two main purposes: protection from corrosion against oxidation or degradation of materials, and for decoration and embellishment of materials and surfaces [54]. Painting industry is one of the largest industries, which is accompanied by much-loaded industrial wastewater that has approximately a generation of 200 m^3/month of wastewater. Paint wastewater presents a big problem in the painting industry, such that this industry uses a large amount of water during washing and mixing tanks in which various organic and inorganic components are present. So, the paint wastewater is consisted of a mixture of unknown chemical structure compounds, which are characterized by a high organic load and a high content of suspended solids [55].

1.3.3.1 Paints Classification

i. Water-based paints consist of complex mixtures of organic and inorganic pigments, cellulosic and non-cellulosic thickeners, dyes, and extenders [56].
ii. Oil-based paints are the oldest form of modern paints and are widely used in decoration and surface protection. Most of them are obtained from vegetable oils, linseed, and soya beans. Natural oil paints are excellent coating materials due to the autoxidation process which occurs only in the presence of atmospheric oxygen to form a thin, solid film to protect the base metal while providing a color. The oil-based paints are used for construction and household purposes which have quickly changed to water-based paints [57].

1.3.3.2 Raw Materials of Paint Manufacturing Industry

There are several ingredients that are used to manufacture the colored liquid paints. Some of these components are solids such as resin, pigment, and additives, and there is just one liquid component that dissolves all of these mentioned solids [58].

1.3.3.2.1 Resin

Resin is the binder that holds all the pigments together, which is the component that identifies the paint. It allows the product to adhere to the painted surface. Resins are classified into two categories: The first is convertible binders which are the most common type of resin. Convertible binders aren't fully polymerized, but they complete its polymerization during the formation of paints. The second one is non-convertible binders that are fully polymerized such as nitro cellulose and cellulose [59, 60].

1.3.3.2.2 Solvents

Solvents act as a carrier that helps holding the pigments and resin together, which give flow ability to the paint to make easy applying [60, 61]. There are two types of solvents, as follows:

i. According to the principle of working
   a. Real solvent (where binder can be dissolved)
   b. Diluent (to increase the binder capacity)
   c. Latent solvent (liquid added to increase binder tolerance for diluent)
   d. Thinner (liquid added to increase fluidity of paint)
   e. Retarder (liquid added to pre-long the evaporation rate)
   f. Front-end solvent (quick drying solvent leaves the paint after applications)
   g. Exempt solvent (solvents don't react with sunlight to form smog)
   h. Middle solvent (medium evaporation rate solvent)

ii. According to its chemical structure
   a. Xylene (used for short alkyd resin, polyurethane, and epoxy)
   b. Toluene (used for air drying vinyl and as a diluent solvent for nitro cellulose)

c. Butyl alcohol (used as slow solvent for amino and acrylic resins and in solvent combination for nitro cellulose)
d. Ethyl alcohol (used as a fast-evaporating solvent for nitro cellulose)
e. Acetone (used as a fast-evaporating solvent for vinyl copolymers and nitro cellulose)
f. Ethylene glycol mono ethyl ether (slow evaporating solvent used as a coalescent agent in emulsion paint)
g. Methyl ethyl ketone (fast evaporating solvent for vinyl copolymers, polyurethane and epoxy)
h. Methyl isobutyl ketone (low evaporating rate solvent)
i. Butyl acetate and ethyl acetate (used as a slow evaporating solvent for nitro cellulose coatings) [60]

1.3.3.2.3 Additives

Additives in low proportions are used in the composition of paints to enhance the properties of the material [59, 60]. There are many types of additives for different applications or to enhance different properties, such as the following:

1. Anti-settling agents (e.g., BYK Anti-Terra 203, Crayvallac, and poly olefins)
2. Anti-skinning agents (e.g., methyl ethyl ketoxime and cyclohexanone oxide)
3. Anti-sagging agents by increasing viscosity (e.g., poly olefins)
4. Anti-foaming agents (e.g., BYK 023, BYK 03, and pyrene)
5. Light stabilizers (e.g., Tinuvin 1130 UV absorber, Tinuvin 1123 HALS, and Tinuvin 1144 HALS)
6. Dryers (e.g., Co, Ca, and Mg)
7. Thickeners (e.g., Benton)
8. Adhesion promoter (e.g., silanes, silicones, titanium compounds, and amides)

1.3.3.2.4 Fillers

Filler is a special type of additive that is added in large quantities to paint formulas, indicating that it is a major component of the paint formula. It serves a variety of objectives, including cost reduction and preventing later layers from being absorbed more quickly than in other places. It comes in a variety of forms, some of which can be used as fillers and others as flame retardants. Alumina trihydrate, calcium carbonate, calcium sulfate, silica, clay, glass structures, and aluminum trihydrate are the most commonly used manufacturing fillers [62, 63].

1.3.3.2.5 Pigments

Pigments are used to add color, luster, opacity, durability, mechanical strength, and wear protection to metal substrates. Pigments are divided into two types: organic pigments used in decorative paints and inorganic pigments (metallic pigments) used in protective paints, as shown in Table 1.3 [62–64].

TABLE 1.3
Types of Pigments with Different Colors

Color	Inorganic Pigments	Organic Pigments
White	TiO_2, ZnO, ZnS, and Sb_2O_3.	
Black	Carbon black and black iron oxide	Aniline black
Brown	Iron oxide brown	Benz imidazole
Yellow	Lead chromate and yellow iron oxide	Aryl amide and di aryl amide
Red	Cadmium red and red iron oxide	Metalized azo red and perylene
Blue	Ferric potassium, ferro cyanide, and cobalt blue	Copper phthalocyanine blue
Green	Chrome green, Cr_2O_3, and hydrated Cr_2O_3	Copper phthalocyanine green

1.3.3.3 Paint Manufacturing Process

1.3.3.3.1 Mixing Operation

The paint components with different properties and homogeneity need the mixing process to achieve homogeneity in the final paint system. The type of mixing depends on the type of mixed components which may be solids or liquids. The mixing technique may be solid-liquid mixing (solids dissolved in liquids by the three processes of suspension, distribution, and drawing down of solids by agitation) or liquid-liquid mixing (there are three processes applied to different liquids to achieve homogeneity, which are coalescence, dispersion, and suspension) [65].

1.3.3.3.2 Milling Operation

The milling process applies if the mixing operation doesn't achieve the required finesse of the paint mixture, which damages the agglomerate particles and disperses them into the paint mixture. Two types of mills are used in the paint industry: rotary mills and ball mills. It's observed that ball mills are more suitable and provide more safety than rotary mills. The rotatory milling process is done in open-air system, which is unsuitable for paints containing volatile solvents. These emissions are harmful to the working team. However, the ball milling process is done under circular cover which prevents solvent emissions [66].

1.3.3.3.3 Filtration Operation

Filtration process is required finally to rest the colloidal particles using several techniques such as screening filtration as a more common technique in the paint industry [59].

The Paint manufacturing process including washing tanks as a basic step in the manufacturing process is shown in Figure 1.10. These washing tanks produce large amounts of wastewater that need to be treated and reused for other purposes. Each stage in the manufacturing process needs to clean and rinse or offers another treatment method to improve the surface in terms of paint adhesion, appearance, and performance [60].

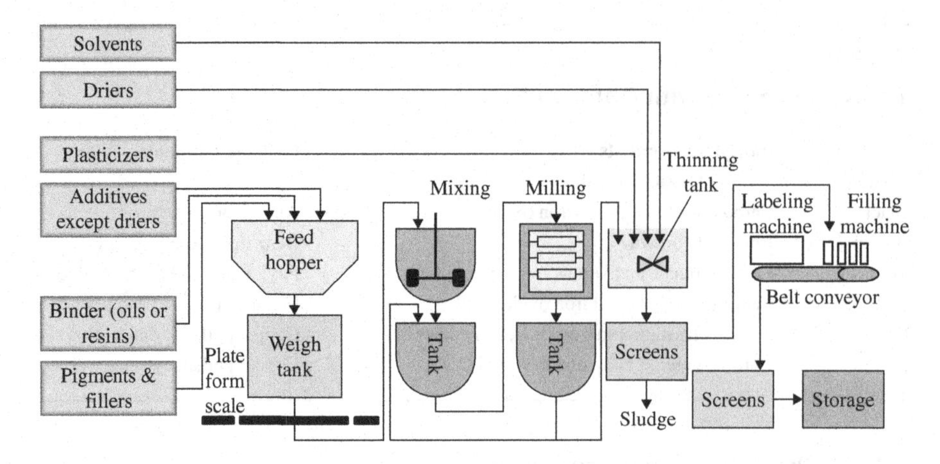

FIGURE 1.10 Paint manufacturing flow diagram.

1.3.3.4 Paint Wastewater

Paint wastewater contaminants include BOD, COD, suspended solids, toxic compounds, and color. The discharge of this wastewater into the natural effluents has resulted in a hazardous environmental pollution that has harmed aquatic life, the food chain, and human health. To achieve a clear effluent from paint wastewater with corresponding COD and BOD removal rates more than 90%, it is required a combination between chemical and biological treatment [67].

1.4 CHARACTERISTICS OF DYES, PAINTS, AND TEXTILE WASTEWATER/EXISTING REGULATIONS

1.4.1 Introduction

Treatment of industrial wastewater begins with the understanding of the wastewater characterization, which may include dye wastewater, textile wastewater, or paint wastewater. Each of the industrial wastewater types has distinct characteristics and properties. Dyes have general characteristics that are similar in all species of dyes but different in their percentages from one type to another and from one environmental industrial area to another. Wastewater should be analyzed before and after treatment with many analytical methods to define the characteristics of this wastewater and identify their physical, chemical, and biological properties. The main parameters that should be analyzed are COD, BOD, TOC, solids, nutrients, and heavy metals [68].

Wastewater from textile industries is subjected to various types of treatment like physical-chemical treatment (such as coagulation-flocculation) and biological treatment (such as aeration). The whole treatment process includes three main steps: the primary, secondary, and tertiary treatment processes. The primary treatment process removes suspended solids, most oils, grease, and coarse materials. The secondary treatment process reduces BOD, phenol, oil in the remaining water, and color control. The tertiary treatment process removes any final pollutants in the wastewater [69].

1.4.2 Types of Textile Fibers and Their Dyes

The textile industry is one of the most important industries in the world, providing workers without the specific skills required and playing a major role in the economies of many countries. The types of dyes and chemicals used in the textile industry vary according to the type of fabric made:

1. Cellulose fibers such as cotton, rayon, linen, ramie, hemp, and lyocell can be stained with reactive, direct, naphthol, and indigo dyes [70–72].
2. Protein fibers like wool, silk, mohair, and cashmere could be dyed with acid dyes [71–74].
3. Synthetic fibers (polyester, nylon, spandex, acetate, acrylic, indigo, and polypropylene) could be dyed with disperse, basic, and direct dyes [71–75].

The textile industry requires many chemicals and a huge amount of water during its production process. About 200 L of water is used to produce 1 kg of textile [76]. Water has been mainly used to mix and rinse the finished products. The wastewater from this process includes dyes and chemicals that contain traces of minerals (Cr, As, Cu, and Zn) which cause so much damage to the environment and human health.

1.4.2.1 Dyes for Cellulose Fibers

Cellulose consists of repeated glucose units. The dyes mainly used for cellulose fibers are reactive dyes and other dyes such as direct dyes, steam, sulfur, indigo, and naphthol.

I. ***Reactive dyes*** (*fiber reactive dyes*): It is the most used pigment type for cellulosic and protein fibers. Reactive dyes have been found to work well on any cellulosic fiber under different temperatures [70, 77]. Reactive dyes have two types: homo bi-functional and hetero bi-functional. Under alkaline conditions, about 50–70% of the dye with one reactive group is fixed onto the fiber, and about 80–95% of the dye with two reactive groups is fixed onto the fiber [78]. The important reactive dye systems are shown as follows:
   i. *Procion MX*: As shown in Figure 1.11, Procion works in a cold-water reactive dye that could be used at normal room temperature [79]. Procion dyes are inactive at alkaline conditions.
   ii. *Cibacron F*: As shown in Figure 1.12, unlike Procion dyes, Cibacron works only in a slightly warmer temperature (40–48°C) [80].
   iii. *Sabracron F*: As shown in Figure 1.13, Sabracron works quite similarly to Procion dyes but can only be used at a slightly warmer temperature (40–48°C) [81].
   iv. *Drimarene K*: As shown in Figure 1.14, they are reactive dyes of stable fibers in liquid and powder form. Drimarene dyes work in wide range of temperature (30–50°C), both cold and warm temperature dyeing [82].
   v. *Remazol*: As shown in Figure 1.15, Remazol works in low pH and cold solution [71].

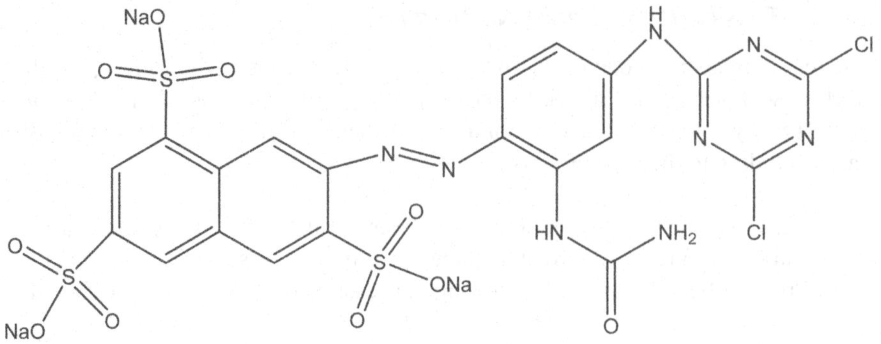

FIGURE 1.11 Chemical structure of Procion MX.

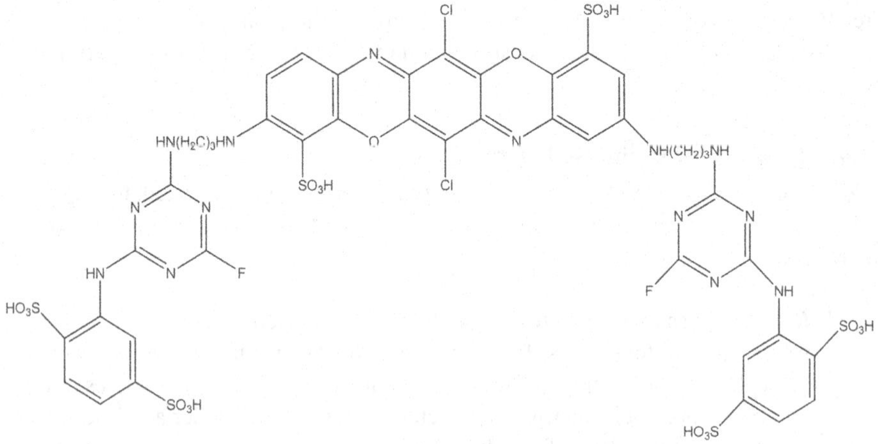

FIGURE 1.12 Chemical structure of Cibacron F.

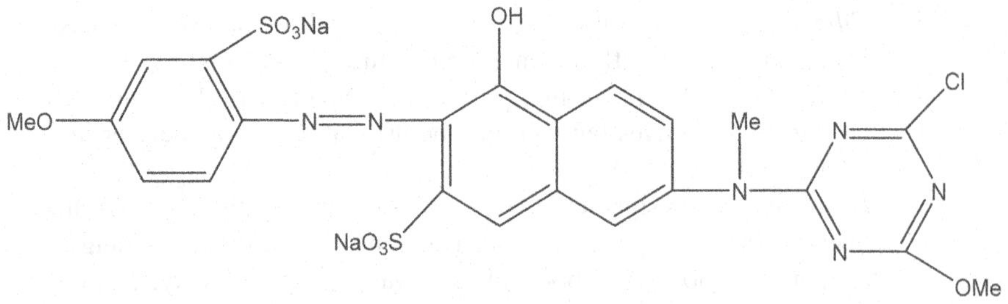

FIGURE 1.13 Chemical structure of Sabracron F.

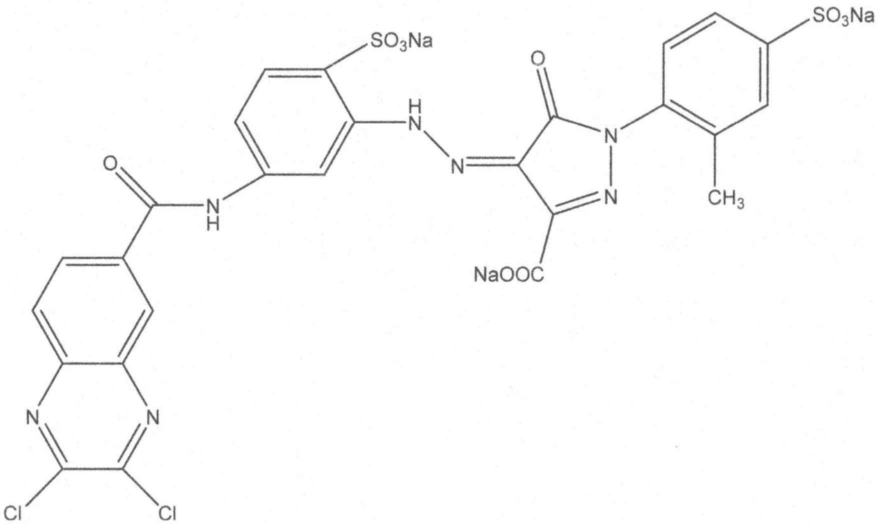

FIGURE 1.14 Chemical structure of Drimarene K.

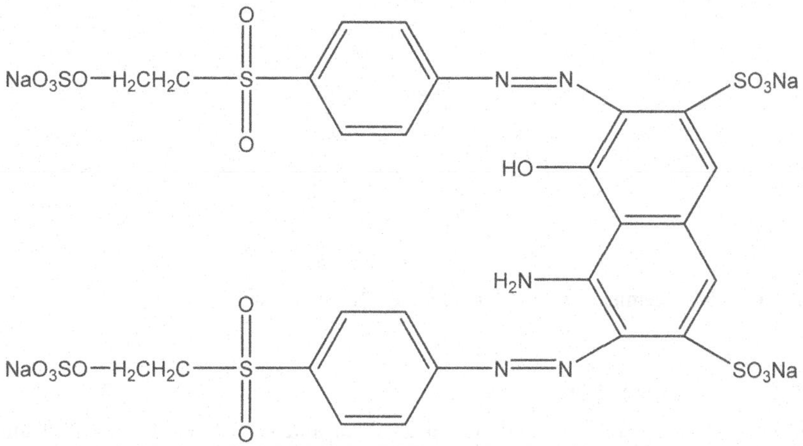

FIGURE 1.15 Chemical structure of 10 Remazol.

II. ***Direct dyes*** (*such as Direct Red 81*): These are applied under an aqueous bath containing electrolytes and ionic salts as shown in Figure 1.16 [70, 71].

III. ***Indigo dyes*** (*such as Dark blue*): As shown in Figure 1.17, indigo dyes are soluble in water at the presence of a reducing agent like sodium hydrosulfite. The soluble indigo dye molecules inside the fiber are then exposed to air, which makes it insoluble again and can be easily locked and held inside the fiber [71, 83].

IV. ***Naphthol*** (*Naphthol AS*): As shown in Figure 1.18, this dye works in a cold-water dye at a temperature from 35°C to 41°C [71, 84].

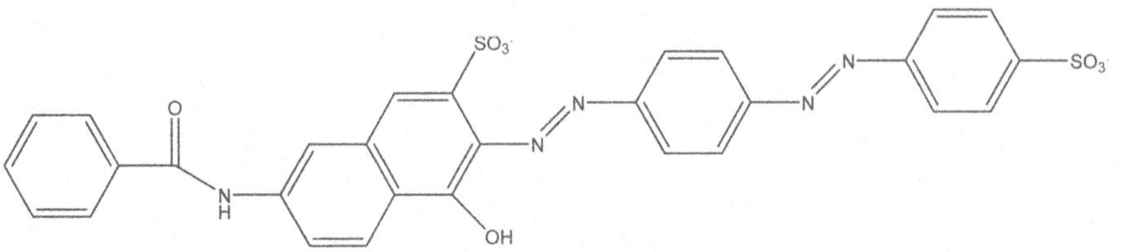

FIGURE 1.16 Chemical structure of direct dyes (Direct Red 81).

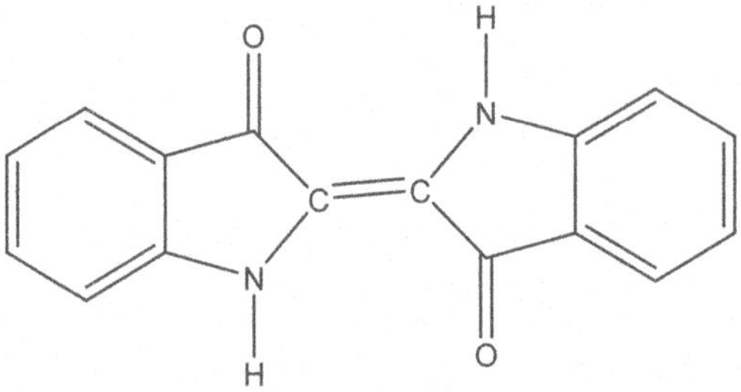

FIGURE 1.17 Chemical structure of indigo (dark blue).

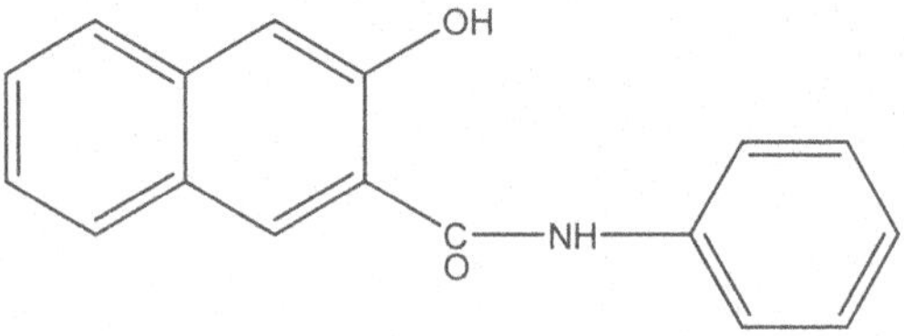

FIGURE 1.18 Chemical structure of Naphthol (Naphthol AS).

1.4.2.2 Dyes for Protein Fibers

Fibers are derived from animals (such as wool and silk). Twenty essential amino acids were found in proteins. These dyes are applied to fabrics with acidic solutions. The chemical reaction between dye and fiber produces the insoluble dye molecule in the fiber [85]. The most used acidic dyes are azo dyes, triaryl methane dyes, and anthraquinone.

i. *Azo dyes*: As mentioned previously, 60–70% of all the dye groups are azo dyes [86]. Azo dyes are classified according to the number of azo-chromophore groups associated with the center of the dye molecule (monohedral, dihedral, triplehedral, tetrahedral, etc.) [72].
ii. *Tri aryl methane dyes* are acidic with at least two SO_3H groups, which are used for wool and silk fibers. On the other hand, basic tri aryl methane dyes are widely used in stamping, writing, and printing [87].

iii. *Anthraquinone dyes* contain a group of sulfonic acids which make them soluble in water. They are used for food, protein fibers, and some nylon fibers under high temperature [71–73, 88]. Unlike other dye groups, these dyes contain a known chromophore system which contains various reactive groups [89].

1.4.2.3 Dyes for Synthetic Fibers

The most commonly used dyes to print the synthetic fibers like polyesters, nylon, and acetates and printing ink are dispersed dyes. Dispersed dyes are classified into three types: the E type (works in low temperature), the S type (works in high temperature), and the other SE type (works in all conditions) [90]. Another classification is as follows: direct dyes and basic dyes. Direct dyes work in a water bath that contains electrolytes and ionic salts and are commonly used to dye proteins and synthetic fibers [71]. Basic dyes (cationic dyes): These positive salts have been found to interact with the anionic surface of the substrate. These dyes prove to be strong dyes for acrylic fibers [91].

1.4.3 Characterization of Textile Wastewater Effluent

The textile manufacture requires large amounts of water to distribute and release chemicals. About 200 L of water is used to produce 1 kg of textile. Water is mainly necessary for (a) chemical applications of fibers and (b) rinsing the products [92]. The production process of textile fibers is mainly classified into two categories: the dry process and the wet process that involve the use of dyes. The dyeing process is an especially important step including a wet method that changes the color of the fabric using dyes. Finishing step as it's named is a final manufacturing process, using various chemicals such as HS-ULTRAPHIL, ECODESIZE-PS-10, and amino silicone fluid to produce fibers to give a better quality [93]. There are many types of textile wastes such as emission gas wastes, liquid wastes, and solid wastes that are produced at the different processes in the textile manufacturing steps as in Table 1.4 [94].

Sizing is one of the primary procedures in the textile business (for example, in the cotton textile industry) that consumes water throughout the wet process:

- sizing (500–8200 L/1000 kg of product)
- de-sizing (2500–21,000 L/1000 kg of product)
- scouring (20,000–45,000 L/1000 kg of product)
- bleaching (2500–25,000 L/1000 kg of product)
- mercerizing (17,000–32,000 L/1000 kg of product)
- dyeing (10,000–300,000 L/1000 kg of product)
- printing (8000–16,000 L/1000 kg of product), and finishing

The synthetic textile industries require much more water than the cotton industry in its wet process [92]. The major contaminants in wastewater are BOD, COD, suspended solids, metals, color, organic toxic compounds, and high pH as a result of the presence of residue NaOH.

TABLE 1.4
Types of Textile Wastes in Different Steps [76]

Process	Gas Wastes	Liquid Wastes	Solid Wastes
Preparation	No waste	No waste	Wastes of fibers and packages
Spinning	No waste	No waste	Wastes of fibers and packages
Sizing	Volatile wastes	Wastewater with high levels of BOD, COD, metals, and size	Wastes of fibers, packages, and chemicals
Weaving	No waste	No waste	Wastes of fibers and packages
Knitting	No waste	No waste	Wastes of fibers and packages
Tufting	No waste	No waste	Wastes of fibers, packages, and manufactured solid materials
De-sizing	Volatile wastes	Wastewater has lubricants, biocides, and anti-static compounds	No waste
Scouring	Volatile wastes	Wastewater has chemical loads of disinfectants, insecticide, detergent oil, lubricants, and high alkalinity	No waste
Bleaching	No waste	H_2O_2 and high alkalinity	No waste
Singeing	Volatile wastes	No waste	No waste
Mercerizing	No waste	High alkalinity	No waste
Heat setting	Volatile wastes	No waste	No waste
Dyeing	Volatile wastes	Wastewater includes metals, chemicals, surfactants, color, BOD, and COD	No waste
Printing	Volatile wastes	Wastewater with suspended solids, solvents, color, and foam	No waste
Finishing	Volatile wastes	Wastewater with suspended solids, toxic materials, and solvents	Wastes of fibers and packages

There are large quantities of dyes and other chemicals that harm the environment and are found in the textile effluent. The chemicals present mainly in the wastewater effluent are used in the wet process in bleaching and dyeing steps. At the bleaching process, the chemicals are wetting agents, such as caustic soda, peroxide, lubricants, stabilizers, and acetic acid, which are used in bleaching process; while at the dyeing process, these chemicals are used with different concentrations, which are lubricants, sequestering agent, light shade, medium shade, dark shade, acetic acid, soap, fixed agent, softening agent, soda ash, sodium chloride, and dye stuff [95]. At the other hand, the presence of dyes in the textile wastewater effluent is coming from textile dyeing process, at which there are always a portion of unfixed dyes that get away by washing with water with different concentrations according to the type of dye and fiber as shown in Table 1.5.

The characteristics of wastewater effluent depend on the type of manufacturing process and the chemicals used. The textile wastewater includes large amounts of some harmful compounds in all the environments [96]. Typical characteristics of an example wastewater effluent are shown in Table 1.6.

TABLE 1.5
Unfixed Dyes Get Away with Washing [76]

Fiber	Dye Type	Unfixed Dye (%)
Wool and nylon	Acid dyes/reactive dyes for wool	7–20
	Pre-metallized dyes	2–7
Cotton and viscose	Azoic dyes	5–10
	Reactive dyes	20–50
	Direct dyes	5–20
	Pigment	1
	Vat dyes	5–20
	Sulfur dyes	30–40
Polyester	Disperse	8–20
Acrylic	Modified basic	2–3

The textile wastewater effluent contains traces of metals like Cr, As, Cu, and Zn, causing environmental damage [97]. The presence of dyes in water can lead to bad color and many diseases [98]. These dyes work to prevent the photosynthesis according to the blocking of penetration of sunlight from passing through the water surface [99]. Dyes also increase the BOD of the receiving water, which decreases the

TABLE 1.6
Characteristics of Untreated Wastewater [97]

Parameter	Range
pH	6–10
Temperature (°C)	35–45
BOD (mg/L)	80–6,000
COD (mg/L)	150–12,000
Total solids (mg/L)	8,000–12,000
Total dissolved solids (mg/L)	2,900–3,100
Total suspended solids (mg/L)	15–8,000
Chlorine (mg/L)	1,000–6,000
Free chlorine (mg/L)	<10
Sodium (mg/L)	70%
Trace elements (mg/L)– Fe, Zn, Cu, As, Ni, B, F, Mg, V, Hg, PO_4, and Cn	<10
Oil/grease (mg/L)	10–30
NO_3-N (mg/L)	<5
Free ammonia (mg/L)	<10
SO_4 (mg/L)	600–1,000
Silica (mg/L)	<15
Total nitrogen (mg/L)	70–80
Color (Pt-Co)	50–2,500

TABLE 1.7
Characteristic of the Effluent of Each Wet Textile Processing Operation [102]

	Parameter		
Source of Effluent Generation	**pH**	**COD (mg/L)**	**BOD (mg/L)**
Process effluent			
De-sizing	5.83–6.50	10,000–15,000	1,700–5,200
Scouring	10–13	1,200–3,300	260–400
Bleaching	8.5–9.6	150–500	50–100
Mercerizing	8–10	100–200	20–50
Dyeing	7–10	1,000–3,000	400–1,200
Wash effluent			
After bleaching	8–9	50–100	10–20
After acid rinsing	6.5–7.6	120–250	25–50
After dyeing (hot wash)	7.5–8.5	300–500	100–200
After dyeing (acid and soap wash)	7.5–8.64	50–100	25–50
After dyeing (final wash)	7–7.8	25–50	
Printing washing	8–9	250–450	115–150
Blanket washing of rotary printer	7–8	100–150	25–50

re-oxygenation process and prevents the growth of photoautotrophic organisms [78]. The presence of suspended solids in the wastewater affects the environment so that when they colloid with oils and oxygen, the transmission mechanism in the air-water interface [79]. Inorganic substances in wastewater convert the water to something unsuitable for use because of the high concentration of TDS and their toxicity to aquatic and marine life [100, 101]. The possible pollutants that can be released at every step of the wet process are presented in Figure 1.11, while Table 1.7 shows the pollution loads from each step of the wet process [102].

The pollution loads of textile wastewater effluent after each processing operation are different with various textile fibers. Table 1.8 illustrates the difference in pollution loads at various textile fibers [103].

1.4.4 Characterization of Paint Wastewater Effluent

Painting industry is one of the most usable industries and has many applications in various fields, which is accompanied by a generation of large amounts of wastewater about 200 m^3/month. Paint industry has many processes (mixing, milling, and filtration), and after each one of these processes, the tanks should be cleaned, washed, and rinsed [59, 60]. Paint wastewater should be collected and treated before discharge in any water flow or reused in other things. So first should be known the characteristics of the paint wastewater, then thinking how to treat it.

The characterization of paint wastewater depends mainly on pH, COD, low biodegradability, and deep color [104, 105]. Paint wastewater has suspended solids which could be decreased easily by conventional methods, and also may contain at least traces

TABLE 1.8
Comparison between Pollution Loads for Processing Different Textile Fibers [103]

Parameter		Textile of 100% Cotton					Textile of 50% Cotton/50% Polyester						Textile of Wool					Textile of Synthetic Fiber				
		De-sizing (Enzyme Starch)	**Scouring** (Unmercerized Greige fabric)	**Mercerizing** (Greige Fabric)	**Bleaching** (Hydrogen peroxide)	**Dyeing** (Fiber Reactive HE dyes)	**De-sizing** (Starch)	**Scouring** (Unmercerized)	**Mercerizing** (Poly/Cotton)	**Dyeing** (Disperse-Vat)	**Printing** (Pigment (Woven))	**Finishing** (Resin Finishing)	**Scouring**	**Dyeing**	**Washing**	**Neutralizing**	**Bleaching**	**Rayon**	**Acetate**	**Nylon**	**Acrylic**	**Polyester**
pH		6–8	12.5	12	9–12	12	6–8	12	–	12	6–8	6–8	9–10.4	4.8–8	7.3–10.3	1.9–9	6	–	–	–	–	–
BOD	kg/1,000 kg of product	45.5	21.5	13	0.5	6	38.5	10.8	3.2	22.8	1.26	–	30–40	0.38–3	4–11.455	0.028	0.39	30	45	45	125	185
COD	kg/1,000 kg of product	91	64.5	39	2	24	–	–	–	68	5	–	–	–	–	–	–	52	78	78	216	320
TS	kg/1,000 kg of product	94	55	153	26	180	97	14.5	82	122	2.63	22	1.129–64.448	2–10	4.83–19.267	1.241–4.83	0.908	155	140	130	187	245
TDS	kg/1,000 kg of product	5	50	148	22	180	20	9.8	77	122	2.5	22	–	–	–	–	–	100	100	100	100	150
TSS	kg/1,000 kg of product	89	5	5	4	–	77	5	5	–	0.13	–	–	–	–	–	–	55	40	30	87	95
Oil & Gas	kg/1,000 kg of product	5	40	10	–	–	–	–	–	–	–	–	–	–	–	–	–	–	–	–	–	–

–: *None*

TABLE 1.9
Characterization of Paint Wastewater Effluent in Different Conditions

Industry	[107] 2002	[107] 2002	[107] 2002	[108] 2012	[109] 2018	[106] 2018	[110] 2011	[111] 2018	[105] 2007
Parameters	**Food**	**Metals**	**Chemicals**	**Paints**	**Meat**	**Decorative**	**Electronics**	**Paints**	**Prints**
pH	5.3	7.9	7.5	8.73	7.29	7.69	787	8	12.2
COD (mg/L)	5796	1152	1590	1711	5293	5970	97,300	–	1201.7
BOD (mg/L)	2530	585	475	748	2554	–	–	–	–
TSS (mg/L)	3450	1471	168	3821	2122	1463	23,000	2690	–
Turbidity (NTU)	–	–	–	–	–	–	13,100	1706.5	–
Oil and grease (mg/L)	77	260	235	–	329	421	–	–	–
Conductivity (μs/cm²)	–	–	–	–	–	–	2.87	1.53	–

of heavy metals from the used pigments [106]. Paints wastewater has high concentrations of oil and grease and phosphorus [107]. Paint wastewater can be highly toxic to the environment and harms aquatic life and wildlife. Table 1.9 illustrates the characterization of paint wastewater with explaining the various painting uses [105–111].

1.4.5 Regulation

1.4.5.1 Laboratory Wastewater Test Understanding

Industrial, institutional, and commercial entities have been required to improve the quality of their process wastewater effluent discharges since the execution of the Clean Water Act and ensuing making of the United States Environmental Protection Agency (USEPA) in the early 1970s. Simultaneously, population and production increases have increased water use, causing a corresponding rise in wastewater amount. This increased process wastewater generation requires more efficient removal of by-products and contaminations that allows for effluent discharge within established environmental regulatory limits. The assurance of wastewater quality set forth in environmental permits has been established since the 1970s in a series of laboratory tests concentrated on four major categories:

1. *Organics*: The concentration of carbon-based (i.e., organic) substances such as BOD, COD, TOC, and oil and grease that are used to determine the relative "strength" of wastewater (O&G).
2. *Solids*: By measuring the concentration of particulate solids that can dissolve or suspend in wastewater as total solids (TS), total suspended solids (TSS), TDS, total volatile solids (TVS), and total fixed solids (TFS).
3. *Nutrients*: The concentration of targeted nutrients measured as nitrogen and phosphorus that can contribute to the acceleration of eutrophication (i.e., the natural aging of water bodies).

4. *Physical properties and other impact parameters*: Varied groups of constituents that directly impact wastewater treatability as temperature, color, pH, turbidity, and odor are measured by analytical tests.

1.4.5.2 Parameters

The pollution is characterized by BOD, COD, suspended solids (SS), bad smell, toxicity (high concentration of nutrients, presence of chlorinated phenolic compounds, sulfur and lignin derivatives, etc.), and especially color [4, 5]. Color is the first contaminant to be recognized in wastewater, and the presence of very small amounts of dyes in water is highly visible and undesirable [112].

Regarding chemical processes, i.e., pre-treatment, dyeing, printing, and finishing, the main problem is water. Many commercial and specialty chemicals are used intensively in the wet processing of textiles.

Strict water discharge criteria, as shown in Table 1.10, are applicable to the Egyptian textile sector in order to improve environmental performance and boost the economic driving force for cleaner manufacturing [113].

The following are the laws concerning environmental protection: Wastewater: Law 9/2009, amended by Decree 1095/2011; Law 93/1962; and Law 48/1982, concerning protection of the River Nile Egypt's waterways from pollution [113].

TABLE 1.10
Egyptian Environmental Legal Requirements for Industrial Wastewater

Parameter (mg/L Unless Otherwise Noted)	Law 9/2009 Discharge Coastal Environment Amended by Decree 1095/2011	Law 93/62 Discharge to Sewer System (as Decree 44/2000)	Law 48/82 Discharge Into			
			Underground Reservoir and Nile Branches/ Canal	Nile (Mainstream)	Drain	
					Municipal	Industrial
BOD (5-day, 20 deg.)	60	600	20	30	60	60
COD	100	1,100	30	40	80	100
pH	6–9	6–9.5	6–9	6–9	6–9	6–9
Oil and grease	15	100	5	5	10	10
Total suspended solids	60	800	30	30	50	60
Total dissolved solids	+5 of the receiving body TDS	–	800	1,200	2,000	2,000
Residual chlorine	–	–	1	1	–	–
Sulfides	–	10	1	1	1	1
Phosphate (inorganic)	–		1	1	–	10
Phenol	0.015	0.05	–	–	–	–
Total nitrogen	10	100	–	–	–	–

1.5 CONCLUSION

In the present work, the importance of water and various sources of industrial wastewater is presented. The manufacturing process of dyes, paints, and textiles is also explained, as it is one of the industries that produce industrial wastewater. The description of industrial wastewater varies according to the source of that water and the materials used during the stages of obtaining that water. The main criteria used to determine the quality of industrial wastewater are also presented. Finally, this chapter presents the Egyptian environmental legal requirements for industrial wastewater for the safe reuse of industrial wastewater in different sectors.

ACKNOWLEDGMENT

This book chapter is based upon the projectsupported by the Egyptian Academy of Science, Research, and Technology (ASRT), Project ID: Call no. 2/2019/ASRT-Nexus\# 4607.

REFERENCES

1. G. Wagh, M. Sayyed, and M. Sayadi, "Evaluating groundwater pollution using statistical analysis of hydrochemical data: a case study from southeastern part of Pune metropolitan city (India)", International Journal of Geomatics and Geoscience, vol. 4, pp. 456–476, 2014.
2. T. Ling, "A global study about water crisis", in: 2021 International Conference on Social Development and Media Communication (SDMC 2021). Atlantis Press, 2022. pp. 809–814.
3. J. E. Miller, "Review of water resources and desalination technologies", Technical Report, Sandia National Laboratories, Albuquerque, NM, vol. 49, pp. 2003–2008, 2003.
4. Widyarani, D. R. Wulan, U. Hamidah, A. Komarulzaman, R. T. Rosmalina, and·N. Sintawardani, "Domestic wastewater in Indonesia: generation, characteristics and treatment", Environmental Science and Pollution Research, 2022. https://doi.org/10.1007/s11356-022-19057-6
5. Z. Hasan, J. Jeon, and S. H. Jhung, "Adsorptive removal of naproxen and clofibric acid from water using metal-organic frameworks", Journal of Hazardous Materials, vol. 209, pp. 151–157, 2012.
6. T. Heberer, "Occurrence, fate, and removal of pharmaceutical residues in the aquatic environment: a review of recent research data", Toxicology Letters, vol. 131, pp. 5–17, 2002.
7. A. M. Deegan, B. Shaik, K. Nolan, K. Urell, M. Oelgemoller, J. Tobin, and A. Morrissey, "Treatment options for wastewater effluents from pharmaceutical companies", International Journal of Environmental Science and Technology, vol. 8, pp. 649–666, 2011.
8. W. W. Ngah, L. Teong, and M. M. Hanafiah, "Adsorption of dyes and heavy metal ions by chitosan composites: a review", Carbohydrate Polymers, vol. 83, pp. 1446–1456, 2011.
9. L. Holzer, B. Münch, M. Rizzi, R. Wepf, P. Marschall, and T. Graule, "3D-microstructure analysis of hydrated bentonite with cryo-stabilized pore water", Applied Clay Science, vol. 47, pp. 330–342, 2010.
10. M. Raouf, N. Maysour, and R. Farag, "Wastewater treatment methodologies, review article", International Journal of Environment & Agricultural Science, vol. 3, pp. 018, 2019.

11. S. A. Ghodke, S. H. Sonawane, B. A. Bhanvase, and I. Potoroko, "Advanced engineered nanomaterials for the treatment of wastewater", Handbook of Nanomaterials for Industrial Applications, pp. 959–970, 2018.
12. S. Valili, G. Siavalas, H. K. Karapanagioti, I. D. Manariotis, and K. Christanis, "Phenanthrene removal from aqueous solutions using well-characterized, raw, chemically treated, and charred malt spent rootlets, a food industry byproduct", Journal of Environmental Management, vol. 128, pp. 252–258, 2013.
13. K. Othmer, Encyclopedia of Chemical Technology, 5th Edition. Wiley-Interscience, vol. 7, 2004.
14. F. M. D. Chequer, G. A. R. de Oliveira, E. R. A. Ferraz, J. C. Cardoso, M. V. B. Zanoni, and D. P. de Oliveira, "Textile dyes: dyeing process and environmental impact", Eco-Friendly Textile Dyeing and Finishing, pp. 152–176, 2013.
15. F. V. Ortega, I. Lagunes, and A. Trigos, "Cosmetic dyes as potential photosensitizers of singlet oxygen generation", Dyes Pigments, vol. 176, pp. 108–248, 2020.
16. E. Guerra, M. Llompart, and C. Garcia-Jares, "Analysis of dyes in cosmetics: challenges and recent developments", Cosmetics, vol. 5, pp. 1–15, 2018.
17. R. Anliker, G. C. Butler, E. A. Clarke, U. Forstner, W. Funke, C. Hyslop, G. Kaiser, C. Rappe, J. Russow, G. Tolg, M. Zander, and V. Zitko, "Anthropogenic compounds", Handbook of Environmental Chemistry, vol. 3, pt. A, pp. 181–216, 1980.
18. A. B. dos Santos, F. J. Cervantes, and J. B. van Lier, "Review paper on current technologies for decolorization of textile wastewaters: perspectives for anaerobic biotechnology", Bioresource Technology, vol. 98, pp. 2369–2385, 2007.
19. C. J. Ogugbue and T. Sawidis, "Bioremediation and detoxification of synthetic wastewater containing triarylmethane dyes by *Aeromonas hydrophile* isolated from industrial effluent", Biotechnology Research International, pp. 1–11, 2011.
20. Q. Y. Sun and L. Z. Yang, "The adsorption of basic dyes from aqueous solution on modified peat-resin particle", Water Research, vol. 37, no. 7, pp. 1535–1544, 2003.
21. S. Benkhaya, S. M'rabet, and A. El Harfi, "Classifications, properties, recent synthesis and applications of azo dyes", Heliyon, vol. 6, pp. e03271, 2020.
22. M. Berradi, R. Hsissou, M. Khudhair, M. Assouag, O. Cherkaoui, A. El Bachiri, and A. El Harfi, "Textile finishing dyes and their impact on aquatic environs", Heliyon, vol. 5, pp. e02711, 2019.
23. Y. Mittal, S. Dash, P. Srivastava, P. M. Mishra, T. M. Aminabhavi, and A. K. Yadav, "Azo dye containing wastewater treatment in earthen membrane based unplanted two chambered constructed wetlands-microbial fuel cells: a new design for enhanced performance", Chemical Engineering Journal, vol. 427, pp. 131856, 2022. https://doi.org/10.1016/j.cej.2021.131856.
24. E. J. Fontenot, Y. H. Lee, R. D. Matthews, G. Zhu, and S. G. Pavlostathis, "Reductive decolorization of a textile reactive dyebath under methanogenic conditions", Applied Biochemistry and Biotechnology, vol. 109, pp. 207–225, 2003.
25. S. Benkhaya, S. Mrabet, and A. El Harfi, "A review on classifications, recent synthesis and applications of textile dyes", Inorganic Chemistry Communications, vol. 115, pp. 1078–1091, 2020.
26. C. P. Pereira, J. P. N. Goldenstein, and J. P. Bassin, "Industrial wastewater contaminants and their hazardous impacts", Biosorption for Wastewater Contaminants, pp. 1–22, 2022.
27. O. J. Hao, H. Kim, and P. C. Chang, "Decolorization of wastewater", Critical Reviews in Environmental Science and Technology, vol. 30, pp. 449–505, 2000.
28. G. Z. Kyzas and K. A. Matis, "Nanoadsorbents for pollutants removal: a review", Journal of Molecular Liquids, vol. 203, pp. 159–168, 2015.
29. W. Konicki, M. Aleksandrzak, and E. Mijowska, "Equilibrium, kinetic and thermodynamic studies on adsorption of cationic dyes from aqueous solutions using graphene oxide", Chemical Engineering Research and Design, vol. 123, pp. 35–49, 2017.

30. P. Bradder, S. K. Ling, S. Wang, and S. Liu, "Dye adsorption on layered graphite oxide", Journal of Chemical & Engineering Data, vol. 56, pp. 138–141, 2011.
31. S. T. Yang, S. Chen, Y. Chang, A. Cao, Y. Liu, and H. Wang, "Removal of methylene blue from aqueous solution by graphene oxide", Journal of Colloid Interface Science, vol. 359, pp. 24–29, 2011.
32. L. Sun, H. Yu, and B. Fugetsu, "Graphene oxide adsorption enhanced by in situ reduction with sodium hydrosulfite to remove acridine orange from aqueous solution", Journal of Hazardous Materials, vol. 203–204, pp. 101–110, 2012.
33. S. Chowdhury and R. Balasubramanian, "Recent advances in the use of graphene-family nano adsorbents for removal of toxic pollutants from wastewater", Advances in Colloid and Interface Science, vol. 204, pp. 35–56, 2014.
34. Y. Yao, F. Xu, M. Chen, Z. Xu, and Z. Zhu, "Adsorption behavior of methylene blue on carbon nanotubes", Bioresource Technology, vol. 101, pp. 3040–3046, 2010.
35. Y. Yao, B. He, F. Xu, and X. Chen, "Equilibrium and kinetic studies of methyl orange adsorption on multiwalled carbon nanotubes", Chemical Engineering Journal, vol. 170, pp. 82–89, 2011.
36. C. H. Wu, "Adsorption of reactive dye onto carbon nanotubes: equilibrium, kinetics and thermodynamics", Journal of Hazardous Materials, vol. 144, pp. 93–100, 2007.
37. C. Y. Kuo, C. H. Wu, and J. Y. Wu, "Adsorption of direct dyes from aqueous solutions by carbon nanotubes: determination of equilibrium, kinetics and thermodynamics parameters", Journal of Colloid Interface Science, vol. 327, pp. 308–315, 2008.
38. Y. Chen, Y. Lin, Y. Liu, J. Doyle, N. He, X. Zhuang, J. Bai, and W. J. Blau, "Carbon nanotube-based functional materials for optical limiting", Journal of Nanoscience and Nanotechnology, vol. 7, pp. 1268–1283, 2007.
39. O. A. Abdel Aziz, K. Arafa, A. S. Abo Dena, and I. M. El-sherbiny, "Superparamagnetic iron oxide nanoparticles (SPIONs): preparation and recent applications", Journal of Nanotechnology & Advanced Materials, vol. 8, pp. 21–29, 2020.
40. S. S. Alaa, A. S. Abo Dena, and I. M. El-Sherbiny, "Matrix-dispersed PEI-coated SPIONs for fast and efficient removal of anionic dyes from textile wastewater samples: applications to triphenylmethanes", Spectrochimica Acta Part A: Molecular and Biomolecular Spectroscopy, vol. 249, pp. 119301, 2021.
41. M. Behera, J. Nayak, S. Banerjee, S. Chakrabortty, and S. K. Tripathy, "A review on the treatment of textile industry waste effluents towards the development of efficient mitigation strategy: an integrated system design approach", Journal of Environmental Chemical Engineering, vol. 9, no. 4, pp. 105277, 2021.
42. H. Zollinger, Color Chemistry: Syntheses, Properties, and Applications of Organic Dyes and Pigments, John Wiley & Sons, UK, 2003.
43. S. R. Couto, "Dye removal by immobilised fungi", Biotechnology Advances, vol. 27, no. 3, pp. 227–235, 2009.
44. W. Przystas, E. Z. Godlewska, and E. G. Sota, "Biological removal of azo and Ttriphenylmethane dyes and toxicity of process by-products", Water, Air, and Soil Pollution, vol. 223, no. 4, pp. 1581–1592, 2012.
45. W. S. Perkins, "A review of textile dyeing processes", American Association of Textile Chemists and Colorists, vol. 23, no. 8, 1991.
46. J. B. Perez, A. G. Arrieta, A. H. Encinas, and A. Queiruga-Dios, "Manufacturing processes in the textile industry. Expert systems for fabrics production", Advances in Distributed Computing and Artificial Intelligence Journal, vol. 6, no. 1, pp. 41–50, 2017.
47. F. Uddin, "Introductory Chapter: Textile Manufacturing Processes", Textile Manufacturing Processes, 2019. DOI: 10.5772/intechopen.87968.
48. I. Bisschops and H. Spanjers, "Literature review on textile wastewater characterization", Environmental Technology, vol. 24, no. 11, pp. 1399–1411, 2003.

49. E. J. Webber and V. C. Stickney, "Hydrolysis kinetics of reactive blue 19-vinyl sulfone", Water Research, vol. 27, no. 1, pp. 63–67, 1993.
50. World Bank Group. "Environmental, health, and safety guidelines for textile manufacturing", IFC, vol. 1, 2007.
51. S. E. Subramani and N. Thinakaran, "Isotherm, kinetic and thermodynamic studies on the adsorption behavior of textile dyes onto chitosan Process", Process Safety and Environmental Protection, vol. 106, pp. 1–10, 2017.
52. T. Adane, A. T. Adugna, and E. Alemayehu. "Textile industry effluent treatment techniques", Journal of Chemistry, vol. 2021, pp. 5314404, 2021.
53. O. Dovletoglou, C. Philippopoulos, and H. Grigoropoulou, "Coagulation for treatment of paint industry wastewater", Journal of Environmental Science and Health, Part A, vol. A37, no. 7, pp. 1361–1377, 2002.
54. M. Malakootian, J. Nouri, and H. Hossaini, "Removal of heavy metals from paint industry's wastewater using Leca as an available adsorbent", International Journal of Environmental Science and Technology, vol. 6, no. 2, pp. 183–190, 2009.
55. D. Krithika and L. Philip, "Treatment of wastewater from water-based paint industries using submerged attached growth reactor", International Biodeterioration & Biodegradation, vol. 107, pp. 31–41, 2016.
56. A. Kumar, P. K. Vemula, P. M. Ajayan, and G. John, "Silver-nanoparticle-embedded antimicrobial paints based on vegetable oil", Nature Materials, vol. 7, pp. 236–241, 2008.
57. B. K. Körbahti, N. Aktas, and A. Tanyolac, "Optimization of electrochemical treatment of industrial paint wastewater with response surface methodology", Journal of Hazardous Materials, vol. 148, pp. 83–90, 2007.
58. A. M. A. Youssef, "Paints industry: raw materials & unit operations & equipment & manufacturing & quality tests", 2019. DOI: 10.13140/RG.2.2.22793.60007.
59. R. Talber, Paint Technology Handbook, 1st Edition. Taylor & Francis Group, London, 2008. https://doi.org/10.1201/9781420017786.
60. M. Martins, R. Oliveira, J. A. P. Coutinho, M. A. F. Faustino, M. G. P. M. S. Neves, D. C. G. A. Pinto, and S. P. M. Ventura, "Recovery of pigments from *Ulva rigida*", Separation and Purification Technology, vol. 255, pp. 117723, 2021.
61. E. M. Angelin, S. França de Sá, M. Picollo, A. Nevin, M. E. Callapez, and M. J. Melo, "The identification of synthetic organic red pigments in historical plastics: developing an in situ analytical protocol based on Raman microscopy", Journal of Raman Spectroscopy, vol. 52, no. 1, pp. 145–158, 2021.
62. N. S. Allen and M. Edge. "Perspectives on additives for polymers. Part 2. Aspects of photostabilization and role of fillers and pigments", Journal of Vinyl and Additive Technology, vol. 27, no. 2, pp. 211–239, 2021.
63. L. M. de Souza Mesquita, M. Martins, L. P. Pisani, S. P. M. Ventura, and V. V. de Rosso, "Insights on the use of alternative solvents and technologies to recover bio-based food pigments", Comprehensive Reviews in Food Science and Food Safety, vol. 10, pp.1–32, 2020. DOI: 10.1111/1541-4337.12685.
64. F. Musabyimana and P. Turabimana, "Development and fabrication of vehicle body paints mixing machine", Engineering Perspective, vol. 1, no. 3, pp. 86–91, 2021.
65. N. S. Allen, M. Edge, C. Hill, and J. M. Kerrod, "Optimization of the ultraviolet–visible absorption properties of nano-particle TiO_2: influence of milling, surface area and surfactants on particle-size distribution, and stability of isocyanate/acrylic paints", Journal of Vinyl and Additive Technology, vol. 28, no. 1, pp. 62–81, 2022.
66. M. A. Aboulhassan, S. Souabi, A. Yaacoubi, and M. Baudu, "Treatment of paint manufacturing wastewater by the combination of chemical and biological processes", International Journal of Science, Environment and Technology, vol. 3, no 5, pp. 1747–1758, 2014.

67. J. Zhang, Y. Shao, G. Liu, L. Qi, H. Wang, X. Xu, and S. Liu, "Wastewater COD characterization: RBCOD and SBCOD characterization analysis methods", Scientific Reports, vol. 11, no 1, pp. 1–10, 2021.
68. M. I. Ejimofor, I. G. Ezemagu, and M. C. Menkiti, "Physiochemical, instrumental and thermal characterization of the post coagulation sludge from paint industrial wastewater treatment", South African Journal of Chemical Engineering, vol. 37, pp. 150–160, 2021.
69. J. Lorimer, T. J. Mason, M. Plattes, S. S. Phull, and D. J. Walton, "Degradation of dye effluent", Pure and Applied Chemistry, vol. 73, pp. 1957–1968, 2001.
70. P. Burch, "About fiber reactive dyes", All About Hand Dyeing, 2013.
71. L. Robert, F. Joseph, and A. Alexander, Fisher's Contact Dermatitis in: Textiles and Shoes, BC Decker Inc., Ontario, pp. 339–401, 2008.
72. V. Moody and H. Needles, Tufted Carpet: Textile Fibers, Dyes, Finishes, and Processes, William Andrew Publishing, 2004.
73. A. Schmidt, E. Bach, and E. Schollmeyer, "The dyeing of natural fibers with reactive disperse dyes in supercritical carbon dioxide", Dyes and Pigments, vol. 56, pp. 27–35, 2002.
74. S. Burkinshaw, Chemical Principles of Synthetic Fiber Dyeing, Blackie Academic & Professional, 1995.
75. A. E. Ghaly, R. Ananthashankar, M. Alhattab, and V. V. Ramakrishnan, "Production, characterization and treatment of textile effluents: a critical review", Journal of Chemical Engineering & Process Technology, vol. 5, 2014.
76. J. V. Valh and A. M. Le Marechal, Decoloration of Textile Wastewaters, Dyes and Pigments: New Research, Nova Science Publishers, Inc., New York, 2009.
77. I. Holme, "Coloration of Technical Textiles", Handbook of Technical Textiles, pp. 187–222, 2000.
78. Jacquard, "Procion MX", 2012.
79. PCD (Pro Chemical & Dye), "PRO sabracron/cibracron F reactive dyes", 2012.
80. PF (Paradise Fibers), "Sabracron F dye sampler", 2012.
81. Batikoetero, "Drimarene K", 2012.
82. V. Prideaux, A Handbook of Indigo Dyeing, Weelwood, Tunbridge Wells, Kent, 2003.
83. Hanu, "Naphthols and bases", 2010. http://dyeingworld1.blogspot.ca/2010/01/naphthols-and-bases.html.
84. E. Valko, The Theory of Dyeing Cellulosic Fibers, Onyx Oil and Chemical Company, Jersey City, NJ, 1957.
85. "Azo dyes", University of Bristol, 2012.
86. Farlex, "Triarylmethane dye", 2012.
87. F. Zhang, A. Yediler, X. Liang, and A. Kettrup, "Effects of dye additives on the ozonation process and oxidation by-products: a comparative study using hydrolyzed C.I. reactive red 120", Dyes Pigments, vol. 60, pp. 1–7, 2003.
88. K. Hunger, Industrial Dyes Chemistry, Properties, Applications, Wiley-VCH, Weinheim, Cambridge, 2003.
89. H. Sunny, "Manufacturer and Exporter of Acid Dye, Solvent Dye, Vat Dye, Direct Dye, Polymer Additive, UV Absorber, Light Stabilizer, Antioxidant, Optical Brightener Agent, Pigments and Fine Chemicals Disperse Dye", 2003.
90. M. Kiron, "Introduction to Mordant Dye/Properties of Mordant Dyes/Mechanism of Mordant Dyeing/Application of Mordant Dyes", 2012.
91. F. Ntuli, I. Omoregbe, P. Kuipa, E. Muzenda, and M. Belaid, "Characterization of effluent from textile wet finishing operations", Proceedings of WCECS, vol. 1, San Francisco, USA, 2009.
92. C. Wang, A. Yediler, D. Lienert, Z. Wang, and A. Kettrup, "Toxicity evaluation of reactive dyestuffs, auxiliaries and selected effluents in textile finishing industry to luminescent bacteria *Vibrio fischeri*", Chemosphere, vol. 46, pp. 339–344, 2002.

93. S. Moustafa, "Process analysis & environmental impacts of textile manufacturing", Dyes and Chemicals, vol. 20, 2008.
94. F. Kiriakidou, D. Kondarides, and X. Verykios, "The effect of operational parameters and TiO_2 – doping on the photocatalytic degradation of azo-dyes", Catalysis Today, vol. 54, pp. 119–130, 1999.
95. D. Arya and P. Kohli, "Environmental impact of textile wet processing, India", Dyes and Chemicals, 2009.
96. S. Eswaramoorthi, K. Dhanapal, and D. Chauhan, "Advanced in textile wastewater treatment: the case for UV-ozonation and membrane bioreactor for common effluent treatment plants in Tirupur, Tamil Nadu, India", Environment with People's Involvement & Co-ordination in India, Coimbatore, India, 2008.
97. T. Nese, N. Sivri, and I. Toroz, "Pollutants of textile industry wastewater and assessment of its discharge limits by water quality standards", Turkish Journal of Fisheries and Aquatic Sciences, vol. 7, pp. 97–103, 2007.
98. M. Laxman, "Pollution and its control in textile industry", Dyes and Chemicals, 2009.
99. M. Tholoana, "Water management at a textile industry: a case study in Lesotho", University of Pretoria, South Africa, 2007.
100. A. Blomqvist, "Food and fashion-water management and collective action among irrigation farmers and textile industrialists in South India", Linkoping University, Studies in Art and Science, vol. 148, pp. 201–213, 1996.
101. DOE (Department of Environment), Guide for Assessment of Effluent Treatment Plants EMP/EIA Reports for Textile Industries, 1st Edition. Ministry of Environment and Forest, Bangladesh, 2008.
102. S. Barclay and C. Buckley, Waste Minimisation Guide for the Textile Industry: A Step Towards Cleaner Production, The Pollution Research Group, University of Natal, Durban, South Africa, 2000.
103. J. Fu and G. Z. Kyzas, "Wet air oxidation for the decolorization of dye wastewater: an overview of the last two decades", Chinese Journal of Catalysis, vol. 35, pp. 1–7, 2014.
104. H. F. Wu, S. H. Wang, H. L. Kong, T. T. Liu, and M. F. Xia, "Performance of combined process of anoxic baffled reactor biological contact oxidation treating printing and dyeing wastewater", Bioresource Technology, vol. 98, pp. 1501–1504, 2007.
105. P. Yapicioglu, "Investigation of environmental-friendly technology for a paint industry wastewater plant in Turkey", Suleyman Demirel University, Journal of Natural and Applied Sciences, vol. 20, 2018. DOI: 10.19113/sdufbed.22148.
106. F. A. El-Gohary, R. A. Wahaab, F. A. Nasr, and H. I. Ali, "Three Egyptian industrial wastewater management programs", Environmentalist, vol. 22, pp. 59–65, 2002.
107. I. E. Mousa, "A full-scale biological aerated filtration system application in the treatment of paints industry wastewater", African Journal of Biotechnology, vol. 11, no. 77, pp. 14159–14165, 2012.
108. P. S. Yapıcıoglu, "Environmental impact assessment for a meat processing industry in Turkey: wastewater treatment plant", Water Practice & Technology, vol. 13, no. 3, pp. 692–704, 2018.
109. M. Mamadiev and G. Yilmaz, "Treatment and recycling facilities of highly polluted water-based paint wastewater", Desalination and Water Treatment. vol. 26, pp. 66–71, 2011.
110. M. C. Menkiti, A. O. Okoani, and M. I. Ejimofor, "Adsorptive study of coagulation treatment of paint wastewater using novel *Brachystegia eurycoma* extract", Applied Water Science, vol. 8, pp. 189, 2018.
111. N. M. Mahmoodi, R. Salehi, M. Arami, and H. Bahrami, "Dye removal from colored textile wastewater using chitosan in binary systems", Desalination, vol. 267, pp. 64–72, 2011.
112. "Best Available Techniques (BAT) for the textile industry in Egypt", 2012.

2 Microbial Bioremediation of Pesticides and Future Scope

Rahul Vikram Singh, Krishika Sambyal, and Pushpa Ruwali

2.1 INTRODUCTION

Pesticides are the chemical agents which prevent the spread of harmful intruders such as insects, rodents, weeds, fungi and pests by killing them, which eliminate the agricultural pests favouring crop productivity and livestock (Bhattacharjee et al., 2020). Due to escalated human population, the usage of pesticides has also increased exponentially to fulfil the soaring demands of foodstuff productivity by diminishing the possible agricultural harvest loss and decreased product quality caused by pests (Hu et al., 2020). The total tropospheric ozone level accounts for about 6% of the pesticides (Coxall, 2014). These pesticides eliminate pest infestations as well as control the disease-causing insect vectors and domestic home gardening pests based on their toxicity and mode of entry (Jayaraj et al., 2016; Rani et al., 2020). Apart from the pesticides mentioned in Figure 2.1, floor surface disinfectants, pool disinfectants, plant defoliants, growth regulators and certain other chemicals are also used in our day-to-day life (Bhattacharjee et al., 2020). Dichloro-diphenyl-trichloroethane (DDT), lindane, endosulfan, dieldrin, aldrin and chlordane are some commonly used organochlorines (Raimondo et al., 2019; Mudhoo et al., 2019) whereas atrazine, simazine, imidacloprid, glyphosate, mecoprop and isoproturon are other frequently used pesticides (Simon-Delso et al., 2015). The intensive use of all these pesticides at unmanageable rates has led to contaminated water bodies through agricultural runoffs, improper disposal and storage (Terzopoulou and Voutsa, 2017; Plattner et al., 2018). The spread of these pollutants indirectly affects the non-targeted species because pesticides can easily enter the food chain by leaching through the soil along with water (García-Galán et al., 2020). Moreover, they gradually bioaccumulate in lower as well as higher trophic level organisms via bioamplification, resulting in serious environmental threats along with animal health hazards (Nie et al., 2020). To overcome the heavy accumulation at pesticide-contaminated sites, few living organisms have established the potential ability to biotransform and bioremediate pesticides, thereby decomposing them through the utilization of a few growth factors, carbon, hydrogen and nitrogen supplied by such chemicals (Bhattacharjee et al., 2020). These living organisms include microbes (actinomycetes, fungi and bacteria) which either detoxify or transform the contaminants,

DOI: 10.1201/9781003354475-2

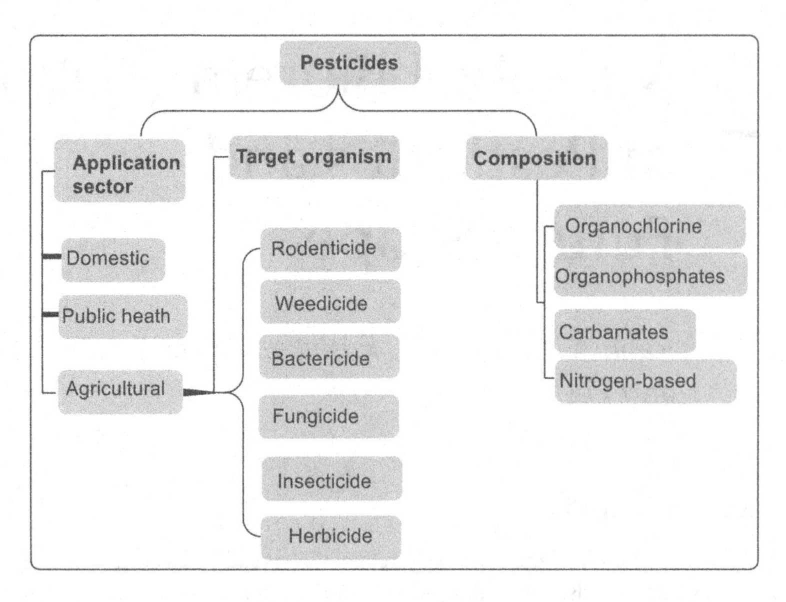

FIGURE 2.1 Classification of pesticides on the basis of their composition, application and target. (Jayaraj et al., 2016.)

and plants (including microalgae) which stabilize or remove contaminants present in the soil (Zhang et al., 2020). The common microbial genera including *Bacillus, Streptomyces, Aspergillus, Serratia, Pseudomonas, Rhizobium, Arthrobacter, Chlorella, Vibrio*, etc., utilize diverse coupled physical and biochemical mechanisms for pesticide degradation such as enzymatic mineralization, hydrolysis of P-O alkyl and aryl bonds, adsorption and photodegradation (Kumar et al., 2018). The methyl detoxification of organochlorines is achieved by the principal mechanism of dehalogenation, resulting in their vulnerability to other biodegradative attacks (Erguven and Koçak, 2019). Another promising tool for the restoration of contaminated soil is through the exploitation of different microbial consortiums (Raimondo et al., 2019). Thus, these remediation technologies convert/degrade the toxic chemicals and agricultural effluents into less/non-toxic constituents under real conditions by utilizing pure bacterial-fungal strains and other microbial consortiums (Góngora-Echeverría et al., 2020). Furthermore, biological decontamination approaches are more advantageous than other conventional methods because these do not produce toxic intermediates while degradation of environmental pollutants (Erguven and Nuhoglu, 2020). The present review focuses on the detrimental outcomes of pesticides on existing life forms and recent advancements in their remediation through various biological approaches.

2.2 EFFECT OF PESTICIDES ON HUMAN HEALTH AND ENVIRONMENT

All pesticides used by farmers for different purposes vary from each other based on their nature, mode of action and entry as displayed in Figure 2.1 (Jayaraj et al., 2016). These contaminants are sufficient and effective on the target species even in very small doses, yet a quantity of several metric tonnes is used globally every year (Malaj et al., 2020). Maximum of the remaining dusted or sprayed pesticides get washed off after application, resulting in an effortless entry into the water table and other water bodies as well as enter into the soil beds consequently invading the food chain. Initially, they affect the lower trophic levels (small fishes, plants and insects), successfully reaching the highest-level including humans, therefore occupying the entire food web (Nie et al., 2020). Approximately 95% of the applied pesticides are estimated to disperse widely in the environment. Plants take up these pesticides through contaminated soil which are then consumed by animals and humans affecting them indirectly (Simeonov et al., 2014). Their continuous use not only leads to loss of biodiversity but also facilitates high pest revival rates due to the resistance towards the chemicals used (Damalas and Eleftherohorinos, 2011). Pests develop this resistance by altering their lifestyle which further causes more hazards to non-targets including humans (Rani et al., 2020). Since soil serves as a sink for storage of such contaminants, the accessibility of already limited clean and portable water is plummeting through indirect pollution (Postigo et al., 2015; Pokhrel et al., 2018). Meat and dairy products such as beef and fish accumulate these pesticides directly from water whereas even a single poisoning event in agricultural commodities can impact an entire population of consumers (Li and Jennings, 2017). Besides this, the microbial environment has also been greatly influenced negatively after the application of volatile and persistent organic pesticides such as polychlorinated biphenyls, organochlorines and fumigants (Wang et al., 2015; García-Galán et al., 2020). The re-emissions of the parent pesticide compounds have been detected in higher concentrations in urban areas when compared to rural areas due to heavily applied amounts (Zhang et al., 2010). The most intensely poisonous type of pesticide is insecticides which pose a danger to the majority of animals (Mahmood et al., 2016). Dieldrin can induce mitochondrial impairment and ubiquitin–proteasomal dysfunction, whereas endosulfan causes developmental and chronic neurodegeneration in rodents. The primary neurotoxicity by organophosphorus compounds is exerted through the inhibition of acetylcholinesterase (Richardson et al., 2019). Such agrochemicals also reduce the population of insect pollinators and are a threat to endangered species and the habitat of birds (Rawtani et al., 2018).

Approximately 98% of pesticides are toxic for the crustaceans, fishes and other aquatic life forms, whereas fertilizers with a high amount of phosphorus contribute towards eutrophication along with improper nutrient cycling in the soil (Bhattacharjee et al., 2020). Most of the chemical pesticides are fat-soluble, thereby getting retained inside the body of organisms (Rawtani et al., 2018). Their accumulation occurs in animal tissue at higher levels than that in the water or soil to which they were applied, which accounts for bioconcentration. Some fishes concentrate certain pesticides in their body fat tissues at 10 million times greater levels than in the water. Carbamates and organophosphorus compounds are the most widely used pesticides affecting birds. Many vulture species have been reported to get accidentally exposed to organochlorines, anticoagulant rodenticides and external antiparasitic drugs (Plaza et al., 2019). The pesticide deliberately used to poison wildlife including vultures is carbofuran (2, 3-dihydro-2, 2-dimethyl-7-benzofuranyl-Nmethylcarbamate) with a broad spectrum of activity as a nematicide, insecticide and acaricide (Alarcón and Lambertucci, 2018). The action mechanism of these three pesticides reversibly inhibits the enzyme acetylcholinesterase, which results in the accumulation of neurotransmitter acetylcholine at the junction of the nerve cell and the receptor sites. This produces alterations such as salivation, tremors and convulsions, ultimately leading to the death of the organism (Richards, 2011).

Besides all the environmental effects, direct exposure through dermal contact, eye, inhalation or ingestion of pesticides through water and foodstuff causes various harmful health issues in humans such as respiratory, neurological and reproductive disorders, diabetes mellitus, cancer and metabolic and developmental toxicity. (Rani et al., 2020) in addition to minor symptoms such as headache, skin rashes, nausea, diarrhoea and dizziness (Boudh and Singh, 2019). The harmful effects of pesticides are also exerted through oxidative stress causing DNA damage turning into malignancies and other disorders (Sabarwal et al., 2018). The vulnerability of a person to high toxicity levels of azoxystrobin and atrazine can initiate carcinogenicity affecting the function of organs and damage DNA at the molecular level leading to neurological diseases (Fatima et al., 2018; Singh et al., 2018) and different types of cancers namely breast, colorectal, brain, prostate, pancreatic and lung cancer (Kim et al., 2017; Mostafalou and Abdollahi, 2017). Organophosphorus components primarily affect the reproductive system of males by reducing sperm activities, damaging sperm DNA and, therefore, inhibiting spermatogenesis (Mehrpour et al., 2014; Kim et al., 2017). The affected mothers can pass on the diseases to children in utero or via breastfeeding. Additionally, endocrine disruptor pesticides have shown genetic toxicity and epigenetic alterations including methylation and acetylation of DNA and its histone (Maqbool et al., 2016). The exposure of humans to pyrethroid-based allethrin and prallethrin mosquito repellents has resulted in decreased plasma cholesterol levels and increased plasma glucose and phospholipid levels (Narendra et al., 2008; Meftaul et al., 2020). Even after all such side effects, the use of pesticides cannot be limited due to high agricultural demands. Gradually, alternate pesticides will enter into the market with high efficiency and consequently have high toxicity levels, causing more environmental pollution and vast human side effects. Therefore, to overcome such issues, eco-friendly techniques must be adapted such as the use of microbes for remediation. They work autonomously either naturally or under mutated conditions to degrade toxic chemicals and pollutants.

2.3 MICROBIAL REMEDIATION OF PESTICIDES

The concept of bioremediation was started 600 BC ago by Romans for the cleaning of wastewater in ancient times. However, modern bioremediation was developed by engineer George M. Robinson in the 1960s; he applied different microbes for the degradation of pollutants. Further his results concluded that microbes have the capability of degrading pollutants (https://opgplus.com/). The removal of pesticides is dependent on the changes in the skeleton structure of the pesticide as well as the optimal environmental conditions favouring its survival and activity, which includes the availability of water, carbon sources, light, pH, temperature, nutrition, surface bonding, biological substrates, oxygen tension and redox potential (Rath, 2012). Microorganisms are known to have enzymes that can act on

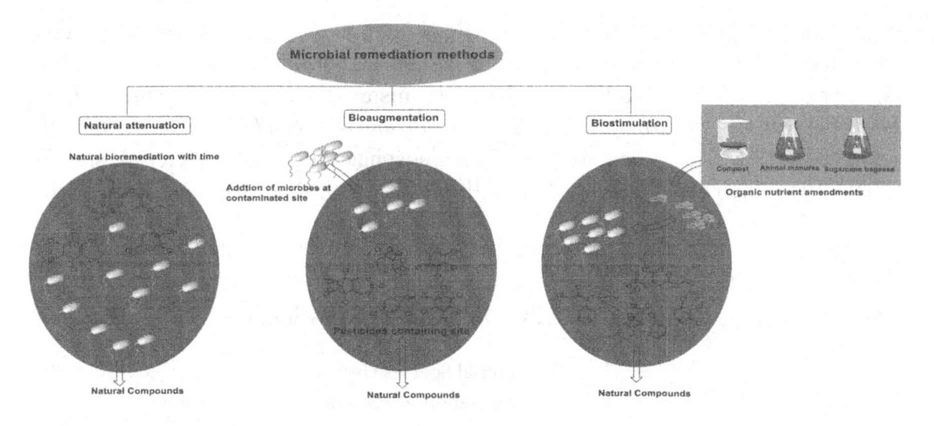

SCHEME 1 Bioremediation methods for pesticides by microbes.

pesticides for their degradation. The basic bioremediation other than natural microbial remediation comprises both bioaugmentation and biostimulation (Scheme 1). Bioaugmentation is the addition of microorganisms on sites contaminated with pesticides so that the bioprocess can be reinforced naturally. Biostimulation is the supply of organic nutrient amendments (animal manures, sugarcane bagasse, straw, sawdust and compost) for the predominantly surviving pesticide degrading microbes at the contaminated site so that the parent population can grow exponentially for their intended work (Varshney, 2019). This helps to increase nutrient content, water-holding capacity, aeration and porosity of soil for microbial diversity (Raimondo et al., 2020). Active substances such as plant hormones and ion carriers are secreted by microbes to promote plant tolerance and resistance to pollutants by their degradation (Hrynkiewicz et al., 2018). Some natural microbial remediation studies have been mentioned below.

2.3.1 Bacterial Remediation of Pesticides

The simple, low molecular weight pesticidal compounds are more vulnerable to degradation than polymers or complex compounds due to distinct features like molecular weight, spatial structure and several substituent groups present in the compound (Chrzanowski et al., 2012). An externally driven energy or immediate physicochemical interaction with chemical structures assists microorganisms for pesticide removal (Nie et al., 2020). After entering the bacterial cell, the pesticide compound breaks down from complex to a simpler non-toxic substance by ensuring a series of pathways with physiological and biochemical enzymatic reactions (Tang et al., 2018) such as N-hydroxylation, S-oxidation, oxidative dehalogenation, oxidative alkylation, quinone reduction and thiocarbamate, followed by a reduction in secondary pollution (Prabha et al., 2017). Mineralization and co-metabolism are other mechanisms involved in the pathway of the degradation process (Arora et al., 2012).

In 1973, a bacterium of *Flavobacterium* sp. gained notoriety for having the potential ability to degrade organophosphorus compounds (Singh and Walker, 2006). Following this, numerous other bacteria, fungi and algae have been reported for the degradation of pesticides. *Bacillus subtilis, Pseudomonas* sp., *Klebsiella* sp., etc., are some of the commonly known bacterial species known for pesticide depletion (Kumar et al., 2018). In addition to these organisms, *Sphingomonas paucimobilis, Microbacterium esteraromaticum, Brevibacterium* sp., *Rhodococcus* sp., *Serratia marcescens,* etc., are some other organophosphate-degrading bacterial species (Table 2.1) (Singh and Walker, 2006; Sidhu et al., 2019; Bhattacharjee et al., 2020).

TABLE 2.1
Bacterial Strains Reported for Bioremediation of Pesticides

Pesticides	Bacterial Species Used for Remediation of Pesticides	Reference
Tetrachlorvinphos	*Vibrio metschinkovii*	Ortiz-Hernández and Sánchez-Salinas (2010)
Diuron	*Micrococcus* sp.	Sharma et al. (2010)
Methyl parathion	*Sphingobium* sp.	Yuanfan et al. (2010)
Fenamiphos	*Pseudomonas putida*	Chanika et al. (2011)
Acephate	*Exiguobacterium* sp. and *Rhodococcus* sp.	Phugare et al. (2012)
Cadusafos	*Pseudomonas putida*	Abo-Amer (2012)
Fluoranthene	*Alcaligenes denitrificans*	Tewari et al. (2012)
Carbofuran and parathion	*Arthrobacter* sp.	
Pentachlorophenol	*Arthrobacter* sp.	
Parathion	*Bacillus sphaericus*	
DH35A Cyclohexylamine	*Brevibacterium oxydans*	
Glyphosate	*Clostridium Quinoline*	
Acetonitrile and carboxylic acid	*Corynebacterium nitrophilus*	
Trichloroethylene	*Dehalococcoides ethenogenes*	
Nitroaromatic compounds	*Desulfovibrio* sp.	
Pentachlorophenol and parathion	*Flavobacterium* sp.	
Arylacetonitrils	*Alcaligenes faecalis*	
Dichlorvos	*Proteus vulgaris, Vibrio* sp. and *Serratia* sp.	Agarry et al. (2013)
Diazinon	*Lactobacillus brevis*	Zhang et al. (2014)
Malathion	*Bacillus amyloliquefaciens* and *Pseudomonas* sp.	Karishma and Sharma (2014)
Diazinon	*Serratia liquefaciens* and *Serratia marcescens*	Iqbal and Bartakke (2014)
Aldrin, chlorpyrifos, coumaphos, dichloro-diphenyl-trichloroethane, diazinon, endosulfan, endrin, hexachlorocyclohexane, methyl parathion, monocrotophos, and parathion	*Pseudomonas* sp.	Verma et al. (2014)
Endosulfan	*Flavobacterium* sp., *Bacillus* sp. and *Alcaligenes* sp.	Kafilzadeh et al. (2015)

(Continued)

TABLE 2.1 *(Continued)*
Bacterial Strains Reported for Bioremediation of Pesticides

Pesticides	Bacterial Species Used for Remediation of Pesticides	Reference
Chlorpyrifos, coumaphos, glyphosate, polycyclic aromatic hydrocarbons, methyl parathion, dieldrin, and endosulfan	*Bacillus* sp.	Upadhyay and Dutt (2017)
Endosulfan	*Stenotrophomonas maltophilia*	Zaffar et al. (2018)
Monocrotophos	*Pseudomonas stutzeri, Bacillus licheniforms* and *Bacillus sonorensis*	Buvaneswari et al. (2018)
Imidacloprid	*Methylobacterium radiotolerans* and *Microbacterium arthrosphaerae*	Erguven and Yildirim (2019)
Atrazine, glyphosate and carbofuran	*Pseudomonas nitroreducens*	Góngora-Echeverría et al. (2020)
Fluazifop-p-butyl	*Bacillus subtilis, Brevibacterium macrolides, Microbacterium chocolatum, Bacillus macrolides, Sphingomonas aquatilis, Ochrobactrum thiophenivorans* and *Sphingomonas melonis*	Erguven and Nuhoglu (2020)
Lindane	*Streptomyces* sp.	Raimondo et al. (2020)
Endosulfan	*Bacillus subtilis*	Ahmad (2020)

Pseudomonas is a bacterial strain that has been efficiently used for cadusafos organothiophosphate degradation. This nematicide is extensively used to protect crops (Abo-Amer, 2012). *Methylobacterium radiotolerans* and *Microbacterium arthrosphaerae* are the two bacterial strains used for the remediation of imidacloprid vegetable insecticide with 98.7% and 88.4% chemical oxygen demand (COD) and biological oxygen demand (BOD) removal rates (Erguven and Yildirim, 2019). Different bacterial strains of *Burkholderia* and *Pseudomonas* can degrade fenitrothion, which is a potent inhibitor of acetylcholinesterase (Tago et al., 2006). The contaminant hexachlorocyclohexane is difficult to be eliminated from the water; therefore a bacterial consortium comprising of *Burkholderia, Pseudomonas, Flavobacterium* and *Vibrio* strains was made to successfully digest hexachlorocyclohexane in liquid culture (Murthy and Manonmani, 2007). Upon biodegradation kinetics and GC-MS analysis, it was revealed that endosulfan can be metabolized into a non-toxic metabolite, i.e. endosulfan diol, by a bacterium isolated from agriculture-contaminated soil identified as *Stenotrophomonas maltophilia* (Zaffar et al., 2018). Multiple exogenous proteins secreting enzymes when combined with single degrading bacterial strains facilitate the process of bioremediation, which is known as the microbial cell's surface technology. To analyse the degradation of monocrotophos, a soil treatment incubation method was adopted in which *Pseudomonas stutzeri*, *Bacillus licheniforms* and *Bacillus sonorensis* were incubated for 45 days with pesticide exposed soil samples. It resulted in the complete degradation of monocrotophos into water-soluble compounds when present in consortium cultures as well as individually (Buvaneswari et al., 2018).

The effective *in situ* treatments for effluent removal also utilize biological beds (bio-beds) in which biomixture (mixture of substrates) and microbes are the keys in pesticide treatment. During bioremediation of agricultural pesticides, the viability of pure bacterial strains and microbial consortium isolated from a biomixture (soil-straw; 1:1, v/v) has been evaluated. The strain *Pseudomonas nitroreducens* was the most abundant (52%) amongst them with the ability to highly degrade atrazine, glyphosate and carbofuran (>90%) in association with others. Therefore, bio-bed efficiency can be increased by using microbial consortium as well as pure strains to remediate contaminated soil (Góngora-Echeverría et al., 2020). A submerged biological aerated filter with bacterial consortium can biologically treat atrazine (0.01–10 mg/L) from wastewater in a pilot-scale trial up to 97.9% (Baghapour et al., 2013; Saleh et al., 2020). For the post-emergence control of grass weeds, a selective phenoxy herbicide, namely Fluazifop-p-butyl ($C_{19}H_{20}F_3NO_4$), is used, which can be broken down in soil using the consortia of *Bacillus subtilis*, *Brevibacterium macrolides*, *Microbacterium chocolatum*, *Bacillus macrolides*, *Sphingomonas aquatilis*, *Ochrobactrum thiophenivorans* and *Sphingomonas melonis*. The pesticide removal rates were determined using COD, BOD and total organic carbon (TOC) parameters, and removal yields of 84%, 87% and 94% were observed, respectively, in 20 cm^3 mixed medium (Erguven and Nuhoglu, 2020). The use of factorial designs has also been done to identify maximum lindane removal by bioaugmentation in different soil types using actinobacteria consortium, enhancing heterotrophic microbial counts and enzymatic activities. Biostimulation using sugarcane filter cake improved soil microbiological properties indicating favourable conditions and is a non-expensive alternative as well. The study showed that *Streptomyces* strains can adapt and grow in different soils, even in the presence of lindane and other pesticides (Raimondo et al., 2020).

The use of recombinant technology can also be performed to develop a plasmid resistant to various environmental conditions, possessing a foreign gene encoding the enzyme mostly involved in the successful degradation of harmful pesticides (Nayak et al., 2018). By the consistent action of three consecutive enzyme genes, namely *Atz A*, *Atz B* and *Atz C* (atrazine chlorohydrolase A, B and C), isolated from *Pseudomonas* sp., complete degradation of atrazine into CO_2 and H_2O has been achieved (Wackett et al., 2002). The degradation of 2, 4-dichlorophenoxyacetic acid (2, 4-D) has already been explored using a plasmid pJP4 with *tfd* gene modules encoding functional enzymes responsible for 2, 4-D degradation. The plasmid-carried donor strain *Pseudomonas putida* SM1443 bioaugmented with glucose-fed granules resulted in the establishment of 2, 4-D degradation (Ledger et al., 2006; Quan et al., 2010).

2.3.2 Fungal Remediation of Pesticides

Different pathways are followed by different fungal strains (Table 2.2) for pesticidal degradation. Due to the involvement of fungi in various biogeochemical cycles and other natural associations, its existence is considered vital for the bioremediation of environmental xenobiotics. Moreover, the extensive network of fungal mycelium and the unspecific nature of catabolite enzymes make it suitable over the other employed microorganisms (Chen et al., 2012). *Zygomycetes, Basidiomycetes, Ascomycetes* and *Mortierella* species primarily follow demethylation and polar hydroxylation

TABLE 2.2
Fungal Species Used for Remediation of Pesticides

Pesticides	Fungal Species Used for Remediation	Reference
Heptachlor epoxide	*Phlebia* sp. and *Phanerochaete chrysosporium*	Xiao et al. (2011)
Lindane	*Fusarium verticillioides, Fusarium solani* and *Fusarium poae*	Sagar and Singh (2011)
2,4-dichlorophenol, 2,4,6-trichlorophenol and pentachlorophenol	*Anthracophyllum discolour*	Diez et al. (2012)
Lindane	*Ganoderma australe*	Hussaini et al. (2013)
Endosulfan and lindane	*Aspergillus niger*	
Lindane and chlorpyrifos	*Ganoderma austral*	Iqbal and Bartakke (2014)
Chlorpyrifos	*Trichosporon* sp. and *Verticillium dahliae*	
β-cypermethrin	*Aspergillus niger*	Deng et al. (2015)
Vydate	*Trichoderma harzianum* and *Trichoderma viride*	Helal and Abo-El-Seoud (2015)
Malathion	*Fusarium oxysporum*	Peter et al. (2015)
Lindane	*Ganoderma lucidum*	Kaur et al. (2016)
Pentachlorophenol	*Rhizopus oryzae*	León-Santiesteban et al. (2016)
Atrazine, heptachlor and metolachlor	*Lentinus subnudus*	Gupta et al. (2017)
Aldrin	*Pleurotus ostreatus, Phanerochaete chrysosporium, Phlebia brevispora* and *Phlebia aurea*	Purnomo et al. (2017)
Acetochlor	*Tolypocladium geodes* and *Cordyceps cicadae*	Erguven (2018)
Endosulfan	*Aspergillus niger*	Ahmad (2020)
Lindane	*Aspergillus fumigatus*	Dey et al. (2020)
Thiencarbazone-methyl	*Aspergillus flavus, Penicillium chrysogenum, Aspergillus niger, Aspergillus terreus* and *Aspergillus fumigatus*	Ahmad et al. (2020)
Diuron, 3,4-dichloroaniline	*Trametes versicolor*	Hu et al. (2020)
Carbofuran and carbaryl	*Acremonium* sp.	Kaur and Balomajumder (2020)
Allethrin	*Fusarium proliferatum*	Bhatt et al. (2020)

for the remediation of phenylurea herbicides such as chlorotoluron, linuron and isoproturon (Ellegaard-Jensen et al., 2013), whereas dehydrogenation, deoxygenation, esterification and hydroxylation are few additional methods utilized by other fungal strains. The bioremediation of endosulfan is also done using varying pathways such as hydroxylation, oxidation, esterification, dechlorination, deoxygenation and dehydrochlorination with the involvement of different enzymatic species (Verma et al., 2011). Pyrethroid insecticide-contaminated environments can be potentially bioremediated using *Aspergillus niger* YAT strain which can transform β-cypermethrin into 3-phenoxybenzoic acid, which gradually transforms into permethric acid, protocatechuic acid, 3-hydroxy-5-phenoxy benzoic acid, gallic acid and phenol along with their effective degradation (Deng et al., 2015). Apart from *Aspergillus niger, Cladosporium cladosporioides, Aspergillus fumigatus, Phanerochaete chrysosporium, Penicillium raistrickii* and *Aspergillus sydowii* have the inherited ability to degrade organophosphates from contaminated samples (Bhattacharjee et al., 2020). The *Fusarium* species, namely *F. verticillioides, F. solani* and *F. poae,* are responsible for lindane degradation by utilizing it as a carbon source (Sagar and Singh, 2011).

Based on response surface methodology, orthogonal central composite design determined optimum biodegradation and biosorption removal conditions of lindane from another effective fungus known as *Ganoderma australe* known to biodegrade 3.11 mg/g (degraded lindane/biomass) pesticide (Dritsa et al., 2009). Many white-rot fungi are regarded as highly effective against pesticide-contaminated sites due to the presence of extracellular enzymatic systems such as lignin peroxide, laccase and manganese peroxidase (Gago-Ferrero et al., 2012). For the degradation of atrazine, heptachlor and metolachlor, the white-rot fungus *Lentinus subnudus* has been used (Nwachukwu and Osuji, 2007; Gupta et al., 2017). Another white-rot fungus *Anthracophyllum discolour* when immobilized on wheat grains produced manganese peroxidase and laccase while degrading 2,4-dichlorophenol, 2,4,6-trichlorophenol and pentachlorophenol in allophanic soil columns activated by acidification (Diez et al., 2012). Another effective species *Trametes versicolor* is known to rapidly degrade diuron (83%) and its major metabolite 3,4-dichloroaniline (100%) in 7 days of incubation using a trickle-bed reactor. A cytochrome P450 enzymatic system was found to catalyse the diuron removal along with its main by-product representing a step forward for bioremediation of real diuron-contaminated waters using such technology (Hu et al., 2020). Carbofuran and carbaryl have been removed from soil by newly isolated *Acremonium* sp. employing a central composite rotatable method. The carbamate-contaminated soils were successfully remediated by carb-PV5 strain (Kaur and Balomajumder, 2020). The presence of mushrooms in the soil also promotes remediation by secreting laccase and manganese-dependent peroxidase (MnP). It promotes higher microbial counts in soil, consequently resulting in remediated co-contaminated soil with heavy metals and pesticides by strains such as *Clitocybe maxima, Pleurotus eryngii* and *Coprinus comatus* through various mechanisms (Zhang et al., 2020).

2.3.3 Algal Role in Remediation of Pesticides

Recently, the utilization of microalgae (Table 2.3) has been demonstrated for the efficient treatment of pesticidal pollution by removal of nutrients, pesticides, toxic

TABLE 2.3
Algal Species Used for Remediation of Pesticides

Pesticides	Algal Species used for Remediation of Pesticides	Reference
Dimethomorph, pyrimethanil, isoproturon	*Scenedesmus quadricauda*	Dosnon-Olette et al. (2010)
and fluroxypyr	*Chlamydomonas reinhardtii*	Zhang et al. (2011)
Organophosphorus	*Synechococcus elongatus*	Subramaniyan (2012)
Organochlorine	*Microcystis aeruginosa*	
Mesotrione	*Pediastrum tetras*, *Ankistrodesmus fusiformis* and *Amphora coffeaeformis*	Valiente Moro et al. (2012)
Bisphenol	*Monoraphidium braunii*	Gattullo et al. (2012)
2,4-Dichlorophenoxyacetic	*Gracilaria verrucosa*	Ata et al. (2012)
Hexachlorocyclohexane	*Sphingobium japonicum*	Cao et al. (2013)
Malathion	*Anabaena oryzae*, *Nostoc muscorum* and *Spirulina platensis*	Ibrahim et al. (2014)
Prometryne, fluroxypyr and isoproturon	*Chlamydomonas reinhardtii*	Chekroun et al. (2014)
Naphthalene	*Dunaliella* sp.	Biswas et al. (2015)
Dichloro-diphenyl-trichloroethane	*Cylindrotheca* sp.	
Diazinon	*C. vulgaris*, *S. obliquus*, *Chlamydomonas Mexicana* and *Chlamydomonas pitschmannii*	Kurade et al. (2016)
Triadimefon	*Scenedesmus obliquus*	Xu and Huang (2017)
Trichlorobenzene	*Scenedesmus quadricauda*	Kováčik et al. (2018)
Hexachlorocyclohexane	*Coccomyxa subellipsoidea*	
Malathion	*Chlorella vulgaris*, *Scenedesmus quadricauda* and *Spirulina platensis*	Abdel-Razek et al. (2019)

elements, pharmaceutical chemicals and oil from wastewater (Ummalyma et al., 2018) mainly due to the ability of microalgae to utilize organic matter as carbon and nutrient sources present in wastewater. In addition, microalgae treat wastewater in a cost-effective way, which ultimately renders more algal cultivation (Barros et al., 2015; Nie et al., 2020). The efficiency of pesticide bioremediation through various microalgae is relatively higher than that of photo-degradation by natural light (Warren et al., 2003). Various studies have also shown that algal cells have the potential to remove radioactive pollutants from waste (Singh et al., 2017). The marine *Chlorella* belonging to Chlorellaceae algal family is an active pesticide decomposer (Huang et al., 2018). *Oscillatoria, Anabaena* and *Nostoc* belong to cyanobacteria, whereas *Scenedesmus* and *Stichococcus* are other potential decomposers of organophosphates from algal genera (Bhattacharjee et al., 2020). To decontaminate agricultural run-off in terms of NH_4^+-N as well as other pesticides in water, a tubular horizontal photobioreactor with a microalgae-based treatment system has been used to eliminate a wide range of priority pesticides under real environmental conditions (García-Galán et al., 2020). Similar mechanisms, namely bioadsorption, bioaccumulation and biodegradation, are utilized by algae to remove pesticides. Around 87–96% of various pesticides (atrazine, simazine, isoproturon, molinate, carbofuran, metolacholar, propanil, dimethoate, pyriproxyfen and pendimethalin) have been demonstrated to be removed via bioadsorption through different mechanisms such as electrostatic interaction, surface complexation, ion exchange, absorption and precipitation in aqueous phase by cultivated algae (Hussein et al., 2017; Bilal et al., 2018). Malathion, an organophosphorus pesticide, showed removal efficiencies of 91%, 65% and 54%, when biodegraded using *N. muscorum, A. oryzae* and *S. platensis,* respectively (Ibrahim et al., 2014). *Chlamydomonas* sp., *Scenedesmus* sp. MM1, *Scenedesmus* sp. MM2, *Stichococcus* sp. and *Chlorella* sp. are select green algae with high detoxification ability for fenamiphos. *Chlorella* sp. can achieve above 99% fenamiphos removal efficiency (Cáceres et al., 2008).

For the bioremediation of aquatic systems contaminated by organochlorine pesticides, several microalgal species have been studied. On exposure to increasing lindane levels, wild-type strain *Scenedesmus intermedius* is known to adapt the lindane contamination, resulting in lindane-resistant cells by rare spontaneous mutations, evident through fluctuation analysis. These cells have a great capability to remove lindane until 99%, presenting an opportunity for clean-up of lindane and other organochlorine polluted habitats (González et al., 2012). Hence, microalgal technology can be efficaciously used by research endeavours and industries for simultaneous degradation and removal of different pesticides along with the production of high-value products (Nie et al., 2020).

2.4 LIMITATIONS OF BIOREMEDIATION

Regardless of the numerous advantages, bioremediation displays limitations. For instance, microbes remediating pollutants under suitable conditions work for their benefit and survival. Therefore, if the environmental state gets inappropriately altered for the microbial growth other than their acceptable conditions, bioremediation can get afflicted (Varshney, 2019). Secondly, some types of environmental modifications are unavoidably required to encourage a few microbes to degrade or take up the pollutant at an acceptable rate (Ummalyma et al., 2018). Moreover, excessive supplementary nutrient and fertilizer supply in the soil may also inhibit soil organisms in situ. Another limitation is that few microbes require low levels of the pollutants initially so that they get a certain period time to induce metabolic pathways needed to digest the pollutant (Singh, 2008). The bioremediated species have stronger adaptability towards the environment, enabling it to reduce the harm of invasive species. But under controlled and suitable conditions, bioremediation through microbes remains the most convenient and cost-effective method for pesticide removal and degradation (Jaiswal et al., 2019).

2.5 FUTURE PROSPECTS FOR BIOREMEDIATION OF PESTICIDES OTHER THAN MICROBIAL REMEDIATION

Besides using microbes, various other remediation techniques such as adsorption and advanced oxidation have also been explored (Rani et al., 2020). The addition of granular repellent substances in pesticide formulations has been tried by some manufacturers to avoid toxic chemical ingestion by non-target animals (Martínez-Haro et al., 2008). An emerging tool for sensing and remediating pesticides is the use of nanotechnology. Various nanotubes (carbon and halloysite) and nanoparticles (metal, bimetallic and metal oxide) can be used for the detection and simultaneously degradation of pesticides. Furthermore, acetylcholinesterase, organophosphate hydrolase and laccase can be used as enzymes in the enzyme-based biosensors for easy pesticide detection (Samsidar et al., 2018; Rawtani et al., 2018). Bioremediation of heavy metals, persistent organic pollutants, petroleum, xenobiotics and acid drainage can be performed using gene editing (Figure 2.2) and systemic biology tools.

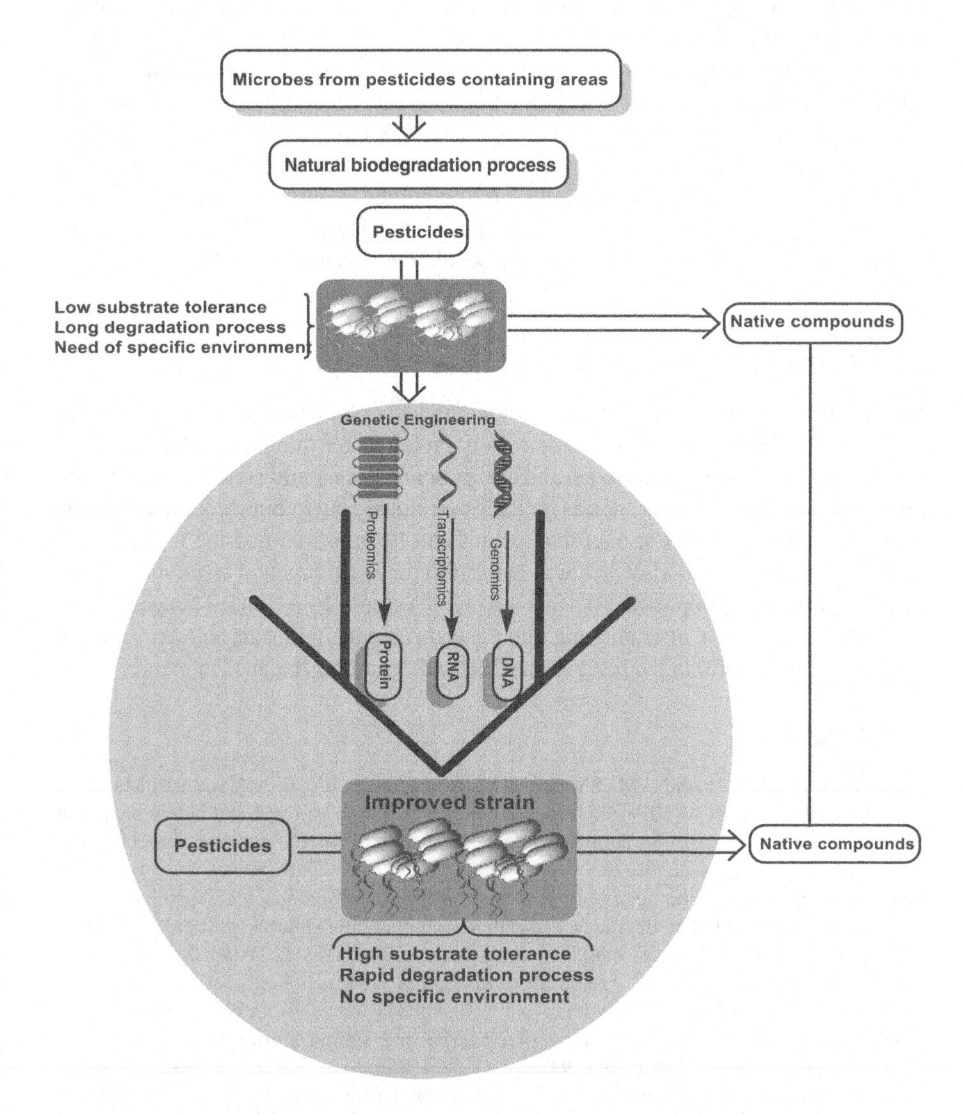

FIGURE 2.2 Overview of natural bioremediation process and proposed scheme representing strain improvement for bioremediation.

A functional gene of interest can be inserted into a microbe designed for the degradation of a particular recalcitrant for improved bioremediation using gene-editing tools like clustered regularly interspaced short palindromic repeats (CRISPR-Cas), transcription activator-like effector nucleases (TALEN) and zinc-finger nucleases (ZFNs) (Singh et al., 2017; Jaiswal et al., 2019, Ruwali et al., 2022).

The utilization of microalgae has also been recently acknowledged by many researchers to eradicate toxic elements, pharmaceutical chemicals and pesticides from wastewater. Genetic engineering can be applied to improve the available microalgae strains against pesticides by genetically modifying them for the removal of contaminants. This aspect has not yet been explored (Nie et al., 2020). To avoid depend on conventional chemical pesticides, eco-friendly bio-pesticides such as *Pseudomonas* sp., *Serratia* sp. and *Bacillus* sp. are a good alternative with highly targeted action mechanisms, no negative impact on surroundings and great cost efficiency (Keswani et al., 2020). Since there is a high demand for agricultural commodities, the prohibition of pesticides cannot be done. But their distribution can be controlled by environmental education producing better outcomes rather than banning them. Therefore, utilization of such advanced techniques for the degradation of toxic pesticides can be a promising aid towards the betterment of the environment and all other life forms.

2.6 CONCLUSION

Bioremediation is a safe environment restoration technology with less risk of secondary pollution. Therefore, microbes are mostly preferred for this process because they utilize pesticides as carbon/nitrogen and energy source. In natural conditions, microbes utilize the low amount of pesticides as their nutritional source but through genomics, proteomics, transcriptomics; microbial strains can be improved and the utilization of pesticides can be increased, which will be helpful for environmental restoration. Thus, there is a need to develop such efficient recombinant strains by advanced biotechnological applications which, after entering into the environment, will start auto-recycling of pesticides, thus helpful in the restoration of natural conditions/balance in nature.

REFERENCES

Abdel-Razek MA, Abozeid AM, Eltholth MM, Abouelenien FA, El-Midany SA, Moustafa NY, Mohamed RA (2019) Bioremediation of a pesticide and selected heavy metals in wastewater from various sources using a consortium of microalgae and cyanobacteria. Slov Vet 56(Suppl 22):61–73.

Abo-Amer AE (2012) Characterization of a strain of *Pseudomonas putida* isolated from agricultural soil that degrades cadusafos (an organophosphorus pesticide). World J Microbiol Biotechnol 28(3):805–14. https://doi.org/10.1007/s11274-011-0873-5

Agarry SE, Olu-Arotiowa OA, Aremu MO, Jimoda LA (2013) Biodegradation of dichlorovos (organophosphate pesticide) in soil by bacterial isolates. Biodegradation 3(8):11–6.

Ahmad KS (2020) Remedial potential of bacterial and fungal strains (*Bacillus subtilis, Aspergillus niger, Aspergillus flavus* and *Penicillium chrysogenum*) against organochlorine insecticide Endosulfan. Folia Microbiol 65:801–10. https://doi.org/10.1007/s12223-020-00792-7

Ahmad KS, Gul P, Gul MM (2020) Efficient fungal and bacterial facilitated remediation of thiencarbazone methyl in the environment. Environ Res 188:109811. https://doi.org/10.1016/j.envres.2020.109811

Alarcón PA, Lambertucci SA (2018) Pesticides thwart condor conservation. Science 360(6389):612. https://doi.org/10.1126/science.aat6039

Arora PK, Sasikala C, Ramana CV (2012) Degradation of chlorinated nitroaromatic compounds. Appl Microbiol Biotechnol 93(6):2265–77. https://doi.org/10.1007/s00253-012-3927-1

Ata A, Nalcaci OO, Ovez B (2012) Macro algae *Gracilaria verrucosa* as a biosorbent: a study of sorption mechanisms. Algal Res 1(2):194–204. https://doi.org/10.1016/j.algal.2012.07.001

Baghapour MA, Nasseri S, Derakhshan Z (2013) Atrazine removal from aqueous solutions using submerged biological aerated filter. J Environ Health Sci Eng 11(1):1–9. https://doi.org/10.1186/2052-336x-11-6

Barros AI, Gonçalves AL, Simões M, Pires JC (2015) Harvesting techniques applied to microalgae: a review. Renew Sustain Energy Rev 41:1489–500. https://doi.org/10.1016/j.rser.2014.09.037

Bhattacharjee G, Gohil N, Vaidh S, Joshi K, Vishwakarma GS, Singh V (2020) Microbial bioremediation of industrial effluents and pesticides. In: Pandey V, Singh V (Eds) Bioremediation of Pollutants. Elsevier, London, pp 287–302

Bhatt P, Zhang W, Lin Z, Pang S, Huang Y, Chen S (2020) Biodegradation of allethrin by a novel fungus *Fusarium proliferatum* strain CF2, isolated from contaminated soils. Microorganisms 8(4):593. https://doi.org/10.3390/microorganisms8040593

Bilal M, Rasheed T, Sosa-Hernández JE, Raza A, Nabeel F, Iqbal H (2018) Biosorption: an interplay between marine algae and potentially toxic elements—a review. Mar Drugs 16(2):65. https://dx.doi.org/10.3390%2Fmd16020065

Biswas K, Paul D, Sinha SN (2015) Biological agents of bioremediation: A concise review. Front Environ Microbiol 1:39–4. https://link.springer.com/article/10.1007/s13201-016-0461-9

Boudh S, Singh JS (2019) Pesticide contamination: environmental problems and remediation strategies. In: Bharagava RN, Chowdhary P (Eds) Emerging and Eco-Friendly Approaches for Waste Management. Springer, Singapore, pp 245–69.

Buvaneswari G, Thenmozhi R, Nagasathya A, Thajuddin N, Kumar P (2018) GC-MS and molecular analyses of monocrotophos biodegradation by selected bacterial isolates. Afr J Microbiol Res 12(3):52–61. https://doi.org/10.5897/AJMR2017.8696

Cáceres TP, Megharaj M, Naidu R (2008) Biodegradation of the pesticide fenamiphos by ten different species of green algae and cyanobacteria. Curr Microbiol 57(6):643–6. https://doi.org/10.1007/s00284-008-9293-7

Cao X, Yang C, Liu R, Li Q, Zhang W, Liu J et al (2013) Simultaneous degradation of organophosphate and organochlorine pesticides by *Sphingobium japonicum* UT26 with surface-displayed organophosphorus hydrolase. Biodegradation 24(2):295–303. https://doi.org/10.1007/s10532-012-9587-0

Chanika E, Georgiadou D, Soueref E, Karas P, Karanasios E, Tsiropoulos NG, Tzortzakakis EA, Karpouzas DG (2011) Isolation of soil bacteria able to hydrolyze both organophosphate and carbamate pesticides. Bioresour Technol 102(3):3184–92. https://doi.org/10.1016/j.biortech.2010.10.145

Chekroun KB, Sánchez E, Baghour M (2014) The role of algae in bioremediation of organic pollutants. Int Res J Public Environ Health 1(2):19–32.

Chen S, Liu C, Peng C, Liu H, Hu M, Zhong G (2012) Biodegradation of chlorpyrifos and its hydrolysis product 3,5,6-trichloro-2-pyridinol by a new fungal strain *Cladosporium cladosporioides* Hu-01. PLoS ONE 7(10):e47205. https://dx.doi.org/10.1371%2Fjournal.pone.0047205

Chrzanowski Ł, Dziadas M, Ławniczak Ł, Cyplik P, Białas W, Szulc A, Lisiecki P, Jeleń H (2012) Biodegradation of rhamnolipids in liquid cultures: effect of biosurfactant dissipation on diesel fuel/B20 blend biodegradation efficiency and bacterial community composition. Bioresour Technol 111:328–35. https://doi.org/10.1016/j.biortech.2012.01.181

Coxall M (2014) Ethical Eating: A Complete Guide to Sustainable Food. In: Caswell G (Ed), 1st edn. Cornelio Books, Spain, pp 1–552.

Damalas CA, Eleftherohorinos IG (2011) Pesticide exposure, safety issues, and risk assessment indicators. Int J Environ Res Public Health 8(5):1402–19. https://doi.org/10.3390/ijerph8051402

Deng W, Lin D, Yao K, Yuan H, Wang Z, Li J, Zou L, Han X, Zhou K, He L, Hu X (2015) Characterization of a novel β-cypermethrin-degrading *Aspergillus niger* YAT strain and the biochemical degradation pathway of β-cypermethrin. Appl Microbiol Biotechnol 99(19):8187–98. https://doi.org/10.1007/s00253-015-6690-2

Dey P, Malik A, Mishra A, Singh DK, von Bergen M, Jehmlich N (2020) Mechanistic insight to mycoremediation potential of a metal resistant fungal strain for removal of hazardous metals from multimetal pesticide matrix. Environ Pollut 262:114255. https://doi.org/10.1016/j.envpol.2020.114255

Diez MC, Gallardo F, Tortella G, Rubilar O, Navia R, Bornhardt C (2012) Chlorophenol degradation in soil columns inoculated with *Anthracophyllum discolor* immobilized on wheat grains. J Environ Manage 95:S83–7. https://doi.org/10.1016/j.jenvman.2010.09.024

Dosnon-Olette R, Trotel-Aziz P, Couderchet M, Eullaffroy P (2010) Fungicides and herbicide removal in Scenedesmus cell suspensions. Chemosphere 79:117–123. https://doi.org/10.1016/j.chemosphere.2010.02.005

Dritsa V, Rigas F, Doulia D, Avramides EJ, Hatzianestis I (2009) Optimization of culture conditions for the biodegradation of lindane by the polypore fungus *Ganoderma australe*. Water Air Soil Pollut 204(1):19–27. https://doi.org/10.1007/s11270-009-0022-z

Ellegaard-Jensen L, Aamand J, Kragelund BB, Johnsen AH, Rosendahl S (2013) Strains of the soil fungus *Mortierella* show different degradation potentials for the phenylurea herbicide diuron. Biodegradation 24(6):765–74. https://doi.org/10.1007/s10532-013-9624-7

Erguven GO (2018) Comparison of some soil fungi in bioremediation of herbicide acetochlor under agitated culture media. Bull Environ Contam Toxicol 100(4):570–5. https://doi.org/10.1007/s00128-018-2280-1

Erguven GÖ, Koçak E (2019) Determining the detoxification potential of some soil bacteria and plants on bioremediation of deltamethrin, fenvalerate and permethrin pesticides. EJAR 3(1):36–47.

Erguven GO, Nuhoglu Y (2020) Bioremediation of fluazifop-p-butyl herbicide by some soil bacteria isolated from various regions of Turkey in an artificial agricultural field. Environ Prot Eng 46(3):5–15. http://dx.doi.org/10.37190/epe200301

Erguven GO, Yildirim N (2019) The evaluation of imidacloprid remediation in soil media by two bacterial strains. Curr Microbiol 76(12):1461–6. https://doi.org/10.1007/s00284-019-01774-w

Fatima SA, Hamid A, Yaqub G, Javed A, Akram H (2018) Detection of volatile organic compounds in blood of farmers and their general health and safety profile. Nat Environ Pollut Technol 17(2):657–60.

Gago-Ferrero P, Badia-Fabregat M, Olivares A, Piña B, Blánquez P, Vicent T, Caminal G, Díaz-Cruz MS, Barceló D (2012) Evaluation of fungal- and photo-degradation as potential treatments for the removal of sunscreens BP3 and BP1. Sci Total Environ 427:355–63. https://doi.org/10.1016/j.scitotenv.2012.03.089

García-Galán MJ, Monllor-Alcaraz LS, Postigo C, Uggetti E, de Alda ML, García J, Díez-Montero R (2020) Microalgae-based bioremediation of water contaminated by pesticides in peri-urban agricultural areas. Environ Pollut 265:114579. https://doi.org/10.1016/j.envpol.2020.114579

Gattullo CE, Bährs H, Steinberg CE, Loffredo E (2012) Removal of bisphenol A by the freshwater green alga *Monoraphidium braunii* and the role of natural organic matter. Sci Total Environ 416: 501–506. https://doi.org/10.1016/j.scitotenv.2011.11.033

Góngora-Echeverría VR, García-Escalante R, Rojas-Herrera R, Giácoman-Vallejos G, Ponce-Caballero C (2020) Pesticide bioremediation in liquid media using a microbial consortium and bacteria-pure strains isolated from a biomixture used in agricultural areas. Ecotoxicol Environ Saf 200:110734. https://doi.org/10.1016/j.ecoenv.2020.110734

González R, García-Balboa C, Rouco M, Lopez-Rodas V, Costas E (2012) Adaptation of microalgae to lindane: a new approach for bioremediation. Aquat Toxicol 109:25–32. https://doi.org/10.1016/j.aquatox.2011.11.015

Gupta S, Wali A, Gupta M, Annepu SK (2017) Fungi: an effective tool for bioremediation. In: Singh D, Singh H, Prabha R (Eds) Plant-Microbe Interactions in Agro-Ecological Perspectives. Springer, Singapore, pp 593–606.

Helal IM, Abo-El-Seoud MA (2015) Fungal biodegradation of pesticide vydate in soil and aquatic system. In: 4th International Conference on Radiation Sciences and Applications, Taba, Egypt, pp 13–17.

Hrynkiewicz K, Złoch M, Kowalkowski T, Baum C, Buszewski B (2018) Efficiency of microbially assisted phytoremediation of heavy-metal contaminated soils. Environ Rev 26(3): 316–32. https://doi.org/10.1139/er-2018-0023

Huang Y, Xiao L, Li F, Xiao M, Lin D, Long X, Wu Z (2018) Microbial degradation of pesticide residues and an emphasis on the degradation of cypermethrin and 3-phenoxy benzoic acid: a review. Molecules 23(9):2313. https://doi.org/10.3390/molecules23092313

Hussaini SZ, Shaker M, Iqbal MA (2013) Isolation of fungal isolates for degradation of selected pesticides. BEPLS 2(4):50–3.

Hussein MH, Abdullah AM, Badr El-Din NI, Mishaqa ES (2017) Biosorption potential of the microchlorophyte *Chlorella vulgaris* for some pesticides. J Fertil Pestic 8(1). https://doi.org/10.4172/2471-2728.1000177

Hu K, Torán J, López-García E, Barbieri MV, Postigo C, de Alda ML, Caminal G, Sarrà M, Blánquez P (2020) Fungal bioremediation of diuron-contaminated waters: evaluation of its degradation and the effect of amendable factors on its removal in a trickle-bed reactor under non-sterile conditions. Sci Total Environ 743:140628. https://doi.org/10.1016/j.scitotenv.2020.140628

Ibrahim WM, Karam MA, El-Shahat RM, Adway AA (2014) Biodegradation and utilization of organophosphorus pesticide malathion by cyanobacteria. BioMed Res Int 2014. https://doi.org/10.1155/2014/392682

Iqbal MA, Bartakke KV (2014) Isolation of pesticide degrading microorganisms from soil. Adv Biores 5(4):164–8.

Jaiswal S, Singh DK, Shukla P (2019) Gene editing and systems biology tools for pesticide bioremediation: a review. Front Microbiol 10:87. https://dx.doi.org/10.3389%2Ffmicb.2019.00087

Jayaraj R, Megha P, Sreedev P (2016) Organochlorine pesticides, their toxic effects on living organisms and their fate in the environment. Interdiscip Toxicol 9(3–4):90–100. https://dx.doi.org/10.1515%2Fintox-2016-0012

Kafilzadeh F, Ebrahimnezhad M, Tahery Y (2015) Isolation and identification of endosulfan-degrading bacteria and evaluation of their bioremediation in Kor River, Iran. Osong Public Health Res Perspect 6(1):39–46. https://dx.doi.org/10.1016%2Fj.phrp.2014.12.003

Karishma B, Sharma HP (2014) Isolation and characterization of organophosphorus pesticide degrading bacterial isolates. Arch Appl Sci Res 6(5):144–9. https://doi.org/10.3844/ojbsci.2015.113.125

Kaur P, Balomajumder C (2020) Bioremediation process optimization and effective reclamation of mixed carbamate-contaminated soil by newly isolated *Acremonium* sp. Chemosphere 249:125982. https://doi.org/10.1016/j.chemosphere.2020.125982

Kaur H, Kapoor S, Kaur G (2016) Application of ligninolytic potentials of a white-rot fungus *Ganoderma lucidum* for degradation of lindane. Environ Monit Assess 188(10):588. https://doi.org/10.1007/s10661-016-5606-7

Keswani C, Singh HB, García-Estrada C, Caradus J, He YW, Mezaache-Aichour S, Glare TR, Borriss R, Sansinenea E (2020) Antimicrobial secondary metabolites from agriculturally important bacteria as next-generation pesticides. Appl Microbiol Biotechnol 104(3):1013–34. https://doi.org/10.1007/s00253-019-10300-8

Kim KH, Kabir E, Jahan SA (2017) Exposure to pesticides and the associated human health effects. Sci Total Environ 575:525–35. https://doi.org/10.1016/j.scitotenv.2016.09.009

Kováčik J, Antoš V, Micalizzi G, Dresler S, Hrabák P, Mondello L (2018) Accumulation and toxicity of organochlorines in green microalgae. J Hazard Mater 347:168–75. https://doi.org/10.1016/j.jhazmat.2017.12.056

Kumar S, Kaushik G, Dar MA, Nimesh S, Lopez-Chuken UJ, Villarreal-Chiu JF (2018) Microbial degradation of organophosphate pesticides: a review. Pedosphere 28(2): 190–208. https://doi.org/10.1016/S1002-0160(18)60017-7

Kurade MB, Kim JR, Govindwar SP, Jeon BH (2016) Insights into microalgae mediated biodegradation of diazinon by *Chlorella vulgaris*: microalgal tolerance to xenobiotic pollutants and metabolism. Algal Res 20:126–34. https://doi.org/10.1016/j.algal.2016.10.003

Ledger T, Pieper DH, González B (2006) Chlorophenol hydroxylases encoded by plasmid pJP4 differentially contribute to chlorophenoxyacetic acid degradation. Appl Environ Microbiol 72(4):2783–92. https://doi.org/10.1128/AEM.72.4.2783-2792.2006

León-Santiesteban HH, Wrobel K, Revah S, Tomasini A (2016) Pentachlorophenol removal by *Rhizopus oryzae* CDBB-H-1877 using sorption and degradation mechanisms. J Chem Technol Biotechnol 91(1):65–71. https://doi.org/10.1002/jctb.4566

Li Z, Jennings A (2017) Worldwide regulations of standard values of pesticides for human health risk control: a review. Int J Environ Res Public Health 14(7):826. https://doi.org/10.3390/ijerph14070826

Mahmood I, Imadi SR, Shazadi K, Gul A, Hakeem KR (2016) Effects of pesticides on environment. In: Hakeem K, Akhtar M, Abdullah S (Eds) Plant, Soil and Microbes. Springer, Cham. pp 253–69.

Malaj E, Liber K, Morrissey CA (2020) Spatial distribution of agricultural pesticide use and predicted wetland exposure in the Canadian Prairie Pothole Region. Sci Total Environ 718:134765. https://doi.org/10.1016/j.scitotenv.2019.134765

Maqbool F, Mostafalou S, Bahadar H, Abdollahi M (2016) Review of endocrine disorders associated with environmental toxicants and possible involved mechanisms. Life Sci 145:265–73. https://doi.org/10.1016/j.lfs.2015.10.022

Martínez-Haro M, Mateo R, Guitart R, Soler-Rodríguez F, Pérez-López M, María-Mojica P, García-Fernández AJ (2008) Relationship of the toxicity of pesticide formulations and their commercial restrictions with the frequency of animal poisonings. Ecotoxicol Environ Saf 69(3):396–402. https://doi.org/10.1016/j.ecoenv.2007.05.006

Meftaul IM, Venkateswarlu K, Dharmarajan R, Annamalai P, Megharaj M (2020) Pesticides in the urban environment: a potential threat that knocks at the door. Sci Total Environ 711:134612. https://doi.org/10.1016/j.scitotenv.2019.134612

Mehrpour O, Karrari P, Zamani N, Tsatsakis AM, Abdollahi M (2014) Occupational exposure to pesticides and consequences on male semen and fertility: a review. Toxicol Lett 230(2):146–56. https://doi.org/10.1016/j.toxlet.2014.01.029

Mostafalou S, Abdollahi M (2017) Pesticides: an update of human exposure and toxicity. Arch Toxicol 91(2):549–99. https://doi.org/10.1007/s00204-016-1849-x

Mudhoo A, Bhatnagar A, Rantalankila M, Srivastava V, Sillanpää M (2019) Endosulfan removal through bioremediation, photocatalytic degradation, adsorption and membrane separation processes: a review. Chem Eng Technol 360:912–28. https://doi.org/10.1016/j.cej.2018.12.055

Murthy HR, Manonmani HK (2007) Aerobic degradation of technical hexachlorocyclohexane by a defined microbial consortium. J Hazard Mater 149(1):18–25. https://doi.org/10.1016/j.jhazmat.2007.03.053

Narendra M, Kavitha G, Kiranmai AH, Rao NR, Varadacharyulu NC (2008) Chronic exposure to pyrethroid-based allethrin and prallethrin mosquito repellents alters plasma biochemical profile. Chemosphere 73(3):360–4. https://doi.org/10.1016/j.chemosphere.2008.05.070

Nayak SK, Dash B, Baliyarsingh B (2018) Microbial remediation of persistent agro-chemicals by soil bacteria: an overview. Microb Biotechnol 2018:275–301. https://doi.org/10.1007/978-981-10-7140-9_13

Nie J, Sun Y, Zhou Y, Kumar M, Usman M, Li J, Shao J, Wang L, Tsang DC (2020) Bioremediation of water containing pesticides by microalgae: mechanisms, methods, and prospects for future research. Sci Total Environ 707:136080. https://doi.org/10.1016/j.scitotenv.2019.136080

Nwachukwu EO, Osuji JO (2007) Bioremedial degradation of some herbicides by indigenous white rot fungus, *Lentinus subnudus*. J Plant Sci 2:619–24. https://dx.doi.org/10.3923/jps.2007.619.624

Ortiz-Hernández M, Sánchez-Salinas E (2010) Biodegradation of the organophosphate pesticide tetrachlorvinphos by bacteria isolated from agricultural soils in México. Revista internacional de contaminación ambiental 26(1):27–38.

Peter L, Gajendiran A, Mani D, Nagaraj S, Abraham J (2015) Mineralization of malathion by *Fusarium oxysporum* strain JASA1 isolated from sugarcane fields. Environ Prog Sustain Energy 34(1):112–6. https://doi.org/10.1002/ep.11970

Phugare SS, Gaikwad YB, Jadhav JP (2012) Biodegradation of acephate using a developed bacterial consortium and toxicological analysis using earthworms (*Lumbricus terrestris*) as a model animal. Int Biodeterior Biodegradation 69:1–9. https://doi.org/10.1016/j.ibiod.2011.11.013

Plattner J, Kazner C, Naidu G, Wintgens T, Vigneswaran S (2018) Pesticide and microbial contaminants of groundwater and their removal methods: a mini review. JJSA 1(1):12–8.

Plaza PI, Martínez-López E, Lambertucci SA (2019) The perfect threat: pesticides and vultures. Sci Total Environ 687:1207–18. https://doi.org/10.1016/j.scitotenv.2019.06.160

Pokhrel B, Gong P, Wang X, Chen M, Wang C, Gao S (2018) Distribution, sources, and air–soil exchange of OCPs, PCBs and PAHs in urban soils of Nepal. Chemosphere 200:532–41. https://doi.org/10.1016/j.chemosphere.2018.01.119

Postigo C, García-Galán MJ, Köck-Schulmeyer M, Barceló D (2015) Occurrence of polar organic pollutants in groundwater bodies of Catalonia. In: Munné A, Ginebreda A, Prat N (Eds) Experiences from Ground, Coastal and Transitional Water Quality Monitoring. Springer, Cham. pp. 63–89.

Prabha R, Singh DP, Verma MK (2017) Microbial interactions and perspectives for bioremediation of pesticides in the soils. In: Malik JA, Goyal MR, Wani KA (Eds) Plant-Microbe Interactions in Agro-Ecological Perspectives. Springer, Singapore, pp 649–71.

Purnomo AS, Nawfa R, Martak F, Shimizu K, Kamei I (2017) Biodegradation of aldrin and dieldrin by the white-rot fungus *Pleurotus ostreatus*. Curr Microbiol 74(3):320–4. https://doi.org/10.1007/s00284-016-1184-8

Quan XC, Tang H, Xiong WC, Yang ZF (2010) Bioaugmentation of aerobic sludge granules with a plasmid donor strain for enhanced degradation of 2,4-dichlorophenoxyacetic acid. J Hazard Mater 179(1–3):1136–42. https://doi.org/10.1016/j.jhazmat.2010.04.002

Raimondo EE, Aparicio JD, Bigliardo AL, Fuentes MS, Benimeli CS (2020) Enhanced bioremediation of lindane-contaminated soils through microbial bioaugmentation assisted by biostimulation with sugarcane filter cake. Ecotoxicol Environ Saf 190:110143. https://doi.org/10.1016/j.ecoenv.2019.110143

Raimondo EE, Aparicio JD, Briceño GE, Fuentes MS, Benimeli CS (2019) Lindane bioremediation in soils of different textural classes by an actinobacteria consortium. J Soil Sci Plant Nutr 19(1):29–41. https://doi.org/10.1007/s42729-018-0003-7

Rani L, Thapa K, Kanojia N, Sharma N, Singh S, Grewal AS, Srivastav AL, Kaushal J (2020) An extensive review on the consequences of chemical pesticides on human health and environment. J Clean Prod 238:124657. https://doi.org/10.1016/j.jclepro.2020.124657

Rath B (2012) Microalgal bioremediation: current practices and perspectives. J Biochem Technol 3(3):299–304.

Rawtani D, Khatri N, Tyagi S, Pandey G (2018) Nanotechnology-based recent approaches for sensing and remediation of pesticides. J Environ Manage 206:749–62. https://doi.org/10.1016/j.jenvman.2017.11.037

Richards N (Ed) (2011) Carbofuran and Wildlife Poisoning: Global Perspectives and Forensic Approaches. John Wiley & Sons, New York pp 251–61. https://doi.org/10.1002/9781119998532.ch9

Richardson JR, Fitsanakis V, Westerink RH, Kanthasamy AG (2019) Neurotoxicity of pesticides. Acta Neuropathol 138:343–62. https://doi.org/10.1007/s00401-019-02033-9

Ruwali P, Pandey N, Ambwani TK, Singh RV (2022) Impact of Cadmium Toxicity on Environment and Its Remedy. In: Kumar V, Garg VK, Kumar S, Biswas JK (Eds) Omics for Environmental Engineering and Microbiology Systems, pp 385–401. CRC Press, Boca Raton. https://doi.org/10.1201/9781003247883-19.

Sabarwal A, Kumar K, Singh RP (2018) Hazardous effects of chemical pesticides on human health–cancer and other associated disorders. Environ Toxicol Pharmacol 63:103–14. https://doi.org/10.1016/j.etap.2018.08.018

Sagar V, Singh DP (2011) Biodegradation of lindane pesticide by non-white-rots soil fungus *Fusarium* sp. World J Microbiol Biotechnol 27(8):1747–54. https://doi.org/10.1007/s11274-010-0628-8

Saleh IA, Zouari N, Al-Ghouti MA (2020) Removal of pesticides from water and wastewater: chemical, physical and biological treatment approaches. Environ Technol Innov 19:101026. https://doi.org/10.1016/j.eti.2020.101026

Samsidar A, Siddiquee S, Shaarani SM (2018) A review of extraction, analytical and advanced methods for determination of pesticides in environment and foodstuffs. Trends Food Sci Technol 71:188–201. https://doi.org/10.1016/j.tifs.2017.11.011

Sharma P, Chopra A, Cameotra SS, Suri CR (2010) Efficient biotransformation of herbicide diuron by bacterial strain *Micrococcus* sp. PS-1. Biodegradation 21(6):979–87. https://doi.org/10.1007/s10532-010-9357-9

Sidhu GK, Singh S, Kumar V, Dhanjal DS, Datta S, Singh J (2019) Toxicity, monitoring and biodegradation of organophosphate pesticides: a review. Crit Rev Environ Sci Technol 49(13):1135–87. https://doi.org/10.1080/10643389.2019.1565554

Simeonov LI, Macaev FZ, Simeonova BG (Eds) (2014) Environmental Security Assessment and Management of Obsolete Pesticides in Southeast Europe. Springer, Netherlands.

Simon-Delso N, Amaral-Rogers V, Belzunces LP, Bonmatin JM, Chagnon M, Downs C, Furlan L, Gibbons DW, Giorio C, Girolami V, Goulson D (2015) Systemic insecticides (neonicotinoids and fipronil): trends, uses, mode of action and metabolites. Environ Sci Pollut 22(1):5–34. https://doi.org/10.1007/s11356-014-3470-y

Singh DK (2008) Biodegradation and bioremediation of pesticide in soil: concept, method and recent developments. Indian J Microbiol 48(1):35–40. https://dx.doi.org/10.1007%2Fs12088-008-0004-7

Singh N, Gupta VK, Kumar A, Sharma B (2017) Synergistic effects of heavy metals and pesticides in living systems. Front Chem 5:70. https://doi.org/10.3389/fchem.2017.00070

Singh M, Pant G, Hossain K, Bhatia AK (2017) Green remediation. Tool for safe and sustainable environment: a review. Appl Water Sci 7(6):2629–35. https://doi.org/10.1007/s13201-016-0461-9

Singh NS, Sharma R, Parween T, Patanjali PK (2018) Pesticide contamination and human health risk factor. In: Oves M, Khan MZ, Ismail IMI (Eds) Modern Age Environmental Problems and their Remediation. Springer, Cham. pp 49–68.

Singh BK, Walker A (2006) Microbial degradation of organophosphorus compounds. FEMS Microbiol Rev 30(3):428–71. https://doi.org/10.1111/j.1574-6976.2006.00018.x

Subramaniyan V (2012) Potential applications of cyanobacteria in industrial effluents – a review. J Bioremediat Biodegrad 3(6). https://doi.org/10.4172/2155-6199.1000154

Tago K, Sekiya E, Kiho A, Katsuyama C, Hoshito Y, Yamada N, Hirano K, Sawada H, Hayatsu M (2006) Diversity of fenitrothion-degrading bacteria in soils from distant geographical areas. Microbes Environ 21(1):58–64. https://doi.org/10.1264/jsme2.21.58

Tang W, Wang DI, Wang J, Wu Z, Li L, Huang M, Xu S, Yan D (2018) Pyrethroid pesticide residues in the global environment: an overview. Chemosphere 191:990–1007. https://doi.org/10.1016/j.chemosphere.2017.10.115

Terzopoulou E, Voutsa D (2017) Study of persistent toxic pollutants in a river basin—ecotoxicological risk assessment. Ecotoxicology 26(5):625–38. https://doi.org/10.1007/s10646-017-1795-2

Tewari L, Saini J, Arti (2012) Bioremediation of pesticides by microorganisms: general aspects and recent advances. In: Maheshwari DK, Dubey RC (Eds) Bioremediation of Pollutants. I.K. International Publishing House Pvt Ltd, New Delhi, pp 24–9.

Ummalyma SB, Pandey A, Sukumaran RK, Sahoo D (2018) Bioremediation by microalgae: current and emerging trends for effluents treatments for value addition of waste streams. In: Varjani SJ, Parameswaran B, Kumar S, Khare SK (Eds) Biosynthetic Technology and Environmental Challenges. Springer, Singapore, pp 355–75.

Upadhyay LS, Dutt A (2017) Microbial detoxification of residual organophosphate pesticides in agricultural practices. In: Patra JK, Vishnuprasad CN, Das G (Eds) Microbial Biotechnology. Springer, Singapore, pp. 225–42.

Varshney K (2019) Bioremediation Of Pesticide Waste At Contaminated Sites. https://www.researchgate.net/publication/333455076_Bioremediation_of_pesticides Accessed on 25 Dec 2020.

Valiente Moro C, Bricheux G, Portelli C, Bohatier J (2012) Comparative effects of the herbicides chlortoluron and mesotrione on freshwater microalgae. Environ Toxicol and Chem, 31:778–786. https://doi.org/10.1002/etc.1749

Verma A, Ali D, Farooq M, Pant AB, Ray RS, Hans RK (2011) Expression and inducibility of endosulfan metabolizing gene in *Rhodococcus* strain isolated from earthworm gut microflora for its application in bioremediation. Bioresour Technol 102(3):2979–84. https://doi.org/10.1016/j.biortech.2010.10.005

Verma JP, Jaiswal DK, Sagar R (2014) Pesticide relevance and their microbial degradation: a-state-of-art. Rev Environ Sci Biotechnol 13(4):429–66. https://doi.org/10.1007/s11157-014-9341-7

Wackett L, Sadowsky M, Martinez B, Shapir N (2002) Biodegradation of atrazine and related s-triazine compounds: from enzymes to field studies. Appl Microbiol Biotechnol 58(1):39–45. https://doi.org/10.1007/s00253-001-0862-y

Wang X, Gong P, Sheng J, Joswiak DR, Yao T (2015) Long-range atmospheric transport of particulate Polycyclic Aromatic Hydrocarbons and the incursion of aerosols to the southeast Tibetan Plateau. Atmos Environ 115:124–31. https://doi.org/10.1016/j.atmosenv.2015.04.050

Warren N, Allan IJ, Carter JE, House WA, Parker A (2003) Pesticides and other micro-organic contaminants in freshwater sedimentary environments—a review. Appl Geochem 18(2):159–94. https://doi.org/10.1016/S0883-2927(02)00159-2

Xiao P, Mori T, Kondo R (2011) Biotransformation of the organochlorine pesticide trans-chlordane by wood-rot fungi. N Biotechnol 29(1):107–15. https://doi.org/10.1016/j.nbt.2011.06.013

Xu P, Huang L (2017) Stereoselective bioaccumulation, transformation, and toxicity of triadimefon in *Scenedesmus obliquus*. Chirality 29(2):61–9. https://doi.org/10.1002/chir.22671

Yuanfan H, Jin Z, Qing H, Qian W, Jiandong J, Shunpeng L (2010) Characterization of a fenpropathrin-degrading strain and construction of a genetically engineered microorganism for simultaneous degradation of methyl parathion and fenpropathrin. J Environ Manage 91(11):2295–300. https://doi.org/10.1016/j.jenvman.2010.06.010

Zaffar H, Sabir SR, Pervez A, Naqvi TA (2018) Kinetics of endosulfan biodegradation by *Stenotrophomonas maltophilia* EN-1 isolated from pesticide-contaminated soil. Soil Sediment Contam 27(4):267–79. https://doi.org/10.1080/15320383.2018.1470605

Zhang S, Qiu CB, Zhou Y, Jin ZP, Yang H (2011) Bioaccumulation and degradation of pesticide fluroxypyr are associated with toxic tolerance in green alga *Chlamydomonas reinhardtii*. Ecotoxicology 20(2):337–47. https://doi.org/10.1007/s10646-010-0583-z

Zhang YH, Xu D, Liu JQ, Zhao XH (2014) Enhanced degradation of five organophosphorus pesticides in skimmed milk by lactic acid bacteria and its potential relationship with phosphatase production. Food Chem 164:173–8. https://doi.org/10.1016/j.foodchem.2014.05.059

Zhang W, Ye Y, Hu D, Ou L, Wang X (2010) Characteristics and transport of organochlorine pesticides in urban environment: air, dust, rain, canopy throughfall, and runoff. J Environ Monit 12(11):2153–60. https://doi.org/10.1039/c0em00110d

Zhang H, Yuan X, Xiong T, Wang H, Jiang L (2020) Bioremediation of co-contaminated soil with heavy metals and pesticides: influence factors, mechanisms and evaluation methods. Chem Eng J 398:125657. https://doi.org/10.1016/j.cej.2020.125657

3 Applied Techniques for Wastewater Treatment

Physicochemical and Biological Methods

Shima Husien, Nagwan G. Mostafa, Alyaa I. Salim, Irene Samy Fahim, Lobna A. Said, and Ahmed G. Radwan

3.1 INTRODUCTION

Wastewater types can be differentiated into three sectors: domestic, agriculture, and industrial wastewater. Lately, rabid industrialization caused an increase in the amount of consumed water, and, as a result, wastewater was discharged into waterways [1]. The paints and textile industries are developed industries that discharge pollutants into water streams. Dyes and heavy metals as a source produced from textile industries are considered coloring agents that were widely applied in the various stages of paints and textile industries [2]. Therefore, discharging paints and textile industries' effluents into water streams cause water coloration. Additionally, they have bad effects on the environment where they decrease visibility even at low concentrations and decrease the penetration of sunlight, resisting biological and physicochemical attacks. Additionally, they had an adverse effect on human health where their degraded products have a mutagenic, carcinogenic, and toxic impact on the health due to the increase in the chemical oxygen demand (COD)/biochemical oxygen demand (BOD) aquatic source levels. Numerous applied technologies were developed to help wastewater treatment and reduction of coloring water, and these technologies were classified as physicochemical and biological techniques. Physicochemical treatment processes are promising approaches that were applied for a long time to remove colloidal particles, floating matters, suspended solids, toxic compounds, and colors from polluted wastewater by using physical forces or chemical reactions. They can be defined as the techniques that may or may not contain chemical changes at different stages, whereas the physical change is always present [3]. These processes include different techniques such as coagulation-flocculation processes, electrocoagulation, sedimentation, flotation, membrane technology, and advanced oxidation processes [4–6]. Physicochemical processes are mainly applied in the preliminary, primary, and tertiary wastewater treatment stages. Selecting the type of the used physicochemical process depends on the contaminants' concentration

DOI: 10.1201/9781003354475-3

in the wastewater. On the other hand, biological processes are those in which biological materials are applied for the treatment process. Biological processes are known for their environmentally friendly properties and low cost. However, they are typically very time-consuming, their design is less flexible, and they are not applicable for several classes of dye removal [7]. Although there is other literature discussing the physicochemical and biological treatment processes, it nearly lacks the collection of all of these reviews in one chapter or review. Therefore, this chapter aims to collect all the literature about the techniques in a simple way that gives a young researcher a basic knowledge of the different steps of wastewater treatment. Different treatment technologies and their impact on the removal of other pollutants and the most promising technologies that can be applied for the pollutant removal process will be presented and discussed in the chapter.

3.2 PHYSICOCHEMICAL TECHNIQUES

Wastewater physicochemical treatment processes mainly include different techniques such as coagulation, sedimentation, flocculation, filtration, etc. The primary target of these processes is to separate the colloidal and suspended particles from the polluted water. For instance, coagulation and flocculation depend on adding the chemical coagulants and flocculants into the colloidal solution, which enables the formation of the floc particles and settles down under gravity during the sedimentation afterward; the solids can be separated by filtration depending upon their particle size.

Physicochemical processes are commonly applied in all stages of wastewater treatment; preliminary, primary, and tertiary treatment stages. Choosing the physicochemical treatment method depends on the concentration and type of pollutant found in the wastewater. For instance, sedimentation is commonly applied in the primary treatment process, where it was reported that 80% of suspended matters from wastewater could be removed by using this process in the immediate treatment stage [8]. Additionally, some researchers combined sedimentation with adsorption and ultra-filtration and achieved 60% and 87% total solid removal of Kraft black liquor [9]. Table 3.1 shows various physicochemical processes that were previously applied. Additionally, the different physicochemical processes will be discussed in the following subsections.

3.2.1 Coagulation-Flocculation Process

The coagulation-flocculation process was considered the backbone technology in most water plants and advanced wastewater treatment technologies. They were applied to enhance the particulate species separation through filtration and sedimentation processes. Traditional use of the coagulation process is the primary treatment of wastewater to remove turbidity of potable water. However, the process showed an effective performance toward many other contaminants' removal such as toxic organic matter, metals, viruses, and radionuclides [19, 20]. In general, the coagulation process could be defined as the process in which chemicals such as aluminum- or iron-based salts or synthetic organic polymers can be added to colloidal solutions. By adding coagulants into the colloidal solutions, the dissolved and suspended solids

TABLE 3.1
Different Physicochemical Methods Applied for Textile Dye Removal

	Wastewater Quality			Operating Conditions				
Wastewater	**Pollutant**	**Initial Conc.**	**Method**	**Dose**	**pH**	**Time (min)**	**RE**	**Ref.**
Winery wastewater	COD	3090–7438 (mg/L)	Chemical precipitation with chelating agents (trimercaptotriazine (TMT)) at pilot scale	0.84 (ml)	8.5	15	9%	[10]
	TSS	213–320 (mg/L)					90%	
	Cu	0.68–1.63 (mg/L)					96%	
	Zn	0.14–1.47 (mg/L)					76%	
Winery wastewater	pH	6.8	Coagulation by chitosan as a natural organic coagulant	20–600 (mg/L)	4–6.8	92	NA	[11]
	COD	1550 (mg/L)					73%	
	TSS	750 (mg/L)					80%	
	Turbidity	180 (NTU)					92%	
Winery wastewater	COD	25, 200–28, 640 (mg/L)	Electrocoagulation (EC) for iron (Fe) electrodes	NA	7	90	47%	[12]
	Color	6500 (Pt–Co)					80%	
	Turbidity	2490 (NTU)					92%	
	pH	5.2					-	
Winery wastewater	COD	25, 200–28, 640 (mg/L)	EC using aluminum (Al) electrodes	NA	5.2	120	49%	[12]
	Color	6500 (Pt–Co)					97%	
	Turbidity	2490 (NTU)					99%	
	pH	5.2					NA	
Winery wastewater	pH	3.2	Nano-filtration (NF) membranes	Feed flow rate = 1999 L h^{-1}	Permeate flux = 7.4–8.1 dm^3 m^{-2} h^{-1}	Fouling index = 0.16–0.22	TPh removal = 74%	[13]
	Reducing sugars	166 (g/l)						
	Turbidity	89 (NTU)						
	TPh	144 (mg/L)						

(Continued)

TABLE 3.1 *(Continued)*

Different Physicochemical Methods Applied for Textile Dye Removal

	Wastewater Quality			Operating Conditions				
Wastewater	**Pollutant**	**Initial Conc.**	**Method**	**Dose**	**pH**	**Time (min)**	**RE**	**Ref.**
Winery wastewater after centrifugation and filtration (Cyprus)	COD	5353 (mg/L)	Reverse osmosis (RO) (pilot scale)	Surface area = 1.2 m^2	P = 5–10 bar	Permeate flux = 10.8–11.4 L m^{-2} h^{-1}	97%	[14]
	TN	10 (mg/L)					67%	
	TSS	66 (mg/L)					94%	
	TS	5040 (mg/L)					96%	
Industrial wastewater	Cu(II)	3.5 (mg/L)	Flotation	Chabazite collector dose 0.5 g/L	5.5	$Fe(OH)_3$ precipitant dose = 30.9, 20.0, and 20.0 (mg/L)	98.26%	[10]
	Ni(II)	2 (mg/L)					98.6	
	Zn(II)	3 (mg/L)					98.6	
	Cu(II)	474 (mg/L)		CTABr collector dose 0.020 g/L	8–10	NA	99.99	[15]
Industrial wastewater	Co(II)	29.47 (mg/L)	Ultra-filtration	YM1 membrane	5–7	NA	100	[16]
	Ni(II)	29.35 (mg/L)					60	
Industrial wastewater	Cu(II)	200 (mg/L)	RO	Pressure (bar) 7	4–11	Polyamide membrane	98	[17]
	Cd(II)						99	
	Cu(II)	NA		Pressure (bar) 4.5	3–5	Sulfonated polysulfone membrane	98	[18]
	Zn(II)						99	

will be removed from the solution subsequently via sedimentation processes [21], whereas flocculation was known as a part of the coagulation process in which the destabilized particles were influenced to form larger aggregates [22]. Hence, coagulation can be divided into three subsequent steps, coagulant formation, destabilization of particles, and inter-particle.

3.2.2 Electrocoagulation

Electrocoagulation was considered one of the well-established water treatment technologies, which is widely applied for drinking water supply treatment and contaminant removal as presented in Figure 3.1 [23]. Furthermore, it was applied for municipal and industrial wastewater treatment, such as inorganic, organic, and biological contaminant removal from the contaminated water. The process of electrocoagulation can be defined as the in situ coagulant generation by electrolytic oxidation of a sacrificial anode, and it was caused by electric current applied through the electrodes [24]. According to the pH of the solution, metal ions generated from the consumable anode electrode solution spontaneously undergo hydrolysis through the solution and form species of coagulants as hydroxide precipitates, which help with dye removal [25]. Aluminum and iron were the most commonly applied materials as electrodes in the electrocoagulation method [24].

Coagulation is performed when the metal cations interact with the harmful particles and are carried toward the anode by electrophoretic motion, including various destabilization mechanisms [26]. At the same time, a beneficial side reaction is noticed, such as hydrogen bubble formation and pH change [27]. The response, in general, was observed when the metallic compounds such as iron and aluminum salts were dissolved into a medium of water from the sacrificial anode. After that,

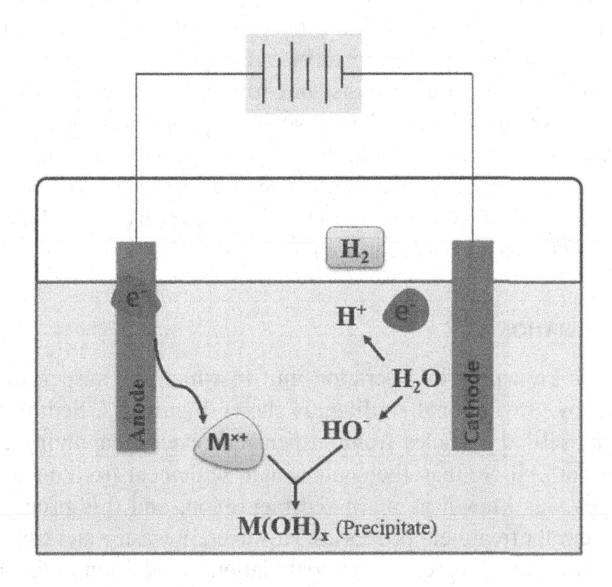

FIGURE 3.1 Electrocoagulation process principle.

the generated metal ions are immediately hydrolyzed to continuously form coagulating polymeric Fe or Al hydroxides near the anode. The electrocoagulation process follows three repetitive phases; at first, in situ coagulant formation by electrolytic oxidation of the sacrificial electrode; secondly, the pollutants and particles destabilization in the colloidal suspension; and finally, the destabilized compounds aggregate to form flocs [28].

Electrocoagulation has many advantages as they are more effective and have rapid organic matter separation in contrast with the coagulation method. Additionally, the reduction was needed for chemical reagents to occur by direct electrogeneration of coagulants with high purity, a lower amount of sludge production, and a reduced risk of secondary pollution [29]. However, on the other hand, the electrocoagulation process has some significant disadvantages, which were its need for maintenance, decomposition of sludge on the electrode, possible anode passivation, sacrificial anode consumption with the need for periodical replacement, in addition to the hydroxides of magnesium, calcium, etc., deposition onto the cathode [30]. Additionally, it was identified that its high costs of electricity and generation of secondary pollutants are the significant limitations [31].

Several mechanisms were reported for the electrocoagulation, such as the diffuse double-layer compression around the charged molecules through its interaction with the ions. These ions were produced by the consumable electrode electro-oxidation [32]. Furthermore, the ionic species was dissociated in water neutralization by counter ions that were generated by sacrificial anode electrochemical dissolution. Additionally, the remaining colloidal matter was entrapment in water by a sludge blanket formed by bridged flocs.

In contrast to chemical coagulation, electrocoagulation enables in situ coagulant delivery without needing other chemicals. Therefore, the related process could be operated in compact treatment facilities, along with adding side reactions (hydrogen bubble formation and pH change) that could also contribute to pollutant destabilizing. Such characteristic was stated to be very useful in enhancing the removal efficiency, reducing the operating costs, in addition to producing less sludge [33]. Moreover, electrocoagulation could remove hydrophobic components and hydrophilic ones, which were not easily removed by conventional coagulation. However, the electrocoagulation process efficiency for water treatment is highly dependent on the water media chemistry, especially pH and conductivity as well as the targeted contaminants' chemical and concentration properties [34].

3.2.3 Sedimentation

Sedimentation is known as an operation unit in which the suspension separation was performed by gravitational settling, as shown Figure 3.2. Sedimentation aims to eliminate the settled particles from suspensions even by applying chemicals as coagulation or not [35]. So that, the contaminant settlement from doldrums without using chemicals was known as plain sedimentation, and this process is usually applied in wastewater treatment processes. However, in wastewater treatment plants, sedimentation is preceded by chemical coagulation; in addition, it can be used as a limited scale in particulate separation from air streams [36].

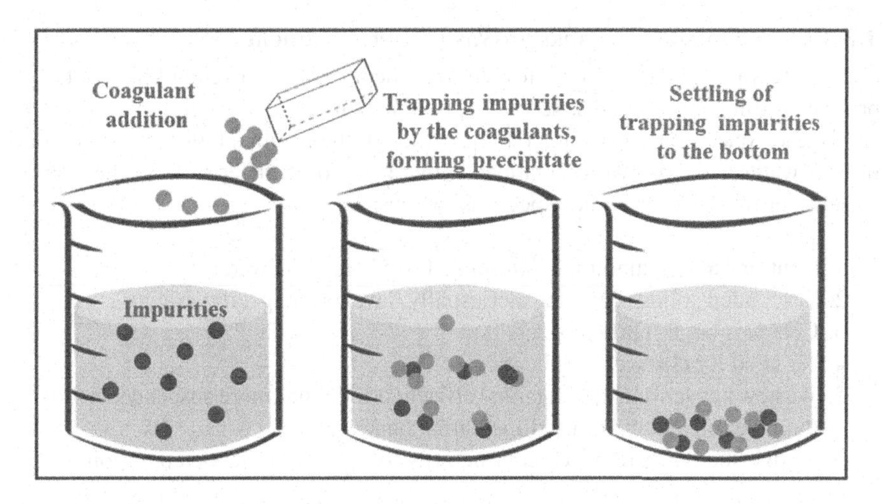

FIGURE 3.2 Coagulation and sedimentation process principle.

A major operation unit is applied in most water and wastewater treatment plants. In the water purification plants, the water turbidity must be reduced to a level as low as possible, where turbidity of 1.5 units was regarded as a desirable target for water to be introduced into filters. This target can be achieved by plain sedimentation or coagulation, which can be followed by sedimentation [37]. However, applying plain sedimentation takes a long time to be effective in water treatment. Consequently, chemicals were added to form flocs that facilitate the settlement process. In most wastewater treatment plants, there is a primary settlement tank that reduces the load of pollutants on the other treatment processes [38]. Additionally, a secondary treatment tank is used to exclude a large amount of the suspended solids after the treatment process of wastewater. This will be carried into the effluent and then to the streams in which the process of discharging effluent was performed [39].

3.2.3.1 Sedimentation Theory

All suspended particles in the solutions, especially of lower density, behave to settle under the gravity influence, which differs from one particle to another. Therefore, the settling rate was known to be dependent on the size, shape, specific gravity of the particles, and the liquid's temperature, velocity, and inactivity. For instance, heavy and coarse particles settle faster than other particles and can be removed by storage in large tanks [40], whereas fine particles with a diameter of 10 μm or less, and a density a little above water density, cannot be economically excluded by the sedimentation process alone. This type of particle requires coagulation through the addition of chemicals to increase the particle size and increase the settling rate. In general, sedimentation could be applied after coagulation and before the filtration process in the surface water treatment process, whereas, in the wastewater treatment processes and plants, sedimentation principles were involved in the grit chamber design through primary settling and secondary or final settling tanks [41].

3.2.3.2 Sedimentation Tanks in Wastewater Treatment

Sedimentation application in wastewater treatment in three different kinds of tanks: primary sedimentation tanks, secondary sedimentation tanks, and gravity thickeners for the concentration of sludge [42]. Although the sedimentation theory is used in both water and wastewater treatments, there are some differences in the state of wastewater treatment in some aspects as follows:

1. Contains a high amount of suspended solids in wastewater
2. Suspended solids of wastewater usually have lower specific gravity
3. Contains larger particle size
4. No need for chemical applications
5. Wastewater sedimentation tanks' effluent usually has more suspended solids than that obtained from sedimentation tanks of water
6. Sludge should be removed continuously to prevent the septic condition

3.2.3.3 Primary Sedimentation Tank

Primary sedimentation tanks were designed to remove readily settleable solids and reduce the load on the different processes of the following biological units. At primary treatment plants, where sedimentation is the only step designed for the treatment process, settleable solids were removed to prevent sludge formation in the receiving waters. By applying a detention time of 120–150 min, the tank's removal efficiency was from 50% to 65% for the removal of suspended solids and 25–40% for BOD, whereas, in secondary treatment plants, primary sedimentation was considered as a preliminary treatment where their tanks usually have fewer detention times from 60 to 90 min. Additionally, primary sedimentation tanks can be applied in stormwater tanks using only 10–30-min detention times. These tanks exclude organic and inorganic solid substantial portion [15].

3.2.3.4 Secondary Sedimentation Tank

Secondary sedimentation tanks were designed to exclude the settleable solids produced after the treatment by biological treatment. It was also known as final sedimentation tanks. This name was attributed to the sedimentation process after the tertiary treatment that was employed in some plants where phosphorous is removed by chemical precipitation by the addition of chemicals. The removal efficiency of suspended solids in the secondary sedimentation tanks is expected to be more than that of primary sedimentation tanks [43]. Types of sedimentation tanks are presented in Figure 3.3, where the tank differs according to some factors such as space availability, engineer experience, installation size, regulatory agencies rules, conditions of the local site, and economic factors.

3.2.3.5 Imhoff Sedimentation Tank

The Imhoff tank is one of the primary treatment technologies applied for raw wastewater treatment. It is designed for the separation of solid-liquid and settled sludge digestion. It composes of a V-shaped settling compartment above a tapering sludge digestion chamber with gas vents. Imhoff tank was an effective settler that reduced the suspended solids to 50% to 70% and COD to 25% to 50%. The compartment

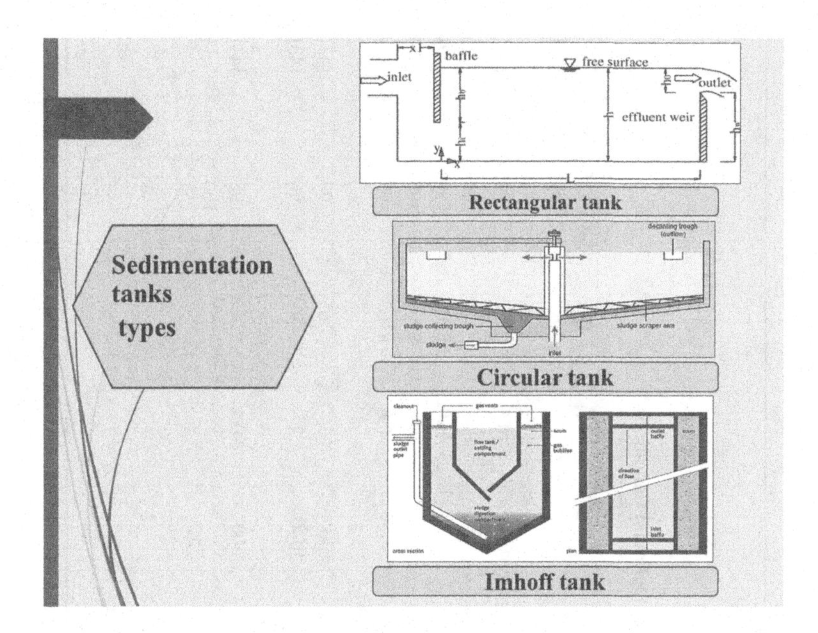

FIGURE 3.3 Different forms of sedimentation tanks.

located in the settling design has a circular or rectangular shape with V-shaped walls and a slot at the bottom, which allows the solids to settle into the digestion compartment. Furthermore, Imhoff tanks prevent foul gas from rising and disturbing the settling process where gas produced in the digestion chamber rises into the gas vents at the edge of the reactor [44].

3.2.4 Flotation

Flotation is employed for solid separation or dispersed liquids from a liquid phase using bubble attachment. The attached particles are separated from the heavy metal colloidal suspension by the effect of bubble rise. Among all types of flotation, dissolved-air flotation is the most applied form of flotation in the pollutant removal from contaminant wastewater [45]. Dissolved air flotation is applied mainly to remove suspended and colloidal solids by flotation by decreasing their apparent density. Wastewater, raw water, and liquid sludge are examples of influent feed liquid. The flotation system comprises four major components: pressurizing pump, air supply, flotation chamber, and retention tank. The solubility of gas as air in the solution increases as pressure increases, and this hypothesis was according to Henry's law.

The separation process depends on foaming produced from the bubbles, which in the following worked in the metal impurities separation. Recently, a new trend has shifted from applying flotation alone to applying a combination of flotation and another type of physicochemical processes treatment as active carbon and filtration [46, 47]. The integrated combination between adsorption, membrane separation, and flotation was successfully applied for Cu(II) and Zn(II) elimination, where the removal efficiency of 97% was achieved by using 60 mg/L initially [48]. Table 3.2 presents

TABLE 3.2
Application of Flotation for Metal Removal

Collector	Pollutant	Precipitant	Optimum Conditions				Efficiency (%)	Ref.
			IC* mg/L	PC** mg/L	CD***	pH		
Chabazite	Zn(II)	$Fe(OH)_3$	2	20	–	–	98.6	[42]
	Ni(II)							
Chabazite	Cu(II)	$Fe(OH)_4$	3.5	30.9	0.5	5.5	98.26	[42]
	Ni(II)		2	20			98.6	
CTABr (cationic collector)	Cu(II)	NA	474	–	0.02	810	99.99	[49]
CTABr (cationic collector)	Ni(II)	NA	3.3	–	0.02	810	98.5	[15]
Zeolite	Cu(II)	NA	60	–	0.8	810	97	[41]
	Zn(II)							
	Zn(II)	Sodium dodecyl sulfate (SDS)	60	40	2	6	99	[43]
		Hexadecyl-trimethyl-ammonium-bromide (HDTMA)	50	2040	2	9	96	[43]
Surfactin-105 and Lichenysin-A	Cr(VI)	Ferric hydroxide	50	600	0.4	4	98	[44]
	Zn(II)						100	

*IC** Initial concentration
*PC**** Precipitant concentration
*CD**** Collector dose

flotation application for various metals' removal from polluted wastewater including the optimum conditions and the removal percent.

3.2.5 Filtration

The filtration aims to extract particulates from polluted water by forcing them to pass through the layer of a porous material where large particles are retained by the staining process and sedimentation, whereas the colloidal particles were retained by other techniques such as adsorption process or coagulation, as well as it can be sedimentation; meanwhile, the biological reaction occurs only by bypassing the polluted water slowly through the porous mass [50]. Three primary filtration forms were reported: vacuum filtration, gravity filtration, and pressure filtration. Through vacuum filtration, the filter was set on the pump suction side, at the same time as the pressure dropping across the filter was limited to the suction lift differential, which could be generated by the pump, usually in the range of 5.5–6.7 m (18–22 ft) of water. The filter is generally performed under a pressure that is less than atmospheric pressure, whereas in the process of pressure filtration, the filter is located on the pump discharge side, whereby dropping the pressure through the filter can be any differential that can be generated by the pump characteristics of usually 3–12 m (10–40 ft) of water [51]. Additionally, if the filter discharge is not directed into the storage tank at atmospheric pressure, the filter pressure appears at a pressure higher than the atmospheric one. Nevertheless, the gravity filtration process starts with water flowing across the filter media under the gravity force. In other words, it can be said that a gravity filter is a type of special pressure filter in which the water was delivered into the filter; in addition, water that appears in the filter of an influent side is at atmospheric pressure [52].

3.2.5.1 Gravity Filtration

Gravity filtration was applied since the 1880s and was improved into a widely generic treatment process in the 1920s. Using the rabid gravity filtration throughout history was combined with solid/liquid processes of separation as coagulation, flocculation, and other subsequent settlements "flotation for example" through the combination with iron or aluminum salts [53].

Additionally, in recent years, the large molecule's weight of charged polymers was investigated to help in the removal of solids, and this will have a significant effect on filters. Gravity filtration was considered one of the most popular applied methods mainly applied in drinking water processes. It plays a vital role in micro-sized particle removal, especially the expected size of bacterial pathogens. Applying gravity filtration to improve the quality of drinking water was more common, and new processes were developed, such as granular activated carbon used to remove pesticides; in addition, intermediate ozone can be added to remove the soluble nitrates, ammonia. Additionally, the optimization process of gravity filtration is essential for contaminant removal where its application reduces the process costs, and energy-consuming procedures [54].

Filtration process is performed with different layer filters; the first layer is the coarser anthracite that helps to remove the floc residuals after the settlement process.

The finer particles that pass across into the finer size of sand media were more susceptible to clogging and led to longer runtime. Clogging can be improved by some surface wash techniques as well as the standard backwashing that was developed by Baylis after the notice that the clogging predominantly occurred in the upper layer of 150 mm through the bed of the filter [55].

3.2.5.2 Membrane Filtration

Membrane filtration was applied for the inorganic effluent treatment since it was able to reject not only the suspended solids and the organic compounds but also the inorganic pollutants such as heavy metals. Depending on the particle size that they tend to retain, different membrane filtration types such as nano-filtration, ultra-filtration, micro-filtration, reverse osmosis, and electrodialysis (ED) can be employed for pollutant removal from the polluted wastewater [52].

3.2.5.3 Ultra-Filtration and Micro-Filtration

Micro-filtration and ultra-filtration are processed pressure driven with usage shared areas, and they have almost the same procedure of separation. The main difference that can be considered between these processes is that the removal solutes by micro-filtration are more extensive than those removed by ultra-filtration [56]. Micro-filtration pore size is large and nearly ranging from 0.1 to10 μm and has the lowest processing pressure from 0.01 to 0.2 MPa and was originally aimed to reduce the microbial load as a pretreatment process in the dairy industry [57]. Additionally, it is performed to separate the suspended particles that appear in the effluent. Meanwhile, ultra-filtration helps remove particles such as dissolved, suspended species, macromolecules, and colloidal materials [45]. Micro-filtration and ultra-filtration can be applied in various industries, such as purification of water in the production process of potable water, different solutions' clarification, and disturbed particle removal. The most important application is the wastewater treatment to reach discharge standards [58, 59]. Choosing the suitable membrane is considered the most challenging part of the micro-filtration and ultra-filtration processes. Membranes generally differ in structure, material, porosity, and micro-filtration, and ultra-filtration membrane types are porous; they can also be classified into symmetric or asymmetric membranes. Symmetric membranes are membranes whose structure does not vary about the membrane cross-section. Micro-filtration membranes are usually symmetric, unlike ultra-filtration membranes, which are asymmetric. Different materials were investigated for micro-filtration, and ultra-filtration designs, such as hydrophobic, hydrophilic, and polymers membranes, were used. Every type of material has advantages and disadvantages. Generally, the hydrophilic type was not applied in the micro-filtration and ultra-filtration due to the water molecules affecting their mechanical and thermal resistance. The micro-filtration and ultra-filtration separation mechanism processes typically depend on the porous membranes' molecular sieving. Larger size particles were rejected because of the pore size constraints, unlike particles with pore size [45]. Since the particles are absorbed onto the membrane surface, a reduction in pore size occurs, so the separated particles are much smaller than the pore size.

3.2.5.4 Reverse Osmosis

Reverse osmosis was first discussed in the 1920s; however, it did not come true in practical part until 30 years later. It can be applied in various industries such as food processing, desalination, water reclamation, pharmaceutical, pulp, paper, textile, and biotechnology [60]. Additionally, it is widely applied in the treatment of a wide variety of wastewater treatment processes [61, 62]. This method was the pressure-driven membrane process and was mainly applied to exclude heavy metal ions. Reverse osmosis works under hydraulic operating pressure, and a semi-permeable membrane that is characterized by it does not contain any distinct pores that can let the particles flow through, so the transition should occur by diffusion [63]. Multivalent and monovalent particles could be separated by the reverse osmosis membrane, even though the multivalent particles are better in removal. The dissolved solute separation in water was also performed through a reverse osmosis membrane that occurs through three basic principles. The first is water adsorption onto the membrane surface from heavy metal concentrated solutions. Meanwhile, the second step is the diffusion through the membrane because of the concentration gradient.

3.2.5.5 Nano-Filtration

Nano-filtration is a new technology designed to facilitate the large molecule's separation by microscopic pores and to overcome conventional techniques' operational deficiencies [64]. Nano-filtration is a "loose." Reverse osmosis process, which has a pore size range from 1 to 10 nm and is characterized by lower processing pressure nearly of 1.5 to 3.0 MPa [65]. Membrane fabrication and modification are considered one of the most important factors in nano-filtration processes due to their capability to affect the permeability of solvents [66]. Nano-filtration membrane creation is possible by some technologies such as nano-particles (NPs), interfacial polymerization, and ultraviolet (UV) incorporation [66]. Nano-filtration is used in eco-friendly and energy-efficient applications such as surface water, groundwater, and purposes related to wastewater treatment [67]. Recently, the nano-filtration treatment method was widely applied in wastewater treatment such as dyes and heavy metals [68]. The whole process was known as liquid phase filtration due to its ability to separate a wide range of organic or uncharged compounds, and inorganic pollutants from contaminated solutions [69]. Pollutant removal by nano-filtration involves three basic steps; the pretreatment step is the first step in which feeding water should be treated before entering the system to decrease the possible formation of fouling. Pretreatment methods can include coagulation, prefiltration, flocculation and filtration, chemical conditions, and ion exchange. The second step is considered the primary membrane separation process, the nano-filtration membranes' general rejection results from a combination of steric Donnan, hindrance, and dielectric effects. The steric hindrance mechanism serves to remove the neutral solutes [70]. Charged components also can be rejected due to the ion size and existing Donnan potential. The third step was known as the post-treatment step, where toxic heavy metals were effectively removed by AFC-80 membranes when treated with nano-filtration aqueous solutions containing lead in an extensive concentration range [66].

3.2.5.6 Electrodialysis

ED was applied in the desalination process since the 1950s and is considered a novel liquid hybrid membrane process [71]. It was applied for the wastewater treatment generated from different industries, water splitting producing acid and base from salts, and fractionation that includes wine stabilization, whey demineralization, etc. Electrodialysis was performed through ion-selective exchange membranes in which cations and anions do not transport simultaneously. Two types of membranes, cation exchange membrane and anion exchange membrane, are stated. The transportation of ions through these membranes occurs because of the electrical potential or concentration gradient. As a result, ED electrode polarity is reversed; the ED process is called ED reversal, which is sometimes considered better than ED. ED reversal lows the fouling and scaling and was distinguished by its higher recovery rates [72]. On the other hand, ED reversal needs more electrical controls and plumbing.

3.2.6 Oil Separation

Recently, a lot of large oil spill incidents were recorded around the world, making the contamination of water with oil a great problem [73, 74]. Therefore, a lot of techniques were applied for the cleanup of the water from the toxic oil spills, such as chemical dispersant addition, skimming, and filtrating medium applying, in addition to various sorbents. Skimming was considered the most popular method that has been used for this purpose; however, it consumes time, costs high, and is inefficient. The second popular method for oil separation is the application of the oil dispersant method that separates the oil into droplets, even though this method causes contamination of the seawater with toxic chemicals and requires additional steps to be removed [75, 76]. The third applied method is the filtrating materials and sorbents, which are considered the most convenient and effective way of cleaning and recollecting oil from contaminated water [77].

Different materials were investigated for oil separation technology; for example, two types of specific wettable materials were suitable for oil/water separation, such as hydrophilic-oleophobic and hydrophobic-oleophilic materials. Using materials with wetting properties will repel the phase of water and absorb oil, which helps separate oil droplets from the oil/water mixture. Super-hydrophilic/super-oleophobic materials are water-removing materials from the oil/water mixture with the reverse selectivity to the oil-absorbing materials. Various materials were reported for the oil/water separation, such as carbon-based material, fabric materials, metallic mesh material, sponge, powder material, or foam until this date [78, 79]. However, mesh and fabric materials have many drawbacks to be applied in the oil/water separation where they cannot be used directly in the sea/oily water that it should collect and then filter, and applying this isn't easy in the scaling up to large scale. The most effective stated materials were foam and sponge-based materials due to their low cost, availability, high porosity, and ability to absorb oil over a hundred times their estimated weight. However, they have less selectivity toward oil or water, where they drink the two of them almost in equal amounts.

3.2.6.1 Filtration-Based Materials

Different materials such as fabric textile-based materials, metallic mesh-based materials, and polymeric-based materials were stated for the oil/separation technologies.

3.2.6.2 Metallic Mesh-Based Materials

Metallic meshes' super-hydrophobicity was investigated for the last few years to improve their selective permeability toward oil and water [80, 81]. The filtering mesh behavior is super-oleophilic and super-hydrophobic, which was prepared and resulted in oil/water separating properties. The first super-oleophilic and super-hydrophobic stainless steel was designed by a dry and spray method [82]. The mesh surface morphology is very rough because it is super-oleophilic and super-hydrophobic; therefore, droplets of water stay in the form of spherical beads on the mesh, and the droplets of oil pass through it. Different mesh materials were prepared, such as copper mesh, which was considered one of another applied substrates for oil/water separation. Accordingly, the copper mesh's super-oleophilicity and super-hydrophobicity were stated and successfully applied in the separation of oil and water mixtures [83].

3.2.6.3 Fabric/Textile-Based Materials

Some of the inorganic porous metal soft films and flexible organic fabric materials were stated as good candidates for the separation of oil/water separation. For instance, super-hydrophobic cotton textile properties are separate oil/water solutions by dip-coating method. Super-hydrophobic and a super-oleophilic nano-fibrous membrane having water/oil separation characteristics was fabricated using electrospun combination poly(m-phenylene isophthalamide) nano-fibers, with the novel in situ polymerized fluorinated polybenzoxazine functional layer that contains SiO_2 NPs. A one-step silicone nano-filament growth was used to fabricate polyester textile with super-hydrophobic and super-oleophobic characteristics onto the fabric by using the chemical vapor deposition method [84]. A couple of robust and durable super-hydrophobic materials were characterized by their excellent chemical, mechanical, and environmental stability. The mixture of oil and water can be separated by applying these simple methods using these textile materials [85].

3.2.6.4 Absorption-Based Separation

Copious advanced materials were applied in the filtration process, where they were characterized by their unique and tunable wetting properties. They were mainly involved in the selective oil/water separation; however, they still have some challenges in their in situ application. Lately, efforts were directed to develop an advanced type of absorbent that has specific properties of wetting and can be selective in the oil-water mixture separation [86].

3.2.6.5 Particle and Powdered Absorbents

Powdered materials such as iron NPs, activated carbons, zeolites, etc., were investigated by various scientists. For instance, powder of oleophilic and super-hydrophobic calcium carbonate, which can easily separate oil/water mixtures, was reported. Additionally, hydrophobic polymer coated with silica particles was stated to be useful for petroleum hydrocarbon removal from polluted/water [87].

3.2.6.6 Sponge-Based Materials

Foam and sponge-based materials were characterized by their excellent properties such as low cost, easy availability, porosity, and high capacity of absorbance, with initial wettable properties. They could absorb copious liquids, oils, and organics and water, making them non-applicable for oils and organics separation from the water phase as they are non-selective and absorb both of them equally. Therefore, researchers tried to fabricate materials with purposeful construction so that their surface modification and topography can fit the functional molecules in addition to unique wetting properties. Various promising routes such as dip coating, in situ growth, polymerized octadecylsiloxane coating, and carbon nanotube/poly(dimethylsiloxane)-based coating were used for developing unique materials' wettability [88]. Sponge-based materials coated by low-surface-energy molecules were the widely applied means of attaining super-hydrophobicity and super-oleophilicity [80, 89]. For instance, polydimethylsiloxane TiO_2 was successfully used for coating super-hydrophobic polyurethane (PU) sponges. Additionally, various carbon-based materials such as graphene, CNT, etc., were applied as hydrophobic coating agents. However, one drawback was associated with most polymeric sponges where they are unstable. These materials cannot withstand the site application situation such as twisting, bending, and pressing. Few groups were identified by their characteristics of super-hydrophobic, super-oleophilic, melamine-based sponge material that was distinguished by their fire resistance and an extensive temperature [90].

3.2.7 Ion Exchange

Ion exchange processes exhibited many advantages, such as high removal efficiency and treatment capacity in addition to fast kinetics that made the highly applied process for pollutant removal [91]. Natural solid and synthetic resins of ion exchange have an enormous ability to exchange their cations with the metals of the suspended pollutants present in the water. However, synthetic resins are commonly preferable to synthetic because they are more applicable and highly efficient for the heavy metal removal from the solution [92]. Furthermore, gel-like ones are highly efficient materials through the ion exchange process and their costs are low. As another type of resin, macropore resins have a sponge-like structure, which affords high physical stability and leads to more stress relief. For instance, gel-like cation exchange resin such as Lewatit S 100 is strongly acidic with uniform size beads, and this resin kind has a maximum ion exchange capacity of 0.39 mmol g–1 toward Cr (III) under optimum conditions and pH of 3.5 [93]. Additionally, DowexHCR S/S resins that were primarily found in sodium form were reported as economically beneficial resins due to their regeneration properties, and the results showed that by using these resins, Ni and Zn were removed with a near percent of 98% at optimum pH of 4 [92]. Another stated resin is the strong basic resin Ambersep 132; its adsorption capacity was recorded at about 92.10 mg/g and was applied for Cr(VI) removal [94]. Table 3.3 represents a variety of natural, synthetic materials applied as a resin in the ion exchange process.

TABLE 3.3
Different Previously Applied Ion Exchange Resin in the Operation of Ion Exchange

			Optimum Conditions					
Ion Exchanger	**Nature**	**Pollutant**	**Dose**	**IC (mg/L)**	**pH**	**AC (mg/g)**	**RE (%)**	**Ref.**
Clinoptilolite	Weak acidic	Ni (II)	10	100	–	2	90	[82]
		Cr(III)				4.1		
		Zn(II)				3.47		
		Cu(II)				5.91		
		Cd(II)				4.61		
Synthetic (NaP1) zeolite	Weak acidic	Ni (II)	2.5	100	–	43.58	100	[82]
		Cr(III)				20.08		
		Zn(II)				32.63		
		Cu(II)				50.48		
		Cd(II)				50.8		
Clinoptilolite	Weak acidic	Ni(II)	100	51.6	–	8.89	75	[83]
Amberlite IR-120	Strongly acidic resin	Zn(II)	–	5.43	–	–	–	[84]
Zeolite	Weak acidic	Ni(II)	0.1	40	5	–	60	[85]
Zeolite	Weak acidic	Cu(II)	0.1	80	5	–	64	[85]
Zeolite tuffs	Weak acidic	Ni(II)	1.2	250	4–4.5	4	–	[86]
Amberjet 1200H	Strong acidic	Cr(III)	0.5	10	–	84.04	100	[87]
			1	10	6–7	188.67		[87]
Lewatit TP 207	Heavy metal		0.4	3.12	4.5	17.73	–	[79]

*IC** Initial concentration
*AC*** Adsorption capacity
*RE**** Removal efficiency

The mechanism of the ion exchange process is based on a reversible interchange of ions between the solid and liquid phases; for instance, the whole process starts with the ion exchange reactions. After that, the ions of heavy metals were physically absorbed, and a complex was formed between them and the functional groups. Finally, hydration occurs at the solution surface or the adsorbent pores. Various factors such as pH, temperature, anions, the adsorbent and sorbet initial concentration, and exposure time affect the ion exchange process [93]. After the removal process, valuable heavy metals can be recovered from inorganic effluents, which were also conducted by the ion exchange process [78]. Metals can be retrieved in a more concentrated form by the elution process with some suitable reagents which follow the loaded resin separation. Resin that has acidic functional groups contains sulfonic acid as a part of its structure; therefore, it can be suggested that, during the metal ion capture, the physicochemical interactions can be explained as follows:

$$nRSO_3^- - H^+ + M^{n+} \leftrightarrows nRSO_3 - M^{n+} + nH^+ \qquad (3.1)$$

where *n* is the reaction species constant relating to the oxidation state of metal ions and –RSO3– denotes the anionic group linked to the ion exchange resin. Ion exchange was distinguished by its high efficiency, producing less sludge after the treatment process in addition to metal recovery, low cost, and high selectivity [81].

3.2.8 Chemical Oxidation

Chemical oxidation is an operation that involves electron transfer from an oxidizing agent to the chemical particles being oxidized or from a reductant "electron donor" into the oxidant "an acceptor." This process results in a contaminant chemical transformation of unstable, reactive species formation "free radicals" that degrade contaminants after that [95]. The chemical oxidation process is widely happening in natural waters and is an essential mechanism in the natural self-purification of the surface waters. For instance, dissolved iron and sulfide pollutants' oxidative removal from aerated water is a prominent example. Additionally, organic waste degradation showed a critical phenomenon that includes water self-purification and was attributed to the microorganisms' presence. Microorganisms help to catalyze a highly dissolved oxygen effective utilization as an oxidant [96]. The microorganisms' catalyzed operation has been optimized and developed to various forms known as a biological process in high concentration organic waste treatment applications. Different types of chemical oxidation can be explained as follows:

3.2.9 Electro-Oxidation Processes

Electrochemical oxidation is considered one of the most important methods in the application of water treatment because of its capability for organic compounds' wide range degradation [97]. Electrochemical oxidation occurs by direct and indirect anodic oxidation, wherein the direct electrolysis pollutants were adsorbed and degraded directly at the electrode. In contrast, in indirect electrolysis, pollutants were degraded in the liquid where it reacts with oxidizing agents that appear on the electrode [98]. Several materials were applied as electrodes in the electro-oxidation process, such as the boron-doped diamond (BDD) electrode, and characterized by its chemical, mechanical, and thermal stability. Other electrodes such as titanium oxide (TiO_2), lead dioxide (PbO_2), and tin oxide (SnO_2) also revealed its ability to treat polluted water [99] in contrast to other inactive electrode substances such as oxide (IrO_2), iridium (VI), or platinum (Pt) [100]. Different operating parameters such as pH, current density, electrode distance, initial concentration, electrolyte type, and temperature were revealed as significant factors that affect electrochemical oxidation.

3.2.10 Photochemical Oxidation

Appearance of contaminants intermediates due to the contaminant's incomplete mineralization as a result of the by-product and intermediate formation. These intermediates could be more persistent and toxic than the parent compounds. In consequence, the treatment time and formed by-products' disposal increase, resulting in high operating costs of photochemical oxidation. Using UV irradiation such as UV lamps or natural sunlight has served to overcome this issue by generating oxidation reactions between the generated free radicals and contaminants that lead to complete mineralization compounds [92]. Lamps' radiation at 254 nanometers (nm), where pollutants absorb the UV energy in the range of 200–300 nm, becomes excited and reacts with contaminants that work as oxidants. Contaminant degradation by UV irradiation can occur by direct photolysis (photooxidation) or indirect photolysis (photochemical oxidation) [93]. The pollutant's degradation resulted from their interaction with the generated active species. For instance, an efficient homolysis of highly fluorinated compounds that contain a CF bond was achieved by applying a wavelength of <190 nm (VUV).

3.2.11 Advanced Oxidation Processes

Various kinds of conventional methods were identified for domestic, agricultural, and industrial wastewater treatment. However, these methods were limited to a certain level [101]. Therefore, pollutant removal like metals and organic and inorganic compounds, in addition to pathogens, became a significant challenge. Hence, researchers did their best to develop a new technique for pollutant removal. These techniques included nano-materials, activated carbon, ozonation, and UV radiation.

3.2.11.1 Ozonation

Ozonation is a type of advanced oxidation process involving the generation of highly reactive oxygen species (ROS). Ozone is an unstable gas encompassing three oxygen atoms; the gas will rapidly degrade and convert to oxygen; during this alteration, ROS form such as a free oxygen atom or free radical. These ROS are highly reactive and short-lived under normal conditions. Additionally, ROS are capable of attacking not only a variety of pathogens but also organic and inorganic pollutants [102]. Thus, the treatment of water with ozone is efficient for disinfection and the degradation of organic and inorganic contaminants. The main advantage of this method is its simple operation, as well as the absence of sludge production [103].

3.2.11.2 Ultraviolet Radiation

UV radiation technology was identified as an efficient technique for water disinfection. The equipment for water treatment contained UV radiation which was applied for domestic and industrial wastewater disinfection processes. Microorganisms were inhibited by using this method where the UV radiation penetrates the microorganism cell wall of the microorganism and damages the genetic mutant. As a result, the radiation prevents the microorganism's cell reproduction and stops their growth in the polluted water. However, applying large UV radiation quantities will destroy nearly 99.9% of microorganisms present within a water stream [104].

3.2.11.3 Fenton's Oxidation

Iron, for instance, can activate H_2O_2 and is responsible for hydroxyl radicals' production in water. The proposed mechanism of the Fenton process starts with H_2O_2 reacting with Fe^{2+} to generate reactive solid species that were conventionally known as "free hydroxyl radical" [105]. These produced hydroxyl radicals are most effective only at acidic conditions of pH. Therefore, Fenton reaction application in wastewater treatment is restricted in practice. However, according to the traditional treatment scheme, three modified Fenton processes were proposed; photo-Fenton system, Fenton-like system, and electro-Fenton system [106].

3.3 BIOLOGICAL TREATMENT PROCESSES

Trace organic contaminants, or micro-pollutants, could be found in the environment in a trace concentration and have a harmful effect on human health, the environment, and aquatic life. These organic contaminants can be produced from household products, industrial chemicals, personal care products, pharmaceuticals, polycyclic aromatic hydrocarbons (PAHs), etc. Pollution caused by this organic contaminant in the water streams may cause collective harmful effects on multigenerational contact in aquatic organisms and destroy human health by becoming a part of the ecosystem [107]. Therefore, different governmental and non-governmental organizations are directed to remove micro- and organic pollutants and protect the environment from their effect [108]. As a result, researchers stated the importance of biological processes for organic contaminant removal from polluted wastewater due to their advantages: their high removal efficiency, low cost, and the process's simple design in contrast to other treatment methods. For instance, biological processes such as aerobic and anaerobic types were applied in most waste water treatment plants (WWTPs) to eliminate organic pollutants from wastewater. Furthermore, aerobic processes are widely used in most paper and pulp mills because their operation is effortless in addition to the relatively low operating costs [109].

Different aerobic technologies, such as aerated lagoons and activated sludge, are commonly applied in the pulp and paper industry [109]. On the other hand, anaerobic processes applied in the different industries of pulp and paper are not standard [110]. On the other hand, aerobic and anaerobic processes have some disadvantages that include the high production of sludge through the aerobic processes and the sensitivity of anaerobic bacteria to toxic materials. Organic materials that appear in the primary effluent that overflows the tanks of primary settling exhibit definite characteristics which require additional treatment types. This organic material consists of dissolved, suspended, and particulate solids that account for the turbid appearance of the primary effluent. Naturally, the dissolved organic material that occurs in the influent will stay in the solution of the liquid flow through the immediate process of treatment. In contrast, the colloidal part was stated with minimal size and mass characteristics, which cannot settle through the primary treatment.

Various methods were developed for wastewater treatments using biological means, known as secondary treatment. The principle of the wastewater treatment biological treatment methods was to introduce contact with bacterial cells that use the organic materials for their feeding that in follow helps the reduction of BOD content, and this is the purpose of the biological treatment process. For instance, wastewater influent usually contains BOD content higher than 200 mg/L; however, the primary settling helps in its reduction to about 150 mg/L as soon as it enters the biological system component "secondary treatment." After the secondary treatment process, BOD content should be no more than 20–30 mg/L. Therefore, after the discharge of these effluents into water streams and diluting in the water bodies, BOD content less than 6–10 mg/L was recorded [111].

3.3.1 Aerobic Treatment Processes

Aerobic treatment processes aim to transform the dissolved unsettled organic particles in the settling tanks into inorganic end products. After that, it is incorporated into a biological floc and settled down in the final clarifier. The process stages start by activating aerobic bacteria and microorganisms using oxygen by exposure to the air produced by the blower. Additionally, it could be through the diffused air system or continuous mixing of liquid wastes. Organic matters and nutrients, such as nitrogen and phosphorus, are removed from polluted water by applying biological treatment.

3.3.1.1 Suspended Growth Treatment Processes

In suspended growth, after the primary treatment, the settled flow was mixed with part of the settled activated sludge obtained from the final clarifier to increase the flocs in the reactor. After that, the resulting mixture known as mixed liquor was activated by oxygen, which can be supplied even by surface aeration or by diffused air compressors. Providing the reaction environment with oxygen allows the accumulation of the aerobic bacteria into flocs where it oxidizes organics exciting in the sewage. Afterward, the formed mixture was transferred to the final clarifier where the flocs will settle sown, and water stays on top of the clarifier. The amount of BOD will be reduced while the settled sludge is transformed again into the aeration tank to keep on the closed system. Suspended growth examples are such as activated oxidation ditches, sludge tanks, sequencing batch reactors, etc. [112, 113].

3.3.1.2 Attached Are Growth Treatment Processes "Biofilm"

Attached growth processes are the second primary type of biological treatment where the flow trickles of the pretreated sewage were fixed over inert media; gravel, ceramic, rocks, or plastic materials. Through the attached growth "biofilm," a thin gelatinous film composed of many microorganisms was connected. Through the process, microorganisms oxidize the suspended and dissolved organic matters, which are found in the sewage, and convert them into inorganic end products and new cell tissue [114]. The attached growth process includes trickling filters and rotating biological contactors.

3.3.2 Hybrid Processes

Suspended growth incorporated with biofilm systems is a reasonable hypothesis where both of the suspended solids' aggregation contains a variety of microorganisms. Modelling these processes to obtain the optimum conditions that achieve high pollutant removal was studied, which covered many aspects of the activated sludge process. However, only limited research was conducted to investigate the diversity of these multiple biomass systems [115, 116]. For instance, the biofilm's variation in the microbes' population was studied. Results showed that the competition between organisms generates oscillations in the abundance of planktonic and sessile microbes' populations. Lately, systems of submerged aerobic biofilm were being promoted in the integration form of fixed film-activated sludge systems. As usual, they were applied to improve the activated sludge nitrifying capacity facilitated by setting the media in existing aeration tanks. The media requires primary sedimentation or fine screening of the influent for clogging prevention. Furthermore, these systems were used to treat industrial and wastewater effluents.

3.3.3 Anaerobic Treatment

Anaerobic processes such as the process of up-flow anaerobic sludge blanket (UASB) reactor and fluidized-bed reactor (FBR) were used to treat paper and pulp wastewaters. UASB was applied for newsprint paper mill wastewater treatment to reduce the COD, and sulfite concentration, where a removal percent of 66% and 73% was achieved for both of them, respectively [117]. Methane and sludge production was monitored through the COD removal process, where the sludge production rate was noticed independent of the duration of the process. The estimated greenhouse gas (GHG) emission through the study was approximately 60 percent of the theoretical one concerning wastewater strength; for example, the anaerobic lagoon was replaced by a UASB reactor for the treatment process of the agro-based pulp and paper mill wastewater, and a COD removal of the 80–93% range was obtained [118]. The new system was characterized by higher COD removal capacity, requirements of small area, and easier biogas recovery; additionally, the anaerobic lagoon has a drawback that, without a biogas recovery system, methane can be released into the atmosphere through his system [119, 120].

Moreover, the UASB system treated both the unbleached and bleached Kraft mill wastewater. Results stated that high removal percent of 79–82% and 71–99% ranges was achieved for both COD and chlorinated organics, respectively [121]. Furthermore, by applying the activated sludge process, the removal efficiency of carbohydrates and volatile fatty acids was in the range of 50–60% after the UASB reactor. Therefore, aerobic post-treatment application was reported as a post-treatment process that improves the treated effluent quality.

Additionally, a packed-bed reactor (PBR) and FBR were applied for the anaerobic pretreatment of the weak black liquor from Kraft pulping wastewater [122]. They tried to improve the treatment process by adding ligninolytic fungi to both reactors to enhance the removal percent of contaminants from polluted water. Results revealed that higher COD and color removal percent in the PBR system were higher in the presence of different fungi in the two types of reactors.

Moreover, anaerobic reactors such as fixed-film, UASB, and fluidized beds were studied and compared with each other concerning their easy operation, low energy consumption, and capital cost [123]. In conclusion, the UASB reactor was characterized by low energy consumption. In contrast, the fixed-film reactor was marked by a lower capital cost, while the removal of contaminants by FBR was higher during pulp and paper wastewater treatment. For instance, an up-flow filter was applied to remove COD, BOD, and Adsorbable organic halides (AOX) from polluted "bleaching" water, with a removal efficiency of 50%, 70%, and 50%, respectively. Additionally, acetates were added as an electron donor to improve the removal percent of AOX, which increased to about 90%. The previous results showed the electron donor cost-effectiveness to the anaerobic treatment process compared to physicochemical or biological processes.

3.3.3.1 Constructed Wetlands

Constructed wetlands are a type of engineered system designed and built to include the natural processes that involve soils, wetland vegetation, and other associated microbial assemblages to help treat wastewaters. They were designed to take advantage of many same processes that happen in the natural wetlands. However, do so within a more controlled environment. Constructed wetlands applying for the treatment of wastewater can be classified according to the life-dominating macrophyte form into systems that contain floating left, free-floating, rooted emergent, and submerged macrophytes [124]. Further classification can be described according to the hydrology of the wetland; "free water surface, and systems of subsurface," and subsurface flow. Constructed wetlands could be classified into the flow direction, even vertical or horizontal [125]. The first tests directed to the possibility of applying wastewater treatment by wetland plants were investigated by Käthe Seidel in Germany in the 1950s at the Max Planck Institute in Plön [126]. After that, numerous experiments were carried out for applying the wetland plants for the treatment of various wastewater types involving phenol, dairy, and livestock wastewater [127, 128]. Most of the experiments were performed in constructed wetlands with horizontal or vertical constructed subsurface flow. However, the first entirely constructed wetland was built with free water surface in the Netherlands in 1967 [129]. Various constructed wetland types can be combined to achieve a higher treatment effect, especially for nitrogen – hybrid systems. In general, all constructed wetland types could be integrated to obtain more complex treatment efficiency [130].

3.4 CONCLUSION

Textile and paint industries are one of the emerged industries that cause adverse bad effects on the environment due to the coloration and toxic dyes; heavy metals appear in the water after discharging their effluents. Therefore different physicochemical methods have been developed to treat their effluents before discharging into water streams. Hence, in this chapter we discussed the different treatment methods applied for COD and BOD reduction such as coagulation-flocculation processes and electrocoagulation in addition to other techniques as sedimentation, flotation, and advanced oxidation processes that have applied for other pollutants' reduction. Biological treatment process is one of the promising processes that have a great role in the trace organic contaminant removal. Different techniques of biological treatment were applied such as aerobic treatment processes, anaerobic treatment processes, and constructed wetlands. As a conclusion from this chapter, it could be stated that wastewater treatment systems started by physical methods to reduce the suspended solids found in wastewater. After that, biological methods could be applied for the removal of trace organic pollutants from wastewater. These processes should be followed by other processes such as adsorption, advanced oxidation processes, and micro- and nano-filtration for the removal of other trace elements from polluted wastewater. This chapter gives a brief summary about most of the applied physical, chemical, and biological techniques applied for the wastewater treatment to help a young researcher identify the wastewater treatment as a whole process.

ACKNOWLEDGMENT

This chapter is based on the work supported by the Egyptian Academy of Science, Research, and Technology (ASRT), Project ID: Call no. 2/2019/ASRT-Nexus # 4607.

REFERENCES

[1] Rajesh Kumar Singh, H Ramalinga Murty, S Kumar Gupta, and A Kumar Dikshit. An overview of sustainability assessment methodologies. *Ecological Indicators*, 9(2): 189–212, 2009.

[2] Jae-Wook Lee, Seung-Phil Choi, Ramesh Thiruvenkatachari, Wang-Geun Shim, and Hee Moon. Evaluation of the performance of adsorption and coagulation processes for the maximum removal of reactive dyes. *Dyes and Pigments*, 69(3):196–203, 2006.

[3] Yahong Li, Bing Zhou, Guoqiang Zheng, Xianhu Liu, Tingxi Li, Chao Yan, Chuanbing Cheng, Kun Dai, Chuntai Liu, Changyu Shen, et al. Continuously prepared highly conductive and stretchable synergistically composited electrospun thermoplastic polyurethane yarns for wearable sensing. *Journal of Materials Chemistry C*, 6(9):2258–2269, 2018.

[4] Jonathan T Alexander, Faisal I Hai, and Turki M Al-Aboud. Chemical coagulation-based processes for trace organic contaminant removal: Current state and future potential. *Journal of Environmental Management*, 111:195–207, 2012.

[5] Ahmad Al-Haj Ali and Ribhi El-Bishtawi. Removal of lead and nickel ions using zeolite tuff. *Journal of Chemical Technology & Biotechnology: International Research in Process, Environmental and Clean Technology*, 69(1):27–34, 1997.

[6] BAM Al-Rashdi, DJ Johnson, and N Hilal. Removal of heavy metal ions by nanofiltration. *Desalination*, 315:2–17, 2013.

[7] S Sadri Moghaddam and MR Alavi Moghaddam. Cultivation of aerobic granules under different pre-anaerobic reaction times in sequencing batch reactors. *Separation and Purification Technology*, 142:149–154, 2015.

[8] Nicole Stadie, Astrid Schroeder, Jenny Postler, Antje Lorenz, Maria Swoboda, Frank Burchert, and Ria De Bleser. Treatment of sentence production in German agrammatism: A multiple single case study. 2005.

[9] Sangita Bhattacharjee, Siddhartha Datta, and Chiranjib Bhattacharjee. Improvement of wastewater quality parameters by sedimentation followed by tertiary treatments. *Desalination*, 212(1–3):92–102, 2007.

[10] G Andreottola, M Cadonna, P Foladori, G Gatti, F Lorenzi, and P Nardelli. Heavy metal removal from winery wastewater in the case of restrictive discharge regulation. *Water Science and Technology*, 56(2):111–120, 2007.

[11] Luigi Rizzo, Giusy Lofrano, and Vincenzo Belgiorno. Olive mill and winery wastewaters pre-treatment by coagulation with chitosan. *Separation Science and Technology*, 45(16):2447–2452, 2010.

[12] Murat Eyvaz. Treatment of brewery wastewater with electrocoagulation: Improving the process performance by using alternating pulse current. *International Journal of Electrochemical Science*, 11(6):4988–5008, 2016.

[13] Roberto Ferrarini, Andrea Versari, and Sergio Galassi. A preliminary comparison between nanofiltration and reverse osmosis membranes for grape juice treatment. *Journal of Food Engineering*, 50(2):113–116, 2001.

[14] LA Ioannou, C Michael, N Vakondios, K Drosou, Nikolaos P Xekoukoulotakis, E Diamadopoulos, and Despo Fatta-Kassinos. Winery wastewater purification by reverse osmosis and oxidation of the concentrate by solar photo-Fenton. *Separation and Purification Technology*, 118:659–669, 2013.
[15] Haiwen Gao and MK Stenstrom. Evaluation of three turbulence models in predicting the steady state hydrodynamics of a secondary sedimentation tank. *Water Research*, 143:445–456, 2018.
[16] C Blöcher, J Dorda, V Mavrov, H Chmiel, NK Lazaridis, and KA Matis. Hybrid flotation—membrane filtration process for the removal of heavy metal ions from wastewater. *Water Research*, 37(16):4018–4026, 2003.
[17] Shigendo Akita, Lourdie P Castillo, Susumu Nii, Katsuroku Takahashi, and Hiroshi Takeuchi. Separation of CO (II)/Ni (II) via micellar-enhanced ultrafiltration using organophosphorus acid extractant solubilized by nonionic surfactant. *Journal of Membrane Science*, 162(1-2):111–117, 1999.
[18] Hani Abu Qdais and Hassan Moussa. Removal of heavy metals from wastewater by membrane processes: A comparative study. *Desalination*, 164(2):105–110, 2004.
[19] E Alvarez-Ayuso, A Garca-Sánchez, and X Querol. Purification of metal electroplating waste waters using zeolites. *Water Research*, 37(20):4855–4862, 2003.
[20] Bilge Alyüz and Sevil Veli. Kinetics and equilibrium studies for the removal of nickel and zinc from aqueous solutions by ion exchange resins. *Journal of Hazardous Materials*, 167(1):482–488, 2009.
[21] Nida Amin, Dominik Schneider, and Michael Hoppert. Bioleaching potential of bacterial communities in historic mine waste areas. *Environmental Earth Sciences*, 77(14):542, 2018.
[22] Vanessa S. Antonin, Mauro C. Santos, Sergi Garcia-Segura, and Enric Brillas. Electrochemical incineration of the antibiotic ciprofloxacin in sulfate medium and synthetic urine matrix. *Water Research*, 83:31–41, 2015.
[23] Junsheng Liu and Xin Wang. Novel silica-based hybrid adsorbents: Lead (II) adsorption isotherms. *The Scientific World Journal*, 2013: 2356–6140, 2013.
[24] Jean Nepo Hakizimana, Bouchaib Gourich, Mohammed Chafi, Youssef Stiriba, Christophe Vial, Patrick Drogui, and Jamal Naja. Electrocoagulation process in water treatment: A review of electrocoagulation modeling approaches. *Desalination*, 404:1–21, 2017.
[25] Sergi Garcia-Segura, Maria Maesia SG Eiband, Jailson Vieira de Melo, and Carlos Alberto Martnez-Huitle. Electrocoagulation and advanced electrocoagulation processes: A general review about the fundamentals, emerging applications and its association with other technologies. *Journal of Electroanalytical Chemistry*, 801:267–299, 2017.
[26] Hossein Shahbeig, Nafiseh Bagheri, Sohrab Ali Ghorbanian, Ahmad Hallajisani, and Sara Poorkarimi. A new adsorption isotherm model of aqueous solutions on granular activated carbon. *World Journal of Modelling and Simulation*, 9(4):243–254, 2013.
[27] Ibrahim Maamoun, Ramadan Eljamal, Omar Falyouna, Khaoula Bensaida, Yuji Sugihara, and Osama Eljamal. Insights into kinetics, isotherms and thermodynamics of phosphorus sorption onto nanoscale zero-valent iron. *Journal of Molecular Liquids*, 328:115402, 2021.
[28] A Shahedi, AK Darban, F Taghipour, and A Jamshidi-Zanjani. A review on industrial wastewater treatment via electrocoagulation processes. *Current Opinion in Electrochemistry*, 22:154–169, 2020.
[29] M Behbahani, MR Alavi Moghaddam, and M Arami. Techno-economical evaluation of fluoride removal by electrocoagulation process: Optimization through response surface methodology. *Desalination*, 271(1-3):209–218, 2011.
[30] Dina T Moussa, Muftah H El-Naas, Mustafa Nasser, and Mohammed J Al-Marri. A comprehensive review of electrocoagulation for water treatment: Potentials and challenges. *Journal of Environmental Management*, 186:24–41, 2017.

[31] Akshaya Kumar Verma, Rajesh Roshan Dash, and Puspendu Bhunia. A review on chemical coagulation/flocculation technologies for removal of colour from textile wastewaters. *Journal of Environmental Management*, 93(1):154–168, 2012.

[32] Drissa Ouattara Kra, Grah Patrick Atheba, Patrick Drogui, and Albert Trokourey. Activated carbon based on acacia wood (auriculeaformis, côte d'ivoire) and application to the environment through the elimination of Pb^{2+} ions in industrial effluents. *Journal of Encapsulation and Adsorption Sciences*, 11(01):18, 2021.

[33] Mustafa Abunowara, Suriati Sufian, Mohamad Azmi Bustam, Muhammad Babar, Usama Eldemerdash, Roberto Bencini, Sami Ullah, Mohammed Ali Assiri, Abdullah G Al-Sehemi, and Ahmad Mukhtar. Experimental and theoretical investigations on kinetic mechanisms of low-pressure CO2 adsorption onto Malaysian coals. *Journal of Natural Gas Science and Engineering*, 88:103828, 2021.

[34] Sousan Hadi, Ensiyeh Taheri, Mohammad Mehdi Amin, Ali Fatehizadeh, and Tejraj M Aminabhavi. Adsorption of 4-chlorophenol by magnetized activated carbon from pomegranate husk using dual stage chemical activation. *Chemosphere*, 270:128623, 2021.

[35] Skelte G Anema. Age gelation, sedimentation, and creaming in UHT milk: A review. *Comprehensive Reviews in Food Science and Food Safety*, 18(1):140–166, 2019.

[36] Majid Rezaei and Roland R Netz. Airborne virus transmission via respiratory droplets: Effects of droplet evaporation and sedimentation. *Current Opinion in Colloid & Interface Science*, 55:101471, 2021.

[37] Jie Ren, Na Li, Hua Wei, Aimin Li, and Hu Yang. Efficient removal of phosphorus from turbid water using chemical sedimentation by fecl3 in conjunction with a starch-based flocculant. *Water Research*, 170:115361, 2020.

[38] CA Johnston. Mechanisms of wetland-water quality interaction. *Constructed Wetlands for Water Quality Improvement*, 293–299, 2020. doi: 10.1201/9781003069997-35.

[39] Emmanuel Kweinor Tetteh and Sudesh Rathilal. Application of organic coagulants in water and wastewater treatment. *Organic Polymer,* 2019. doi: 10.5772/intechopen.84556.

[40] Haojie Lai, Hongwei Fang, Lei Huang, Guojian He, and Danny Reible. A review on sediment bioflocculation: Dynamics, influencing factors and modeling. *Science of the Total Environment*, 642:1184–1200, 2018.

[41] Juanjuan Cao, Liang Zhang, Jiaying Hong, Jianliang Sun, and Feng Jiang. Different ferric dosing strategies could result in different control mechanisms of sulfide and methane production in sediments of gravity sewers. *Water Research*, 164:114914, 2019.

[42] Jinte Zou, Yaqiang Tao, Jun Li, Shuyun Wu, and Yongjiong Ni. Cultivating aerobic granular sludge in a developed continuous-flow reactor with two-zone sedimentation tank treating real and low-strength wastewater. *Bioresource Technology*, 247:776–783, 2018.

[43] M Zamani, A Montazeri, M Gheibi, A Fathollahi, and Kourosh Behzadian Moghadam. An efficient design of primary sedimentation tanks using a combination of response surface, metaheuristic and scenario building. *International Journal of Environmental Science and Technology*, 20(2):1215–1246. doi: 10.1007/s13762-022-04076-0. 2022.

[44] Imhoff Tank, Hawley's Condensed Chemical Dictionary, Mar. 2007. doi: 10.1002/9780470114735.hawley08835.

[45] Ganesh Kumar Rajahmundry, Chandrasekhar Garlapati, Ponnusamy Senthil Kumar, Ratna Surya Alwi, and Dai-Viet N. Vo. Statistical analysis of adsorption isotherm models and its appropriate selection. *Chemosphere*, 276:130176, 2021.

[46] Show-Chu Huang, Tsair-Wang Chung, and Hung-Ta Wu. Effects of molecular properties on adsorption of six-carbon vocs by activated carbon in a fixed adsorber. *ACS Omega*, 6(8):5825–5835, 2021.

[47] Jing Xia, Yanxin Gao, and Gang Yu. Tetracycline removal from aqueous solution using zirconium-based metal-organic frameworks (Zr-MOFs) with different pore size and topology: Adsorption isotherm, kinetic and mechanism studies. *Journal of Colloid and Interface Science*, 590:495–505, 2021.

[48] Junxiong Lin and Lan Wang. Comparison between linear and non-linear forms of pseudo-first-order and pseudo-second-order adsorption kinetic models for the removal of methylene blue by activated carbon. *Frontiers of Environmental Science & Engineering in China*, 3(3):320–324, 2009.

[49] J Rubio and F Tessele. Removal of heavy metal ions by adsorptive particulate flotation. *Minerals Engineering*, 10(7):671–679, 1997.

[50] Thomas Benzing and David Salant. Insights into glomerular filtration and albuminuria. *New England Journal of Medicine*, 384(15):1437–1446, 2021.

[51] Abbas Tcharkhtchi, Navideh Abbasnezhad, M Zarbini Seydani, Nader Zirak, Sedigheh Farzaneh, and Mohammadali Shirinbayan. An overview of filtration efficiency through the masks: Mechanisms of the aerosols penetration. *Bioactive Materials,* 6(1):106–122, 2021.

[52] Björg Bjarnadóttir, Margrét Ásta Bjarnadóttir, Snærós Axelsdóttir, and Bing Wu. Direct membrane filtration for wastewater treatment and resource recovery: A review. *Science of the Total Environment*, 710:136375, 2020.

[53] Wouter Pronk, An Ding, Eberhard Morgenroth, Nicolas Derlon, Peter Desmond, Michael Burkhardt, Bing Wu, and Anthony G Fane. Gravity-driven membrane filtration for water and wastewater treatment: A review. *Water Research*, 149:553–565, 2019.

[54] Irish Valerie Maggay, Yung Chang, Antoine Venault, Gian Vincent Dizon, and Chien-Jung Wu. Functionalized porous filtration media for gravity-driven filtration: Reviewing a new emerging approach for oil and water emulsions separation. *Separation and Purification Technology*, 259:117983, 2021.

[55] Feng Chen, Zhongli Ji, and Qiangqiang Qi. Effect of pore size and layers on filtration performance of coalescing filters with different wettabilities. *Separation and Purification Technology*, 201:71–78, 2018.

[56] Ayushi Verma, Meenu Agarwal, Shweta Sharma, Neetu Singh. Monitoring environmental nanotechnology, and management. Competitive removal of cadmium and lead ions from synthetic wastewater using *Kappaphycus striatum. Environmental Nanotechnology, Monitoring & Management*, 15:100449, 2021.

[57] BG Carter, N Cheng, R Kapoor, GH Meletharayil, and MA Drake. Invited review: Microfiltration-derived casein and whey proteins from milk. *Journal of Dairy Science*, 104(3):2465–2479, 2021.

[58] A Hashem, GM Taha, AJ Fletcher, LA Mohamed, SH. Highly efficient adsorption of Cd (II) onto carboxylated camelthorn biomass: Applicability of three-parameter isotherm models, kinetics, and mechanism. *Journal of Polymers and the Environment*, 29(5):1630–1642, 2021.

[59] S Mahendran, R Gokulan, A Aravindan, H Joga Rao, G Kalyani, S Praveen, T Bhagavathi Pushpa, M Senthil. Production of ulva prolifera derived biochar and evaluation of adsorptive removal of reactive red-120: Batch, isotherm, kinetic, thermodynamic and regeneration studies. *Biomass Conversion and Biorefinery*, 1–12, 2021.

[60] Péter Sipos. Searching for optimum adsorption curve for metal sorption on soils: Comparison of various isotherm models fitted by different error functions. *SN Applied Sciences*, 3(3):1–13, 2021.

[61] F Brouers. Generalized fractal kinetics in complex systems (application to biophysics and biotechnology). *Physica A: Statistical Mechanics and its Applications*, 368(1): 165–175, 2006.

[62] Sami Guiza, François Brouers, Mohamed Bagane. Fluoride removal from aqueous solution by montmorillonite clay: Kinetics and equilibrium modeling using new generalized fractal equation. *Environmental Technology & Innovation*, 21:101187, 2021.

[63] F. Brouers. The fractal (BSf) kinetics equation and its approximations. *Journal of Modern Physics*, 5(16):1594, 2014.

[64] François Brouers, O Sotolongo, F Marquez, Jean-Paul. Microporous and heterogeneous surface adsorption isotherms arising from levy distributions. *Physica A: Statistical Mechanics and Its Applications*, 349(1-2):271–282, 2005.

[65] K Smith. Membrane processing. Development of Membrane Processes. *Dairy and Beverage Application*, 3–14, 2013.

[66] Francois Brouers. Dubinin isotherms versus the Brouers–Sotolongo family isotherms: A case study, *Francisco Technology*, 34(9-10):552–564, 2016.

[67] Malgorzata Wasilewska, Adam Wojciech Marczewski, Anna Deryło-Marczewska, and Dariusz Sternik. Nitrophenols removal from aqueous solutions by activated carbon–temperature effect of adsorption kinetics and equilibrium. *Journal of Environmental Chemical Engineering*, 9:105459, 2021.

[68] AA Chiniforush, M Gharib. A unified moisture sorption-desorption isotherm for engineered wood. *Journal of Materials in Civil Engineering*, 33, 2021.

[69] Flor R Siperstein, Carlos Avendaño, Jordan J Ortiz, and Alejandro Gil-Villegas. Analytic expressions for the isosteric heat of adsorption from adsorption isotherm models and two-dimensional SAFT-VR equation of state. *AIChE Journal* 67(3):e17186, 2021.

[70] Yassine Saji and Mohammed Barkatou. A discrete bat algorithm based on lévy flights for Euclidean traveling salesman problem. *Expert Systems with Applications*, 172:114639, 2021.

[71] Chijioke Elijah Onu, Joseph T Nwabanne, Paschal E Ohale, and Christian O Asadu. Comparative analysis of RSM, ANN and ANFIS and the mechanistic modeling in eriochrome black-t dye adsorption using modified clay. *South African Journal of Chemical Engineering*, 36:24–42, 2021.

[72] HR Zakeri, M Yousefi, AA Mohammadi, M Baziar, SA Mojiri, S Salehnia, and A Hosseinzadeh. Chemical coagulation-electro Fenton as a superior combination process for treatment of dairy wastewater: Performance and modelling. *International Journal of Environmental Science and Technology*, 8:1–14, 2021.

[73] Pamila Ramesh, Vasanthi Padmanabhan, Praveen Saravanan, and Bhagavathi Pushpa Thillainayagam. Batch studies of turquoise blue dye (tb) adsorption onto activated carbon prepared from low-cost adsorbents: An ANN approach. *Biomass Conversion and Biorefinery*, 13(4): 1–14, 2021.

[74] Susan SA Alkurdi, Raed A Al-Juboori, Jochen Bundschuh, Les Bowtell, and Alla Marchuk. Inorganic arsenic species removal from water using bone char: A detailed study on adsorption kinetic and isotherm models using error functions analysis. *Journal of Hazardous Materials*, 405:124112, 2021.

[75] Fahad A Alabduljabbar, Sajjad Haider, Abdulaziz A Alghyamah, Adnan Haider, Rawaiz Khan, Waheed A Almasry, Raj Patel, Iqbal M Mujtaba, and Fekri Abdulraqeb Ahmed. Ethanol amine functionalized electrospun nanofibers membrane for the treatment of dyes polluted wastewater. *Applied Nanoscience*, 7:1–14, 2021.

[76] Bożena Czech, Magdalena Kończak, Magdalena Rakowska, and Patryk Oleszczuk. Engineered biochars from organic wastes for the adsorption of diclofenac, naproxen and triclosan from water systems. *Journal of Cleaner Production*, 288:125686, 2021.

[77] AA Fodeke and OJ Ayejuyone. Adsorption of methylene blue on corncob charcoal: Thermodynamic studies. *Ife Journal of Science*, 23(1):131–144, 2021.

[78] Jana Braniša, Klaudia Jomová, Ľubomír Lapčík, and Mária Porubská. Testing of electron beam irradiated sheep wool for adsorption of Cr (III) and Co (II) of higher concentrations. *Polymer Testing*, 10:107191, 2021.

[79] María Ángeles Lobo-Recio, Caroline Rodrigues, Thamires Custódio Jeremias, Flávio Rubens Lapolli, Isabel Padilla, and Aurora López-Delgado. Highly efficient removal of aluminum, iron, and manganese ions using Linde type – a zeolite obtained from hazardous waste. *Chemosphere*, 267:128919, 2021.

[80] Junfeng Liu, Dingfang Li, Yun Wu, and Dedi Liu. Lion swarm optimization algorithm for comparative study with application to optimal dispatch of cascade hydropower stations. *Applied Soft Computing*, 87:105974, 2020.

[81] Sodabeh Ebrahimpoor, Vahid Kiarostami, Morteza Khosravi, Mehran Davallo, and Abdolmohammad Ghaedi. Bees metaheuristic algorithm with the aid of artificial neural networks for optimization of acid red-27 dye adsorption onto novel polypyrrole/ $SrFe_{12}O_{19}$/graphene oxide nanocomposite. *Polymer Bulletin*, 76(12):6529–6553, 2019.

[82] Feixiang Zhao, Mingzhe Liu, Kun Wang, Tao Wang, and Xin Jiang. A soft measurement approach of wastewater treatment process by lion swarm optimizer-based extreme learning machine. *Measurement*, 179:109322, 2021.

[83] Shahrzad Saremi, Seyedali Mirjalili, and Andrew Lewis. Grasshopper optimisation algorithm: Theory and application. *Advances in Engineering Software*, 105:30–47, 2017.

[84] Xiwang Xiang, Xin Ma, Yizhu Fang, Wenqing Wu, and Gaoxun Zhang. A novel hyperbolic time-delayed grey model with grasshopper optimization algorithm and its applications. *Ain Shams Engineering Journal*, 12(1):865–874, 2021.

[85] Feng Wang, Heng Zhang, and Aimin Zhou. A particle swarm optimization algorithm for mixed-variable optimization problems. *Swarm and Evolutionary Computation*, 60: 100808, 2021.

[86] Salvador Asensio-Delgado, Fernando Pardo, Gabriel Zarca, and Ane Urtiaga. Absorption separation of fluorinated refrigerant gases with ionic liquids: Equilibrium, mass transport, and process design. *Separation and Purification Technology*, 276: 119363, 2021.

[87] Lu Lu, Hui Zheng, Jing Jie, Miao Zhang, and Rui Dai. Reinforcement learning-based particle swarm optimization for sewage treatment control. *Complex & Intelligent Systems*, 11:1–12, 2021.

[88] Deepak Sinwar, Monika Saini, Dilbag Singh, Drishty Goyal, and Ashish Kumar. Availability and performance optimization of physical processing unit in sewage treatment plant using genetic algorithm and particle swarm optimization. *International Journal of System Assurance Engineering and Management*, 1–12, 2021.

[89] Yang Liu, Zihang Zhang, Lei Bo, and Dongxu Zhu. Multi-objective optimization of a mine water reuse system based on improved particle swarm optimization. *Sensors*, 21(12):4114, 2021.

[90] Kung-Jeng Wang, Pei-Shan Wang, and Phuc Hong Nguyen. A data-driven optimization model for coagulant dosage decision in industrial wastewater treatment. *Computers & Chemical Engineering*, 8:107383, 2021.

[91] Shuzhen Zhang, Zhigang Ge, and Yanhua Lai. Application of genetic algorithm in optimizing a chemical adsorption bed with cacl2/expanded graphite adsorbent. *Procedia Engineering*, 205:1828–1834, 2017.

[92] Qiurong Yang, Rongjie Xu, Pan Wu, Jian He, Changjun Liu, and Wei Jiang. Three-step treatment of real complex, variable high-cod rolling wastewater by rational adjustment of acidification, adsorption, and photocatalysis using big data analysis. *Separation and Purification Technology*, 270:118865, 2021.

[93] Bo Ke, Hoang Nguyen, Xuan-Nam Bui, Hoang-Bac Bui, Yosoon Choi, Jian Zhou, Hossein Moayedi, Romulus Costache, and Thao Nguyen-Trang. Predicting the sorption efficiency of heavy metal based on the biochar characteristics, metal sources, and environmental conditions using various novel hybrid machine learning models. *Chemosphere*, 276:130204, 2021.

[94] Jimei Qi, Yu Hou, Jiwei Hu, Wenqian Ruan, Yiqiu Xiang, and Xionghui Wei. Decontamination of methylene blue from simulated wastewater by the mesoporous RgO/Fe/Co nanohybrids: Artificial intelligence modeling and optimization. *Materials Today Communications*, 24:100709, 2020.

[95] Jason Sahl and Junko Munakata-Marr. The effects of in situ chemical oxidation on microbiological processes: A review. *Remediation Journal: The Journal of Environmental Cleanup Costs, Technologies & Techniques*, 16(3):57–70, 2006.

[96] Mohamed Ksibi. Chemical oxidation with hydrogen peroxide for domestic wastewater treatment. *Chemical Engineering Journal*, 119(2-3):161–165, 2006.

[97] Seef Saadi Fiyadh, Mohammed Abdulhakim AlSaadi, Wan Zurina Jaafar, Mohamed Khalid Al Omar, Sabah Saadi Fayaed, Nuruol Syuhadaa Mohd, Lai Sai Hin, and Ahmed El-Shafie. Review on heavy metal adsorption processes by carbon nanotubes. *Journal of Cleaner Production*, 230:783–793, 2019.

[98] E Sanmuga Priya and P Senthamil. Water hyacinth (*Eichhornia crassipes*)—An efficient and economic adsorbent for textile effluent treatment—A review. *Arabian Journal of Chemistry*, 10:S3548–S3558, 2017.

[99] Khaled S. Abou-El-Sherbini and Mohamed M. Hassanien. Study of organically-modified montmorillonite clay for the removal of copper(II). *Journal of Hazardous Materials*, 184(1):654–661, 2010.

[100] Shuying Jia, Zhen Yang, Kexin Ren, Ziqi Tian, Chang Dong, Ruixue Ma, Ge Yu, and Weiben Yang. Removal of antibiotics from water in the coexistence of suspended particles and natural organic matters using amino-acid-modified-chitosan flocculants: A combined experimental and theoretical study. *Journal of Hazardous Materials*, 317:593–601, 2016.

[101] David B Miklos, Christian Remy, Martin Jekel, Karl G Linden, Jörg E Drewes, and Uwe Hübner. Evaluation of advanced oxidation processes for water and wastewater treatment—A critical review. *Water Research*, 139:118–131, 2018.

[102] Chaohai Wei, Fengzhen Zhang, Yun Hu, Chunhua Feng, and Haizhen Wu. Ozonation in water treatment: The generation, basic properties of ozone and its practical application. *Reviews in Chemical Engineering*, 33(1):49–89, 2017.

[103] Yifei Guo, Li Yang, X Cheng, and X Wang. The application and reaction mechanism of catalytic ozonation in water treatment. *Journal of Environmental and Analytical Toxicology*, 2(7):2161–0525, 2012.

[104] William H Glaze, Joon-Wun Kang, and Douglas H Chapin. The chemistry of water treatment processes involving ozone, hydrogen peroxide and ultraviolet radiation. *The Journal of the International Ozone Association*, 9(4):335–352 1987.

[105] Christos Comninellis, Agnieszka Kapalka, Sixto Malato, Simon A Parsons, Ioannis Poulios, and Dionissios Mantzavinos. Advanced oxidation processes for water treatment: Advances and trends for R&D. *Journal of Chemical Technology & Biotechnology: International Research in Process, Environmental & Clean Technology*, 83(6):769–776, 2008.

[106] EG Solozhenko, NM Soboleva, and VV Goncharuk. Decolourization of azodye solutions by Fenton's oxidation. *Water Research*, 29(9):2206–2210, 1995.

[107] Christian G Daughton. Pharmaceutical ingredients in drinking water: Overview of occurrence and significance of human exposure. In *Contaminants of Emerging Concern in the Environment: Ecological and Human Health Considerations*, pp. 9–68, 2010. doi: 10.1021/bk-2010-1048.ch002.

[108] Liping Li, Dongbin Wei, Guohua Wei, and Yuguo Du. Transformation of cefazolin during chlorination process: Products, mechanism and genotoxicity assessment. *Journal of Hazardous Materials*, 262:48–54, 2013.

[109] Ibtissem Ghorbel-Abid and Malika Trabelsi-Ayadi. Competitive adsorption of heavy metals on local landfill clay. *Arabian Journal of Chemistry*, 8(1):25–31, 2015.

[110] Alaa H. Hawari and Catherine N. Mulligan. Biosorption of lead (II), cadmium (II), copper (II) and nickel (II) by anaerobic granular biomass. *Bioresource Technology*, 97(4):692–700, 2006.

[111] Jixian Yang, Wei Wei, Shanshan Pi, Fang Ma, Ang Li, Dan Wu, and Jie Xing. Competitive adsorption of heavy metals by extracellular polymeric substances extracted from *Klebsiella* sp. J1. *Bioresource Technology*, 196:533–539, 2015.

[112] Eun-Ju Kim, Chung-Seop Lee, Yoon-Young Chang, and Yoon-Seok Chang. Hierarchically structured manganese oxide-coated magnetic nanocomposites for the efficient removal of heavy metal ions from aqueous systems. *ACS Applied Materials & Interfaces*, 5(19):9628–9634, 2013.

[113] Chih-Huang Weng and CP Huang. Adsorption characteristics of Zn(II) from dilute aqueous solution by fly ash. *Colloids and Surfaces A: Physicochemical and Engineering Aspects*, 247(1):137–143, 2004.

[114] Ronak Rahimian. A review of studies on the removal of methylene blue dye from industrial wastewater using activated carbon adsorbents made from almond bark. *Progress in Chemical and Biochemical Research*, 3(3):251–268, 2020.

[115] Yifei Li, Lee Nuang Sim, Jia Shin Ho, Tzyy Haur Chong, Bing Wu, and Yu Liu. Integration of an anaerobic fluidized-bed membrane bioreactor (MBR) with zeolite adsorption and reverse osmosis (RO) for municipal wastewater reclamation: Comparison with an anoxic-aerobic MBR coupled with RO. *Chemosphere*, 245:125569, 2020.

[116] Vincenzo Naddeo, Mona Freda N Secondes, Laura Borea, Shadi W Hasan, Florencio Ballesteros Jr, and Vincenzo Removal of contaminants of emerging concern from real wastewater by an innovative hybrid membrane process—Ultrasound, adsorption, and membrane ultrafiltration (USAME®). *Ultrasonics Sonochemistry*, 68:105237, 2020.

[117] Jiangkun Du, Sang Hoon Kim, Muhammad Azher Hassan, Sana Irshad, Jianguo Bao. Application of biochar in advanced oxidation processes: Supportive, adsorptive, and catalytic role. *Environmental Science*, 11:1–27, 2020.

[118] Danyan Li, Haidong Zhou, Liping Huang, Jingyuan Zhang, Jinyu Cui, Xin Li. Role of adsorption during nanofiltration of sulfamethoxazole and azithromycin solution. *Separation Science and Technology*, 15:1–15, 2020.

[119] Ling Zhang, Weiya Niu, Jie Sun, and Qi Zhou. Efficient removal of Cr (VI) from water by the uniform fiber ball loaded with polypyrrole: Static adsorption, dynamic adsorption and mechanism studies. *Chemosphere*, 248:126102, 2020.

[120] Hao Peng and Jing. Removal of chromium from wastewater by membrane filtration, chemical precipitation, ion exchange, adsorption electrocoagulation, electrochemical reduction, electrodialysis, electrodeionization, photocatalysis and nanotechnology: A review. *Environmental Chemistry Letters*, 1–14, 2020.

[121] Saeed Farrokhpay and Bradshaw, Dee. *Effect of Clay Minerals on Froth Stability in Mineral Flotation. A review, XXVI. IMPC*, New Delhi, India, *Paper*, (313), 2012.

[122] SM Yakout and G Sharaf El-Deen. Characterization of activated carbon prepared by phosphoric acid activation of olive stones. *Arabian Journal of Chemistry*, 9:S1155–S1162, 2016.

[123] FBKG Bergaya and G Lagaly. *General Introduction: Clays, Clay Minerals, and Clay Science*, volume 5, pages 1–19. Elsevier, 2013.

[124] Colin M Sayers and Lennert D. Den Boer. The elastic anisotropy of clay minerals. *Geophysics* 81(5):C193–C203, 2016.

[125] Mirela Rožić, Štefica Cerjan-Stefanović, Stanislav Kurajica, V Vančina. Ammoniacal nitrogen removal from water by treatment with clays and zeolites. *Water Research*, 34(14):3675–3681, 2000.

[126] Prashant Pandey and Vipin Kumar. Pillared interlayered clays: Sustainable materials for pollution abatement. *Environmental Chemistry Letters*, 17(2):721–727, 2019.

[127] Ali Tor, Nadide Danaoglu, Gulsin Arslan, and Yunus Cengeloglu. Removal of fluoride from water by using granular red mud: Batch and column studies. *Journal of Hazardous Materials*, 164(1):271–278, 2009.

[128] Mohammad Kashif Uddin. A review on the adsorption of heavy metals by clay minerals, with special focus on the past decade. *Chemical Engineering Journal*, 308:438–462, 2017.

[129] Maria Fernanda Oliveira, Victor M. de Souza, Meuris G. C. da Silva, and Melissa G. A. Vieira. Fixed-bed adsorption of caffeine onto thermally modified Verde-lodo bentonite. *Industrial & Engineering Chemistry Research*, 57(51):17480–17487, 2018.

[130] Júlia R. de Andrade, Maria F. Oliveira, Meuris G. C. da Silva, and Melissa G. A. Vieira. Adsorption of pharmaceuticals from water and wastewater using nonconventional low-cost materials: A review. *Industrial & Engineering Chemistry Research*, 57(9): 3103–3127, 2018.

4 Adsorption as an Emerging Technology and Its New Advances of Eco-Friendly Characteristics

Isotherm, Kinetic, and Thermodynamic Analysis

Reem M. Eltaweel, Shima Husien, Alyaa I. Salim, Nagwan G. Mostafa, Khloud Ahmed, Irene Samy Fahim, Lobna A. Said, and Ahmed G. Radwan

4.1 INTRODUCTION

The adsorption process is the adherence of soluble liquids and gases to adsorbents. The pollutant surface "adsorbate" interacts and adheres with the surface of the adsorbent [1]. Therefore, adsorption quality is determined by the amount of adsorbate that accumulates on the adsorbent surface. The amount adsorbed is calculated by isotherm studies that are discussed in this chapter [2]. Adsorption mechanism could be returned to two main types: physisorption and chemisorption; in physisorption, the adsorption mechanism is returned to the van der Waals force, whereas in the chemisorption process, the adsorbate molecules are attached to the adsorbent surface by chemical bonding such as ion exchange and complexation [3, 4]. Adsorption techniques are efficient in the removal of both organic and inorganic pollutants from domestic, agricultural, and industrial wastewater [5]. Furthermore, adsorption operation is free from sludge [6]. Additionally, adsorption is characterized by the availability of different synthetic and natural materials that can be used as adsorbents [7].

Many adsorbents were stated for their ability to remove pollutants such as manganese oxides, agriculture wastes, granular biomasses, sewage sludge ash, activated carbon, minerals, and extracellular polymeric substances [8–12]. Adsorbents that were derived from carbon were investigated as a potential adsorbent material for dyes' removal from polluted wastewater [13, 14]. Nano-sorbent materials, metal oxides, carbon, graphene oxide, metal oxides, and carbon, were used for heavy metal removal from industrial wastewater [15, 16]. The iron oxide modified by mesoporous

DOI: 10.1201/9781003354475-4

silica had a significant influence on methyl orange and lead elimination from polluted water because of their high surface area characteristics [17]. Cotton waste as an agriculture waste type was used for the removal of basic dyes with adsorption capacity of 875 and 277 mg/g for safranin and methylene blue, respectively [18]. Additionally, coconut husks were stated as a promising adsorbent material for bisphenol A (BPA) from polluted aqueous solution with 72% at 20 mg/L concentration [19]. Sugarcane bagasse (SCB)-activated carbon was used for phenol removal from natural streams with an adsorption capacity of 101 mg/g at a range of concentration of 100 to 1600 mg/g [20]. Water hyacinth plant was reported as a promising adsorbent material for different types of heavy metals with high removal efficiency as Cd, Cu, Pb, and Zn that achieved removal efficieny of 98%, 99%, 98%, and 84% [21, 22].

4.2 FACTORS AFFECTING ADSORPTION PROCESS

The adsorption process is affected by different environmental parameters such as time, pH, initial pollutant concentration, and dose of adsorbent material. Therefore, these parameters must be optimized to obtain high adsorption capacity for the pollutant from contaminated wastewater.

4.2.1 Effect of Contact Time

Adsorption contact time is an essential factor that can influence the adsorption process, where it affects its economic efficiency and the adsorption kinetics [23]. The optimum time that helps reach the equilibrium in the adsorption experiment can be determined using a contact time experiment. This equilibrium time can be later applied to the designs of wastewater treatment plants. In contrast to the other adsorption parameters, contact time does not follow a determined general trend [24]. The adsorption most probably occurs rapidly in the first few minutes of adsorption contact time and then flocculates gradually until reaching the equilibrium time. Equilibrium time is when no change in the pollutant removal percent is noticed [25]. The previous explanation was noticed in the safranin dye removal behaviour by magnetic mesoporous clay adsorbent material [26]. Different previous studies such as application of Fe_3O_4/chitosan composite in the adsorption process achieved high removal efficiency of 86% after 60–150 min contact time [27]. In a different study, the contact time effect on the safranin removal was demonstrated from 0 to 90 min, where results demonstrated that safranin removal was performed through two-stage processes. The first stage was fast; initially, safranin was adsorbed in the first 10 min and followed by the second stage, the slow adsorption, and the total adsorption capacity was 10.7 mg/g. Then safranin removal remained constant through time increasing. Researchers suggested that the fast initial uptake of safranin was owed to the rabid dye diffusion of the aqueous solution onto the external surface of mixed metal oxide composite (MMC) composite. Adsorption sites on the MMC composite surface are gradually incubated by safranin dye, where the dye molecules transfer from the bulk phase to the MMC pores' inner sphere. However, the steady equilibrium was achieved after 30 min contact time using the shaking model [24]. The effect of contact time on the Congo red removal amount using kaolin-bentonite as an adsorbent material was investigated [28]. Results indicated that contact time did not significantly influence the Congo red removal percent where the adsorption was rapid until nearly the first 50 min, and after 80-min contact time, no significant change was noticed. Another comparable study demonstrated the effect of contact time on the dyes' adsorption. DO34 uptake by hydrocarbon (HC), for example, was investigated through time intervals from 5 to 90 min. The DO34 removal increased by increasing the time until 60 min contact time was reached [29]. The initial rabid adsorption was attributed to large numbers of adsorption sites available at the beginning of the adsorption process of DO34 molecules. After that, the adsorption capacity remained constant because of the adsorption site saturation process. Finally, from the previous studies, it can be concluded that the adsorption process occurs rapidly in the first few contact time minutes, followed by a gradual increase until equilibrium is reached; after that, the dyes' removal percentage remains constant.

4.2.2 Effect of Agitation Rate and Time

The agitation rate significantly affects the removal of pollutants from polluted water. By increasing the agitation rate, the pollutant removal percent increases. This could be attributed to the reduction in film boundary layer around the sorbent particles. Additionally, agitation may also increase the external mass transfer coefficient and the rate of ion uptake [30]. For instance, the dyes' removal by risk husk was increased by increasing the agitation rate. After 15 min, crystal violet removal percent was 70% and increased to 82.5% after 60 min of agitation time [21, 31]. In contrast, direct orange and magenta removal were 47% and 54% in 15 min and increased up to 77% and 84% in 45 min agitation time, respectively. The increase in the removal percent by increasing the agitation time was attributed to the formation of the monolayer on the adsorbents' surface that is controlled by the adsorbate transport rate of the pollutant molecules from the external sites into the interior sites of the adsorbents [32].

4.2.3 Effect of Temperature

The effect of temperature is a physicochemical parameter on the adsorption process where it causes a change in the adsorption capacity of the adsorbent. Two states were stated for the adsorption process [33]; the first state is the endothermic process. Through the endothermic process, the adsorption capacity increases by increasing the temperature, and this was attributed to the increase in the dyes molecules' mobility that increases the number of active sites. In contrast, the second state is the exothermic process, where the degree of adsorption capacity decreases by increasing the temperature. Increasing the temperature decreases the adsorptive forces between the active sites of the adsorbent surface and the dye molecules, which in follow decreases adsorption amount [34]. Energy and entropy values are used to determine the adsorption process type [35]. In addition, thermodynamic parameter values are used to determine the process's practical applicability. The thermodynamic parameters affect the adsorption process, and they are usually obtained from the experimental data of different temperatures. The parameters of thermodynamic involved entropy (S) are evaluated according to Eqs. (4.1) and (4.2). The enthalpy variation (ΔH) and Gibbs free energy variation (ΔG) are calculated by using the following equation [36]:

$$\Delta G = -RT \ln Ka, \tag{4.1}$$

where ΔG is the sorption free energy (kJ/mol), T is the temperature in Kelvin (K), R is the constant of the universal gas (8.314 J/mol/K), and Ka is the constant of sorption equilibrium.

Additionally, Ka can be expressed in terms of enthalpy change (ΔH) and entropy change (ΔS) as a function of temperature as given by [37]

$$\ln Ka = \frac{\Delta H}{RT} + \frac{\Delta S}{RT}, \tag{4.2}$$

where ΔH is the sorption heat (kJ/mol) and ΔS is the change standard entropy (kJ/mol/K). Results obtained from the previous equation can be implied to determine the process type of the reaction. The DH has positive values in the endothermic adsorption process, indicating that dyes' adsorption nature is physical adsorption with weak interaction forces. ΔS positive values refer to an improved chance of the solid solution interface with some sorption system, structural changes, and an attraction to the adsorbate. Additionally, it refers that the adsorbed molecules' degree of freedom risen [38]. The temperature usually has a significant effect on the reaction rate where by increasing temperature, the rate of chemical reaction increases. The temperature determines the reaction process type as explained previously [39]. The malachite green adsorbed quantity by using clays as an adsorbent material increased by increasing the temperature from 10 to 50°C, and the methyl orange adsorbed quantity decreased significantly by increasing the temperature [40]. The temperature less influenced the adsorption of methylene blue as the adsorbed amount decreased slightly by increasing the temperature. By applying the thermodynamic parameters and calculating constants, it was demonstrated that the adsorption process was exothermic in the case of methylene blue (MB) and methyl orange [40]. At the same time, it was endothermic in the case of malachite green [40]. Another study demonstrated the effect of temperature for diazo dye uptake by modified bentonite where different temperature values of 25, 30, 40, 50, and 60°C were applied. A minor decrease in the diazo dye by increasing the temperature indicated that the adsorption process was exothermic [41].

4.2.4 Effect of pH

pH value is an important parameter that affects the removal process due to its effect on the pollutants' solubility and the binding sites of the adsorbent material [42]. pH is defined as the measurements of hydrogen ion concentrations in the medium and the variability of the aqueous solution pH where the adsorption process can take place [40]. In general, pH significantly affects pollutant adsorption due to the change in the characteristics of the adsorbent surface and the pollutant chemistry. Furthermore, the pH of the solution estimates the OH^- exchange sites' protonation as well as the adsorbate protonation. pH determines the exchange site's specific charge, and, in consequence, it improves the substrate adsorption tendency [43]. Generally, at low pH values (acidic conditions), the removal of the anionic contaminants increases because of the electrostatic attraction between the positive charges of the adsorbent surface and the anionic dye. Nevertheless, at high pH values (basic conditions), electrostatic repulsion is generated between the negative adsorbent surface charge and the anionic dye, which in consequence reduces the adsorption capacity and percentage removal of anionic pollutants [44, 45].

Several studies investigated the pH effect on dyes' adsorption; for example, the adsorption capacity of organo-clay adsorbent material between 200 and 400 mmol/L methylene blue, crystal violet, and reactive blue solutions increased by increasing pH [46]. This process was owed to the decrease in the positive charges' surface area by increasing the pH. Additionally, it increases the surface negatively charged sites. As a result, the clay surface's negative charges favour the cationic dye's uptake due to the electrostatic attraction [47]. Moreover, it was noticed that in acidic conditions, dyes' adsorption decreased due to the electrostatic force of repulsion where the protonated surface of the adsorbent. After modification, the surface of the adsorbent became positive, giving a 7.8 pHpzc value, and pH 9 achieved the maximum removal capacity [48]. According to the pHpzc result, the surface of the clay is opposing, and cationic dyes are positive, generating an electrostatic attraction between the opposing clay surface and the cationic dye molecules. This result revealed that the adsorbent surface's characteristics influence its adsorption capacity [49]. It was reported that anionic dyes' adsorption strongly depended on the pH value, where low ranges of pH were ideal for their adsorption, especially by using metal oxides as adsorbents [50]. Furthermore, the removal of azo dyes, Direct Fast Scarlet 4BS, and weak acid dark blue 5R by using acid-activated kaolinite with TiO_2 exhibited that maximum removal of these dyes was in the acidic medium than basic medium [51]. The authors reported that these findings resulted from the TiO_2 nanoparticle's surface determination by pH and the dyes' ionic nature [46]. The previous expectation cannot be suggested when using metal oxides as an adsorbent, although oxide-modified adsorbents under acidic conditions increase the dyes' removal capacity. The previous findings were attributed to the attraction force of electrostatic between the positively charged Ti^-OH^{2+}. On the other side, in the case of basic conditions, groups of OH on the TiO_2 dissociated and formed TiO^-. Therefore, the surfaces of TiO_2 nanoparticles became negatively charged, causing electrostatic repulsion between the negatively charged TiO_2 nanoparticles and the dye anions, which resulted in dyes' adsorption. Other reports suggested that adsorption processes do not always depend on the pH value. For instance, the Basic Red 46 removal by using bentonite clay did not affect by pH [52]. The study's findings demonstrated that pH ranged from 2 to 6; the Basic Red 46 adsorbed amount recorded nearly 220 mg/g, and no considerable impact was stated on that adsorption capacity. pH influences the adsorption of La by different adsorbents [53].

4.2.5 Effect of Adsorbent Dose

The dose of the adsorbent material is directly proportional to the pollutant removal percent, where an increase in pollutant uptake is observed by increasing the adsorbent material dose [54]. Increasing the adsorbent dose leads to increase the active exchangeable sites of the adsorbent surface that make the taking of ions more available. Additionally, low doses of adsorbents accelerate the adsorption rate due to the readily available active sites. In the case of a high adsorbent dose, the dye ions cannot readily access the adsorption sites easily until the equilibrium is achieved [55]. Generally, the overall solute adsorption per adsorbent unit weight can be decreased by increasing the adsorbent dose because of the interference that resulted from the active site interactions of the adsorbent [56, 57].

The increasing of Fe_3O_4@Ca-Alg beads' dose from 2 to 6 g/L did not enhance the La adsorption [58].

Raw bentonite and CTMAB-bentonite were applied for Acid Blue-93, Acid Blue-25, and Acid Golden Yellow removal with different doses to study the effect of its dose on these dyes' removal percent [29]. Results suggested that the dyes' removal percent increased by increasing the amount of the clay, where 80% removal was achieved for Acid Blue-93, and the Acid Blue dose was 0.02 g [59]. Acid Golden Yellow G recorded rapid adsorption with the equilibrium state being achieved at 0.05 g surface [60].

The brilliant green dye uptake by red clay was investigated by varying the dose from 0.3 to 1.5 g/L, and the results showed that adsorption increased from 67 to 94% by increasing the dose [61]. Nevertheless, adsorption capacity decreased from 117.3 to 31.4 mg/g. A similar performance was noticed by Mane and Babu [62], where the uptake of the dye increased by increasing the amount of red clay due to an increase in the available adsorption sites that were noticed.

4.2.6 Effect of Initial Pollutant Concentration

The adsorption capacity of pollutant removal, such as dyes, is highly influenced by the initial concentration of the pollutant. Furthermore, it depends on the immediate relation between the pollutant concentration and the available surface-active sites of the adsorbent. Generally, the pollutant removal decreased by increasing the initial concentration due to the adsorption sites' saturation with the pollutant [54, 63]. However, other studies revealed that increasing the concentration helps increase the removal percent, and this may be due to the high driving force for the mass transfer at high initial pollutant concentration [42, 64]. The effect of initial dyes' concentration on methylene blue adsorption using pine leaves [65]. Results suggested that by increasing the concentration from 10 to 90 mg/L, the removal of the dye decreased from 96.5% to 40.9% after 240 min contact time [66].

A further study investigated the effect of initial concentration in the state of dyes' removal using kaolin-bentonite. The results suggested that removal efficiency increased by increasing the initial concentration [28]. They attributed the previous result to the increasing in the concentration, which cause a rising mass gradient between the solution and the adsorbent, which acted as a driving force. This driving force helped transfer dyes' ions from the solution into the surface of the adsorbent. Moreover, the increase in the proportional dye adsorption was owed to the shift in equilibrium during the adsorption process of clay [28].

4.2.7 Effect of Ionic Strength

The availability of these salts in the solution records high ionic strength that has a significant effect on the whole adsorption process [67]. Usually, the concentration of salts, such as variation of KCl, NaCl, and $CaCl_2$, has a noticeable effect on the acidic dye adsorption. NaCl was applied as a simulator for the dying processing. Congo red removal was assessed using raw bentonite surfactant-modified bentonite, and the positively charged "CTABMBn". Moreover, the negatively charged bentonite was improved by adding the following salts in order NaCl < KCl < $CaCl_2$ [68]. This result was attributed to the ionic strength, which increased in the solution and caused the diffuse double-layer compression on the adsorbent and thus facilitated the electrostatic attraction and, in follows, the adsorption process [69,70].

PAA/PVA/clay composite hydrogel was investigated towards crystal violet uptake, and the effect of NaCl concentration on their adsorption capacity was demonstrated. Results stated the adsorption capacity decreased by increasing NaCl concentration. This was due to the increase in the ionic strength by adding NaCl concentration, which increases the negative charge of the adsorbent surfaces. In consequence, the attractive electrostatic force was reduced, and in consequence, the adsorption of the cationic dye decreased [71].

Moreover, the authors concluded that the addition of NaCl deteriorates the performance of the prepared hydrogel towards dyes' adsorption [72]. A possible illustration of the previous result is that increasing ionic strength leads to the reduction in the electrostatic attraction between the surface of the adsorbent and the dyes' molecules [73, 74].

4.2.8 Effect of Surface Area and Porosity

The surface area significantly affects adsorption capacity where by increasing the surface area, the adsorption capacity increases. This was illustrated by the assumption that increasing the surface area increases the available active sites, which increases the bindings between the pollutant molecules and the adsorbent's surface. In the state of porosity, pore size distribution plays a vital role in the adsorption rate by porous adsorbents. The optimum-size pores positively affected the adsorbate removal process where macropores had a tiny contribution to the surface area of the adsorbate. Nevertheless, the micropores and mesoporous contributed significantly and proportionally from 5% to 95% of the total surface area [75].

For instance, pore volumes from small pore size had a significant role in carbon dioxide adsorptions at 273 K. In conclusion, small pore size is very effective for high-performance adsorbents. For example, activated carbon size < 1.0 nm was found to be more effective for carbon dioxide uptake in addition to the specific surface area, and sulphur doping [76].

4.3 ADSORBENT TYPES

Different natural and synthetic materials are identified as potent adsorbent materials for organic and inorganic pollutant removal from industrial wastewater [77]. Synthetic adsorbents are materials whose origin is chemically not from living organisms' nature, such as metal ions' syncretization and nanoparticles. On the other hand, the nature adsorbents are the natural sorbents that originated from nature, such as agriculture wastes, microorganisms, and clays, and these materials are distinguished by their low costs [78]. Additionally, adsorbents can be classified into commercial and low-cost materials, indicating the materials obtained from natural sources.

4.3.1 Natural Adsorbents

Different clay minerals were applied for pollutant removal from polluted wastewater (see Table 4.1).

4.3.1.1 Clay Minerals

Clay and its minerals were identified as a promising adsorbent material due to its characteristic properties that include its high ion exchangeability and its mechanical and chemical stability. Additionally, clay has a large surface area characterized by its layered structure, non-toxic, available, affordable, and sorption and complexation ability. Therefore, many researchers investigated clay for pollutant adsorption, especially heavy metals from industrial wastewater, and considered it an outstanding alternative for traditional-commercial adsorbents [92].

Clay minerals majority belong to the group of phyllosilicate, and they can be classified according to the structure of their layers [93]. The first 1:1 clay mineral type consists of one octahedral and one tetrahedral sheet, and this structure appears in serpentine groups and kaolinite. The second type is 2:1 that consists of one sheet of tetrahedral located between two sheets of octahedral, and this structure can be found in chlorite, vermiculite, talc, montmorillonite, smectite, saponite groups, and mica [94–96]. Clay minerals are found in various sources such as soils, argillaceous shale rocks, and marine sediments formed due to weathering, especially for aluminium silicate rocks, sedimentation, and hydrothermal actions. In general, clay minerals exhibited the same physical properties, which can be examined using a variety of methods such as Fourier transform infrared spectroscopy (FT-IR), X-ray diffraction (XRD), scanning/tanning electron microscopy (SEM/TEM), and differential thermal analysis. Clay structure is characterized by its plastic properties on hardness as well as wetting in the drying or firing processes due to the silicate availability in its constituents [97]. Furthermore, the units of the plate-like structural layers convey high flexible anisotropy to clay minerals due to the appearance of strong covalent bonds through the layers of the sheet and the weaker electrostatic interactions. Moreover, the clay particle's irregular morphology gives it the property of the fine colloidal particles in watery suspensions [98]. They demonstrated more extensive particle size distribution and maintained a system of an isometric crystal feature. Clay particles' stability in colloidal suspensions depends on their structural characteristics such as charges, surface area, layer flexibility, and cation exchange capacity. Additionally, colloidal stability is affected by characteristics of suspensions such as

TABLE 4.1
Previous Studies for Different Minerals of Clay Application for Dyes' Removal

Clay	Pollutant	Factors	AC (mg/g)	References
Montmorillonite	Methylene blue	–	99.47 (%)	[79]
	Methyl green		68.35 (%)	
Bentonite	Congo red	–	95 (%)	[41]
Montmorillonite	Acid Scarlet G	–	85.7	[80, 81]
	Basic Red 18		530.645	
Bentonite	Crystal violet	–	496	[82]
Bentonite	Acid Blue-193	–	740.5	[83]
Ball clay	Methylene blue	Dose = 1 g/L Conc. = 30–300 mg/L Time = 100 min pH = 4–12	142	[84]
Bentonites	Methyl orange Methylene blue	Dose = 0.4 g/L Conc. = 50–2000 mg/L Time = 12–15 h pH = 2–11	141 183	[85]
Natural clay	Methylene blue	Dose = 1 g/L Conc. = 600 mg/L Time = 10 h pH = 7.14	330	[86]
Kaolinite	Methylene blue	Dose = 1 g/L; Conc. = 25–400 mg/L Time = 29 h pH = 3–13	143.47	[87]
Bentonite	Methylene blue	Dose = 1 g/L conc. = 25–500 mg/L Time = 30 h pH = 3–11	756.79	[88]
Malaysia clay	Methylene blue	Dose = 2 g/L Conc. = 30–300 mg/L Time = 12 h pH = 3–12	223.19	[88]
Moroccan clay	Methylene blue	Dose = 2g/L Conc. = 100–1200 mg/L Time = 2 h pH = 5.6	500	[89]
Zeolite	Goldgelb GL EC	Conc. = 20–200 m day^{-3} Time = 4–10 days	14.91	[90]
Zeolite	Basic dye – Maxilon Schwarz FBL-01	Conc. = 50–500 m day^{-3} Time = 4–10 days	55.86	[90]
Activated clay	Basic Red-18	Time = 2 h	157	[91]

ionic strength, pH, and surfactants [99, 100]. Many researchers successfully applied raw and modified clays for uptake by organic compounds, metals, and dyes. For instance, organophilic modified bentonite was stated as an impure clay. It consists of montmorillonite and other crystalline structures and was effectively applied for 2,4-dichlorophenoxyacetic acid removal [96]. The mechanism that is responsible for the adsorption process was hydrophobic interactions between the bisphenol A and the surface of the adsorbent [101]. For instance, Ni(II), Co(II), and Zn(II) ions could be adsorbed from polluted solutions using natural Jordanian clay. Additionally, Ni(II) adsorption from an aqueous medium was demonstrated under optimum experimental conditions where clay adsorption capacity was found at 19.4 mg g^{-1} for Ni(II) [102]. Moreover, chemical treatment of clay, such as hydrochloric acid, causes an increase in the clay surface area, which in follow increases the heavy metal removal capacity of the clay [103, 104]. Therefore, clay minerals are applied as potent adsorbent materials.

Clay minerals could be classified into natural and modified clays. The natural clay minerals are montmorillonite, kaolinite, bentonite, sepiolite/paly gorskite, and illite [105–109], whereas the modified clay is the type of clay that was modified to improve its capacity to remove pollutants from polluted wastewater. There are different forms of modification such as clay modification with activated carbon or surfactant [110, 111]. These methods were identified by their higher performance for improving the clay adsorption capacity towards dyes and other pollutants [112].

4.3.1.2 Zeolites

Zeolites were applied in metal ion uptake from polluted streams due to their structural characteristics and valuable properties that strongest their ability as an ion exchange material towards the metal cations. In addition to that, zeolites were distinguished by their high specific surface area, low prices, and high ion exchange capability. Furthermore, most zeolites are abundant and were characterized by their high selectivity for certain pollutants such as heavy metals [113]. For instance, the cation exchange capacity of Brazilian natural zeolites towards different heavy metals; Ni(II), Cd(II), Cr(III), and Mn(II), in a synthetic medium, were investigated, and the results revealed that Brazilian natural zeolite recorded a good removal performance towards these metals [114]. Additionally, it was reported that modified zeolites with Cu and Zn gel were more effective for As(V) removal from contaminated wastewater than natural zeolites [115]. As clay, zeolites could be classified into natural and modified zeolites. There are over 40 types of natural zeolite in the world, where the most common forms are mordenite, clinoptilolite, chabazite, analcime, phillipsite, stilbite, and laumontite [116, 117]. Nevertheless, the main zeolite modification was performed by the simple methods of acid/base treatment and modifications by surfactants. Simple washing of the zeolites with acid removes the impurities that block the pores of the zeolite material, which facilitates the entering process of cations into the pores and, as a result, enhances the removal rate of cations from polluted wastewater [118, 119].

4.3.2 Industrial By-product

Industrial activity is widely spread and recognized by its generation for a considerable quantity of solid wastes and by-products that hurt the environment. Various solid wastes such as fly ash, metal hydroxide "sludge," biosolids, waste slurry, and red mud are identified as potent adsorbents. They were characterized by their low cost and local availability, which facilitate their application in different pollutants' removal [120]. Fly ash is a material originated from plant and waste combustion processes and was applied as an adsorbent for different pollutants and showed a suitable adsorbent for phenolic compounds. Phenol's maximum adsorption capacity of 27.9 mg/g was recorded in fly ash and 108.0 mg/g by applying granular activated carbon using a phenol concentration of 100 mg/L [121].

Sludge, or metal hydroxide waste, is a type of industrial by-product that was applied for azo dye removal, and its contents are insoluble metal hydroxides and some salts.

Red mud is another type of industrial by-product produced through alumina production, and it was applied for dye removal and heavy metals such as arsenic [122, 123]. Other industrial by-products, such as steel and fertilizer industries, were applied as adsorbents for cationic dye removal from polluted wastewater. Table 4.2 shows a comparison of low-cost industrial by-product materials for different pollutants' removal.

TABLE 4.2
Different Applied Industrial By-Product Wastes as Adsorbents for Various Pollutants

Adsorbent Material	Pollutant	Factors	Adsorption Capacity (mg/g)	References
Fly ash	Methylene blue	–	4.47	[133]
Acid-treated fly ash	Methylene blue	–	7.99	[133]
Sludge	Reactive red-120	30°C	45.87	[134, 135]
	Reactive red-2	pH = 8–9	61.73	
Sludge	Congo red dye	30°C Initial pH = 10.4	270.8	
Metal hydroxides	Reactive dye Remazol Brilliant blue	25°C pH = 7	91	[136]
Fly ash	Methylene blue	pH = 6.75 Dose = 900 mg/L Initial concentration = 65 mg/L	58.24	[137]
Fly ash	Methylene blue	–	1.91	[138]
Red mud	Methylene blue	–	2.49	[139]
Activated sludge	Basic red-18	–	285.71	[140]
Activated sludge	Basic blue-9	–	256.41	[141]
Metal hydroxide sludge	Congo red	–	271	[134]
Red mud	Rhodamine B	–	92.5	[141]
Carbon slurry of fertilizer industry	Ethyl orange Metanil yellow Acid blue-113	Initial concentration = 5.10^{-4} M t Time = 45 min	198 211 219	[142]

4.3.3 Agriculture Wastes

Agriculture wastes are applied in pollutant removal due to their potential components such as hemicellulose, polysaccharides, lipids, lignin, proteins, simple sugars, hydrocarbons, water, and starch. These components afford a variety of functional groups, which enhance the capacity of adsorbents towards pollutant removal [124, 125]. The mechanism of adsorption is the cation exchange, which occurs between the functional groups like "hydroxylic, phenolic or carboxylic" and the metal ions "pollutants" [126]. The physicochemical characteristics of the agricultural wastes make them a potential adsorbent for various pollutants [127]. Using agriculture wastes as adsorbents affords twofold advantages to the environment, where first, it effectively reduces the waste materials' volume; secondly, their low cost [92]. Moreover, agricultural wastes were identified as a rich source for activated carbon production "Biochar" because of their low ash content and reasonable hardness [128]. There are different sources for the agricultural solid wastes such as almond shell, poplar, walnut sawdust, hazelnut shell, sawdust, orange peel, rice husk, sugarcane bagasse, coconut Burch waste, and papaya seed [129–132].

4.3.4 Biological Biomasses

These biological materials or biomasses are applied in dead and live forms a for dye removal from polluted wastewater [143, 144]. Biomasses, also known as biosorbents, were identified as a more selective material for pollutants than traditional adsorbents in addition to their ability to reduce dyes to low-level concentrations [145]. Biosorption is a developed technique characterized by a competitive, effective, and low-priced approach to remove pollutants from polluted wastewater. The adsorption capacity of the biosorbent materials was returned to the available binding sites. This could be attributed to the functional groups' availability such as carboxyl, sulphate, hydroxyl, and amino groups on algal composition [53]. Different biological biomasses that were employed as biosorbent materials are represented in Table 4.3.

4.3.4.1 Fungi

Fungi as adsorbent materials play an essential role in the biosorption process due to the content of their cell wall that exhibited an excellent metal binding capacity agent [146]. For instance, the fungus cell wall consisted of chitins, glucans, and mannans as well as lipids, polysaccharides, and pigments, e.g. melanin [147, 148]. In consequence, these different contents provide copious functional groups; carboxyl, nitrogen-containing ligands, and phosphate that contributed to the process of pollutant binding to the adsorbents [149]. For instance, other geniuses of Aspergillus and Penicillium were reported for their effectiveness in pollutant removal from industrial wastewater [150].

4.3.4.2 Bacteria

The bacterial cell was stated as a strong biosorbent due to the cell wall composition of various functional groups, which firmly were responsible for the pollutant binding mechanism [151]. Gram-positive bacteria were stated as having a more vital

TABLE 4.3
Previously Applied Agriculture By-Product Wastes as Adsorbents for Various Pollutants

Adsorbent Material	Pollutant	Factors	AC (mg/g)	References
Orange peel	Direct red-23 Direct red-80	Initial concentrations = 50, 75, 100, and 125 mg/L 25°C pH = 2 15 min contact time	10.72 21.05	[131]
Cedar sawdust	Methylene blue	Initial concentration = 40 mg/L 20°C	142.36	[161]
Rice husk	Safranin Methylene blue	–	838 312	[125]
Modified sunflower stalks	Congo red Direct blue	–	191.0 216.0	[162]
Pre-treated risk husk (sodium hydroxide)	Cd(II)	Initial concentration = 50 mg/L 240 min	20.24	[163]
Pre-treated risk husk (sodium carbonate)	Cd(II)	Initial concentration = 50 mg/L 60 min	16.18	[163]
Rice husk ash	Methylene blue	Initial concentration = 50 mg/L 600 min	690	[164]
Tea waste	Cu	Initial concentration = 100 mg/L 90 min	48	[165]
Beech sawdust	Methylene blue	Initial concentration = 14 mg/L	9.78	[166]
Beech sawdust (H_2SO4)	Methylene blue Basic red-22	Initial concentration = 14 and 21 mg/L	30.5 24.10	[167]
Pine sawdust	Acid -yellow 132 Acid blue-256	–	398.8 280.3	[168]
Egyptian bagasse pith	Basic blue-69	–	168	[169]
Chitosan	Acid orange-10 Acid red-73 Acid red-18 Acid green-25	–	922.9 728.2 693.2 645.1	[170]
Modified pinecone	Methylene blue	–	142.24	[171]
Mango seed kernel	Methylene blue	–	142.86	[11]
Pinecone	Methylene blue	–	109.89	[172]
Modified saw dust	Methylene blue	–	111.46	[173]

(Continued)

TABLE 4.3 *(Continued)*
Previously Applied Agriculture By-Product Wastes as Adsorbents for Various Pollutants

Adsorbent Material	Pollutant	Factors	AC (mg/g)	References
Green alga	Reactive red 5	–	555.6	[174]
Caulerpa lentillifera	Methylene blue	–	417	[175]
Sargassum muticum	Methylene blue	–	279.2	[176]
Duckweed	Methylene blue	–	144.93	[177]
Rhizopus arrhizus	Reactive orange-16 Reactive red 4	–	190 150	[178]
Spirodela polyrhiza	Basic blue-9	–	144.93	[177]
Living fungus	Methylene blue	–	1.17	[179]
Dead fungus	Methylene blue	–	18.54	[179]
Algal waste	Methylene blue	–	104	[180]
Algae Gelidium	Methylene blue	–	171	[180]

AC, Adsorption capacity.

biosorption capacity that can achieve high pollutant uptake because of the glycoprotein presence in their cell wall [152], whereas, Gram-negative bacteria showed a lower pollutant uptake capacity due to the phospholipids' appearance in the cell wall [153, 154]. Many bacterial species were stated as powerful biosorbents, such as *Agrobacterium* sp. and *Bacillus* sp. [155].

4.3.4.3 Yeast

Yeasts were little studied for pollutant removal. However, they are considered versatile microorganisms, which developed in both aerobic and anaerobic environments and are cultivated at low-cost nutritional requirements such as *Saccharomycetes* sp. [156]. In addition, various yeast species were reported by their high removal efficiencies, such as *Cyberlindnera fabiani*, *Candida intermedia,* and *Pichia guilliermondii* ATCC 201911 (L2) [157, 158].

4.3.4.4 Algae

Algae were recorded as potential biosorbents because of their high binding ability, surface area, easy handling, low costs, and availability [159]. For instance, algal cell wall composition contains chitin, and polysaccharides facilitate the biosorption process due to different functional groups; hydroxyl, amine, sulphate, carboxyl, and amino. Algal cells could be classified into two types, macro- and micro-algal cells [160].

4.3.5 Nano-Sorbents

Nano-material particles with sizes ranging from 1 to 100 nm were stated as potent adsorbent materials for pollutant removal from polluted wastewater [181]. Table 4.6 shows the different applied nanomaterials previously applied for pollutant removal.

TABLE 4.4
Comparison between Top-Down and Bottom-Up Approaches of Nano-Sorbent synthesis processes

Comparison Item	Top-Down Approach	Bottom-Up Approach
Definition	Conventional methods were applied for the nano-sorbent methods and followed by ball milling to reduce the particle size into nanoscale	It is the newest applied method, and it depends on building the substance from the bottom atom by atom, and molecule by molecule
Disadvantages	Damage of crystallographic and surface structure during particle size reduction	Chemical purification is required, and large-scale production is difficult
Advantages	Plays an important role in the nano-sorbent synthesis process	More preferable because it can produce materials with specific properties tailored to the remediation need depending on the route chosen to fabricate them
Examples	Mechanical alloying, reactive milling, and high-energy ball milling	Molecular self-assembly, sol–gel, and chemical/physical vapour deposition

4.3.5.1 Nano-Adsorbent Synthesis

The nano-sorbent synthesis process was dependent on two main approaches: bottom-up and top-down processes; comparison between the two approaches was performed in Table 4.4 [182]. Additionally, various chemical and physical ways were applied for the nano-sorbent material synthesis with specific morphology and size. However, these methods were stated as unsafe methods for the environment due to using of poisonous non-chemical agents through the synthesis processes [183]. Therefore, biological synthetic methods as a new approach for the nano-sorbent material synthesis were identified using biological agents such as bacteria, fungi, algae, and plant extract [184]. Table 4.5 shows the different applied physical, chemical, and biological approaches for nano-sorbent synthesis.

Different nano-material forms; nano-based metal oxides with their different types, carbon nanotubes, composites, and polymers, were applied for the pollutant removal process, and they can be discussed as follows:

4.3.5.2 Nano-Metal Oxides

Different nanosized metal oxides such as manganese, iron, titanium, and aluminium oxides were widely applied for pollutant adsorption from polluted wastewater. The most critical factors for determining the adsorption performance of nanosized metal oxides are their shape and size. Several characterization methods were used for nano-materials metal oxides cauterization as morphology, size, crystal structure, specific surface area, and the pH of zero point of charges (pHpzc), such as FT-IR, XRD, SEM, TEM, and DLS.

TABLE 4.5
Different Chemical, Physical, and Biological Synthesis Approaches of Nano-Sorbents

	Synthesis Method	Advantages	Disadvantages	References
Chemical methods	Coprecipitation method	Easy approach for producing gram scale magnetic nanoparticles, and the main diameter of the particles is up to 50 nm Additionally, reaction process is faster and has a significant yield and economical cost of precursor	Poor crystalline particles were produced due to requirements of size sorting methods In addition to that, polymeric matrix comprises of particle clusters due to the use of polymers for the stability of magnetite collides	[185]
	Sol–gel synthesis	In contrast to other synthesis methods, sol–gel can traditionally disperse in aqueous as well as in polar solvents due to the hydrophilic ligands Sol–gel synthesis can produce magnetic nanoparticles that have extensive crystallinity and saturation magnetization because of the higher process temperature	The process consumes higher alcohol content at the calcination time Weak bonding, high permeability, and high safety measures are required for the process	[186]
	Hydrothermal/ solvent thermal	Hydrothermal synthesis could form crystalline phases that are not steady at the melting point in contrast to other crystal growth types Additionally, iron oxide shape and size can be controlled by applying this method	However, this process requires maintaining the pressure >2000 psi, and temperature >200°C	[146]
	Aerosol/vapour phase	High-quality products can be obtained by applying this method that was further purified through lowering down the gas contaminants and maintaining the heating time as well as gas concentration	Larger aggregates can be formed, and low yield was obtained	[187]
	Sono-chemical decomposition	Mixing uniformity and crystal growth reduction result in acceleration effects in chemical dynamics and reaction rates Growth of particles can be controlled using polymers, organic agents, etc.	Not valuable to realize the iron oxide nanoparticle production Cavitation is caused in aqueous phase due to the use of ultrasonic irradiation	[34]

(Continued)

TABLE 4.5 *(Continued)*
Different Chemical, Physical, and Biological Synthesis Approaches of Nano-Sorbents

	Synthesis Method	Advantages	Disadvantages	References
	Electrochemical	This method was responsible for production of Fe_2O_3 and Fe_3O_4 nanoparticles through electrochemical decomposition under oxidizing environment	Inadequate reproducibility	[17]
	Flow injection	Laminar and plug-flow conditions validate extensive nanoparticle reproducibility in addition to the extensive mixing homogeneity Furthermore, produced nanoparticles' size such as magnetite ranged from 2 to 7 nm	Require continuous/segmented mixing of reagents under laminar flow regime in capillary reactor	[188]
Physical methods	Electron beam lithography	Well-maintained interparticle spacing	Requires complex machines, and expensive	[77]
	High-energy ball milling	Time-saving approach, cost-effective, facile approach, and particle size maintaining	Requires highly complex machines	[192]
	Gas-phase deposition	Convenient approach	Particle size maintaining is difficult	[189]
	Laser-induced pyrolysis	Saves time, the uniform shape of the produced nanoparticles, direct, and high production	Costs high, requires installation process, and requires expensive reagents and laser sources	[190]
Biological methods	Microbial route	Eco-friendly and costs low High production yield was obtained and requires low energy and temperature	Takes more time to obtain the desired product	[191]
	Plant-mediated extract route	Clean, non-toxic, environment-friendly, costs low, high yield production, and requires low time over the microbial route	–	[192]

4.3.5.2.1 Nano-Zerovalent Iron

Nano-zerovalent iron (nZVI) particles shows high efficiency towards pollutants, such as dyes, heavy metals, organochlorine pesticides, removal from polluted wastewater [193]. The core-shell nanoparticle's reactivity of the nano zerovalent iron was attributed to the higher intrinsic reactivity and greater density of their surface sites. The FeO core oxidation drove it in contrast to their microscale counterparts. Additionally, nano zerovalent iron was stated as an efficient adsorbent for three azo dyes: Acid Blue A, Sunset yellow, and methyl orange [194, 195].

4.3.5.2.2 Magnetic Nanoparticles

The magnetic properties of the nano-materials can be obtained by their shape, particle size, crystallite degree, type, chemical composition, and the curation regarding the non-homogeneous structure [196]. They were characterized by their non-toxic characteristics and were reusable and highly recyclable [197]. The nano-structural of metal oxide was known as magnetic nanoparticles and had properties such as an insulator, semiconductor, insulator, or metallic due to its electronic morphology [198]. Nanoscale materials with distinctive size have a significant impact on nanotechnology and provide a promising platform for the different types of wastewater treatment [199]. Among the different types of nanoparticles, magnetic nanoparticles were characterized by their most straightforward and most cost-effective way of wastewater purification of the wastewater [200, 201]. Additionally, they can be quickly recovered after the treatment process by applying a simple magnet. For instance, magnetite (Fe_3O_4) magnetic nanoparticles showed a promising path for the removal of metals and pollutants [202]. Magnetic nanoparticles were identified as a strong adsorbent due to their high surface area to volume ratio and their magnetic behaviour [203].

4.3.5.2.3 Magnesium Nanoparticles

As one of the metal oxide types, Mn-oxides had a tremendous specific interest because of their wealthy structural and various structural composition; MnO, Mn_2O_3, Mn_5O_8, MnO_2, and Mn_3O_4. They could be applied in different fields such as molecular sieves, batteries, catalysts, solar cells, magnetic materials, and water treatment and purification. Furthermore, they were identified as a less toxic material, cost-effective, with high specific capacitance, and environmentally compatible with other nanoparticle compounds [204].

4.3.5.2.4 Carbon Nanotubes

Carbon nanotubes are carbon nano-structures, whose shape is cylindrically divided into single-walled or multiwalled nanotubes according to their synthesis process. Carbon nanotubes are distinguished by their high surface area, tunable surface chemistry, and adsorption sites. The carbon nanotube's hydrophobic surface properties stabilize by the aggregation of particles in the aqueous solution to obviate the decrease in surface activity. The synthetic carbon nanotubes were applied for pollutant removal and determination of the pre-concentrate and reveal pollutants [205]. Furthermore, it was stated that boron addition to the prepared sponge enhances its absorption capacity to oil and water. Then the oil can be retained from the sponge and recovered for further application [206]. Applying carbon nanotubes had various advantages over activated carbon, such as its low cost through the application on the industrial scale to treat large wastewater amounts in contrast to activated carbon [207].

4.3.5.3 Polymeric Nano-Sorbents

Polymeric adsorbents are a growing type of nano-adsorbents, which are considered an alternative method for traditional adsorbents. They have various properties such as high surface area, surface chemistry, adaptability, distribution of their pore size, and mechanical rigidity. They could be classified into four major categories based on their composition. The dendrimers' nano-adsorbent type is a branched molecule considered a perfect example for polymeric nano-adsorbent application on organic and inorganic contaminant removal from contaminated wastewater. Additionally, it was stated that heavy metals were adsorbed through the dendrimers' external branches, whereas organic material adsorption was through the internal hydrophobic fractions. For instance, it was reported that combining dendrimers with ultrafiltration is a technique that succeeded in the elimination of all copper cations from polluted wastewater. However, few recent companies produced dendrimers due to their complex multistage synthesis process [208].

4.3.5.4 Composite Material Nano-Sorbents

Composites are identified as a blend of two or more different constituents to impart outstanding characteristics in the final product. They can be fabricated by different methods such as fictionalization, grafting, and cross-linking to enhance their properties. Composites have various characteristics such as fictionalization characteristics, surface area, and chemical accessibility. Additionally, composites' material comprises two phases; the first is continuous, known as matrix, and the second is discontinuous, which was identified by the dispersed phase [209].

4.4 THE ISOTHERMAL MODELS OF THE ADSORPTION PROCESS

Industrial wastewater contains toxic heavy metals such as Cr, Ni, Cd, Pb, Hg, Zn, Co, and Cu [227]. These solids have high toxic properties. Nano-materials can be used as adsorbents. The nano-sorbents should be non-toxic and have a high capacity at a low concentration of solids; the solids can be removed easily from the adsorbent surface and can be recycled [218]. In the adsorption process, a porous solid medium is used to attract a mixture of multicomponent fluid (liquid or gas) by chemical or physical bonds [228]. So, the adsorption is called chemisorption or a physisorption process. The physisorption happens because of the weak electrostatic charges such as van der Waals interactions, where the bands can be easily broken, and it produces a multilayer of the adsorbate on the adsorbent [229]. The chemisorption occurs covalent bond forms between adsorbates and adsorbent by sharing or transferring an electron which produces a monolayer of the adsorbate on the adsorbent [230]. The physisorption is reversible, and the chemisorption is irreversible. The differences between the

TABLE 4.6
Different Nano-Materials Previously Applied for Pollutant Removal

Adsorbent Material	Pollutant	Factors	Removal Efficiency	References
TiO_2	Crystal violet Methylene blue	Conc. = 10 ppm Dose = 50 mg pH = neutral Time = 60 min	92.65% 94.77%	[210]
TiO_2/Fe_2O_3/PAC	Cyanide	Conc. = 300 mg/L Dose = 1.4 g/L pH = 10.0 Time = 180 min	97%	[211]
$MgFe_2O_4$ $CaFe_2O_4$ $BaFe_{12}O_{19}$ $CuFe_2O_4$	Methylene blue	Conc. = 10 mg/L Dose = 0.1 g pH = neutral Time = 30 min	96%	[212]
$ZnFe_2O_4$	Methylene blue	Conc. = 10 mg/L Dose = 0.1 g pH = neutral Time = 30 min	85%	[213]
ZnO/SnO_2	Methyl orange	Conc. = 20 ppm Dose = 0.2 g pH = neutral Time = 100 min	56%	[214]
Magnetic activated carbon composite	Methylene blue Methyl orange Rhodamine B	Conc. = 4 mg/L (methylene orange), 45 mg/L (methylene blue), and 4 mg/L (rhodamine B) Dose = 0.44 g/L pH = 2.5	>90%	[215]
$TiO_2/ZnTiO_3/Fe_2O_3$	Methylene blue Methyl orange	Conc. = 1000 mg/L Dose = 10 g/L pH = 2.0 Time = 15 min	>95%	[216]
Nano-TiO_2	Reactive red Reactive yellow	Conc. = 40 mg/L Dose = 160 mg/L (Reactive red) and 500 mg/L (Reactive yellow) pH = 5.5 (Reactive red) and 7.4 (Reactive yellow) Time = 95 min	100%	[217]
TiO_2 and ZnO	Reactive black 5 Reactive orange 4	Conc. = 25 mg/L Dose = 1.25 g/L (Reactive black 5) and 1 g/L (Reactive orange 4) pH = 4.0 (Reactive black 5) and 11.0 (Reactive orange 4) Time = 7 min	75% 62%	[218]

(Continued)

TABLE 4.6 *(Continued)*
Different Nano-Materials Previously Applied for Pollutant Removal

Adsorbent Material	Pollutant	Factors	Removal Efficiency	References
Multiwalled carbon nanotubes	Methyl orange	Conc. = 10 mg/L Dose = 0.2 g/L Time = 140 min	52.9 mg/g	[86]
Iron oxide/silica gel nanomaterial	Sulphur dyes	Conc. = 100 mg/L Dose = 0.20 g pH = 4–5 Time = 20 min	11.1 mg/g	[150]
Silica nanomaterial	Basic violet	Conc. = 50 mg/L Time = 60 min	416 mg/g	[219]
Metal oxide nano-adsorbents Amine-modified nanoceria	Fenalan Yellow G	Conc. = 10 mg/L Dose = 0.1 g pH = 2.0 Time = 210 min	25.58 mg/g	[188]
Zinc oxide	Congo red	Conc. = 250 mg/L Dose = 1.2 g/L pH = 2.0–10 Time = 10 min	208 mg/g	[220]
Zinc oxide	Malachite green	Conc. = 10 mg/L 80 mg/L pH = 8.0 Time 120 min	310.5 mg/g	[221]
Ferric oxide with carbon composite	Methylene blue	pH = 2 0.1 mg/L Time = 3 h	117 mg/g	[222]
Cobalt oxide/ Silicon dioxide composite	Methylene blue	Conc. = 20 mg/L Dose = 0.8 g/L pH = 6–7 Time = 30 min	53.9 mg/g	[223]
Manganese oxide, ferric oxide, and ionic polymer composite	Crystal violet	Conc. = 50 mg/L Dose = 1.5 g/L Time = 20–60 min	1200%	[224]
Ferric oxide with montmorillonite composite	Methylene blue	Conc. = 120 mg/L Dose = 2.5 g/L pH = 7.37 Time = 25 min	106.4%	[225]
Magnesium oxide/Graphene oxide	Methylene blue	Conc. = 20 mg/L Dose = 0.1–1 g/L pH = 7.0 Time = 20 min	833%	[226]

TABLE 4.7
Comparison between Chemisorption and Physisorption Adsorption Process [266]

Chemisorption	Physisorption
Monolayer	Multilayer
Covalent bond forms between the adsorbent and the adsorbate by sharing or transferring of electrons	Weak electrostatic (van der Waals interactions)
Irreversible	Reversible
It has a high enthalpy and occurs at all temperatures	It has a low enthalpy and occurs at low temperature below the adsorbate boiling point

chemisorption and the physisorption are summarized in Table 4.7. The adsorption isothermal models are used to describe the equilibrium performance of the adsorbents at a constant temperature [16]. A comparison between different isothermal models is shown in Table 4.22.

4.4.1 One-Parameter Isotherm

4.4.1.1 Henry's Isotherm Model

Henry's isotherm model is the simplest model, where the partial pressure of the adsorptive gas is directly proportional to the amount of surface adsorbate [231]. For the liquid phase, the pressure is replaced by the adsorbate concentration [16]. It is used to fit adsorption at low concentration, and it is given by

$$q_e = K_{\mathrm{HE}} C_e, \tag{4.3}$$

where q_e is the amount of adsorbate on the adsorbent at equilibrium, C_e is the adsorbate concentration at equilibrium, and K_{HE} is Henry's constant [16].

4.4.2 Two-Parameter Isotherm

4.4.2.1 Langmuir Model

The Langmuir model is an empirical model that assumes monolayer adsorption, where the adsorption process happens at identical sites [232]. There are no side interactions, even on the adjacent sites, between the adsorbed molecules. The adsorption is assumed to be homogenous in that model [7]. The Langmuir model is given by

$$q_e = \frac{q_{\max} k_L C_e}{1 + k_L C_e}, \tag{4.4}$$

where q_e is a fraction of the filled sites, which shows how many molecules are on the surface at equilibrium, and C_e is the equilibrium concentration of the adsorbates, k_L is the Langmuir equilibrium constant, and $q_{\max}$ is the maximum adsorption capacity [14].

The model characteristics can be determined by the separation factor R_L which is a dimensionless constant represented as follows [126]:

$$R_L = \frac{1}{1 + K_L C_o}, \tag{4.5}$$

where C_o is the initial concentration of the adsorbate. The separation factor R_L values indicate the adsorption nature to be linear ($R_L = 1$), unfavourable ($R_L < 1$), irreversible ($R_L = 0$), and favourable ($0 < R_L < 1$). The Langmuir model is converted to Henry's model at high concentrations [233].

In inorganic arsenic species removal from water using bone char, four different isothermal models were investigated. The models were Langmuir, Freundlich, Redlich–Peterson, and Sips models [23]. The statistical analysis was implemented to find the best fit model to the data that has the minimum error and a higher R^2 value. For As(V) removal on bone char, the data best fitted the non-linear Langmuir model with R^2 of 0.977. So, the result explained a monolayer homogeneous surface coverage of As(V) on the surface [23].

4.4.2.2 Freundlich Model

The Freundlich model describes the reversible adsorption process where multilayer adsorption is possible, and the adsorption is heterogeneous [234]. This model fits the systems with heterogeneous surfaces in the gas phase. It provides a narrow range of pressures because at low pressure, it has improper behaviour towards Henry's law [235]. The Freundlich model has an empirical equation, and it has non-linear and linear models; the non-linear model is represented as follows:

$$q_e = K_F C_e^{1/n}, \tag{4.6}$$

where q_e is the amount of adsorbate on the adsorbent at equilibrium and C_e is the adsorbate concentration at equilibrium. K_F and n depend on the temperature, where $1/n$ is the intensity of the adsorption. When $(0 < (1/n) < 1)$ the adsorption is desirable, it is undesirable if $(1 < (1/n))$, and it is irreversible if $(1/n = 1)$. The linear form is represented as follows:

$$\ln q_e = \ln k_F + \frac{\ln c_e}{n}. \tag{4.7}$$

For the treatment of dye-polluted wastewater using ethanol-amine-functionalized electrospun nano-fibre membrane, Langmuir and Freundlich models were used in [14]. The data best fitted the linear Freundlich with $R^2 = 0.998$, for the adsorption of MB onto ethanolamine (EA). Freundlich model suggested multilayer adsorption due to physical adsorption [14].

4.4.2.3 Dubinin–Radushkevich Model

The Dubinin–Radushkevich model describes the adsorption mechanism with the distribution of Gaussian energy onto the heterogeneous surfaces [236]. This model usually fits systems with vapours and gases adsorbed in microporous sorbents. The Dubinin–Radushkevich model assumes multilayer adsorption occurs by van der Waal's forces, and this model can be used for the physical adsorption processes. This model can be used to distinguish between chemical and physical adsorption. Unlike the Langmuir and Freundlich models, the Dubinin–Radushkevich model is a semiempirical equation [16]. The Dubinin–Radushkevich model has both non-linear and linear models, as shown in Table 4.8. The theoretical isotherm saturation

TABLE 4.8
Illustration of the Linear Forms, the Non-Linear Forms, and the Plot of the Isotherm Models

Isotherm Model	Non-Linear Form	Linear Form	Plot
Langmuir	$q_e = \frac{q_{max} k_L C_e}{1 + k_L C_e}$	$\frac{C_e}{q_e} = \frac{C_e}{q_{max}} + \frac{1}{q_{max} k_L}$	$\frac{C_e}{q_e}$ vs. C_e
Dubinin–Radushkevich	$q_e = q_s e^{-k_{ad}\varepsilon^2}$	$\ln(q_e) = \ln(q_s) - k_{ad}\varepsilon^2$	$\ln(q_e)$ vs. ε^2
Temkin	$q_e = \frac{RT}{b}\ln(K_T C_e)$	$q_e = \frac{RT}{b}\ln(K_T) + \frac{RT}{b}\ln(C_e)$	q_e vs. $\ln(C_e)$
Flory–Huggins	$\frac{\theta}{C_O} = K_{FH}(1-\theta)^{n_{FH}}$	$\log\left(\frac{\theta}{C_O}\right) = \log(K_{FH}) + n_{FH}\log(1-\theta)$	$\log\left(\frac{\theta}{C_O}\right)$ vs. $\log(1-\theta)$
Hill	$q_e = \frac{q_{sH} C_e^{n_H}}{K_D + C_e^{n_H}}$	$\log\left(\frac{q_e}{q_{sH} - q_e}\right) = n_H \log(C_e) - \log(K_D)$	$\log\left(\frac{qe}{qsH - qe}\right)$ vs. $\log(C_e)$
Halsey	$q_e = e^{\frac{\ln K_H - \ln C_e}{n_H}}$	$\ln q_e = \left(\frac{1}{n_H}\right)\ln K_H - \left(\frac{1}{n_H}\right)\ln C_e$	$\ln q_e$ vs. $\ln C_e$
Harkins–Jura	$q_e = \left(\frac{A_{HJ}}{B_{HJ} - \log C_e}\right)^{1/2}$	$\frac{1}{q_e^2} = \frac{B_{HJ}}{A_{HJ}} - \left(\frac{1}{A_{HJ}}\right)\log C_e$	$\frac{1}{q_e^2}$ vs. $\log C_e$
Elovich	$\frac{q_e}{q_m} = K_e C_e e^{-(q_e/q_m)}$	$\ln\left(\frac{q_e}{C_e}\right) = \ln K_e q_m - \frac{q_e}{q_m}$	$\ln\left(\frac{q_e}{C_e}\right)$ vs. q_e
Kiselev	$K_1 C_e = \frac{\theta}{(1-\theta)(1+K_n\theta)}$	$\frac{q_m}{C_e(q_m - q_e)} = \frac{K_1 q_m}{q_e} + K_1 K_n$	$\frac{q_m}{C_e(q_m - q_e)}$ vs. $\frac{q_m}{q_e}$

(Continued)

TABLE 4.8 ***(Continued)***
Illustration of the Linear Forms, the Non-Linear Forms, and the Plot of the Isotherm Models

Isotherm Model	Non-Linear Form	Linear Form	Plot
Redlich–Peterson	$q_e = \frac{K_R C_e}{1 + a_R C_e^g}$	$\ln\left(\frac{K_R C_e}{q_e} - 1\right) = g \ln C_e + \ln a_R$	$\ln\left(\frac{C_e}{q_e}\right)$ vs. $\ln C_e$
Toth	$q_e = \frac{K_T C_e}{(a_T + C_e)^{1/t}}$	$\ln\left(\frac{q_e}{K_T}\right) = \ln(C_e) - \frac{1}{t}\ln(a_T + C_e)$	$\ln\left(\frac{q_e}{K_T}\right)$ vs. $\ln C_e$
Sips	$q_e = \frac{K_s C_e^{\beta}}{1 + a_s C_e^{\beta}}$	$\ln\left(\frac{K_s}{q_e}\right) = \ln a_s - \beta \ln C_e$	$\ln\left(\frac{K_s}{q_e}\right)$ vs. $\ln C_e$
Kahn	$q_e = \frac{q_{\max} b_k C_e}{(1 + b_k C_e) a_k}$		
Koble–Corrigan		$\frac{1}{q_e} = \left(\frac{1}{A_K C_e^P}\right) + \frac{B_K}{A_K}$	
Radke–Prausniiz	$q_e = \frac{q_{\mathrm{MRP}} K_{RP} C_e}{(1 + K_{RP} C_e)^{\mathrm{MRP}}}$		
Langmuir–Freundlich	$q_e = \frac{q_{\mathrm{MLF}} (K_{LF} C_e)^{\mathrm{MLF}}}{1 + (K_{LF} C_e)^{\mathrm{MLF}}}$		
Jossens	$C_e = \frac{q_e}{H} e^{F q_e^P}$	$\ln\left(\frac{C_e}{q_e}\right) = -\ln(H) + F q_e^P$	$\ln\left(\frac{C_e}{q_e}\right)$ vs. q_e
Fritz–Schlunder	$q_e = \frac{q_{mFS5} K_{FS} C_e}{1 + q_m C_e^{MFS}}$		
Baudu	$q_e = \frac{q_m b_0 C_e^{1+x+y}}{1 + b_0 C_e^{1+x}}$		
Weber–Van Vliet	$C_e = p_1 q_e^{(p_2 q_e^{p_3} + p_4)}$		
Marczewski–Jaroniec	$q_e = q_{MMJ} \left(\frac{(K_{MJ} C_e)^{nMJ}}{1 + (K_{MJ} C_e)^{nMJ}}\right)^{M_{MJ}/n_{MJ}}$		
Fritz–Schlunder (five-parameter)	$q_e = \frac{q_{mFS5} K_1 C_e^{\alpha_{FS}}}{1 + K_2 C_e^{\beta_{FS}}}$		

capacity is represented by q_s, ε is the Polanyi potential, and K_{ad} is a constant referring to the average free energy of adsorption. ε can be calculated by the following equation:

$$\varepsilon = RT\ln\left(1+\frac{1}{C_e}\right), \tag{4.8}$$

where C_e is the concentration of adsorbate. This model may not generate accurate mean free energy for distinguishing the type of the adsorption, either chemical or physical adsorption, in the case of a solid or liquid adsorption system. The Dubinin–Radushkevich model is used to know if the adsorption is either chemisorption or physisorption of the adsorbate ions on the adsorbent by calculating the mean free energy E [145]:

$$E = \frac{1}{\sqrt{2K_{\text{ad}}}}. \tag{4.9}$$

The free energy values define the adsorption mechanism, physisorption if $E \leq 8$, and chemisorption if E is more than that [145].

For the adsorption of diclofenac, naproxen, and triclosan from water systems using engineered biochars from organic wastes, four different isothermal models were used in [237]. Langmuir, Freundlich, Temkin, and Dubinin–Radushkevich (DR) models were investigated, and the data best fitted the DR with the highest R^2. The data that fitted by the DR model was the DCF adsorption onto BCSLW_{N2}; therefore, the adsorption occurred on heterogeneous surfaces [237].

4.4.2.4 Temkin Model

The Temkin model assumes that the binding energies are uniformly distributed, and it takes into account the indirect reaction between the adsorbent and the adsorbate [238]. The Temkin model assumes that the heat of the adsorption decreases linearly when the surface coverage increases. This model ignores the low and high concentrations; it is applicable only for an intermediate range of ion concentration [41]. It has non-linear and linear models, as shown in Table 4.8. The Temkin isotherm constant is represented by K_T, T is the absolute temperature, and R is the molar gas constant. This model does not provide an appropriate fit for an aqueous phase adsorption system [16].

For the adsorption of MB on corncob charcoal, Langmuir, Freundlich, and Temkin models were used in [79]. The linear Temkin model was the best to fit the adsorption of MB on acid-treated TCC with $R^2 = 0.956$ [79].

4.4.2.5 Flory–Huggins Model

The Flory–Huggins model defines the surface coverage degree characteristics of the adsorbate on the adsorbent by the equation shown in Table 4.8 [252]. The degree of the surface coverage is represented by θ, n_{FH} is the number of sites occupied by the adsorbate molecules, and K_{FH} is Flory–Huggins equilibrium constant.

For the adsorption of Cr(III) and Co(II) using the electron beam irradiated sheep wool, Langmuir, Freundlich, Dubinin–Radushkevich, Temkin, Flory–Huggins, Halsey, Harkins–Jura, Jovanovic, Elovich, and Redlich–Peterson models were used to study this case. The Co(II) adsorption best fitted the linear Flory–Huggins isotherm [193].

4.4.2.6 Hill Model

The Hill model assumes that the adsorption is cooperative, where the adsorbates can bind at one site on the adsorbent [239]. The non-linear and the linear models are shown in Table 4.8. The constants are represented by n_H, K_D, and q_{sH}. If $n_H = 1$ means that the binding is non-cooperative if $n_H > 1$, the binding is a positive cooperation, and if $n_H < 1$, the binding is a negative cooperation [16].

4.4.2.7 Halsey Model

Like Freundlich model, the Halsey model is used to describe the multilayer adsorption on the heterogeneous surfaces where the heat of the adsorption is non-uniformly distributed [240]. Halsey model is represented by both linear model and non-linear model as shown in Table 4.8. Halsey isotherm constants are represented by K_H and n_H [123].

4.4.2.8 Harkins–Jura Model

Harkins–Jura model describes the multilayer adsorption on the adsorbent surface, where the pore distribution is heterogeneous [241]. The non-linear and linear forms are shown in Table 4.8. Harkins–Jura constants are represented by A_{HJ} and B_{HJ} [145].

4.4.2.9 Elovich Model

The Elovich model adopts that the adsorption sites increase exponentially with the adsorption [242]. This model is applied to multilayer adsorption, and it was first used to describe the kinetics of chemisorption of gas onto solids. The non-linear and the linear models are shown in Table 4.8. The Elovich maximum capacity is represented by q_m, and K_e is the Elovich constant [41].

4.4.2.10 Kiselev Model

The Kiselev model is used to describe the monomolecular layer [112]. The linear and non-linear models are shown in Table 4.8. The Kiselev constant is represented by K_1, and K_n is the equilibrium constant of complex formation between adsorbed molecules [112]. It is also applicable only for surface coverage $\theta > 0.68$ [41].

4.4.3 Three-Parameter Isotherm

4.4.3.1 Redlich–Peterson Model

The Redlich–Peterson model is a hybrid model that combines between Langmuir and Freundlich models [243]. The numerator is based on the Langmuir model that can be decreased to Henry's model at infinite. The Redlich–Peterson model is an empirical isotherm that contains three parameters. It is applied in heterogeneous and homogeneous systems. The linear and non-linear equations are shown in Table 4.8. Redlich–Peterson isotherm constants are represented by K_R and a_R, and g is an exponent [169].

4.4.3.2 Toth Model

The Toth model is a combination of Henry's model and the Langmuir model. It is applicable for heterogeneous adsorption [106]. The non-linear and the linear equations are shown in Table 4.8. The parameter that measures the heterogeneity of the adsorption system is represented by t, K_T is Toth isotherm constant, and a_T is a constant that depends on the temperature. When $t = 1$, the model reduces to the Langmuir model, and if it does not equal unity, the sorption is heterogeneous [16].

For the adsorption of aluminium Al^{3+} ions using Linde type-A zeolite obtained from hazardous waste, the Sips, Toth, and Freundlich models were implemented in [143]. The Al^{3+} adsorption using synthetic monocomponent solutions was best fitted to Toth model with $R^2 = 0.9719$ [143].

4.4.3.3 Sips Model

The Sips model is the derivative of the Langmuir model; at high adsorbate concentration, it shows the monolayer adsorption characteristics of the Langmuir model. It is reduced to the Freundlich model at low adsorbate concentration [235]. The non-linear and linear expressions are shown in Table 4.8. The exponent is represented by β; a_s and K_s are the Sips isotherm constants [16].

For the adsorption of manganese Mn^{2+} ions using Linde type-A zeolite obtained from hazardous waste, the Sips, Toth, and Freundlich models were implemented in [143]. The Mn^{2+} adsorption using synthetic monocomponent solutions was best fitted to Sips model with $R^2 = 0.9978$ [143].

4.4.3.4 Kahn Model

Kahn model is a generalized model that is used to describe the adsorption of the adsorbate from pure solutions [41]. The model equation is given in Table 4.8. The maximum adsorption capacity is represented by $q_{\max}$; a_k and b_k are the Kahn constants [157].

4.4.3.5 Koble–Corrigan Model

Koble–Corrigan model integrates both Langmuir and Freundlich models to represent the equilibrium adsorption data [244]. The linear expression is shown in Table 4.8. Koble–Corrigan isotherm constants are represented by A_K, B_K, and P. It is reduced to the Freundlich model at a high concentration of the adsorbate. This model is only applicable when the constant P is greater than or equal to unity. When P is less than unity, it indicates that the model cannot define the experimental data [169].

4.4.3.6 Radke–Prausnitz Model

The Radke–Prausnitz model reduces to a linear isotherm at low adsorbate concentration [210]. It converts to Freundlich isotherm at high adsorbate concentration, and when $M_{RP} = 0$, it converts to the Langmuir isotherm. So, it fits well at a wide range of adsorbate concentration, which makes it preferred in most adsorption systems with low concentration. The non-linear model is shown in Table 4.8. Radke–Prausnitz maximum adsorption capacity is represented by q_{MRP}; K_{RP} is Radke–Prausnitz constant and MRP is the model exponent [210].

4.4.3.7 Langmuir–Freundlich Model

Langmuir–Freundlich model is used to describe the adsorption energy distribution onto the heterogeneous surface of the adsorbent [245]. It is reduced to Freundlich isotherm at low adsorbate concentration, while it reduces to Langmuir model at high adsorbate concentration. This model has non-linear expression as given in Table 4.8. Langmuir–Freundlich maximum adsorption capacity is represented by q_{MLF}, K_{LF} is the equilibrium constant for heterogeneous solid, and MLF is a heterogeneous parameter that varies between 0 and 1 [174].

4.4.3.8 Jossens Model

The Jossens isotherm assumes a simple model equation that depends on the energy distribution of adsorbate-adsorbent interactions at adsorbent sites [246]. It assumes heterogeneous surfaces on the adsorbent. The non-linear and linear expressions are shown in Table 4.8. Jossens isotherm constants are represented by H, P, and F [98].

4.4.4 Four-Parameter Isotherm

4.4.4.1 Fritz–Schlunder Model

Fritz–Schlunder model is an empirical equation that fits a wide range of experimental data due to a large number of coefficients [14]. It has a non-linear expression that is shown in Table 4.8. The Fritz–Schlunder maximum adsorption capacity is represented by q_{mFS}, K_{FS} is the Fritz-Schlunder equilibrium constant, and MFS is the Fritz–Schlunder model exponent. It becomes the Langmuir model when MFS = 1, and at high adsorbate concentration, the model reduces to Freundlich [14].

4.4.4.2 Baudu Model

Baudu found that the estimation of the Langmuir coefficients by tangent measurements at different equilibrium concentrations shows that they are not constant in the broad concentration range. It has a non-linear model, as shown in Table 4.8 [174]. It is applicable only in the range of $(1+x+y)<1$ and $(1+x)<1$. It is reduced to the Freundlich model when surface coverage is low, and it is given by

$$q_e = \frac{q_m b_0 C_e^{1+x+y}}{1+b_0}, \tag{4.10}$$

where q_m is Baudu maximum adsorption capacity, b_o is the equilibrium constant, and x and y are Baudu parameters [14].

4.4.4.3 Weber–Van Vliet Model

Weber–Van Vliet assumed an empirical equation that contains four parameters that are used to describe a wide range of adsorption data [247]. It has a non-linear expression, as shown in Table 4.8. Equilibrium concentration is represented by C_e, q_e is the adsorption capacity, and p_1, p_2, p_3, and p_4 are Weber–Van Vliet parameters [14].

4.4.4.4 Marczewski–Jaroniec Model

The Marczewski–Jaroniec model is also called the four-parameter general Langmuir equation [14]. This model has a non-linear expression, as shown in Table 4.8. The parameters that describe the heterogeneity of the adsorbent surface are represented by n_{MJ}; M_{MJ}. M_{MJ} defines the spreading of distribution in the path of higher adsorption energy, and n_{MJ} defines the spreading of distribution in the path of less adsorption energy. It is reduced to Langmuir isotherm when n_{MJ} and $M_{MJ} = 1$, and if $M_{MJ} = n_{MJ}$, then it is reduced to Langmuir–Freundlich model [248].

4.4.5 Five-Parameter Isotherm

4.4.5.1 Fritz–Schlunder Model

Fritz and Schlunder generated an empirical equation with five parameters that can simulate the model variations over a wide range of equilibrium data. It has a non-linear expression, as shown in Table 4.8 [14]. The Fritz–Schlunder maximum adsorption capacity is represented by q_{mFS5}; K_1, K_2, α_{FS}, and β_{Fs} are Fritz–Schlunder parameters. It becomes the Langmuir model when the values of both exponents α_{FS} and β_{Fs} equal to 1, and at high adsorbate concentrations, it is reduced to Freundlich isotherm [193].

4.5 GENERALIZED ISOTHERMAL MODEL

4.5.1 The Generalized Brouers-Sotolongo Isotherm

Broures and Sotolongo developed the generalized Broures and Sotolongo (GBS) isotherm based on statistical and mathematical considerations, as given by [248]

$$\frac{W_{GBS}(x)}{W_m} = 1 - \left(1 + c\left(\frac{x}{b}\right)^{\alpha}\right)^{-(1/c)}, \tag{4.11}$$

where x is the sorbate concentration or pressure, $W_{GBS}(x)$ is the uptake, W_m is the equilibrium uptake, c and α are form parameters, and b is a scale parameter. The Burr–Singh–Maddala cumulative probability distribution function is the right-hand side of the equation [248]. The GBS is used for the isothermal analysis of fluoride removal from an aqueous solution by montmorillonite clay, as given by

$$\frac{Qe_{GBS}(x)}{Qe_m} = 1 - \left(1 + c\left(\frac{x}{b}\right)^{\alpha}\right)^{-(1/c)}, \tag{4.12}$$

where Qe_{GBS} is the uptake and Qe_m is the equilibrium uptake [167]. At different values of c and α, different empirical and semiempirical isotherms can be driven. The Jovanovich isotherm is driven from the GBS when $c = 0$ and $\alpha = 1$; the expression is given by [167]

$$\frac{Qe_J(x)}{Qe_m} = 1 - \exp\left(-\frac{x}{b}\right). \tag{4.13}$$

The Broures and Sotolongo isotherm is driven at $c = 0$, and the expression is given by [167]

$$\frac{Qe_{BS}(x)}{Qe_m} = 1 - \exp\left(\left(-\frac{x}{b}\right)^{\alpha}\right). \tag{4.14}$$

The Sips isotherm is driven at $c = 1$, and the expression is given by [167]

$$\frac{Qe_S(x)}{Qe_m} = \frac{\left(\frac{x}{b}\right)^{\alpha}}{1+\left(\frac{x}{b}\right)^{\alpha}}. \tag{4.15}$$

The Langmuir isotherm is driven at $c = 0$ and $\alpha = 1$, and the expression is given by [167]

$$\frac{Qe_L(x)}{Qe_m} = \frac{\left(\frac{x}{b}\right)}{1+\left(\frac{x}{b}\right)}. \tag{4.16}$$

The Freundlich isotherm is driven from the GBS, BS, and Sips isotherms when $x \to 0$ and $\alpha < 1$, and the expression is given by [248]@@@@[249]

$$\frac{W_F(x)}{W_m} = \left(\frac{x}{b}\right)^{\alpha}. \tag{4.17}$$

4.6 ADSORPTION ISOTHERM ANALYSIS

Isothermal models are applied to the experimental data of P sorption onto NZVI [33], as shown in Tables 4.9–4.13. The curve fitting tool in MATLAB(R2019a) is used to fit the data to the models. The purpose of the analysis is to define the type of adsorption and to get the capacity of the adsorbent. The best isothermal model that fits the experimental data is selected depending on the value of the correlation factor R^2. The adsorption parameters are calculated after the curve fitting for each model as shown in Table 4.14. The adsorption parameters of the Langmuir linear and non-linear models, $q_{\max}$ and K_L, are calculated by comparing the equations resulting from curve fitting in Table 4.9 to the linear and non-linear model forms shown in Table 4.8, respectively. The adsorption parameters of the Freundlich linear and non-linear model, K_F and n, are evaluated by comparing the models generated from curve fitting in Table 4.9 to the models in Table 4.8, and the values are given in Table 4.14.

TABLE 4.9
Langmuir Isotherm Models

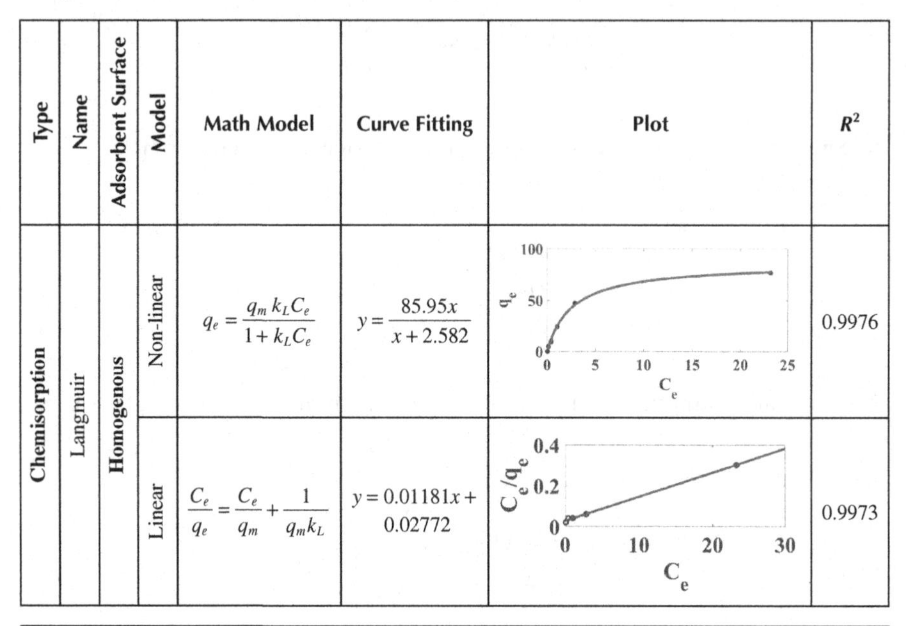

Type	Name	Adsorbent Surface	Model	Math Model	Curve Fitting	Plot	R^2
Chemisorption	Langmuir	Homogenous	Non-linear	$q_e = \frac{q_m k_L C_e}{1 + k_L C_e}$	$y = \frac{85.95x}{x + 2.582}$	q_e vs C_e	0.9976
			Linear	$\frac{C_e}{q_e} = \frac{C_e}{q_m} + \frac{1}{q_m k_L}$	$y = 0.01181x + 0.02772$	C_e/q_e vs C_e	0.9973

TABLE 4.10
Temkin Isotherm Models

Type	Name	Adsorbent Surface	Model	Math Model	Curve Fitting	Plot	R^2
Chemisorption	Temkin		Non-linear	$q_e = \frac{RT}{b}\ln(K_T C_e)$	$y = 12.75\ln(11.35x)$	q_e vs C_e	0.933
			Linear	$q_e = \frac{RT}{b}\ln(K_T) + \frac{RT}{b}\ln(C_e)$	$y = 15.91x + 36.82$	q_e vs $\ln(C_e)$	0.9557

TABLE 4.11
Freundlich Isothermal Models' Comparison

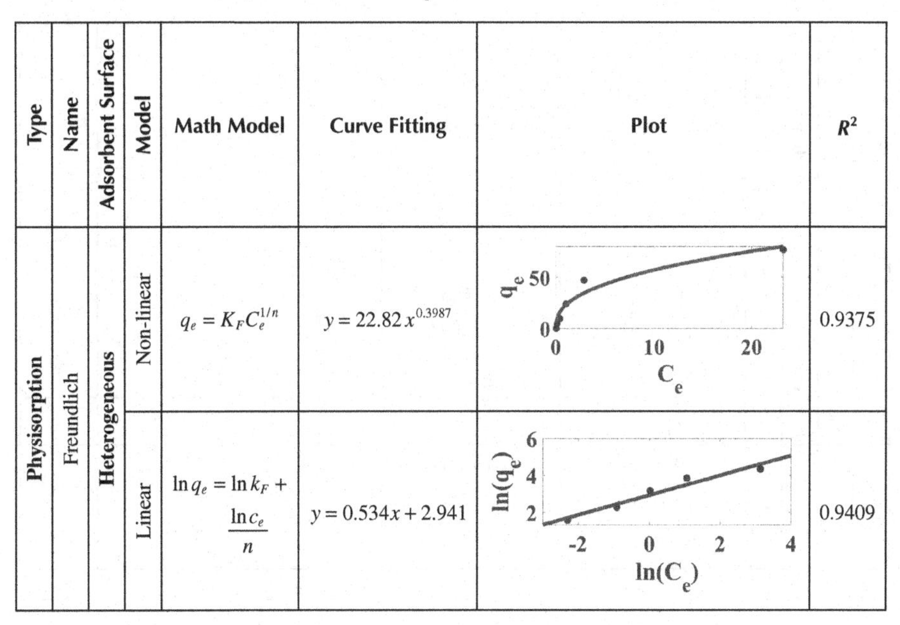

Type	Name	Adsorbent Surface	Model	Math Model	Curve Fitting	Plot	R^2
Physisorption	Freundlich	Heterogeneous	Non-linear	$q_e = K_F C_e^{1/n}$	$y = 22.82\,x^{0.3987}$		0.9375
			Linear	$\ln q_e = \ln k_F + \frac{\ln c_e}{n}$	$y = 0.534x + 2.941$		0.9409

TABLE 4.12
Dubinin–Radushkevich Isothermal Models' Comparison

Type	Name	Adsorbent Surface	Model	Math Model	Curve Fitting	Plot	R^2
Physisorption	Dubinin–Radushkevich	Heterogeneous	Non-linear	$q_e = q_m e^{-kad[RT\ln(1+(1/Ce))]^2}$	$y = 71.13e^{-2.675\ln^2(1+(1/x))}$		0.963
			Linear	$\ln(q_e) = \ln(q_m) - k_{ad}\left[RT\ln\left(1+\frac{1}{Ce}\right)\right]^2$	$y = 0.315\times10^{-5}x + 5.938$		0.9649

TABLE 4.13
Harkins–Jura Isotherm Model

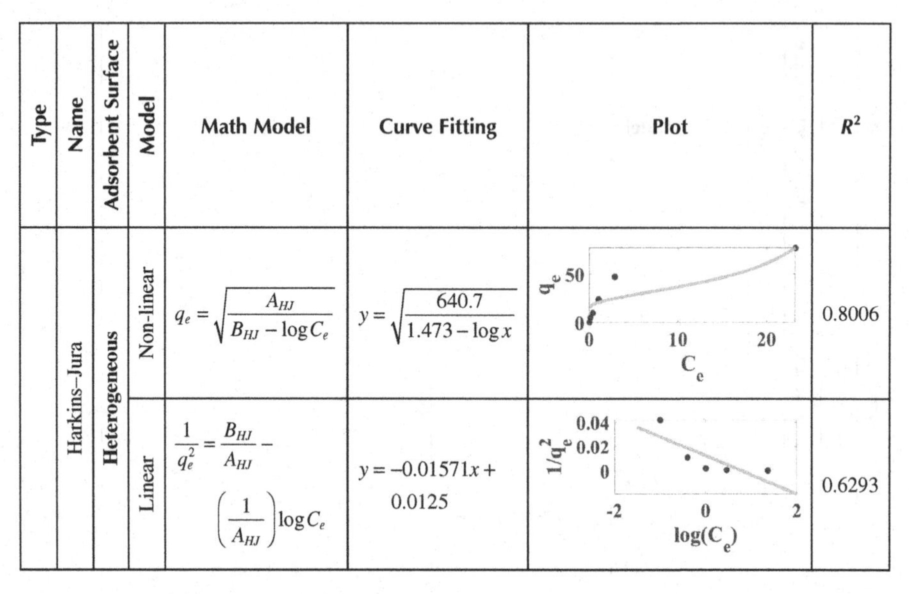

Type	Name	Adsorbent Surface	Model	Math Model	Curve Fitting	Plot	R^2
	Harkins–Jura	Heterogeneous	Non-linear	$q_e = \sqrt{\frac{A_{HJ}}{B_{HJ} - \log C_e}}$	$y = \sqrt{\frac{640.7}{1.473 - \log x}}$		0.8006
			Linear	$\frac{1}{q_e^2} = \frac{B_{HJ}}{A_{HJ}} - \left(\frac{1}{A_{HJ}}\right)\log C_e$	$y = -0.01571x + 0.0125$		0.6293

TABLE 4.14
Isothermal Adsorption Parameters for Different Models

	Adsorption Parameters		
		Value	
Isothermal Model	**Parameters**	**Linear**	**Non-Linear**
Langmuir	q_{max} (mg/g)	84.67	85.95
	K_L (L/mg)	0.426	0.387
Freundlich	n	1.872	2.508
	K_F (L/g)	18.93	22.82
Dubinin–Radushkevich	q_s (mg/g)	379.18	71.13
	K_{ad} (mol^2/kJ^2)	1.315×10^{-5}	3.286×10^{-7}
	E (kJ/mol)	194.95	1233.51
Temkin	β_T	15.91	12.75
	K_T (L/g)	10.115	11.35
Harkins–Jura	A_{HJ} (J/mol)	63.654	640.7
	B_{HJ} (L/g)	0.796	1.473

The Dubinin–Radushkevich model plots are shown in Table 4.11, and the model parameters are evaluated by comparing the generated models in Table 4.11 to the models in Table 4.8. So, K_{ad} and q_s are given in Table 4.14, and the free energy E is evaluated; it defines that the mechanism is chemisorption.

The model parameters of the Temkin model are evaluated by comparing the generated models in Table 4.9 to the models in Table 4.8. So, B_T and K_T are shown in Table 4.14, where $B_T = RT/B$.

The model parameters of the Harkins–Jura model are evaluated by comparing generated models in Table 4.13 to the models in Table 4.8, and the results are given in Table 4.14.

Finally, the best model that fits the experimental data is the non-linear Langmuir model with $R^2 = 0.9976$. The P sorption onto NZVI is chemisorption, and $q_{\text{max}} = 84.67\,\text{mg/g}$.

4.7 THE KINETIC MODELS OF THE ADSORPTION PROCESS

The kinetic model describes the mechanism of metal adsorption and the main parameters governing sorption kinetics [150]. The effect of contact time is studied to identify the possible application and to analyse the adsorption kinetics. Initially, the adsorption process accelerates, which can be because of surface complexation or immediate electron transfer. Then, the adsorption looks constant, showing the gradual saturation of adsorption sites on the outside surface [250]. The comparison between different kinetic models is shown in Table 4.23

4.7.1 First-Order Model

The first-order equation is based on the semiunimolecular reaction that contains one reactant, and the other reactant has a zero-order, as given by [33]

$$\frac{dC_t}{dt} = -K_1 C_t. \tag{4.18}$$

After integration with the initial condition C_0 and C_t at time t, the result is given by [33]

$$-\ln\left(\frac{C_t}{C_0}\right) = K_1 t, \tag{4.19}$$

where C_t is the concentration of the reactant and K_1 is the first-order kinetic constant that can be obtained from the linear slope of the plot of $-\ln(C_t/C_0)$ vs. t.

4.7.2 Second-Order Model

The second-order rate equation describes the concentrations of two reactants within the reaction, as given by [33]

$$\frac{dC_t}{dt} = -K_2 C_t^2. \tag{4.20}$$

The result of the integration with the initial condition C_0 and C_t at time t is given by [33]

$$\frac{1}{C_t} - \frac{1}{C_0} = K_2 t, \tag{4.21}$$

where K_2 is the second-order kinetic constant that can be evaluated from the linear slope of the plot of $(1/C_t) - (1/C_0)$ vs. time.

4.7.3 Pseudo-First-Order Model

In pseudo-first-order kinetic, a higher order reaction is considered a first-order reaction. Reaction rate equals $K[A][B]$; it is supposed to be second order. However, B is constant, so the rate will be $k'[A]$. The pseudo-first-order expression is given by [150]

$$\frac{dq_t}{dt} = k_1 (q_e - q_t), \tag{4.22}$$

where q_e and q_t are the amount of the metal adsorbed at equilibrium and at time t and k_1 is the pseudo-constant rate for the adsorption process.

The integration of (4.22), can be written as follows, with the initial condition q_t equals is given by:

$$\int_0^{q_t} \frac{dq_t}{(q_e - q_t)} = \int_0^t k_1 \, dt. \tag{4.23}$$

The linear form is given by

$$\ln(q_e - q_t) = \ln q_e - k_1 t, \tag{4.24}$$

and the non-linear model is given by [63]

$$q_t = q_e \left(1 - e^{-K_1 t}\right). \tag{4.25}$$

4.7.4 Pseudo-Second-Order Model

Pseudo-second-order model assumes that the sorption follows second-order chemisorption, as given by [54]

$$\frac{dq_t}{dt} = k_1 (q_e - q_t)^2. \tag{4.26}$$

The integration of the kinetic expression is given by

$$\int_0^{q_t} \frac{dq_t}{(q_e - q_t)^2} = \int_0^t k_1 dt. \tag{4.27}$$

and the non-linear form for the kinetic model is given by

$$q_t = \frac{k_1 t q_e^2}{1 + k_1 t q_e}. \tag{4.28}$$

and the linear form is given by

$$\frac{t}{q_t} = \frac{1}{k_1 q_e^2} + \frac{t}{q_e}. \tag{4.29}$$

4.7.5 Intra-Particle Diffusion Model

The intra-particle diffusion model describes the adsorption kinetics when the adsorption rate depends on the diffusion rate of the adsorbate within the pores of the adsorbent particles, as given by [33]

$$q_t = K_{\text{int}} \sqrt{t} + C_{\text{int}}, \tag{4.30}$$

where K_{int} (min^{-1}) is the intra-particle diffusion rate constant and C_{int} is the intercept.

4.7.6 Avrami Model

The Avrami model describes the kinetics of phase transformation, and this model equation is used for the kinetic and thermal decomposition modelling. The non-linear model is given by [251]

$$q_t = q_e \left(1 - e^{-(K_{\text{AV}} t)^{n\text{AV}}}\right), \tag{4.31}$$

where K_{AV} (min^{-1}) and n_{AV} are the Avrami kinetic constants.

4.7.7 Elovich Model

The Elovich model was usually used for gas-solid adsorption systems, and now, it is used for the sorption process of soluble pollutants from aqueous solutions, as given by [33]

$$\frac{dq_t}{dt} = a e^{-b q_t}. \tag{4.32}$$

The results of the integration with an initial condition q_t equal to 0, which is given by

$$e^{bq_t} = abt + 1. \tag{4.33}$$

The linear model is given by [33]:

$$q_t = \frac{1}{b}\ln\left((ab)t + 1\right). \tag{4.34}$$

Simplification is made such that $(ab)t \gg 1$, so the linear model can be expressed as given:

$$q_t \cong \frac{1}{b}\ln(ab) + \frac{1}{b}\ln(t), \tag{4.35}$$

where a and b are constants that can be evaluated from the linear plot of q_t vs. $\ln t$. This model is good at describing the kinetics of systems with heterogeneous surfaces [141].

4.8 GENERALIZED KINETIC MODEL

4.8.1 The Brouers–Sotolongo Kinetic

Brouers and Sotolongo developed a universal function for complex systems kinetics that is characterized by stretched exponential and power law behaviours. This model generalizes the previous theoretical attempts to describe the fractal kinetic. A complex system consists of *A*, *B*, *C*,.... reacting molecules, which can be given by [56]

$$\frac{dc_A}{dt} = KC_A^{\alpha}C_B^{\beta}C_C^{\gamma}\ldots\ldots, \tag{4.36}$$

where k is the rate coefficient, $\alpha, \beta, \delta, \ldots$ relate to the concentrations of chemical species *A*, *B*, *C*,.... that exist in the reaction; the sum of the rate coefficients is the order of the reaction. The concentrations $C_B, C_C, \ldots\ldots$ can be considered as constants, so the Eq. (4.36) for the reactant *A* reduces to [56]

$$-\frac{dc_n(t)}{dt} = k_n c_n(t)^n, \tag{4.37}$$

then n is the order of the reaction. The solution of this differential equation is given by [56]

$$c_n(t) = c_n(0)\left[1 + (n-1)c_n(0)^{n-1} k_n t\right]^{-(1/(n-1))}. \quad (4.38)$$

According to the definition introduced by Tsallis given in [252].

$$\exp_n(x) = \left(1 - (n-1)x\right)^{-(1/(n-1))}. \quad (4.39)$$

if $\left(1-(n-1)x\right) > 0$ and 0 otherwise. So the solution can be written in a more compact form as follows [252]:

$$c_n(t) = c_n(0)\exp_n\left(-\frac{t}{\tau_n}\right), \quad (4.40)$$

with a characteristic time τ_n equal to [252]

$$\tau_n = \left(c_n(0)^{n-1} k_n\right)^{-1}. \quad (4.41)$$

After using the n-Weibull function introduced by Picoli, which was also used in the theory of relaxation, the solution can be expressed as follows [56]:

$$c_{n,\alpha}(t) = c(0)\left(1 + (n-1)\left(\frac{t}{\tau_{n,\alpha}}\right)^{\alpha}\right)^{-(1/(n-1))}. \quad (4.42)$$

The Brouers–Sotolongo fractal kinetic was used for the kinetic analysis of fluoride removal from aqueous solution by montmorillonite clay, and the expression is given by [167]

$$BSf(n,\alpha)(t) = q_m\left(1 + (n-1)\left(\frac{t}{\tau}\right)^{\alpha}\right)^{-(1/(n-1))}, \quad (4.43)$$

where $q(t)$ is the sorbet quantity, q_m is the maximum sorbet quantity, τ is a characteristic time, α is the fractal time exponent, and n is the fractional reaction order. The adsorption kinetic equations can be driven from the general $BSf(n,\alpha)$ at different values of n and α. The pseudo-first-order kinetic is driven when $n = 1$ and $\alpha = 1$; the expression is given by [253]

$$c_1(t) = c_1(0)\exp(-K_1 t). \quad (4.44)$$

The Avrami equation is driven when $n = 1$ and $\alpha \neq 1$; the expression is given by [253]

$$c_{1,\alpha}(t) = c_{1,\alpha}(0)\exp\left(-K_{1,\alpha}t^{\alpha}\right). \tag{4.45}$$

The pseudo-second-order kinetic is driven when $n = 2$ and $\alpha = 1$; the expression is given by [141]

$$-\frac{dc_2(t)}{dt} = k_2 c_n(t)^2 -> \frac{1}{c_2(t)} - \frac{1}{c_2(0)} = k_2 t. \tag{4.46}$$

The Hill kinetic equation is driven when $n = 2$ and $\alpha \neq 1$; the expression is given by [141]

$$c_{2,\alpha}(t) = c_{2,\alpha}(0)\left(1 + c_{2,\alpha}(0)k_{2,\alpha}t^{\alpha}\right)^{-1}. \tag{4.47}$$

4.9 KINETIC ANALYSIS

The kinetic models fit the experimental data of P sorption onto NZVI [33], as shown in Tables 4.15–4.19. The best kinetic model that fits the experimental data is selected based on the R^2 values. The curve fitting is done by using MATLAB (R2019a) to

TABLE 4.15
Comparison between First-Order and Second-Order Kinetic Models

Name	Model	Math Model	Curve Fitting	Plot	R^2
1st order	Linear	$-\ln\left(\frac{C_t}{C_0}\right) = K_1 t$	$y = 0.01739x + 0.212$	-ln[c_t/c_0] vs t (0, 100, 200)	0.9583
2nd order	Linear	$\frac{1}{C_t} - \frac{1}{C_0} = K_2 t$	$y = 0.00209x - 0.02096$	[1/c_t]-[1/c_0] vs t (0, 100, 200)	0.9378

TABLE 4.16
Comparison between Pseudo-First-Order Kinetic Models

Name	Model	Math Model	Curve Fitting	Plot	R^2
Pseudo-First Order	Non-linear	$q_t = q_e\left(1 - e^{-K_1 t}\right)$	$y = 47.32\left(1 - e^{-0.02967x}\right)$		0.9987
	Linear	$\ln(q_e - q_t) = \ln q_e - k_1 t$	$y = -0.00969x + 1.636$		0.9948

TABLE 4.17
Comparison between Pseudo-Second-Order Kinetic Models

Name	Model	Math Model	Curve Fitting	Plot	R^2
Pseudo-second order	Non-linear	$q_t = \dfrac{k_1 t q_e^2}{1 + k_1 t q_e}$	$y = \dfrac{58.18x}{x + 33.46}$		0.9949
	Linear	$\dfrac{t}{q_t} = \dfrac{1}{k_1 q_e^2} + \dfrac{t}{q_e}$	$y = 0.01723x + 0.5976$		0.9947

find different kinetic models. All the parameters of the kinetic models are evaluated by comparing the generated models from curve fitting to the math model in Tables 4.15, 4.17, and 4.19. Table 4.20 shows the values of the evaluated parameters. Finally, the best model that fits the experimental data was the non-linear pseudo-first order with $R^2 = 0.9987$.

TABLE 4.18
Comparison between Elovich Kinetic Models

Name	Model	Math Model	Curve Fitting	Plot	R^2
Elovich	Non-linear	$q_t = \frac{1}{b}\ln\left((ab)t+1\right)$	$y = (15.09)\ln((0.1573x)+1)$	q_t: 0, 20, 40; t: 0, 100, 200	0.982
	Linear	$q_t = \frac{1}{b}\ln(ab) + \frac{1}{b}\ln(t)$	$y = 12.48x - 15.55$	q_t: 0, 50; ln(t): 2, 4, 6	0.99

TABLE 4.19
Intra-Particle Diffusion Kinetic Model

Name	Model	Math Model	Curve Fitting	Plot	R^2
Intra-particle diffusion	Non-linear	$q_t = K_{\text{int}}\sqrt{t} + C_{\text{int}}$	$y = 3.904\sqrt{x} + 1.845$	q_t: 0, 20, 40; t: 0, 100, 200	0.939
	Linear	$q_t = K_{\text{int}}\sqrt{t} + C_{\text{int}}$	$y = 4.068x + 0.7137$	q_t: 0, 20, 40; sqrt(t): 0, 5, 10, 15	0.9582

4.10 THERMODYNAMIC ANALYSIS

The thermodynamic analysis is used to study the effect of temperature on the adsorption of the adsorbate on the adsorbent [254]. Gibbs free energy ($\Delta G°$), standard enthalpy ($\Delta H°$), and standard entropy ($\Delta S°$) are thermodynamic parameters that were evaluated by using the following Van't Hoff equation [44]:

$$\Delta G° = -RT \ln K_d, \tag{4.48}$$

TABLE 4.20
Kinetic Parameters for Different Models

Kinetic Model	Kinetic Parameters	Value: Linear	Value: Non-Linear
1st order	K_1(1/min)	0.01739	–
2*nd* order	K_2(L/mg min)	0.00209	–
Pseudo *1st* order	q_e(mg/g)	5.134	47.32
	K_1(1/min)	0.00969	0.02967
Pseudo 2*nd* order	q_e(mg/g)	58	58.18
	K_1(1/min)	0.000497	0.0005
Elovich	*a* (g/mg)	3.59	2.38
	b (mg/gmin)	0.0801	0.066
Intra-particle diffusion	K_{int}(mg/g min$^{0.5}$)	4.068	3.904
	C_{int}(mg/g)	0.7137	1.845

where R (8.314 J/mol K) is the universal gas constant, T (K) is the temperature in Kelvin, and K_d is the distribution coefficient that is given by

$$K_d = \frac{C_{Ae}}{C_e}, \tag{4.49}$$

where C_{Ae} is the amount of the adsorbate on the adsorbent, and C_e is the equilibrium concentration.($\Delta H°$) and ($\Delta S°$) can be evaluated by

$$\ln K_d = \frac{\Delta S°}{R} - \frac{\Delta H°}{RT}, \tag{4.50}$$

where the values of $\Delta S°$ and $\Delta H°$ are evaluated from the intercept and the slope of the plot of $\ln K_d$ vs. $1/T$, respectively [44]. The change in Gibbs free energy ($\Delta G°$) shows the spontaneity of the adsorption process; the negative value of ($\Delta G°$) means that the adsorption process of the adsorbate on the adsorbent is spontaneous.

The temperature effect on P-adsorption by NZVI was studied at different ranges of temperature (15, 25, 40, 55, and 70°C) in [33]. Figure 4.1 shows a plot of $\ln K_d$ vs. $1/T$ of P-adsorption onto NZVI experimental data [33]. The linear curve fitting for this experimental data is implemented by using MATLAB (R2019a); the result is

$$y = -8176x + 29.42, \tag{4.51}$$

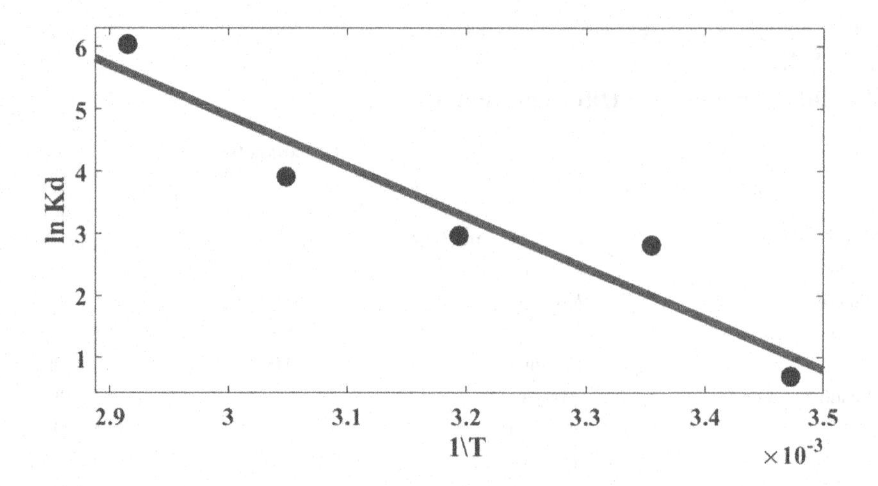

FIGURE 4.1 Thermodynamic Van't Hoff plot of *P*-adsorption by NZVI.

and the R^2 value equals 0.9033. The values of ($\Delta H°$) and ($\Delta S°$) are evaluated by comparing Eqs. (4.50) to (4.51), as shown in Table 4.21. The negative values of $\Delta G°$ indicate the spontaneous nature of the *P*/NZVI reaction. When the temperature increases, the $\Delta G°$ values increase. The positive value of the enthalpy $\Delta H°$ means that the adsorption is endothermic, indicating the improvement of *P*-removal by raising the operating temperature. The positive sign of entropy change means that randomness during the *P*-adsorption increases.

Usually, the activation energies in the range of 5–40 kJ/mol indicate the physisorption process, while higher energies between 40 and 800 kJ/mol indicate chemisorption [44]. So, the magnitude of $\Delta G°$ at different temperatures was in the range 2–20 kJ/mol, which indicates that the physisorption process is involved. However, the magnitude of enthalpy change is >20 kJ/mol and less than 80 kJ/mol, which

TABLE 4.21
The Gibbs Free Energy ($\Delta G°$), Standard Enthalpy ($\Delta H°$), and Standard Entropy ($\Delta S°$) Values

Temperature (Kelvin)	$\Delta G°$ (kJ/mol)	$\Delta H°$ (kJ/mol)	$\Delta S°$ (J/molK)
288.15	−1.68	67.975	244.598
298.15	−6.97		
313.15	−7.72		
328.15	−10.66		
343.15	−17.23		

indicates that the adsorption may involve both the physisorption and chemisorption processes. The thermodynamic analysis results confirm the results of the isothermal and the kinetic models according to the adsorption mechanism of *P* onto NZVI [33] as shown in Table 4.22. If the sign of ΔH° is negative, that means that the adsorption is exothermic. The negative values of ΔS° mean that the randomness during the adsorption process decreases [44].

TABLE 4.22
Comparison between Isotherm Models

Adsorbent	Adsorbate	Isotherm Models	Best Fit	References
Inorganic arsenic (As)	Bone char pyrolysed at 900°C	Langmuir Freundlich Redlich–Peterson Sips	Sips model for As(III) Langmuir model for As(V)	[23]
N-[2-(methylamino) ethyl] ethane-1 2-diaminated acrylic fibre (DAAF)	Hg (II)	Langmuir Freundlich Temkin	Langmuir	[255]
Novel bacterium *Chelatococcus* sp. biomass	Pb (II) ions	Langmuir Freundlich Temkin	Langmuir (non-linear)	[256]
Natural safiot clay	Industrial dyes	Langmuir Freundlich Dubinin–Radushkevich	Langmuir	[257]
Nanoscale zerovalent iron	Phosphorus	Langmuir Freundlich Dubinin–Radushkevich Temkin Harkins–Jura	Langmuir	[33]
Low-cost sugarcane waste	Pb (II)	Langmuir Freundlich Temkin	Freundlich	[258]
SiO_2–MoO_3	Lead ions	Langmuir Freundlich Temkin	Langmuir	[259]
Modified sawdust	Chromium (VI) ions	Langmuir Freundlich Temkin	Freundlich	[260]
Natural and acid-activated nano-bentonites	Ni (II)	Langmuir Freundlich Temkin Dubinin–Radushkevich Flory-Huggins	Langmuir	[261]

(Continued)

TABLE 4.22 ***(Continued)***
Comparison between Isotherm Models

Adsorbent	Adsorbate	Isotherm Models	Best Fit	References
Activated carbon: BAC and SAC	*p*-Nitrophenol	Langmuir Freundlich Jovanovic Redlich–Peterson Koble–Corrigan Brouers-Sotolongo	Langmuir, Redlich–Peterson, and Brouers-Sotolongo show the best fit of experimental data of p-nitrophenol onto BAC Freundlich shows the best fit of experimental data of p-nitrophenol onto SAC	[262]
Magnetic kaolinite clay	Methylene blue	Langmuir Freundlich Temkin Elovich Flory–Huggins	Freundlich	[263]
Mesoporous cerium-aluminium binary oxide nano-materials	Fluoride	Langmuir Freundlich Temkin	Freundlich at higher fluoride concentrations Langmuir at low fluoride concentrations	[264]
Natural biosorbent (walnut husks)	Titan Yellow	Langmuir Freundlich Temkin Jovanovich Halsey	Freundlich and Halsey	[265]
Graphene oxide/ Lanthanum Aluminate nano-composites ($RGO - LaAlO_3$) and Lanthanum Aluminate ($LaAlO_3$)	Anionic dye	Langmuir Freundlich Dubinin–Radushkevich	Freundlich	[266]
Modified expanded graphite Fe_3O_4 composite	Methylene blue	Langmuir Freundlich Redlich–Peterson	Redlich–Peterson	[267]

TABLE 4.23
Comparison between Kinetic Models

Adsorbent	Adsorbate	Kinetic Models	Best Fit	References
Cocoa pod activated carbon prepared by ortho-phosphoric acid activation	Anti-malaria drug	PFO PSO Elovich Intra-particle diffusion	PSO	[268]
Hexavalent chromium	Royal poinciana pod-derived biosorbent	PFO PSO Elovich Intra-particle diffusion	PSO	[269]
Cu–Ca–Al–layered double hydroxide modified by itaconic acid	Anionic dye	PFO PSO Elovich	PSO	[270]
Black mustard husk ash	Fluoride	PFO PSO Intra-particle diffusion	PFO	[271]
PANI@ZnO nano-composite	Congo red and methylene blue dyes	PFO PSO	PSO	[272]
Rape straw biomass fibre/$\beta - CD/Fe_3O_4$	Ibuprofen	PFO PSO	PSO	[273]
Microporous and hierarchical LTA zeolites	CO_2	PFO PSO Elovich	PFO	[274]
Nano-fibrous membrane of polyacrylonitrile (NMP)	Cadmium ions	PFO PSO Elovich	PSO	[275]
A superabsorbent hydrogel ("c-hydrogel")	Lead ion (Pb (II))	PFO PSO	PFO	[276]
Activated charcoal	Triphenylmethane (crystal violet)	PFO PSO Elovich Intra-particle diffusion	PSO	[277]

4.11 CONCLUSION

Adsorption is a new emerging technique that was applied lately for pollutant removal of polluted wastewater. Pollutant adsorbed to the adsorbent surface via ion exchange or complexation. Different adsorbent materials have been applied, and these materials could be sorted into natural eco-friendly materials, industrial and agriculture waste by-products, biomasses, and nano-materials. Different factors have been identified with their effect on adsorption process such as pH, time, dose, and concentration. Additionally, for the purpose of adsorption mechanism explanation, isotherm and kinetics with their different methods were applied. In this chapter, adsorption process, the different types of adsorbent materials, in addition to the factors affecting adsorption, and different isothermal and kinetics models have been discussed.

ACKNOWLEDGEMENT

This chapter is based on the work supported by the Egyptian Academy of Science, Research, and Technology (ASRT), Project ID: Call no. 2/2019/ASRT-Nexus # 4607.

REFERENCES

[1] Hany H Abdel Ghafar, Gomaa AM Ali, Osama A Fouad, and Salah A Makhlouf. Enhancement of adsorption efficiency of methylene blue on Co_3O_4/SiO_2 nanocomposite. *Desalination and Water Treatment*, 53(11):2980–2989, 2015.

[2] Ola Abdelwahab, Ahmed El Nemr, Amany El Sikaily, and Azza Khaled. Use of rice husk for adsorption of direct dyes from aqueous solution: A case study of direct F. Scarlet. *Journal of Aquatic Research*, 31(1):1–11, 2005.

[3] Omran Abdi, and Mosstafa A review study of biosorption of heavy metals and comparison between different biosorbents. *Journal of Materials and Environmental Science*, 6(5):1386–1399, 2015.

[4] Ornella Abollino, Maurizio Aceto, Mery Malandrino, Corrado Sarzanini, and Edoardo. Adsorption of heavy metals on NA-montmorillonite. Effect of pH and organic substances. *Water Research*, 37(7):1619–1627, 2003.

[5] Khaled S. Abou-El-Sherbini, and Mohamed M. Hassanien. Study of organically-modified montmorillonite clay for the removal of copper(II). *Journal of Hazardous Materials*, 184(1):654–661, 2010.

[6] Mustafa Abunowara, Suriati Sufian, Mohamad Azmi Bustam, Muhammad Babar, Usama Eldemerdash, Roberto Bencini, Sami Ullah, Mohammed Ali Assiri, Abdullah G Al-Sehemi, and Ahmad Mukhtar. Experimental and theoretical investigations on kinetic mechanisms of low-pressure CO_2 adsorption onto Malaysian coals. *Journal of Natural Gas Science and Engineering*, 88:103828, 2021.

[7] Aderonke Ajibola Adeyemo, Idowu Olatunbosun Adeoye, and Olugbenga Solomon Bello. Adsorption of dyes using different types of clay: A review. *Applied Water Science*, 7(2):543–568, 2017.

[8] Oluranti Agboola, Patricia Popoola, Rotimi Sadiku, Samuel Eshorame Sanni, Sunday Ojo Fayomi, and Olawale Samuel Fatoba. *Nanotechnology in Wastewater and the Capacity of Nanotechnology for Sustainability*, 1–45. Springer International Publishing, Cham, 2020.

[9] Chi K Ahn, Donghee Park, Seung H Woo, and Jong M Park. Removal of cationic heavy metal from aqueous solution by activated carbon impregnated with anionic surfactants. *Journal of Hazardous Materials*, 164(2–3):1130–1136, 2009.

[10] MA Akl, AM Youssef, and MM Al-Awadhi. Adsorption of acid dyes onto bentonite and surfactant-modified bentonite. *Journal of Analytical & Bioanalytical Techniques*, 4(4):3–7, 2013.

[11] Ahmad Akrami, and Ali Niazi. Synthesis of maghemite nanoparticles and its application for removal of titan yellow from aqueous solutions using full factorial design. *Desalination and Water Treatment*, 57(47):22618–22631, 2016.

[12] Zümriye Aksu, and Sevilay Tezer. Biosorption of reactive dyes on the green alga chlorella vulgaris. *Process Biochemistry*, 40(3):1347–1361, 2005.

[13] Z Aksu, and J Yener. The usage of dried activated sludge and fly ash wastes in phenol biosorption/adsorption: Comparison with granular activated carbon. *Journal of Environmental Science and Health, Part A*, 34(9):1777–1796, 1999.

[14] Fahad A Alabduljabbar, Sajjad Haider, Abdulaziz A Alghyamah, Adnan Haider, Rawaiz Khan, Waheed A Almasry, Raj Patel, Iqbal M Mujtaba, and Fekri Abdulraqeb Ahmed Ali. Ethanol amine functionalized electrospun nanofibers membrane for the treatment of dyes polluted wastewater. *Applied Nanoscience*, 12:3153–3166, 2022.

[15] Yahya S Al-Degs, Musa I El-Barghouthi, Ayman A Issa, Majeda A Khraisheh, and Gavin M. Sorption of Zn (II), Pb (II), and Co (II) using natural sorbents: Equilibrium and kinetic studies. *Water Research*, 40(14):2645–2658, 2006.

[16] Mohammad A Al-Ghouti, and Dana A Da'ana. Guidelines for the use and interpretation of adsorption isotherm models: A review. *Journal of Hazardous Materials*, 393:122383, 2020.

[17] Attarad Ali, Muhammad Zia Hira Zafar, Ihsan ul Haq, Abdul Rehman Phull, Joham Sarfraz Ali, and Altaf Hussain. Synthesis, characterization, applications, and challenges of iron oxide nanoparticles. *Nanotechnology, Science and Applications*, 9:49, 2016.

[18] Imran Ali, Khalaf AlGhamdi, and Fahd T Al-Wadaani. Advances in iridium nano catalyst preparation, characterization and applications. *Journal of Molecular Liquids*, 280:274–284, 2019.

[19] Nauman Ali, Amir Zada, Muhammad Zahid, Ahmed Ismail, Misbha Rafiq, Aaisha Riaz, and Adnan. Enhanced photodegradation of methylene blue with alkaline and transition-metal ferrite nanophotocatalysts under direct sun light irradiation. *Journal of the Chinese Chemical Society*, 66(4):402–408, 2019.

[20] Wajid Ali, Hameed Ullah, Amir Zada, Muhammad Khalid Alamgir, Wisal Muhammad, Muhammad Jawad Ahmad, and Akhtar Nadhman. Effect of calcination temperature on the photoactivities of ZnO/SnO_2 nanocomposites for the degradation of methyl orange. *Materials Chemistry and Physics*, 213:259–266, 2018.

[21] S Aliramaji, A Zamanian, and Z Sohrabijam. Characterization and synthesis of magnetite nanoparticles by innovative sonochemical method. *Procedia Materials Science*, 11:265–269, 2015.

[22] Mahir Alkan, Özkan Demirbas, Sermet Çelikçapa, and Mehmet Doğan. Sorption of acid red-57 from aqueous solution onto sepiolite. *Journal of Hazardous Materials*, 116(1–2):135–145, 2004.

[23] Susan SA Alkurdi, Raed A Al-Juboori, Jochen Bundschuh, Les Bowtell, and Alla Marchuk. Inorganic arsenic species removal from water using bone char: A detailed study on adsorption kinetic and isotherm models using error functions analysis. *Journal of Hazardous Materials*, 405:124112, 2021.

[24] Sh Husien, A Labena, EF El-Belely, Hamada M Mahmoud, and Asmaa S Hamouda. Adsorption studies of hexavalent chromium [Cr (VI)] on micro-scale biomass of sargassum dentifolium, seaweed. *Journal of Environmental Chemical Engineering*, 7(6): 103444, 2019.

[25] Tarek Mansour Mohammed, Ahmed Labena, Nabila Ahmed Maziad, and Shimaa Husien. Radiation copolymerization of PVA/Malic acid/HEMA/Macro-algal (*Sargassum* sp.) biomass for removal of hexavalent chromium. *Egyptian Journal of Chemistry*, 63(6):2019–2035, 2020.

[26] A Allwar. Characteristics of micro- and mesoporous structure and surface chemistry of activated carbons produced by oil palm shell. In *International Conference on Chemical, Ecology and Environmental Sciences Proceedings*, pages 138–141.

[27] Karima Almashhori, Tarek T Ali, Abdu Saeed, Reem Alwafi, Magda Aly, and Faten E Al-Hazmi. Antibacterial and photocatalytic activities of controllable (anatase/rutile) mixed phase TiO_2 nanophotocatalysts synthesized via a microwave-assisted sol–gel method. *New Journal of Chemistry*, 44(2):562–570, 2020.

[28] Sibel Klnç Alpat, Özge Özbayrak, Senol Alpat, and Hüsamettin Akçay. The adsorption kinetics and removal of cationic dye, Toluidine Blue O, from aqueous solution with Turkish zeolite. *Journal of Hazardous Materials*, 151(1):213–220, 2008.

[29] Tamer M Alslaibi, Ismail Abustan, Mohd Azmier Ahmad, and Ahmad Abu. A review: Production of activated carbon from agricultural byproducts via conventional and microwave heating. *Journal of Chemical Technology & Biotechnology*, 88(7):1183–1190, 2013.

[30] Erol Alver and Aysegül Ü Metin. Anionic dye removal from aqueous solutions using modified zeolite: Adsorption kinetics and isotherm studies. *Chemical Engineering Journal*, 200:59–67, 2012.

[31] BMWPK Amarasinghe, and Richard A Williams. Tea waste as a low cost adsorbent for the removal of Cu and Pb from wastewater. *Chemical Engineering Journal*, 132(1–3): 299–309, 2007.

[32] TS Anirudhan, and M Ramachandran. Surfactant-modified bentonite as adsorbent for the removal of humic acid from wastewaters. *Applied Clay Science*, 35(3):276–281, 2007.

[33] TS Anirudhan, and M Ramachandran. Adsorptive removal of basic dyes from aqueous solutions by surfactant modified bentonite clay (organoclay): Kinetic and competitive adsorption isotherm. *Process Safety and Environmental Protection*, 95:215–225, 2015.

[34] Hossein Shahbeig, Nafiseh Bagheri, Sohrab Ali Ghorbanian, Ahmad Hallajisani, and Sara Poorkarimi. A new adsorption isotherm model of aqueous solutions on granular activated carbon. *World Journal of Modelling and Simulation*, 9(4):243–254, 2013.

[35] Mokhtar Arami, Nargess Yousefi Limaee, Niyaz Mohammad Mahmoodi, and Nooshin Salman. Removal of dyes from colored textile wastewater by orange peel adsorbent: Equilibrium and kinetic studies. *Journal of Colloid and Interface Science*, 288(2):371–376, 2005.

[36] Mehmet Emin Argun, Sukru Dursun, Mustafa Karatas, and Metin Gürü. Activation of pine cone using fenton oxidation for Cd(II) and Pb(II) removal. *Bioresource Technology*, 99(18):8691–8698, 2008.

[37] Bulent Armagan, Orhan Ozdemir, Mustafa Turan, and Mehmet S. Çelik. Adsorption of negatively charged azo dyes onto surfactant-modified sepiolite. *Journal of Environmental Engineering*, 129(8):709–715, 2003.

[38] AM Atta, HA Al-Hodan, SA Al-Hussain, AO Ezzat, AM Tawfik, and YA El-Dosary. Preparation of magnetite and manganese oxide ionic polymer nanocomposite for adsorption of a textile dye in aqueous solutions. *Digest Journal of Nanomaterials and Biostructures*, 11(3):909–919, 2016.

[39] Pratibha Attri, Sangeeta Garg, and Jatinder Kumar Ratan. Kinetic modelling and proposed mechanistic pathway for photocatalytic degradation of 4-aminopyridine using cuprous oxide nanoparticles. *Research on Chemical Intermediates*, 47(4):1535–1562, 2021.

[40] M Auta and BH Hameed. Acid modified local clay beads as effective low-cost adsorbent for dynamic adsorption of methylene blue. *Journal of Industrial and Engineering Chemistry*, 19(4):1153–1161, 2013.

[41] Nimibofa Ayawei, Augustus Newton Ebelegi, and Donbebe Wankasi. Modelling and interpretation of adsorption isotherms. *Journal of Chemistry*, 2017:3039817, 2017.

[42] AH Aydin, and Y Bulut. Acid dyes removal using low cost adsorbents. *International Journal of Environment and Pollution*, 21(1):97–104, 2004.

[43] Roop Chand Bansal and Meenakshi Goyal. *Activated Carbon Adsorption*. CRC Press, Boca Raton, FL, 2005.

[44] P Baskaralingam, M Pulikesi, D Elango, V Ramamurthi, and S Sivanesan. Adsorption of acid dye onto organobentonite. *Journal of Hazardous Materials*, 128(2):138–144, 2006.

[45] FA Batzias, and DK Sidiras. Dye adsorption by prehydrolysed beech sawdust in batch and fixed-bed systems. *Bioresource Technology*, 98(6):1208–1217, 2007.

[46] FA Batzias, and DK Sidiras. Dye adsorption by calcium chloride treated beech sawdust in batch and fixed-bed systems. *Journal of Hazardous Materials*, 114(1–3):167–174, 2004.

[47] Bjorn Larsen, Dalia MSA Salem, Mohammed AE Sallam, Morcos M Mishrikey, and Ali I Beltagy. Characterization of the alginates from algae harvested at the egyptian red sea coast. *Carbohydrate Research*, 338(22):2325–2336, 2003.

[48] Hülya Genç-Fuhrman, Jens Christian Tjell, and David McConchie. Adsorption of arsenic from water using activated neutralized red mud. *Environmental Science & Technology*, 38(8):2428–2434, 2004.

[49] Sachin Girdhar Shinde, Maheshkumar Prakash Patil, Gun-Do Kim, and Vinod Shankar Shrivastava. Multi-doped ZnO photocatalyst for solar induced degradation of indigo carmine dye and as an antimicrobial agent. *Journal of Inorganic and Organometallic Polymers and Materials*, 30(4):1141–1152, 2020.

[50] N Belhouchat, H Zaghouane-Boudiaf, and César Viseras. Removal of anionic and cationic dyes from aqueous solution with activated organo-bentonite/sodium alginate encapsulated beads. *Applied Clay Science*, 135:9–15, 2017.

[51] Assia Benhouria, Md Azharul Islam, H. Zaghouane-Boudiaf, M Boutahala, and BH Hameed. Calcium alginate–bentonite–activated carbon composite beads as highly effective adsorbent for methylene blue. *Chemical Engineering Journal*, 270:621–630, 2015.

[52] F Bergaya, and G Lagaly. General introduction: Clays, clay minerals, and clay science. In *Developments in Clay Science*, vol. 5, 1–19. Elsevier, 2013.

[53] S Mahendran, R Gokulan, A Aravindan, H Joga Rao, G Kalyani, S Praveen, T Bhagavathi Pushpa, M Senthil. Production of Ulva prolifera derived biochar and evaluation of adsorptive removal of reactive red 120: Batch, isotherm, kinetic, thermodynamic and regeneration studies. *Biomass Conversion and Biorefinery*, 10:1–12, 2021.

[54] Krishna G Bhattacharyya, and Susmita Sen Gupta. Kaolinite, montmorillonite, and their modified derivatives as adsorbents for removal of Cu (II) from aqueous solution. *Separation and Purification Technology*, 50(3):388–397, 2006.

[55] Jana Braniša, Klaudia Jomová, L'ubomr Lapčk, and Mária Porubská. Testing of electron beam irradiated sheep wool for adsorption of Cr (III) and Co (II) of higher concentrations. *Polymer Testing*, 99:107191, 2021.

[56] F Brouers and O Sotolongo-Costa. Generalized fractal kinetics in complex systems (application to biophysics and biotechnology). *Physica A: Statistical Mechanics and its Applications*, 368(1):165–175, 2006.

[57] F Brouers. The fractal (BSf) kinetics equation and its approximations. *Journal of Modern Physics*, 5(16):1594–1601, 2014.

[58] Yasemin Bulut, and Haluk Aydın. A kinetics and thermodynamics study of methylene blue adsorption on wheat shells. *Desalination*, 194(1):259–267, 2006.

[59] Bingdi Cao, Ruilu Wang, Weijun Zhang, Hanjun Wu, and Dongsheng Wang. Carbon-based materials reinforced waste activated sludge electro-dewatering for synchronous fuel treatment. *Water Research*, 149:533–542, 2019.

[60] PJM Carrott, PAM Mourao, MML Ribeiro Carrott, and EM Gonçalves. Separating surface and solvent effects and the notion of critical adsorption energy in the adsorption of phenolic compounds by activated carbons. *Langmuir*, 21(25):11863–11869, 2005.

[61] Islem Chaari, Mongi Feki, Mounir Medhioub, Jalel Bouzid, Emna Fakhfakh, and Fakher Jamoussi. Adsorption of a textile dye "Indanthrene Blue RS (C.I. Vat Blue 4)" from aqueous solutions onto smectite-rich clayey rock. *Journal of Hazardous Materials*, 172(2):1623–1628, 2009.

[62] Venkat S Mane, and PV Vijay Babu. Studies on the adsorption of brilliant green dye from aqueous solution onto low-cost naoh treated saw dust. *Desalination*, 273(2–3): 321–329, 2011.

[63] Islem Chaari, Bechir Moussi, and Fakher Jamoussi. Interactions of the dye, C.I. direct orange 34 with natural clay. *Journal of Alloys and Compounds*, 647:720–727, 2015.

[64] LS Chan, WH Cheung, SJ Allen, and G McKay. Error analysis of adsorption isotherm models for acid dyes onto bamboo derived activated carbon. *Chinese Journal of Chemical Engineering*, 20(3):535–542, 2012.

[65] Sathy Chandrasekhar, and PN Pramada. Rice husk ash as an adsorbent for methylene blue—effect of ashing temperature. *Adsorption*, 12(1):27–43, 2006.

[66] Jiali Chang, Jianchao Ma, Qingliang Ma, Duoduo Zhang, Nannan Qiao, Mengxiao Hu, and Hongzhu Ma. Adsorption of methylene blue onto Fe_3O_4/activated montmorillonite nanocomposite. *Applied Clay Science*, 119:132–140, 2016.

[67] Sneha Chawla, Himani Uppal, Mohit Yadav, Nupur Bahadur, and Nahar Singh. Zinc peroxide nanomaterial as an adsorbent for removal of Congo red dye from waste water. *Ecotoxicology and Environmental Safety*, 135:68–74, 2017.

[68] Buning Chen, Chi Wai Hui, and Gordon McKay. Film-pore diffusion modeling and contact time optimization for the adsorption of dyestuffs on pith. *Chemical Engineering Journal*, 84(2):77–94, 2001.
[69] Daimei Chen, Jian Chen, Xinlong Luan, Haipeng Ji, and Zhiguo Xia. Characterization of anion–cationic surfactants modified montmorillonite and its application for the removal of methyl orange. *Chemical Engineering Journal*, 171(3):1150–1158, 2011.
[70] Ying Chen, Zeting Zhang, Dong Chen, Yong Chen, Qiang Gu, and Haoran Liu. Removal of coke powders in coking diesel distillate using recyclable chitosan-grafted Fe_3O_4 magnetic nanoparticles. *Fuel*, 238:345–353, 2019.
[71] Keun J Choi, Sang G Kim, Chang W Kim, and Seung H Kim. Effects of activated carbon types and service life on removal of endocrine disrupting chemicals: Amitrol, nonylphenol, and bisphenol-A. *Chemosphere*, 58(11):1535–1545, 2005.
[72] GE Christidis, PW Scott, and AC Dunham. Acid activation and bleaching capacity of bentonites from the islands of Milos and Chios, Aegean, Greece. *Applied Clay Science*, 12(4):329–347, 1997.
[73] Wang Chunfeng, Li Jiansheng, Sun Xia, Wang Lianjun, and Sun Xiuyun. Evaluation of zeolites synthesized from fly ash as potential adsorbents for wastewater containing heavy metals. *Journal of Environmental Sciences*, 21(1):127–136, 2009.
[74] Inês F Cruz, Cristina Freire, João P Araújo, Clara Pereira, and André M Pereira. Multifunctional ferrite nanoparticles: From current trends toward the future. In *Magnetic Nanostructured Materials*, 59–116. Elsevier, 2018.
[75] Sandra Maria Dal Bosco, Ricardo Sarti Jimenez, and Wagner Alves Carvalho. Removal of toxic metals from wastewater by Brazilian natural scolecite. *Journal of Colloid and Interface Science*, 281(2):424–431, 2005.
[76] Nilanjana Das, and Devlina Das. Recovery of rare earth metals through biosorption: An overview. *Journal of Rare Earths*, 31(10):933–943, 2013.
[77] RFW Pease. Electron beam lithography. *Contemporary Physics*, 22(3):265–290, 1981.
[78] Nilanjana Das, R Vimala, and P Karthika. Biosorption of heavy metals–an overview. *Indian Journal of Biotechnology* 7(2):159–169, 2008.
[79] AA Fodeke, and OJ Ayejuyone. Adsorption of methylene blue on corncob charcoal: Thermodynamic studies. *Life Journal of Science*, 23(1):131–144, 2021.
[80] Keng Yuen Foo, and Bassim H Hameed. Insights into the modeling of adsorption isotherm systems. *Chemical Engineering Journal*, 156(1):2–10, 2010.
[81] Elvis Fosso-Kankeu. *Nano and Bio-Based Technologies for Wastewater Treatment: Prediction and Control Tools for the Dispersion of Pollutants in the Environment*. John Wiley and Sons, UK, 2019. doi: 10.2166/wqrj.2000.006.
[82] Yuzhu Fu, and T Viraraghavan. Removal of a dye from an aqueous solution by the fungus *Aspergillus niger. Water Quality Research Journal*, 35(1):95–112, 2000.
[83] GM Gadd, and AJ Griffiths. Effect of copper on morphology of aureobasidium pullulans. *Transactions of the British Mycological Society*, 74(2):387–392, 1980.
[84] Jixian Yang, Wei Wei, Shanshan Pi, Fang Ma, Ang Li, Dan Wu, and Jie Xing. Competitive adsorption of heavy metals by extracellular polymeric substances extracted from *Klebsiella* sp. *Bioresource Technology*, 196:533–539, 2015.
[85] Eun-Ju Kim, Chung-Seop Lee, Yoon-Young Chang, and Yoon-Seok Chang. Hierarchically structured manganese oxide-coated magnetic nanocomposites for the efficient removal of heavy metal ions from aqueous systems. *ACS Applied Materials & Interfaces*, 5(19):9628–9634, 2013.
[86] AK Golder, AN Samanta, and S Ray. Anionic reactive dye removal from aqueous solution using a new adsorbent—sludge generated in removal of heavy metal by electrocoagulation. *Chemical Engineering Journal*, 122(1):107–115, 2006.

[87] Sami Guiza, François Brouers, and Mohamed Bagane. Fluoride removal from aqueous solution by montmorillonite clay: Kinetics and equilibrium modeling using new generalized fractal equation. *Environmental Technology and Innovation*, 21:101187, 2021.
[88] Osman Gulnaz, Aysenur Kaya, Fatih Matyar, and Burhan Arikan. Sorption of basic dyes from aqueous solution by activated sludge. *Journal of Hazardous Materials*, 108(3):183–188, 2004.
[89] VK Gupta, and Suhas. Application of low-cost adsorbents for dye removal – a review. *Journal of Environmental Management*, 90(8):2313–2342, 2009.
[90] VK Gupta, Suhas, Imran Ali, and VK Saini. Removal of rhodamine B, fast green, and methylene blue from wastewater using red mud, an aluminum industry waste. *Industrial and Engineering Chemistry Research*, 43(7):1740–1747, 2004.
[91] Sousan Hadi, Ensiyeh Taheri, Mohammad Mehdi Amin, Ali Fatehizadeh, and Tejraj M Aminabhavi. Adsorption of 4-chlorophenol by magnetized activated carbon from pomegranate husk using dual stage chemical activation. *Chemosphere*, 270:128623, 2021.
[92] Y Hai, X Li, H Wu, S Zhao, W Deligeer, and S Asuha. Modification of acid-activated kaolinite with TiO_2 and its use for the removal of azo dyes. *Applied Clay Science*, 114: 558–567, 2015.
[93] Y Haldorai, A Rengaraj, T Ryu, J Shin, YS Huh, and Y-K Han. Response surface methodology for the optimization of lanthanum removal from an aqueous solution using a Fe_3O_4/chitosan nanocomposite. *Materials Science and Engineering: B*, 195:20–29, 2015.
[94] Oualid Hamdaoui. Batch study of liquid-phase adsorption of methylene blue using cedar sawdust and crushed brick. *Journal of Hazardous Materials*, 135(1–3):264–273, 2006.
[95] BH Hameed, DK Mahmoud, and AL Ahmad. Equilibrium modeling and kinetic studies on the adsorption of basic dye by a low-cost adsorbent: Coconut (*Cocos nucifera*) bunch waste. *Journal of Hazardous Materials*, 158(1):65–72, 2008.
[96] BH Hameed. Removal of cationic dye from aqueous solution using jackfruit peel as non-conventional low-cost adsorbent. *Journal of Hazardous Materials*, 162(1):344–350, 2009.
[97] Zhao-Xiang Han, Zhen Zhu, Dan-Dan Wu, Jue Wu, and Yu-Rong Liu. Adsorption kinetics and thermodynamics of acid blue-25 and methylene blue dye solutions on natural sepiolite. *Synthesis and Reactivity in Inorganic, Metal-Organic, and Nano-Metal Chemistry*, 44(1):140–147, 2014.
[98] A Hashem, GM Taha, AJ Fletcher, LA Mohamed, and SH Samaha. Highly efficient adsorption of Cd (II) onto carboxylated camelthorn biomass: Applicability of three-parameter isotherm models, kinetics, and mechanism. *Environmental Science and Pollution Research*, 29(5):1630–1642, 2021.
[99] Daniel P Hashim, Narayanan T Narayanan, Jose M Romo-Herrera, David A Cullen, Myung Gwan Hahm, Peter Lezzi, Joseph R Suttle, Doug Kelkhoff, E Muñoz-Sandoval, Sabyasachi Ganguli, Ajit K Roy, David J Smith, Robert Vajtai, Bobby G Sumpter, Vincent Meunier, Humberto Terrones, Mauricio Terrones, and Pulickel M Ajayan. Covalently bonded three-dimensional carbon nanotube solids via boron induced nanojunctions. *Scientific Reports*, 2(1):363, 2012.
[100] Alaa H Hawari, and Catherine N. Mulligan. Biosorption of lead(II), cadmium (II), copper (II) and nickel (II) by anaerobic granular biomass. *Bioresource Technology*, 97(4):692–700, 2006.
[101] Abolghasem Hedayatkhah, Mariana Silvia Cretoiu, Giti Emtiazi, Lucas J Stal, and Henk. Bioremediation of chromium contaminated water by diatoms with concomitant lipid accumulation for biofuel production. *Journal of Environmental Management*, 227: 313–320, 2018.

[102] Mahdi Heidarizad, and S Sevinç. Synthesis of graphene oxide/magnesium oxide nanocomposites with high-rate adsorption of methylene blue. *Journal of Molecular Liquids*, 224:607–617, 2016.

[103] Mihir Herlekar, Siddhivinayak Barve, and Rakesh Kumar. Plant-mediated green synthesis of iron nanoparticles. *Journal of Nanoparticles*, 2014:140614, 2014.

[104] Aldahir Alberto Hernández-Hernández, Giaan Arturo Álvarez Romero, Araceli Castañeda-Ovando, Yucundo Mendoza-Tolentino, Elizabeth Contreras-López, Carlos A Galán-Vidal, and María E Physics. Optimization of microwave-solvothermal synthesis of Fe_3O_4 nanoparticles. Coating, modification, and characterization. *Materials Chemistry and Physics*, 205:113–119, 2018.

[105] AR Hidayu, and N Muda. Preparation and characterization of impregnated activated carbon from palm kernel shell and coconut shell for CO_2 capture. *Procedia Engineering*, 148:106–113, 2016.

[106] Show-Chu Huang, Tsair-Wang Chung, and Hung-Ta Wu. Effects of molecular properties on adsorption of six-carbon vocs by activated carbon in a fixed adsorber. *ACS Omega*, 6(8):5825–5835, 2021.

[107] Ramona L Johnson, Amy J Anschutz, Jean M Smolen, Matt F Simcik, and R Lee Penn. The adsorption of perfluorooctane sulfonate onto sand, clay, and iron oxide surfaces. *Journal of Chemical and Engineering Data*, 52(4):1165–1170, 2007.

[108] Sushil Kumar Kansal, Navjeet Kaur, and Sukhmehar Singh. Photocatalytic degradation of two commercial reactive dyes in aqueous phase using nanophotocatalysts. *Nanoscale Research Letters*, 4(7):709, 2009.

[109] Ekta Khosla, Satindar Kaur, and Pragnesh N Dave. Tea waste as adsorbent for ionic dyes. *Desalination and Water Treatment*, 51(34–36):6552–6561, 2013.

[110] M Kobya, E Demirbas, E Senturk, and M Ince. Adsorption of heavy metal ions from aqueous solutions by activated carbon prepared from apricot stone. *Bioresource Technology*, 96(13):1518–1521, 2005.

[111] DDE Koyuncu, and M Okur. Removal of AV 90 dye using ordered mesoporous carbon materials prepared via nanocasting of KIT-6: Adsorption isotherms, kinetics and thermodynamic analysis. *Separation and Purification Technology*, 257:117657, 2021.

[112] Drissa Ouattara Kra, Grah Patrick Atheba, Patrick Drogui, Albert Trokourey, et al. Activated carbon based on acacia wood (*Auriculeaformis*, Côte d'Ivoire) and application to the environment through the elimination of Pb^{2+} ions in industrial effluents. *Journal of Encapsulation and Adsorption Sciences*, 11(1):18, 2021.

[113] Upendra Kumar, and Manas Bandyopadhyay. Sorption of cadmium from aqueous solution using pretreated rice husk. *Bioresource Technology*, 97(1):104–109, 2006.

[114] Kurt Lackovic, Michael J Angove, John D Wells, and Bruce B Johnson. Modeling the adsorption of Cd (II) onto muloorina illite and related clay minerals. *Journal of Colloid and Interface Science*, 257(1):31–40, 2003.

[115] Sophie Laurent, Delphine Forge, Marc Port, Alain Roch, Caroline Robic, Luce Vander Elst, and Robert N. Muller. Magnetic iron oxide nanoparticles: Synthesis, stabilization, vectorization, physicochemical characterizations, and biological applications. *Chemical Reviews*, 108(6):2064–2110, 2008.

[116] Jieun Lee, Sanghyun Jeong, and Zongwen Liu. Progress and challenges of carbon nanotube membrane in water treatment. *Critical Reviews in Environmental Science and Technology*, 46(11–12):999–1046, 2016.

[117] Jie Li, Xiangxue Wang, Guixia Zhao, Changlun Chen, Zhifang Chai, Ahmed Alsaedi, Tasawar Hayat, and Xiangke Wang. Metal–organic framework-based materials: Superior adsorbents for the capture of toxic and radioactive metal ions. *Chemical Society Reviews*, 47(7):2322–2356, 2018.

[118] Min Li, Jian Feng, Kun Huang, Si Tang, Ruihua Liu, Huan Li, Fuye Ma, and Xiaojing Meng. Amino group functionalized SiO_2 graphene oxide for efficient removal of Cu(II) from aqueous solutions. *Chemical Engineering Research and Design*, 145:235–244, 2019.

[119] Qian Li, Qin-Yan Yue, Hong-Jian Sun, Yuan Su, and Bao-Yu Gao. A comparative study on the properties, mechanisms and process designs for the adsorption of non-ionic or anionic dyes onto cationic-polymer/bentonite. *Journal of Environmental Management*, 91(7):1601–1611, 2010.

[120] Wei Li, Li-bo Zhang, Jin-hui Peng, Ning Li, and Xue-yun Zhu. Preparation of high surface area activated carbons from tobacco stems with K_2CO_3 activation using microwave radiation. *Industrial Crops and Products*, 27(3):341–347, 2008.

[121] Junxiong Lin, and Lan Wang. Comparison between linear and non-linear forms of pseudo-first-order and pseudo-second-order adsorption kinetic models for the removal of methylene blue by activated carbon. *Frontiers of Environmental Science and Engineering in China*, 3(3):320–324, 2009.

[122] Darja Lisjak, and Alenka Mertelj. Anisotropic magnetic nanoparticles: A review of their properties, syntheses and potential applications. *Progress in Materials Science*, 95:286–328, 2018.

[123] Junsheng Liu, and Xin Wang. Novel silica-based hybrid adsorbents: Lead (II) adsorption isotherms. *The Scientific World Journal*, 2013:897159, 2013.

[124] Niyaz Mohammad Mahmoodi, Adonis Maghsoudi, Farhood Najafi, Mojtaba Jalili, and Hojjat Kharrati. Primary–secondary amino silica nanoparticle: Synthesis and dye removal from binary system. *Desalination and Water Treatment*, 52(40–42):7784–7796, 2014.

[125] Benamar Makhoukhi, Mohamed Djab, and Mohamed Amine Didi. Adsorption of telon dyes onto bis-imidazolium modified bentonite in aqueous solutions. *Journal of Environmental Chemical Engineering*, 3(2):1384–1392, 2015.

[126] Afshin Maleki, Bagher Hayati, Farhood Najafi, Fardin Gharibi, and Sang Woo Joo. Heavy metal adsorption from industrial wastewater by pamam/TiO_2 nanohybrid: preparation, characterization and adsorption studies. *Journal of Molecular Liquids*, 224:95–104, 2016.

[127] John Geraldine Sandana Mala, Dhanasingh Sujatha, and Chellan Rose. Inducible chromate reductase exhibiting extracellular activity in bacillus methylotrophicus for chromium bioremediation. *Microbiological research*, 170:235–241, 2015.

[128] Ashraf N Malhas, Ramadan A Abuknesha, and Robert G Price. Removal of detergents from protein extracts using activated charcoal prior to immunological analysis. *Journal of immunological methods*, 264(1–2):37–43, 2002.

[129] S Matali, SA Khairuddin, ASAK Sharifah, and AR Hidayu. Removal of selected gaseous effluent using activated carbon derived from oil palm waste: An overview. In IEEE Symposium on Business, Engineering, and Industrial Applications. *Procedia Engineering*, 148: 106–113, 2016.

[130] G McKay, JF Porter, and GR Prasad. The removal of dye colours from aqueous solutions by adsorption on low-cost materials. *Water, Air, and Soil Pollution*, 114(3):423–438, 1999.

[131] Maryam Mehrabi, and Vahid Javanbakht. Photocatalytic degradation of cationic and anionic dyes by a novel nanophotocatalyst of $TiO_2/ZnTiO_3/\alpha Fe_2O_3$ by ultraviolet light irradiation. *Journal of Materials Science: Materials in Electronics*, 29(12):9908–9919, 2018.

[132] Surya Kant Mehta, and Jai Prakash. Characterization and optimization of Ni and Cu sorption from aqueous solution by *Chlorella vulgaris*. *Ecological Engineering*, 18(1): 1–13, 2001.

[133] El Hadj Mekatel, Samira Amokrane, Asma Aid, Djamel Nibou, and Mohamed Trari. Adsorption of methyl orange on nanoparticles of a synthetic zeolite NaA/CuO. *Comptes Rendus Chimie*, 18(3):336–344, 2015.

[134] José D Méndez-Díaz, Mahmoud M Abdel daiem, José Rivera-Utrilla, Manuel Sánchez-Polo, and Isidora Bautista-Toledo. Adsorption/bioadsorption of phthalic acid, an organic micropollutant present in landfill leachates, on activated carbons. *Journal of Colloid and Interface Science*, 369(1):358–365, 2012.

[135] JA Menéndez-Díaz, and I Martín-Gullón. Types of carbon adsorbents and their production. *Interface Science and Technology*, 7:1–47, 2006.

[136] V Meshko, L Markovska, M Mincheva, and AE Rodrigues. Adsorption of basic dyes on granular activated carbon and natural zeolite. *Water Research*, 35(14):3357–3366, 2001.

[137] Shuying Jia, Zhen Yang, Kexin Ren, Ziqi Tian, Chang Dong, Ruixue Ma, Ge Yu, and Weiben Yang. Removal of antibiotics from water in the coexistence of suspended particles and natural organic matters using amino-acid-modified-chitosan flocculants: A combined experimental and theoretical study. *Journal of Hazardous Materials*, 317: 593–601, 2016.

[138] Jie Xie, Yan Lin, Chunjie Li, Deyi Wu, and Hainan Kong. Removal and recovery of phosphate from water by activated aluminum oxide and lanthanum oxide. *Powder Technology*, 269:351–357, 2015.

[139] Cristina M Monteiro, Paula ML Castro, and F Xavier Malcata. Metal uptake by microalgae: Underlying mechanisms and practical applications. *Biotechnology Progress*, 28(2):299–311, 2012.

[140] Duane Milton Moore, and Robert C Reynolds Jr. *X-ray Diffraction and the Identification and Analysis of Clay Minerals*. Oxford University Press, UK, 1989. doi: 10.1016/j.chemosphere.2020.128919.

[141] JM Morgan-Sagastume, B Jiménez, and A Noyola. Tracer studies in a laboratory and pilot scale UASB reactor. *Environmental Technology*, 18(8):817–825, 1997.

[142] Khanidtha Marungrueng, and Prasert Pavasant. High performance biosorbent (*Caulerpa lentillifera*) for basic dye removal. *Bioresource Technology*, 98(8):1567–1572, 2007.

[143] Mara Ángeles Lobo-Recio, Caroline Rodrigues, Thamires Custódio Jeremias, Flávio Rubens Lapolli, Isabel Padilla, and Aurora López-Delgado. Highly efficient removal of aluminum, iron, and manganese ions using linde type-A zeolite obtained from hazardous waste. *Chemosphere*, 267:128919, 2021.

[144] Jun Ma, Yongzhong Jia, Yan Jing, Ying Yao, and Jinhe Sun. Kinetics and thermodynamics of methylene blue adsorption by cobalt-hectorite composite. *Dyes and Pigments*, 93(1):1441–1446, 2012.

[145] Ibrahim Maamoun, Ramadan Eljamal, Omar Falyouna, Khaoula Bensaida, Yuji Sugihara, and Osama Eljamal. Insights into kinetics, isotherms and thermodynamics of phosphorus sorption onto nanoscale zero-valent iron. *Journal of Molecular Liquids*, 328:115402, 2021.

[146] PAM Mourao, PJM Carrott, and MML Ribeiro. Application of different equations to adsorption isotherms of phenolic compounds on activated carbons prepared from cork. *Carbon*, 44(12):2422–2429, 2006.

[147] Swapna Mukherjee. *Clays: Industrial Applications and Their Determinants*, 113–122. Springer, Dordrecht, Netherlands, 2013.

[148] Sumona Mukherjee, Soumyadeep Mukhopadhyay, Mohd Ali Hashim, and Bhaskar Sen Gupta. Contemporary environmental issues of landfill leachate: Assessment and remedies. *Critical Reviews in Environmental Science and Technology*, 45(5):472–590, 2015.

[149] Haydn H Murray. *Applied Clay Mineralogy: Occurrences, Processing and Applications of Kaolins, Bentonites, Palygorskitesepiolite, and Common Clays*. Elsevier, UK 2006.

[150] S Netpradit, P Thiravetyan, and S Towprayoon. Application of 'waste' metal hydroxide sludge for adsorption of azo reactive dyes. *Water Research*, 37(4):763–772, 2003.

[151] Tholiso Ngulube, Jabulani Ray Gumbo, Vhahangwele Masindi, and Arjun Maity. An update on synthetic dyes adsorption onto clay based minerals: A state-of-art review. *Journal of Environmental Management*, 191:35–57, 2017.

[152] Bernd Nowack, Harald F Krug, and Murray Height. 120 years of nanosilver history: Implications for policy makers. *Environmental Science and Technology*, 45(4):1177–1183, 2011.

[153] T O'Mahony, E Guibal, and JM Tobin. Reactive dye biosorption by rhizopus arrhizus biomass. *Enzyme and Microbial Technology*, 31(4):456–463, 2002.

[154] SO Obaje, JI Omada, and UA Dambatta. Clays and their industrial applications: Synoptic review. *International Journal of Science and Technology*, 3(5):264–270, 2013.

[155] OT Ogunmodede, AA Ojo, E Adewole. Adsorptive removal of anionic dye from aqueous solutions by mixture of kaolin and bentonite clay: Characteristics, isotherm, kinetic and thermodynamic studies. *Iranian (Iranica) Journal of Energy & Environment*, 6(2):147–153, 2015.

[156] Maria Fernanda Oliveira, Victor M de Souza, Meuris GC da Silva, and Melissa GA Vieira. Fixed-bed adsorption of caffeine onto thermally modified Verde-lodo bentonite. *Industrial and Engineering Chemistry Research*, 57(51):17480–17487, 2018.

[157] Safir Ouassel, Salah Chegrouche, Djamel Nibou, Abderahmane Aknoun, Redouane Melikchi, Sihem Khemaissia, and Sabino De Gisi. Adsorption of uranium (VI) onto natural algerian phosphate: Study of influencing factors, and mechanism. *Arabian Journal for Science and Engineering*, 46:6645–6661, 2021.

[158] Mahmut Özacar, and I. Ayhan Sengil. Adsorption of metal complex dyes from aqueous solutions by pine sawdust. *Bioresource Technology*, 96(7):791–795, 2005.

[159] Prashant Pandey, and Vipin Kumar. Pillared interlayered clays: Sustainable materials for pollution abatement. *Environmental Chemistry Letters*, 17(2):721–727, 2019.

[160] A. Safa Özcan, Bilge Erdem, and Adnan Özcan. Adsorption of acid blue 193 from aqueous solutions onto Na–bentonite and dtma–bentonite. *Journal of Colloid and Interface Science*, 280(1):44–54, 2004.

[161] Karina I Paredes-Páliz, Enrique Mateos-Naranjo, Bouchra Doukkali, Miguel A Caviedes, Susana Redondo-G6mez, Ignacio D Rodríguez-Llorente, and Eloísa Pajuelo. Modulation of *Spartina densiflora* plant growth and metal accumulation upon selective inoculation treatments: A comparison of gram negative and gram positive *rhizobacteria*. *Marine Pollution Bulletin*, 125(1–2):77–85, 2017.

[162] E Sanmuga Priya and P Senthamil. Water hyacinth (*Eichhornia crassipes*)–an efficient and economic adsorbent for textile effluent treatment–a review. *Arabian Journal of Chemistry*, 10:S3548–S3558, 2017.

[163] Arivalagan Pugazhendhi, Desika Prabakar, Jaya Mary Jacob, Indira Karuppusamy, and Rijuta Ganesh Saratale. Synthesis and characterization of silver nanoparticles using gelidium amansii and its antimicrobial property against various pathogenic bacteria. *Microbial Pathogenesis*, 114:41–45, 2018.

[164] Eric Kristia Putra, Ramon Pranowo, Jaka Sunarso, Nani Indraswati, and Suryadi Ismadji. Performance of activated carbon and bentonite for adsorption of amoxicillin from wastewater: Mechanisms, isotherms and kinetics. *Water Research*, 43(9):2419–2430, 2009.

[165] Xiaolei Qu, Pedro JJ Alvarez, and Qilin Li. Applications of nanotechnology in water and wastewater treatment. *Water Research*, 47(12):3931–3946, 2013.

[166] Khyle Glainmer Quiton, Bonifacio Doma Jr, Cybelle M Futalan, and Meng-Wei Wan. Removal of chromium (VI) and zinc (II) from aqueous solution using kaolin-supported bacterial biofilms of Gram-negative *E. coli* and Gram-positive *Staphylococcus epidermidis*. *Sustainable Environment Research*, 28(5):206–213, 2018.

[167] Ronak Rahimian, and S. Zarinabadi. A review of studies on the removal of methylene blue dye from industrial wastewater using activated carbon adsorbents made from almond bark. *Progress in Chemical and Biochemical Research*, 3(3):251–268, 2020.

[168] Jothi Ramalingam Rajabathar, Arun K Shukla, Aldalbahi Ali, and Hamad A Al-Lohedan. Silver nanoparticle/r-graphene oxide deposited mesoporous-manganese oxide nanocomposite for pollutant removal and supercapacitor applications. *International Journal of Hydrogen Energy*, 42(24):15679–15688, 2017.

[169] Ganesh Kumar Rajahmundry, Chandrasekhar Garlapati, Ponnusamy Senthil Kumar, Ratna Surya Alwi, and Dai-Viet N Vo. Statistical analysis of adsorption isotherm models and its appropriate selection. *Chemosphere*, 276:130176, 2021.

[170] Nithya Rajarathinam, Thirunavukkarasu Arunachalam, Sivashankar Raja, and Rangabhashiyam. Fenalan Yellow G adsorption using surface-functionalized green nanoceria: An insight into mechanism and statistical modelling. *Environmental Research*, 181:108920, 2020.

[171] Mohamed Nageeb Rashed. Adsorption technique for the removal of organic pollutants from water and wastewater. *Organic Pollutants-Monitoring, Risk and Treatment*, 7:167–194, 2013.

[172] Tim Robinson, Geoff McMullan, Roger Marchant, and Poonam Nigam. Remediation of dyes in textile effluent: A critical review on current treatment technologies with a proposed alternative. *Bioresource Technology*, 77(3):247–255, 2001.

[173] J Romero-González, JC Walton, JR Peralta-Videa, E Rodríguez, J Romero. and JL Gardea. Modeling the adsorption of Cr (III) from aqueous solution onto *Agave lechuguilla* biomass: Study of the advective and dispersive transport. *Environmental Research*, 161(1):360–365, 2009.

[174] M Rožić, Š Cerjan-Stefanović, S Kurajica, V Vančina, and E Hodžić. Ammoniacal nitrogen removal from water by treatment with clays and zeolites. *Water Research*, 34(14):3675–3681, 2000.

[175] Eugenia Rubin, Pilar Rodriguez, Roberto Herrero, Javier Cremades, Ignacio Barbara, and Manuel E Sastre de Vicente. Removal of Methylene Blue from aqueous solutions using as biosorbent *Sargassum muticum*: An invasive macroalga in Europe. *Journal of Chemical Technology & Biotechnology*, 80(3):291–298, 2005.

[176] Yusra Safa, and Haq Nawaz Bhatti. Kinetic and thermodynamic modeling for the removal of Direct Red-31 and Direct Orange-26 dyes from aqueous solutions by rice husk. *Desalination*, 272(1):313–322, 2011.

[177] Bedabrata Saha, Sourav Das, Jiban Saikia, and Gopal Das. Preferential and enhanced adsorption of different dyes on iron oxide nanoparticles: A comparative study. *The Journal of Physical Chemistry C*, 115(16):8024–8033, 2011.

[178] Papita Saha, and S. Datta. Assessment on thermodynamics and kinetics parameters on reduction of methylene blue dye using flyash. *Desalination and Water Treatment*, 12(1–3):219–228, 2009.

[179] Mohamad Amran Mohd Salleh, Dalia Khalid Mahmoud, Wan Azlina Wan Abdul Karim, and Azni Idris. Cationic and anionic dye adsorption by agricultural solid wastes: A comprehensive review. *Desalination*, 280(1):1–13, 2011.

[180] Melvin S. Samuel, and Ramalingam Chidambaram. Hexavalent chromium biosorption studies using *Penicillium griseofulvum* MSR1 a novel isolate from tannery effluent site: Box–Behnken optimization, equilibrium, kinetics and thermodynamic studies. *Journal of the Taiwan Institute of Chemical Engineers*, 49:156–164, 2015.

[181] Slvia CR Santos, Alvaro FM Oliveira, and Rui AR Boaventura. Bentonitic clay as adsorbent for the decolourisation of dyehouse effluents. *Journal of Cleaner Production*, 126:667–676, 2016.

[182] Sílvia CR Santos, Vítor JP Vilar, and Rui AR Boaventura. Waste metal hydroxide sludge as adsorbent for a reactive dye. *Journal of Hazardous Materials*, 153(3):999–1008, 2008.

[183] Kailasam Saranya, Arumugam Sundaramanickam, Sudhanshu Shekhar, Moorthy Meena, Rengasamy Subramaniyan Sathishkumar, and Thangavel Balasubramanian. Biosorption of multi-heavy metals by coral associated phosphate solubilising bacteria *Cronobacter muytjensii* $KSCAS_2$. *Journal of Environmental Management*, 222:396–401, 2018.

[184] Gautam Kumar Sarma, Susmita Sen Gupta, and Krishna G Bhattacharyya. Retracted: Adsorption of Crystal violet on raw and acid-treated montmorillonite, K10, in aqueous suspension. *Journal of Environmental Management*, 171:1–10, 2016.

[185] R Elmoubarki, FZ Mahjoubi, H Tounsadi, J Moustadraf, M Abdennouri, A Zouhri, A El Albani, and N Barka. Adsorption of textile dyes on raw and decanted Moroccan clays: Kinetics, equilibrium and thermodynamics. *Water Resources and Industry*, 9:16–29, 2015.

[186] Colin M Sayers, and Lennert D. The elastic anisotropy of clay minerals. *Geophysics*, 81(5):C193–C203, 2016.

[187] P Senthil Kumar, S Ramalingam, C Senthamarai, M Niranjanaa, P Vijayalakshmi, and S. Sivanesan. Adsorption of dye from aqueous solution by cashew nut shell: Studies on equilibrium isotherm, kinetics and thermodynamics of interactions. *Desalination*, 261(1):52–60, 2010.

[188] Shifa MR Shaikh, Mustafa S Nasser, Ibnelwaleed Hussein, Abdelbaki Benamor, Sagheer A Onaizi, and Hazim Qiblawey. Influence of polyelectrolytes and other polymer complexes on the flocculation and rheological behaviors of clay minerals: A comprehensive review. *Separation and Purification Technology*, 187:137–161, 2017.

[189] Areeb Shehzad, Mohammed JK Bashir, Sumathi Sethupathi, and Jun-Wei Lim. An overview of heavily polluted landfill leachate treatment using food waste as an alternative and renewable source of activated carbon. *Process Safety and Environmental Protection*, 98:309–318, 2015.

[190] A Sheikhhosseini, M Shirvani, and H. Shariatmadari. Competitive sorption of nickel, cadmium, zinc and copper on palygorskite and sepiolite silicate clay minerals. *Geoderma*, 192:249–253, 2013.

[191] Ying Shen, Wenzhe Zhu, Huan Li, Shih-Hsin Ho, Jianfeng Chen, Youping Xie, and Xinguo Shi. Enhancing cadmium bioremediation by a complex of water-hyacinth derived pellets immobilized with *Chlorella* sp. *Bioresource Technology*, 257:157–163, 2018.

[192] Silke Karcher, Anja Kornmüller, and Martin Jekel. Screening of commercial sorbents for the removal of reactive dyes. *Dyes and Pigments*, 51(2–3):111–125, 2001.

[193] Péter Sipos. Searching for optimum adsorption curve for metal sorption on soils: Comparison of various isotherm models fitted by different error functions. *SN Applied Sciences*, 3(3):1–13, 2021.

[194] D. Sivakumar. Biosorption of hexavalent chromium in a tannery industry wastewater using fungi species. *Global Journal of Environmental Science and Management*, 2(2):105–124, 2016.

[195] G Skodras, IR Diamantopoulou, A Zabaniotou, G Stavropoulos, and GP Sakellaropoulos. Enhanced mercury adsorption in activated carbons from biomass materials and waste tires. *Fuel Processing Technology*, 88(8):749–758, 2007.

[196] Varsha Srivastava, and Mika Sillanpää. Synthesis of malachite clay nanocomposite for rapid scavenging of cationic and anionic dyes from synthetic wastewater. *Journal of Environmental Sciences*, 51:97–110, 2017.

[197] Vicki Stone, Bernd Nowack, Anders Baun, Nico van den Brink, Frank von der Kammer, Maria Dusinska, Richard Handy, Steven Hankin, Martin Hassellöv, Erik Joner, and Teresa F Fernandes. Nanomaterials for environmental studies: Classification, reference material issues, and strategies for physico-chemical characterisation. *Science of the Total Environment*, 408(7):1745–1754, 2010.

[198] Mar-Yam Sultana, Abu Khayer Md Muktadirul Bari Chowdhury, Michail K Michailides, Christos S Akratos, Athanasia G Tekerlekopoulou, and Dimitrios V. Integrated Cr(VI) removal using constructed wetlands and composting. *Journal of Hazardous Materials*, 281:106–113, 2015.

[199] Christos Tapeinos. Magnetic nanoparticles and their bioapplications. In *Smart Nanoparticles for Biomedicine*, 131–142. Elsevier, UK, 2018.

[200] J Suresh, R Yuvakkumar, M Sundrarajan, and Sun Ig Hong. Green synthesis of magnesium oxide nanoparticles. *Advanced Materials Research*, 952:141–144, 2014.

[201] Mehdi Taghdiri. Selective adsorption and photocatalytic degradation of dyes using polyoxometalate hybrid supported on magnetic activated carbon nanoparticles under sunlight, visible, and UV irradiation. *International Journal of Photoenergy*, 2017:8575096, 2017.

[202] Naser Tavassoli, Reza Ansari, and Zahra Mosayebzadeh. Synthesis and application of iron oxide/silica gel nanocomposite for removal of sulfur dyes from aqueous solutions. *Archives of Hygiene Sciences*, 6(2):214–220, 2017.

[203] Vaishali Tomar, Surendra Prasad, and Dinesh Kumar. Adsorptive removal of fluoride from aqueous media using citrus limonum (lemon) leaf. *Microchemical Journal*, 112:97–103, 2014.

[204] Ru-Ling Tseng. Mesopore control of high surface area NaOH-activated carbon. *Journal of Colloid and Interface Science*, 303(2):494–502, 2006.

[205] Maciej Tulinski, and Mieczyslaw Jurczyk. Nanomaterials Synthesis Methods, 75–98. 2017.

[206] Fatma Tümsek, and Özlem Avcı. Investigation of kinetics and isotherm models for the Acid Orange-95 adsorption from aqueous solution onto natural minerals. *Journal of Chemical & Engineering Data*, 58(3):551–559, 2013.

[207] Faheem Uddin. Clays, nanoclays, and montmorillonite minerals. *Metallurgical and Materials Transactions A*, 39(12):2804–2814, 2008.

[208] Mohammad Kashif Uddin. A review on the adsorption of heavy metals by clay minerals, with special focus on the past decade. *Chemical Engineering Journal*, 308:438–462, 2017.

[209] EI Unuabonah, KO Adebowale, BI Olu-Owolabi, LZ Yang, and LX Kong. Adsorption of Pb (II) and Cd (II) from aqueous solutions onto sodium tetraborate-modified Kaolinite clay: Equilibrium and thermodynamic studies. *Hydrometallurgy*, 93(1–2):1–9, 2008.

[210] Ayushi Verma, Meenu Agarwal, Shweta Sharma, Monitoring Singh, and Neetu Singh. Competitive removal of cadmium and lead ions from synthetic wastewater using *Kappaphycus striatum*. *Environmental Nanotechnology, Monitoring & Management*, 15:100449, 2021.

[211] VK Verma, and AK Mishra. Kinetic and isotherm modeling of adsorption of dyes onto rice husk carbon. *Global NEST Journal*, 12(2):190–196, 2010.

[212] Vítor JP Vilar, Cidália MS Botelho, and Rui AR Boaventura. Methylene blue adsorption by algal biomass based materials: Biosorbents characterization and process behaviour. *Journal of Hazardous Materials*, 147(1):120–132, 2007.

[213] Rafeah Wahi, Luqman Abdullah Chuah, Thomas Shean Yaw Choong, Zainab Ngaini, and Mohsen Mobarekeh Nourouzi. Oil removal from aqueous state by natural fibrous sorbent: An overview. *Separation and Purification Technology*, 113:51–63, 2013.

[214] En-Jie Wang, Zhu-Yin Sui, Ya-Nan Sun, Zhuang Ma, and Bao-Hang. Effect of porosity parameters and surface chemistry on carbon dioxide adsorption in sulfur-doped porous carbons. *Langmuir*, 34(22):6358–6366, 2018.

[215] Shaobin Wang, Y Boyjoo, A Choueib, and ZH Zhu. Removal of dyes from aqueous solution using fly ash and red mud. *Water Research*, 39(1):129–138, 2005.

[216] Shaobin Wang, Y Boyjoo, and A Choueib. A comparative study of dye removal using fly ash treated by different methods. *Chemosphere*, 60(10):1401–1407, 2005.

[217] Shaobin Wang, and Yuelian Peng. Natural zeolites as effective adsorbents in water and wastewater treatment. *Chemical Engineering Journal*, 156(1):11–24, 2010.

[218] Xiangtao Wang, Yifei Guo, Li Yang, Meihua Han, Jing Zhao, and Xiaoliang Cheng. Nanomaterials as sorbents to remove heavy metal ions in wastewater treatment. *Journal of Environmental and Analytical Toxicology*, 2(7):154–158, 2012.

[219] Chih-Huang Weng, and CP. Huang. Adsorption characteristics of Zn(II) from dilute aqueous solution by fly ash. *Colloids and Surfaces A: Physicochemical and Engineering Aspects*, 247(1):137–143, 2004.

[220] Chih-Huang Weng, and Yi-Fong Pan. Adsorption of a cationic dye (methylene blue) onto spent activated clay. *Journal of Hazardous Materials*, 144(1):355–362, 2007.

[221] Muhammad Saif Ur Rehman, Muhammad Munir, Muhammad Ashfaq, Naim Rashid, Muhammad Faizan Nazar, Muhammad Danish, and Jong-In Han. Adsorption of brilliant green dye from aqueous solution onto red clay. *Chemical Engineering Journal*, 228: 54–62, 2013.

[222] Silvan Wick, Bart Baeyens, Maria Marques Fernandes, and Andreas Voegelin. Thallium adsorption onto illite. *Environmental Science & Technology*, 52(2):571–580, 2017.

[223] KK Wong, CK Lee, KS Low, and MJ Haron. Removal of Cu and Pb by tartaric acid modified rice husk from aqueous solutions. *Chemosphere*, 50(1):23–28, 2003.

[224] YC Wong, YS Szeto, WH Cheung, and G McKay. Adsorption of acid dyes on chitosan—equilibrium isotherm analyses. *Process Biochemistry*, 39(6):695–704, 2004.

[225] Dongbei Wu, Jing Zhao, Ling Zhang, Qingsheng Wu, and Yuhui Yang. Lanthanum adsorption using iron oxide loaded calcium alginate beads. *Hydrometallurgy*, 101(1): 76–83, 2010.

[226] Pingxiao Wu, Qian Zhang, Yaping Dai, Nengwu Zhu, Zhi Dang, Ping Li, Jinhua Wu, and Xiangde Wang. Adsorption of Cu (II), Cd (II) and Cr (III) ions from aqueous solutions on humic acid modified Ca-montmorillonite. *Geoderma*, 164(3–4):215–219, 2011.

[227] Mohammad Hadi Dehghani, Daryoush Sanaei, Imran Ali, and Amit Bhatnagar. Removal of chromium (VI) from aqueous solution using treated waste newspaper as a low-cost adsorbent: Kinetic modeling and isotherm studies. *Journal of Molecular Liquids*, 215:671–679, 2016.

[228] Victor Saldanha Carvalho, Arthur Luiz Baião Dias, Karina Pantoja Rodrigues, Tahmasb Hatami, Lucia Helena Innocentini Mei, Julian Martnez, and Juliane Viganó. Supercritical fluid adsorption of natural extracts: Technical, practical, and theoretical aspects. *Journal of CO_2 Utilization*, 56:101865, 2022.

[229] Anton Kokalj. Corrosion inhibitors: physisorbed or chemisorbed? *Corrosion Science*, 196:109939, 2022.

[230] Yameng Yu, Boya Ma, Xinbang Jiang, Chen Guo, Zhuang Liu, Nan Li, Lichun Wang, Yunzheng Du, Biao Wang, Wenzhong Li, et al. Amphiphilic shell nanomagnetic adsorbents for selective and highly efficient capture of low/-density lipoprotein from hyperlipidaemia serum. *Journal of Materials Chemistry B*, 10, 2022.

[231] Xiujun Tang, Aocheng Cao, Yi Zhang, Xinhua Chen, Baoqiang Hao, Jin Xu, Wensheng Fang, Dongdong Yan, Yuan Li, and Qiuxia Wang. Soil properties affect vapor-phase adsorption to regulate dimethyl disulfide diffusion in soil. *Science of the Total Environment*, 825:154012, 2022.

[232] Saad SM Hassan, Ehab M Abdel Rahman, Gehan M El-Subruiti, Ayman H Kamel, and Hanan M Diab. Removal of uranium/-238, thorium/-232, and potassium/-40 from wastewater via adsorption on multiwalled carbon nanotubes. *ACS Omega*, 7:12342–12353, 2022.

[233] Mona M Naim, Nouf F Al-Harby, Mervette El Batouti, and Mahmoud M Elewa. Macro-reticular ion exchange resins for recovery of direct dyes from spent dyeing and soaping liquors. *Molecules*, 27(5):1593, 2022.

[234] Urjinlkham Ryenchindorj, Qammer Zaib, Agusta Samodra Putra, and Hung-Suck Park. Production and characterization of cost-effective magnetic pine bark biochar and its application to remove tetracycline from water. *Environmental Science and Pollution Research*, 29:62382–62392, 2022.

[235] Jing Xia, Yanxin Gao, and Gang Yu. Tetracycline removal from aqueous solution using zirconium-based metal-organic frameworks (Zr-MOFs) with different pore size and topology: Adsorption isotherm, kinetic and mechanism studies. *Journal of Colloid and Interface Science*, 590:495–505, 2021.

[236] Ondřej Kadlec. The history and present state of Dubinin's theory of adsorption of vapours and gases on microporous solids. *Adsorption Science and Technology*, 19(1): 1–24, 2001.

[237] Sodabeh Ebrahimpoor, Vahid Kiarostami, Morteza Khosravi, Mehran Davallo, and Abdolmohammad Ghaedi. Bees metaheuristic algorithm with the aid of artificial neural networks for optimization of acid red-27 dye adsorption onto novel polypyrrole/ $SrFe_{12}O_{19}$/graphene oxide nanocomposite. *Polymer Bulletin*, 76(12):6529–6553, 2019.

[238] Lisdelys González-Rodrguez, Julio Omar Prieto Garca, Lien Rodrguez-López, Yoan Hidalgo-Rosa, Manuel A Treto-Suaréz, Mixary Garcia Enriquez, and Ángel Mollineda Trujillo. Cassava husk powder as an eco-friendly adsorbent for the removal of nickel (II) ions. In *Proceedings of the 3rd International Conference on BioGeoSciences*, 21–38. Springer, 2022.

[239] S Keerthanan, Chamila Jayasinghe, Nanthi Bolan, Jörg Rinklebe, and Meththika Vithanage. Retention of sulfamethoxazole by cinnamon wood biochar and its efficacy of reducing bioavailability and plant uptake in soil. *Chemosphere*, 297:134073, 2022.

[240] Özge Demir, Asl Gök, and Şah İsmail Kırbaşlar. Optimization of protocatechuic acid adsorption onto weak basic anion exchange resins: kinetic, mass transfer, isotherm, and thermodynamic study. *Biomass Conversion and Biorefinery*, 8:1–17, 2022.

[241] Abdallah Tageldein Mansour, Ahmed E Alprol, Khamael M Abualnaja, Hossam S El-Beltagi, Khaled MA Ramadan, and Mohamed Ashour. Dried brown seaweed's phytoremediation potential for methylene blue dye removal from aquatic environments. *Polymers*, 14(7):1375, 2022.

[242] Salima Tlemsani, Zoubida Taleb, Laurence Piraúlt-Roy, and Safia Taleb. Temperature and pH influence on Diuron adsorption by Algerian Mont-Na Clay. *International Journal of Environmental Analytical Chemistry*, 8:1–18, 2022.

[243] Nilly Ahmed Kawady, Ebrahim Abd El Gawad, and Amal Essam Mubark. Modified grafted nano cellulose based bio-sorbent for uranium (VI) adsorption with kinetics modeling and thermodynamics. *Korean Journal of Chemical Engineering*, 39:408–422, 2022.

[244] Pratibha Attri, Sangeeta Garg, Jatinder Kumar Ratan, and Ardhendu Sekhar Giri. Silver nanoparticles from *Tabernaemontana divaricate* leaf extract: Mechanism of action and bio-application for photo degradation of 4-aminopyridine. *Environmental Science and Pollution Research*, 7:1–20, 2022.

[245] Xiangyi Ma, Zhen Chen, Yawen Sun, Zhihan Cai, Fang Cheng, and Wei Ma. Effect on kinetics and energy distribution of riboflavin adsorption from magnetic nano-carbon composites with adsorbed water layer. *Separation and Purification Technology*, 292: 120995, 2022.

[246] Asma S Al-Wasidi, Ibtisam IS AlZahrani, Hotoun I Thawibaraka, Ahmed M Naglah, Mohamed G El-Desouky, and Mohamed A El-Bindary. Adsorption studies of carbon dioxide and anionic dye on green adsorbent. *Journal of Molecular Structure*, 1250: 131736, 2022.

[247] Tanweer Ahmad, and Mohammed Danish. A review of avocado waste-derived adsorbents: Characterizations, adsorption characteristics, and surface mechanism. *Chemosphere*, 296:134036, 2022.

[248] Francois Brouers, and Francisco Marquez-Montesino. Dubinin isotherms versus the brouers–sotolongo family isotherms: A case study. *Adsorption Science and Technology*, 34(9–10):552–564, 2016.

[249] Yingze Cao, Xiaoyong Zhang, Lei Tao, Kan Li, Zhongxin Xue, Lin Feng, and Yen Wei. Mussel-inspired chemistry and Michael addition reaction for efficient oil/water separation. *ACS Applied Materials & Interfaces*, 5(10):4438–4442, 2013.

[250] Hifsa Khurshid, Muhammad Raza Ul Mustafa, and Mohamed Hasnain Isa. A comprehensive insight on adsorption of polyaromatic hydrocarbons, chemical oxygen demand, pharmaceuticals, and chemical dyes in wastewaters using biowaste carbonaceous adsorbents. *Adsorption Science & Technology*, 2022:9410266, 2022.

[251] Ling Zhang, Weiya Niu, Jie Sun, and Qi Zhou. Efficient removal of Cr (VI) from water by the uniform fiber ball loaded with polypyrrole: Static adsorption, dynamic adsorption and mechanism studies. *Chemosphere*, 248:126102, 2020.

[252] BK Nandi, A Goswami, and MK Purkait. Removal of cationic dyes from aqueous solutions by kaolin: Kinetic and equilibrium studies. *Applied Clay Science*, 42(3):583–590, 2009.

[253] Sami Guiza, François Brouers, and Mohamed Bagane. Fluoride removal from aqueous solution by montmorillonite clay: Kinetics and equilibrium modeling using new generalized fractal equation. *Environmental Technology & Innovation*, 21:101187, 2021.

[254] Manman Wei, Fatma Marrakchi, Chuan Yuan, Xiaoxue Cheng, Ding Jiang, Fatemeh Fazeli Zafar, Yanxia Fu, and Shuang Wang. Adsorption modeling, thermodynamics, and DFT simulation of tetracycline onto mesoporous and high-surface-area NaOH-activated macroalgae carbon. *Journal of Hazardous Materials*, 425:127887, 2022.

[255] A Hashem, A Okeil, M Fikry, A Aly, and CO Aniagor. Isotherm and kinetics parametric studies for aqueous Hg(II) uptake onto *N*-[2-(methylamino) ethyl] ethane-1,2-diaminated acrylic fibre. *Arabian Journal for Science and Engineering*, 46(7):6703–6714, 2021.

[256] VK Chintalpudi, RKSL Kanamarlapudi, UR Mallu, and S Muddada. Isolation, identification, biosorption optimization, characterization, isotherm, kinetic and application of novel bacterium *Chelatococcus* sp. biomass for removal of Pb (II) ions from aqueous solutions. *International Journal of Environmental Science and Technology*, 19:1531–1544, 2022.

[257] AE Kassimi, Youness Achour, ME Himri, My Rachid Laamari, and ME Haddad. High efficiency of natural Safiot Clay to remove industrial dyes from aqueous media: Kinetic, isotherm adsorption and thermodynamic studies. *Biointerface Research in Applied Chemistry*, 11:12717–12731, 2021.

[258] A Hashem, CO Aniagor, Ghada M Taha, and M Fikry. Utilization of low-cost sugarcane waste for the adsorption of aqueous Pb (II): Kinetics and isotherm studies. *Current Research in Green and Sustainable Chemistry*, 4:100056, 2021.

[259] Jasmina Sulejmanović, Neira Kovač, Mustafa Memić, Elma Šabanović, Sabina Begić, and Farooq Sher. Selective removal of lead ions from aqueous solutions using SiO_2–MoO_3: Isotherm, kinetics and thermodynamic studies. *Case Studies in Chemical and Environmental Engineering*, 3:100083, 2021.

[260] Rupa Chakraborty, Renu Verma, Anupama Asthana, S Sree Vidya, and Ajaya Kumar Singh. Adsorption of hazardous chromium (VI) ions from aqueous solutions using modified sawdust: Kinetics, isotherm and thermodynamic modelling. *International Journal of Environmental Analytical Chemistry*, 101(7):911–928, 2021.

[261] Zahra Ashouri Mehranjani, Majid Hayati-Ashtiani, and Mehran Rezaei. Isotherm and selectivity studies of Ni (II) removal using natural and acid-activated nanobentonites. *Water Science and Technology*, 84:2394–2405, 2021.

[262] Daniela Roxana Popovici, Mihaela Neagu, Cristina Maria Dusescu-Vasile, Dorin Bombos, Sonia Mihai, and Elena-Emilia Oprescu. Adsorption of *p*-nitrophenol onto activated carbon prepared from fir sawdust: Isotherm studies and error analysis. *Reaction Kinetics, Mechanisms and Catalysis*, 133:483–500, 2021.

[263] Victor O Shikuku, and Trilochan Mishra. Adsorption isotherm modeling for methylene blue removal onto magnetic kaolinite clay: A comparison of two-parameter isotherms. *Applied Water Science*, 11(6):1–9, 2021.

[264] Rumman Zaidi, Saif Ullah Khan, IH Farooqi, and Ameer Azam. Investigation of kinetics and adsorption isotherm for fluoride removal from aqueous solutions using mesoporous cerium–aluminum binary oxide nanomaterials. *RSC Advances*, 11(46): 28744–28760, 2021.

[265] Hanadi K Ibrahim, Mahmood A Albo Hay Allah, and A Muneer. Adsorption of titan yellow using walnut husks: Thermodynamics, kinetics and isotherm studies. *Annals of the Romanian Society for Cell Biology*, 25(6):12576–12587, 2021.

[266] H Alrobei, MK Prashanth, CR Manjunatha, CB Pradeep Kumar, CP Chitrabanu, Prasanna D Shivaramu, K Yogesh Kumar, and MS Raghu. Adsorption of anionic dye on eco-friendly synthesised reduced graphene oxide anchored with lanthanum aluminate: Isotherms, kinetics and statistical error analysis. *Ceramics International*, 47(7):10322–10331, 2021.

[267] Kuo-Hui Wu, Wen-Chien Huang, Wei-Che Hung, and Chih-Wei Tsai. Modified expanded graphite/Fe_3O_4 composite as an adsorbent of methylene blue: Adsorption kinetics and isotherms. *Materials Science and Engineering: B*, 266:115068, 2021.

[268] Abimbola Oluyomi Araoye, Oluwatobi Samuel Agboola, and Olugbenga Solomon Bello. Insights into chemically modified cocoa pods for enhanced removal of an anti-malaria drug. *Chemical Data Collections*, 36:100775, 2021.

[269] Arvind Singh, Neha Srivastava, Maulin Shah, Abeer Hashem, Elsayed Fathi Abd Allah, and Dan Bahadur Pal. Investigation on hexavalent chromium removal from simulated wastewater using royal poinciana pods-derived bioadsorbent. *Biomass Conversion and Biorefinery*, 15:1–12, 2021.

[270] Shirin Shabani, and Mohammad Dinari. Cu-Ca-Al-layered double hydroxide modified by itaconic acid as an adsorbent for anionic dye removal: Kinetic and isotherm study. *Inorganic Chemistry Communications*, 133:108914, 2021.

[271] Akash Sitaram Jadhav and Madhukar Vinayak Jadhav. Utilization of black mustard husk ash for adsorption of fluoride from water. *Korean Journal of Chemical Engineering*, 38(10):2082–2090, 2021.

[272] Imane Toumi, Halima Djelad, Faiza Chouli, and A. Benyoucef. Synthesis of PANI@ ZnO hybrid material and evaluations in adsorption of Congo red and Methylene blue dyes: Structural characterization and adsorption performance. *Journal of Inorganic and Organometallic Polymers and Materials*, 32:112–121, 2022.

[273] Guangyu Wu, Qi Liu, Jingyi Wang, Siye Xia, Hualong Wu, Jiaxiang Zong, Jiangang Han, and Weinan Xing. Facile fabrication of rape straw biomass fiber/ β-Cd/Fe_3O_4 as adsorbent for effective removal of ibuprofen. *Industrial Crops and Products*, 173:114150, 2021.

[274] Diogo PS Silva, Alef T Santos, Thas RS Ribeiro, Julyane RS Solano, Roberta KBC Cavalcanti, Bruno JB Silva, Paulo HL Quintela, and Antonio OS Silva. Monosodium glutamate-mediated hierarchical porous formation in LTA zeolite to enhance CO_2 adsorption performance. *Journal of Sol-Gel Science and Technology*, 100:360–372, 2021.

[275] MT Amin, AA Alazba, and M Shafiq. Nanofibrous membrane of polyacrylonitrile with efficient adsorption capacity for cadmium ions from aqueous solution: Isotherm and kinetic studies. *Current Applied Physics*, 40:101–109, 2021.

[276] Chukwunonso O Aniagor, MA Afifi, and A Hashem. Rapid and efficient uptake of aqueous lead pollutant using starch-based superabsorbent hydrogel. *Polymer Bulletin*, 79:6373–6388, 2022.

[277] Mohamed R Hassan, Sobhy M Yakout, Ahmed A Abdeltawab, and Mohamed I Aly. Ultrasound facilitates and improves removal of triphenylmethane (crystal violet) dye from aqueous solution by activated charcoal: A kinetic study. *Journal of Saudi Chemical Society*, 25(6):101231, 2021.

5 Potential of Algae in the Phyco-Remediation of Industrial Wastewater and Valorization of Produced Biomass

Sayed Rashad, Imran Ahmad, and Ghadir A. El-Chaghaby

5.1 INTRODUCTION

Industrial wastewater treatment is a global environmental challenge, and many treatment methods are being adopted. Among the industrial wastewater treatment techniques, "Phyco-remediation" is rising as an environmentally friendly way that employs algae for wastewater remediation. Algae easily multiply in wastewater by utilizing micronutrients already present in it. Algae have been used to reduce toxic and recalcitrant pollutants in effluents. Toxins can be removed by microalgae in a variety of ways, including biosorption, bioaccumulation, and biodegradation. Algae have been discovered to be capable of eliminating toxins from a variety of sources, including household effluents, agricultural run-off, textile, leather, pharmaceutical, and electroplating industries (Xin et al. 2021). Algae use micronutrients in effluents to help them absorb carbon dioxide during their growth phase. In this context, microalgae are thought to be a possible option for the elimination of harmful compounds because of their numerous means of nutrition, including heterotrophy, mixotrophy, and phototrophy. By accumulating, adsorbing, and metabolizing these hazardous components, microalgae transform them to a significant degree (Mustafa et al. 2021). The purpose of this chapter is to give an overview of algae and their uses in wastewater treatment. The role of algae in industrial wastewater treatment is emphasized, and the possible applications of the resulting algal biomass are also highlighted.

5.2 ALGAE CULTIVATION FOR BIOMASS PRODUCTION

In aquatic habitats, algae are the major producers. Algae are a diverse collection of organisms that play crucial roles in ecosystems, and their biomass is a valuable biological resource. Algae are photosynthetic organisms that can be found in a variety of aquatic environments such as lakes, ponds, rivers, oceans, and even wastewater.

DOI: 10.1201/9781003354475-5

They can grow alone or in association with other species and can endure a wide range of temperatures, salinities, and pH values; varying light intensities; and environments in reservoirs or deserts (Khan et al. 2018). Algal biomass can thrive in seawater, saline-alkali land, and even poor-quality water settings due to its rapid growth, adaptability, and high consumption rate and is occasionally used as sewage purifiers (Zhang et al. 2021).

Algae either macroalgae or microalgae can be cultivated through different environments, natural or aqua-cultured. Aqua-cultured and wild seaweed are the two kinds of macroalgae production. Aqua-cultured seaweeds are farmed either on land (in ponds or tanks) or in the sea. Pond cultivation, like in the case of microalgae farming, is seen as a feasible means of expansion. It entails establishing a pilot plant, which includes a hatchery and a growing area. Filtration of seawater and air supply, suitable lighting, chiller units, tanks, and storage are all necessities for the hatchery (Campbell et al. 2019).

On the other hand, microalgae cultivation systems are divided into two categories: open and closed. Algae are exposed to nature in open cultivation systems (Jerney and Spilling 2018). A good example is an open-pond cultivation. A pond of adequate depth and exposure to sun radiation has been established here. The water-carrying nitrogen and phosphorus are channelled to supply nutrients. Temperature, land, and algae strains are all taken into account when designing the pond (Khan et al. 2018). Closed cultivation techniques, on the other hand, work with carefully controlled conditions that promote algae development. Photobioreactors (PBRs) are extremely efficient, producing high biomass content while occupying smaller spaces and requiring lower operating costs (Bošnjaković and Sinaga 2020).

The conditions under which algae are grown have a big impact on their biomass production and structure. Among these conditions, the medium composition represents a key factor in algae cultivation. The inclusion of all critical elements in the algal growth media is required for optimal biomass development. The most important nutrients for algae growth are carbon, nitrogen, and phosphorus (Panahi et al. 2019). The pH of the medium is another important factor that influences algal composition and growth. High pH inhibits algal development, most likely due to inorganic carbon (HCO_3 and CO_2) restrictions, resulting in decreased organic cell carbon content due to plentiful carbonate ions (CO_3^{2-}) (Raven et al. 2020). Furthermore, excessive energy consumption for internal acid-base balance reduces the amount of available energy for ATP synthesis, resulting in a decrease in development. pH changes have a negative impact on intracellular metabolic enzymes as well (Liao et al. 2019).

5.3 ROLE OF ALGAE IN THE REMOVAL OF HEAVY METALS AND EMERGING CONTAMINANTS

Algae may provide a natural alternative for the remediation of an industrial effluent containing heavy metals. Algal-based remediation is known as phyco-remediation, and it not only solves the problems that traditional treatment methods have, but it's also a cost-effective and environmentally beneficial treatment alternative (Bwapwa et al. 2017). The use of algae to adsorb heavy metals is an environmentally friendly way to clean contaminated water that has sparked a lot of interest in the scientific community (Lin et al. 2020). Biosorption is also used by the algal population to remove heavy metals from contaminated water. Algal cell walls naturally contain a variety of chemical groups, including amino, carboxy, imidazole, phosphate, phenolic, thioether, and sulfhydryl moieties, which selectively attach heavy metals to them. To remove heavy metals from effluents, algae can act as anionic and cationic exchangers, chelating agents, pH-based precipitating agents, and form complexes via electrostatic or covalent interactions (Giarikos et al. 2021; Rebello et al. 2021). Apart from living algae, preparations of brown algae containing bioactive polysaccharides and their butanedioic anhydride derivatives have been reported to be useful in heavy metal bioadsorption. There has been a brief examination of the many features of algal heavy metal clean-up (Rebello et al. 2021). As part of phyco-remediation, several researchers are investigating biological mechanisms to remove heavy metals. (Mohammed Danouche et al. 2021). Phyco-remediation is the process of using algae's inherent ability to absorb nutrients, collect heavy metals, and break down organic contaminants through a symbiotic relationship with aerobic bacteria. Algae have pigments that are similar to those of higher plants with higher photosynthetic efficiency; as a result, algae released more oxygen into the aquatic system and induced aerobic breakdown of organic molecules (Dayana Priyadharshini et al. 2021). Algae have been discovered to have the potential to use waste as a nutritional source while also reducing contaminants via metabolic and enzymatic activities. Through algae, metabolic pathways, xenobiotics, and heavy metal pollutants can be detoxified, converted, and volatilized. Biosorption is the most common process for heavy metal uptake by active algal biomass, or algae (Ahmad, Pandey, and Pathak 2020). Biosorption, unlike ion-exchange resins, requires several processes such as chelation, partial adsorption, complexation, microprecipitation, and so on. It is solely dependent on the interaction between the biomass and the metal ion; yet it is similar to the binding of metal ions via ion-exchange resins (Aryal 2021). The biosurf method employing algae has several advantages, including lower capital and operating costs, true destruction of organic and inorganic pollutants, oxidation of a wide range of organic compounds, and removal of reduced inorganic compounds (Almomani and Bohsale 2021). In recent years, research into nitrogen and phosphorus absorption and transformation by the symbiotic system of algae and fine bacteria has attracted a lot of interest (Mao et al. 2021). Table 5.1 shows some of the algae species that have been used to remove heavy metals and emerging contaminants.

5.4 REACTOR CONFIGURATIONS FOR WASTEWATER TREATMENT USING ALGAE

Various reactor set-ups have been described to date in order to maximize the development of microalgae in contact with wastewater. Their objective is to obtain high pollutant removal yields while accommodating huge amounts of wastewater while ensuring optimal microalgae productivity. In this respect, two types of reactor configurations were proposed, namely "suspended" and "immobilized" (Plöhn et al. 2021). Suspended cultures allow tiny algae cells to move freely throughout a body of water in dilute quantities. On the other hand, the immobilization technique creates a fixed system in which live cells develop in a limited space and are unable to move freely in the system's aqueous environment. Both types of reactors are

TABLE 5.1
Some Algae Species Used for Removal of Heavy Metals and Emerging Contaminants

Algae	Source of the Wastewater	Pollutant	Reference
Spirulina platensis	Wastewater	Cu^{2+}	(Anastopoulos and Kyzas 2015)
Pterocladia capillacea	Wastewater	Cr^{3+}	(El Nemr et al. 2015)
Chlorella pyrenoidosa	Wastewater	Nitrate, phosphate, and chemical oxygen demand (COD)	(Kumari et al. 2020)
Chlorella pyrenoidosa	Wastewater	Phosphate, nitrate, and biological oxygen demand (BOD)	(Pathak et al. 2015)
Microalgal	Wastewater	Phosphate, ammoniacal nitrogen, and COD	(Gatamaneni Loganathan et al. 2020)
Chlorella vulgaris	Aqueous medium	Ethidium bromide	(de Almeida et al. 2020)
Botryococcus sp.	Domestic	Total phosphorus (TP), Total nitrogen (TN), and total organic carbon (TOC)	(Gani et al. 2016)
Diatom	Wastewater	Copper, cadmium, chromium, and lead	(Marella et al. 2020)
Aphanothece sp., *Chlorella vulgaris, C. ellipsoidea, C. sorokiniana, C. pyrenoidosa, Scenedesmus dimorphus, S. obliquus,* and *Chlamydomonas reinhardtii*	Wastewater	Copper (Cu(II)), chromium (Cr(VI)), lead (Pb(II)), and cadmium (Cd(II))	(Danouche et al. 2020)
Cladophora fracta	Mine gallery waters	Au and Ag	(Topal et al. 2020)
Chlorella minutissima	Wastewater	Total dissolved solids (TDS), N, P, K, BOD, and COD	(Malla et al. 2015)
Ulva prolifera	Wastewater	Bisphenol A	(Zhang et al. 2019)
Scenedesmus sp.	Tannery wastewater	(Cr-Cu- Pb-Zn)	(Ajayan et al. 2015)
Chlorella pyrenoidosa	Wastewater	As(III) and As(V)	(Podder and Majumder 2015)
Nostoc muscorum, Anabaena oryzae, and *Spirulina platensis*	Olive mill wastewater	Phenolic	(Mostafa et al. 2019)

subdivided into the open or closed system (Figure 5.1) (Whitton et al. 2015). The advantages and limitations of suspended and immobilized systems for wastewater treatment using algae have been documented by several authors. According to Lahin, Sarbatly, and Suali (2016), the advantages of immobilization of microalgae cells include high biomass concentrations, the avoidance of any additional filtration

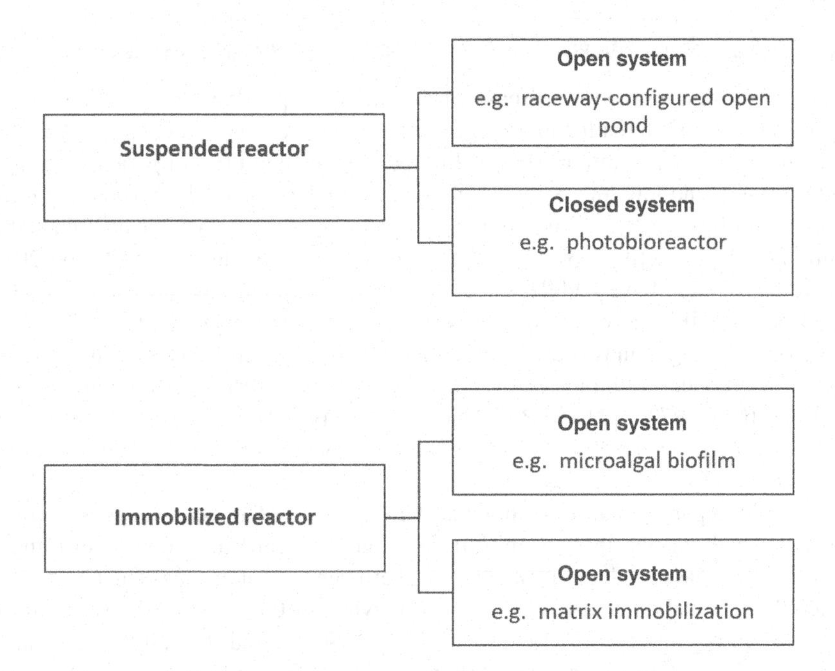

FIGURE 5.1 Different reactor set-ups.

procedure, the potential to entrap more than one microbe, and the fact that it is a simple technique that can be performed even by non-professionals. However, the key challenges in microalgae immobilization are the system's efficacy in removing contaminants, the cost of ingredients, and the cost of the immobilization process itself. Also, Christenson and Sims (2011) determined that an algae immobilized system, as opposed to suspended cultures, can better combine production, harvesting, and dewatering activities, thereby leading to a more streamlined process with lower downstream processing costs. It could be concluded that the suspended reactor type may be more suitable for small scale wastewater treatment such as laboratory experiments, whereas the immobilized culture systems are more effective for large-scale applications and offer better biomass harvesting.

5.4.1 Photobioreactors Used in the Bioremediation of Wastewater

PBRs are divided into open and closed systems, each with its own set of benefits and drawbacks. Open ponds, raceway ponds, scrubbers, and tanks are examples of open systems, while closed systems include tubular PBR (bubble and airlift mechanism) and flat plate PBRs (Ahmad et al. 2021). Microalgae's suitability for wastewater treatment necessitates further technological advancements in PBRs. Microalgae systems for wastewater treatment can be divided into two types: (i) suspended microalgae systems and (ii) immobilized microalgae systems (Wollmann et al. 2019).

5.4.2 Suspended Microalgae Systems for Wastewater Treatment

Because of their low cost and ease of construction, pond systems are extensively utilized reactor systems for microalgae production and wastewater treatment. They do, however, have limitations such as limited light availability, temperature changes, poor mixing, and low microalgal biomass output. Because of the carbon and nitrogen and phosphorus C:N:P imbalance, CO_2 supply is not homogeneous and stable, limiting the productivity of microalgal biomass (Park, Craggs, and Shilton 2011). However, as they have paddle wheel stirrers and enough gas intrusion, high-rate algal ponds (HRAPs) can help overcome some of these constraints by providing improved mixing. Improved aeration and CO_2 supply can boost biomass productivity and reduce pollutant removal rates. Microalgae suspended systems remove between 10 and 97% of N and P, depending on the type of microalgae, culture mode, wastewater characterization, tank size, and process parameters (El Hafiane and El Hamouri 2005).

When compared to an open-pond system, closed systems offer superior options for efficient light distribution and mixing (bubble or airlift systems), resulting in increased biomass productivity and removal efficiency. Matamoros et al. (2015) evaluated the removal of various electric conductivities (ECs) using a pilot scale HRAP fed with an organic loading rate of 7–29 g COD/m^3/d, and the results were classified as (1) Class-A having removal >90 percent (caffeine and ibuprofen); (2) Class-B having removal 60–90 percent (naproxen, bisphenol A, and tributyl phosphate); and (3) Class-C having (methyl paraben, 2,4-D).

5.4.3 Immobilized Microalgae Systems for Wastewater Treatment

Microalgae operate as filters for N and P, heavy metals (HMs), and ECs in wastewater. Because biomass concentrations are sometimes quite low, harvesting and downstream processing can be costly. Immobilization of microalgal cells has been proposed in this context as a way to avoid the harvest problem while keeping the high-value algal biomass for subsequent processing (Mallick 2020).

5.4.4 Microalgae Turf Scrubber

It treats urban and agricultural wastewater with periphyton, a community of microalgae, cyanobacteria, and other microbes. A substrate liner is given for periphyton on a raceway with a mild slope, according to the mechanism. The wastewater is then pumped through a growing biomass, where contaminants are digested or filtered.

Fixed bed systems for microalgal immobilization use stationary metrics (porous matrix or fibres). High surface to volume ratios, hydrophobicity, and the surface material of the matrix utilized will influence the removal of contaminants present in the wastewater stream and are critical in the growth, stable creation, and adhesion strength of the biofilm. Using microalgae and cyanobacteria on a horizontal flat panel PBR, Sukačová, Trtílek, and Rataj (2015) reported removal rates of around 92%.

5.4.5 Fluidized Bed Systems

Microalgal biomass is immobilized on a floating substratum, which increases the surface-to-volume ratios and improves light distribution by boosting mixing capacity. To hold the microalgal biomass, they usually use carrageen beads, chitosan, or alginate. The cells infiltrate and proliferate inside the porous matrix of the beads.

Fluidized bed reactors can be combined with other PBRs (bubble column and stirred tank reactors) to achieve synergistic growth and treatment effects. The rate of growth and removal cannot be compared to fixed bed systems because it is dependent on the kind and material of immobilization matrix, microalgae type, and wastewater parameters. Fixed bed systems and carrageen-immobilized chlorella vulgaris cells were found to remove 82–100 percent of contaminants from wastewater in studies (Lau, Tam, and Wong 1998).

Leading companies from Australia, the United States, and the United Kingdom have been working diligently for the past few years to create algal biomass from various wastewater sources. Microbio-(US) Engineering's RNEW technology used CO_2-aerated and mechanically agitated raceway ponds to treat municipal wastewater having high levels of N and P while also creating algal biomass, which was then used to make biofuels (Fon Sing et al. 2013). Oswald Technologies (US) invented Advanced Integrated Wastewater Pond Systems (AIWPS), which is based on a symbiotic consortium of bacteria and algae and is used to remove inorganic and organic contaminants from agricultural, municipal, and industrial wastewater (Green, Lundquist, and Oswald 1995).

Algae Systems (USA) developed a low-cost offshore PBR (floating system) that operates on CO_2 and natural light to absorb nutrients and pollutants from the original source. The efficiency was rather high, since it cleansed roughly 50,000 gallons of city wastewater per day, while the algal biomass generated is hydrothermally liquefied (HTL) and transported ashore to produce biofuels and biofertilizers (Green et al. 2003).

Immobilized microalgae PBR systems are another alternative. The Algae Wheel system, a cutting-edge algal-fixed film technology, was created by One Water. The Algae Wheels biofilm ecology contains a wide collection of bacteria and algae, and their combined synergetic impact improves the system's removal effectiveness. The microalgae absorb sunlight to repair CO_2, which is produced by bacteria. Photosynthesis produces polysaccharides, which bacteria use as a source of nutrition.

The bacteria can then utilize photosynthetically produced oxygen, resulting in a self-regulating, stable, and environmentally sustainable wastewater treatment (WWT) system (Johnson et al. 2018). Gross-Wen Technologies' revolving algal biofilm (RAB) technology consists of an algae biofilm attached to vertically oriented rotating conveyor belts. The connected microalgae fix N and P from the nutrient-rich liquid while undergoing photoautotrophic growth in the gaseous phase. The algal biomass of the RAB system may be scraped off the surface with ease, saving money on harvesting procedures. The RAB system enhanced algal biomass productivity while effectively removing contaminants such as sulphate (46%) TP (80%) and TN (87%) (Gross et al. 2013). Table 5.2 gives the details of some of the companies involved in the bioremediation of wastewaters.

TABLE 5.2
Companies Involved in the Bioremediation of Wastewaters Using Microalgae

Technologies/Companies Using Microalgae for Wastewater Treatment	Mechanism/PBR	Pollutants/ Wastewater	Utilization of Biomass Produced	Ref.
RNEW® Technology (Microbio-Engineering), the USA	Mechanically mixed, CO_2 gassed raceway ponds	N- and P-rich municipal wastewater	Biofuels	(Craggs, Lundquist, and Benemann 2013)
Advanced Integrated Wastewater Pond System (AIWPSR) or *Energy Ponds*™ (Oswald Green Technologies), the USA	Employing bacterial algal consortium	Organic/inorganic pollutants from agricultural and industrial wastewaters	Biofertilizer Bioplastics Animal feed	(Green, Lundquist, and Oswald 1995)
Algae Systems, the USA	Offshore floating PBR using natural light and CO_2 conditions	50,000 gallons/d municipal wastewaters (75% TN, 93% TP, and 93% BOD)	Biofuels Biofertilizers	(Green et al. 2003)
Algae Enterprises, Australia	Closed PBR system	Municipal, agricultural, and industrial	Methane electricity	(Montingelli, Tedesco, and Olabi 2015)
Algal Turf Scrubber (ATS) by *Hydro Mentia*	Immobilized system	Aquaculture wastewater and river water treatment	Biofertilizer and biofuels	(Kangas et al. 2017)
Algae Wheel® system by *One Water*	Immobilized system	Different sources of wastewater	Biofuels	(Johnson et al. 2018)
RAB system of Gross-Wen Technologies	Immobilized system	Sulphate-contained mining wastewater	Bioplastics	(Zhao et al. 2018)

5.5 ALGAE-BACTERIA INTERACTION FOR WASTEWATER BIOREMEDIATION

Algae have piqued the interest of academics in recent years as a potential alternative feedstock. Algae cultivation in wastewaters to bioremediate nutrient-rich effluent while maintaining a high biomass output is a more cost-effective and environmentally friendly solution (Khoo et al. 2021). The inclusion of algal-bacterial interaction exposes the symbiotic link between microorganisms, with algae serving as primary producers of organic compounds from CO_2 and heterotrophic bacteria serving as secondary consumers, digesting the organic compounds created by algae (Yong et al. 2021). Many recent research has found that algal bacterial symbiosis has a favourable impact on algal growth, flocculation processes, spore germination, pathogen resistance, and morphogenesis, all of which are important factors in achieving long-term algae biotechnology (Lian et al. 2018).

One of the green technologies that have been successfully used to eliminate hazardous contaminants from the environment is the algae-bacteria consortium for wastewater treatment. As a result of interactions between algae and bacteria, significant progress is being made in the field of vitality creation (Morais et al. 2021).

Algal-microbe interactions in wastewater and the utilization of consortia for increased nutrient recovery and biomass production have recently become hot topics (Fallahi et al. 2021). At this time, research into the use of algae and bacteria for wastewater or effluent treatment has progressed, with the focus now being on coupling algae and bacteria with the development of monetarily viable inventive bioproducts, influencing the circular economy, and bringing a multipurpose arrangement to fruition. The bioeconomy and the fabrication of unusual bio-products are currently the focus of financing opportunities, which will aid in the growth of this exploration field (Fallahi et al. 2021; Mu et al. 2021).

The algal-bacterial symbiosis has been proven to have significant effects on the treatment of numerous pollutants in the environment (Ji 2021). Aside from aiding microalgae formation, related bacteria can also assist the algae in performing more sophisticated tasks with a wide range of uses. In wastewater treatment, for example, algae and bacteria collaborate to remove organic and inorganic waste as well as harmful compounds more quickly and efficiently (Qi et al. 2021; Arora et al. 2021). Bacterial and viral infections, in turn, can damage or destroy the algal cell wall, which is a necessary step in algal-based chemical extraction and might potentially be used to combat toxic algae blooms early on (Dayana Priyadharshini et al. 2021). The algal-bacterial consortium system was chosen by researchers because of its unique characteristics, such as low power usage and biomass definability. Its principle is to improve the effect of heterotrophic bacteria removing organic pollutants from wastewater and algae removing N and P by utilizing the link between algal bacteria and community structure (Sánchez-zurano et al. 2021; Qi et al. 2021) Aeration benefits COD-degrading bacteria by providing them with greater oxygen. For bacteria to remove COD, the oxygen provided by algal photosynthesis is clearly insufficient. As a result, bacteria in the COD elimination efficiency group have a dominant position in the population. High algae/sludge ratios, for example, form algal-dominated algal-bacterial consortia that remove nitrogen and phosphorus by algae absorption (Qi et al. 2021).

The advantages of algal-bacterial systems were tested using various wastewaters (Qi et al. 2021). Immobilized algal-bacterial cells were used to treat municipal wastewater, marine microalgal-bacterial consortia were used in fixed bed PBRs for single-stage saline wastewater treatment, and domestic wastewater sludges were treated with algal-bacterial biofilm bioreactors. Chemicals are removed from pharmaceutical wastewater by algal-bacterial consortia by bioaccumulation or biodegradation, whereas heavy metals are removed from metal-containing wastewater through biosorption, bioconvention, and bioaccumulation (Saravanan et al. 2021). Table 5.3 illustrates some algal-bacterial symbiotic interactions that are used in wastewater treatment to remove nutrients.

TABLE 5.3
Some Algal-Bacterial Interactions Employed in Wastewater Treatment

Algae	Bacteria	Source of the Wastewater	Reference
Selenastrum bibraianum	Activated sludge bacteria	Printing and dyeing industrial wastewater	(Lin et al. 2019)
C. vulgaris	*B. licheniformis*	Synthetic wastewater	(Ji et al. 2018)
Navicula sp.	Comamonadaceae and Nitrosomonadaceae	Synthetic wastewater	(Meng et al. 2019)
C. vulgaris	*Pseudomonas putida*	Synthetic municipal wastewater	(Mujtaba et al. 2017)
Scenedesmus acuminatus	Filamentous bacteria	Milk whey processing wastewater	(Marazzi et al. 2020)
C. vulgaris	Activated sludge bacteria	Municipal and industrial wastewater	(Sepehri et al. 2020)
C. vulgaris	*B. licheniformis*	Municipal wastewater	(Ji et al. 2019)
Chlorella sp.	*Bacillus firmus* and *Beijerinckia fluminensis*	Sludge wastewater	(Huang et al. 2020)
C. sorokiniana	Activated sludge bacteria	Domestic wastewater	(Fan et al. 2020)
Chlorella sp.	*Beijerinckia fluminensis*	Vinegar production wastewater	(Huo et al. 2020)
Chlorella sp., *Nannochloropsis* sp., *Scenedesmus bijugatus*, *Chlamydomonas reinhardtii*, and *Oscillatoria*	Bacteria from activated sludge in a municipal wastewater treatment plant	Municipal wastewater	(Sharma et al. 2020)
Dunaliella sp., *Chlorella* sp., *Scenedesmus* sp., *Nannochloropsis* sp., and *Neochloris* sp.	Wastewater bacteria	Tannery wastewater	(Kumar et al. 2022)
Scenedesmus sp.	Activated sludge bacteria	Sewage wastewater	(Lei et al. 2018)
Ulva lactuca	Gram-negative bacteria	Industrial wastewater	(Aly and Amasha 2019)

5.6 ALGAL WASTEWATER TREATMENT: CONCEPT OF CIRCULAR BIOECONOMY

Circular economy is the optimized utilization of resources in a close loop, thus opposing the linear way of utilization in petroleum-based economy (Ahmad, Abdullah, Iwamoto, et al. 2021). Microalgae-based circular bioeconomy instigates the complete utilization of cultivated biomass by establishing a balance between sustainability and ecology (Nagarajan et al. 2020). The most dominant example is the bioremediation

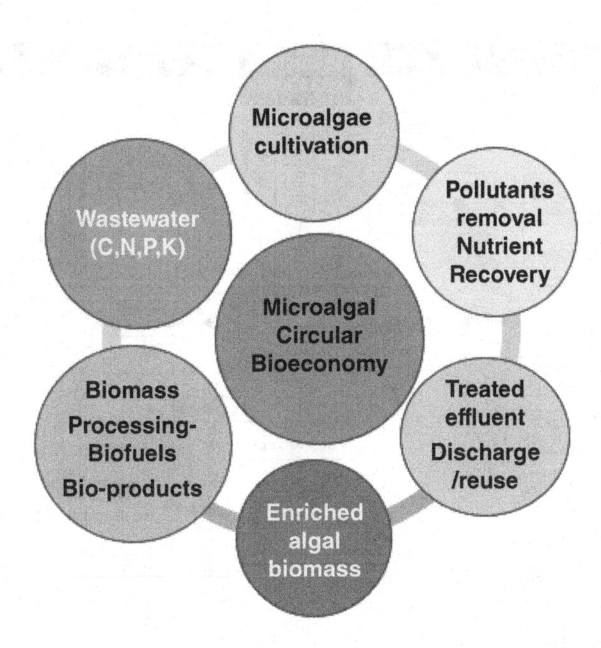

FIGURE 5.2 Concept of circular bioeconomy.

of wastewater using microalgae as it serves two purposes, i.e. (i) nutrient uptake and pollutant removal and (ii) production of enriched biomass (Ahmad, Abdullah, Koji, et al. 2021). The closed loop concept of microalgae-based bioremediation of wastewater is shown in Figure 5.2.

5.7 ALGAL BIOREFINERY CONTRIBUTING TO BIOECONOMY

The biorefinery concept includes the objective of attaining biofuels and other biocompounds from the cultivated biomass. The generation of biofuels and value-added compounds in bulk can contribute to the overall process economically viable (Bhattacharya and Goswami 2020). The production of various biofuels and bioproducts after the harvesting of cultivated algal biomass is shown as a flow diagram in Figure 5.3. Various applications of microalgae via biorefinery are elaborated in the subsections.

5.8 END-USE APPLICATIONS OF MICROALGAE CULTIVATED IN WASTEWATERS

Nowadays, several opportunities are appearing for the use of microalgal biomass grown on wastewater in several areas. Effluents of (dyes, pharmaceutical residues, pesticides, metal processing, and others) are examples of industrial effluents that include a wealth of value-added goods, making their recovery vital (Ali et al. 2021). Algae-based nutrient removal processes aid wastewater treatment by recycling nutrients, reducing environmental footprints associated with conventional wastewater

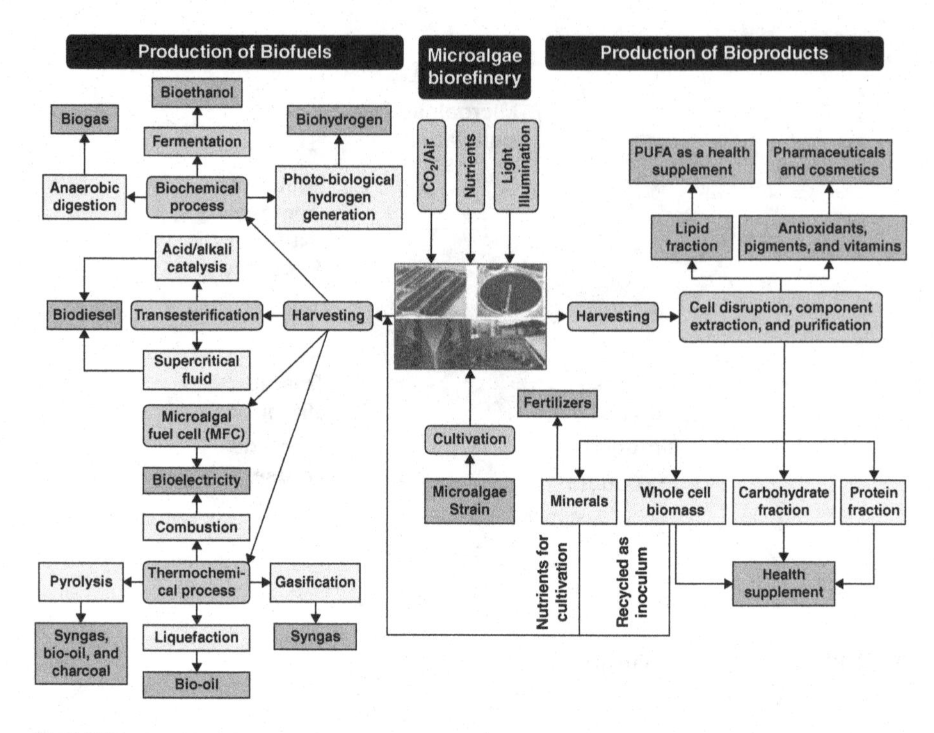

FIGURE 5.3 Biorefinery concept in microalgae cultivation to biomass valorization.

treatment, preventing eutrophication, sequestering CO_2, and producing useful biomass, which is a storehouse of energy-rich compounds with applications in biocrude, biofuels, biodiesel, biomethane, biochar, biopolymers, exopolysaccharides, biofertilizers, aquaculture, and agriculture (Roles et al. 2020; Renuka et al. 2021).

Bioenergy generation from microalgae biomass feedstock in the form of biodiesel, biomethane, biochar, and other products has piqued researchers' interest over the last decade. Microalgae biofuels have the potential to increase energy security and reduce major environmental concerns associated with the use of fossil fuels if they are used successfully in transportation and power generation (Awogbemi et al. 2021). There have been numerous investigations on the manufacture of biofuels from wastewater-grown biomass. Among the various biofuels, biodiesel production combined with wastewater treatment has proven to be a potential method for lowering production costs. Combining wastewater treatment with algal biomass processing could be one of the most cost-effective and ecologically friendly strategies to produce bioenergy and bio-products (Xu et al. 2020). Under natural stress conditions of wastewater, research studies on the use of wastewater to produce microalgae biomass for biodiesel production have revealed that wastewater provides a cheap source of nutrients and can increase the accumulation of lipids; however, this can vary depending on the cultivation conditions (Ge et al. 2018; Do et al. 2019).

The development of liquid biofuel technology is part of the broader scientific and technological research and development programme of the industrial world. Alternative liquid biofuels made from microalgae are without a doubt the driving force behind the surge in research activity that has occurred since the turn of the century. Algae have a higher tendency of producing lipids when they are grown in harsh stress conditions, and these lipids are later converted to biodiesel (Naruka et al. 2019).

Microalgal biomass obtained from wastewater treatment can be used to manufacture biohydrogen using a variety of processes, including direct and indirect photolysis of water and dark fermentation, resulting in the synthesis of various volatile fatty acids as well as hydrogen. The rate of hydrogen production is influenced by a number of factors, including carbon and nitrogen supplies and their relative ratios, pH, temperature, light/dark cycles, microalgal strain, culture set-up, and pre-treatment (Mustafa et al. 2021).

Microalgae offer a promising way to convert CO_2 into biofuels and high-value products. Algae contribute to about 40% of the global carbon fixation, with marine algae accounting for the majority of this (Sydney et al. 2019). The integration of microalgae biorefineries to produce value-added products may shift the balance in favour of low-cost biofuel generation. Bioethanol, biodiesel, biooil, biomethane, biohydrogen, and biobutanol are the significant microalgae biofuels (Arun et al. 2020).

Bioethanol production has already been shown to be a feasible biodiesel manufacturing solution on a big scale (Ahmad, Yuzir, et al. 2021). Several carbohydrate components found in microalgal species such as *Dunaliella* sp., *Chlorella* sp., *and Scenedesmus* sp. could be used to ferment carbohydrates from *Spirogyra* sp. extract into bioethanol (Özçimen et al. 2020); for example, *Saccharomyces cerevisiae* and *Zymomonas mobilis* were used to ferment carbohydrates from a Spirogyra sp. extract into bioethanol. The chemical conversion process, biochemical conversion, thermochemical conversion, and direct combustion are some of the biofuel-generating processes developed so far (Culaba et al. 2020).

When cultured under nitrogen limitation, some algae species, such as *Chlorella protothecoides*, have been shown to gather up to 55% of their biomass in lipids. Cultivation, harvesting, drying, extraction of microalgal lipids, and subsequently transesterification by using alcohol with catalysts are all stages of the microalgal to biofuel conversion process (Rodionova et al. 2017). According to some research, H_2 gas can be produced at a rate of up to 300 mol H_2 (mg Chl/hr) in *Scenedesmus obliquus* via the reversible hydrogenase pathway. As a result, we emphasize how easily microalgae can be included into the process of circular economy that reduces carbon emissions, cleans wastewater, and utilizes waste as a source of producing bioenergy/biofuel to contribute for energy recovery (Serrà i Ramos et al. 2020).

Chemical fertilizers are used in agriculture to assure good yields, but their widespread use causes a slew of environmental problems. Biofertilizers have evolved as a more environmentally friendly alternative to chemical fertilizers for lowering their impact. Photosynthetic microorganism-based formulations stand out among the many biofertilizers studied for their capacity to boost soil fertility and crop yield (Guo et al. 2020).

The use of wastewater grown microalgae biomass as a soil amendment or biofertilizer is another possible practice. Microalgae biomass in the soil could improve soil nitrogen and phosphorus levels, as well as a variety of other plant-required minerals (Ca, K, Fe, Mn, etc.) (Reda et al. 2020). Heavy metals might be adsorbed by microalgae as well; however, the concentrations of heavy metals in the biomass, or anaerobic-digested biomass, were lower than the permitted values. Biomass of microalgae growing in wastewater was employed as a biofertilizer in small-scale trials to grow wheat, corn, cucumber, barley, tomato, and other crops. Microalgae biomass can also be used as a slow-release biofertilizer. In addition, using microalgae as a biofertilizer could improve soil organic content. Depending on the species, microalgae biomass could also introduce a variety of plant-stimulating chemicals. wastewater-grown microalgae biomass, on the other hand, may be accompanied by pathogens and other micropollutants, raising concerns about its use as a soil addition (Kang et al. 2021; Al-Jabri et al. 2021). Table 5.4 shows how microalgae grown in wastewater can be used for a variety of applications.

TABLE 5.4
Some Examples for Algal Biomass Applications

Algae	Product	Source of the Wastewater	Reference
Chlorella vulgaris	Biodiesel and biogas	Sludge wastewater	(Sakarika and Kornaros 2019)
Desmodesmus sp.	Biodiesel	Municipal wastewater	(Do et al. 2019)
Microalgal consortia	Biodiesel	Municipal wastewater	(Sharma et al. 2020)
Spirulina sp.	Biofertilization	Aquaculture wastewater	(Wuang et al. 2016)
Cyanobacteria	Biofertilization	Olive milling	(Rashad et al. 2018)
Chlorella pyrenoidosa	Biofertilization	Rice mill wastewater	(Umamaheswari and Shanthakumar 2021)
Spirulina platensis	Phytohormones	Dairy wastewater	(Zapata et al. 2021)
Spirulina platensis	Carbohydrate	Dairy wastewater	(Pereira et al. 2019)
Chlorella sorokiniana	Biodiesel	Dairy wastewater	(Hamidian and Zamani 2021)
Nannochloropsis oculata and *Tetraselmis suecica*	Bioethanol	Municipal	(Reyimu and Özçimen 2017)
Botryococcus sp.	Hydrocarbons	Domestic	(Gani et al. 2017)
Chlorella vulgaris	Source of nutrients	Aquaculture wastewater	(Mtaki et al. 2021)
Spirulina sp.	Biofertilization	Aquaculture wastewater	(Wuang et al. 2016)
Chlorella variabilis and *Scenedesmus obliquus*	Lutein	Dairy wastewater	(Loganathan et al. 2021)
Tetradesmus obliquus	Biooil, biochar, and biogas	Brewery	(Ferreira et al. 2020)
Chlamydomonas reinhardtii UTEX 2243 and *Chlorella sorokiniana UTEX 2714*	Biohydrogen	Acetate rich wastewater	(Hwang and Lee 2021)
Chlorella vulgaris	Biodiesel	Textile	(Fazal et al. 2021)
Microalgae consortium	Biodiesel	Dairy	(Chandra et al. 2021)

To test rice growth and productivity, Dineshkumar et al. (2018) used varied dosages of algal biomass. The usage of algal biomass had a good impact on the plant's performance, according to the researchers, resulting in a 21% boost in yield. Jochum, Moncayo, and Jo (2018) analysed rice and discovered that plants fed microalgae in a polyculture performed similarly to or better than plants treated with urea. Microalgae's capability to sequestrate atmospheric CO_2 is another important factor that influences productivity.

Microalgae viability and propagation are significantly correlated with CO_2 concentration in the atmosphere; the more is the CO_2 concentration, the greater is the viability and propagation, allowing for superior crop development and output. Carbon enrichment improves the quality and fertility of soil in bull and rhizosphere soils while also preserving ecology and enhancing micro- and macronutrients. According to one study, inoculation treatments enhanced the biomass of a specific microalgae employed as a biofertilizer for wheat by up to 67% when compared to controls (Renuka et al. 2018).

Together with providing direct nutrient enrichment, microalgae can also produce compounds that work as accelerators for plant growth. Mutale-Joan et al. (2020) examined the biostimulatory effects of 18 crude biological extracts obtained from microalgae on tomato plant growth and observed that there was an increase in root and shoot length, as well as dry weight. Rachidi et al. (2020), used microalgae polysaccharides for growing tomatoes as plant biostimulants.

These researchers found that as compared to the non-inoculated control dry weight, the number of nodes and length of the aerial section increased, as did the amount of chlorophyll, carotenoids, and proteins in the leaves. Treatments with *Chlorella sorokiniana* filtrate increased total dry biomass by 30%, subsurface biomass by 51%, and aerial biomass by 22%, in wheat, according to the study (Kholssi et al. 2019).

Apart from promoting nutrient availability and plant growth, research has focused on improving soil quality, specifically strengthening soil structure, and reclaiming damaged land (Roncero-Ramos et al. 2020). Adding microalgal suspensions to the soil increased eukaryotic and prokaryotic biomass as well as heterotrophic microbe activity, according to Marks, Montero, and Rad (2019). Biofertilizer effects of on micro- and macroaggregate stability were studied by Yilmaz and Sönmez (2017). The researchers discovered that biofertilizer treatments enhanced macro- and microaggregate stability and that this impact was amplified when biofertilizer and vermicompost treatments were used together. These findings back up the theory that microalgae play a key role in arid soil stabilization by producing soil biocrust (Kumar and Singh 2020).

Microalgae cultures offer a practicable substitute to physical and chemical procedures for eliminating dangerous pollutants from the environment, in addition to all of the benefits listed above. Microalgal use is justified by its renewable nature and ability to partially reduce and replace the use of chemical fertilizers in the soil, which are the source of causing deleterious impact to the environment and soil (Ali et al. 2021). Baglieri et al. (2016) studied the development of microalgae in hydroponic tomato-growing effluent and discovered that the microalgae were capable of degrading all of the agrochemicals tested (iprodione, metalaxyl, triclopyr, and fenhexamid).

According to Nie et al. (2020), *C. mexicana, C. vulgaris, A. oryzae, S. obliquus, C. pitschmannii, N. muscorum, C. reinhardtii and S. platensis* are the microalga species implicated in pesticide removal. Microalgae as biofertilizers have been shown to be a practical method for enhancing plant growth, soil fertility, aiding pest, and disease management and contributing to remove the toxic substances in agronomic operations, and can help meet the demands of a sustainable development (Kholssi et al. 2021).

5.9 CONCLUSIONS

Algae seem to be a technologically and environmentally viable biological wastewater treatment alternative. Combining wastewater treatment with algal biomass production is an innovative and vibrant strategy from both an economic and environmental aspect. Algal biomass grown on wastewater represents a source of valuable compounds that may be utilized to produce many products via innovative routes. Algal biomass could be used to produce soil compost, fine chemicals, biofuels, and aquaculture animal feed. These techniques lower wastewater treatment costs and promote the adoption of zero-liquid wastewater discharge algae bloom technologies. The considerable literature on nutrient uptake by diverse microalgal species has demonstrated unequivocally that these organisms may be employed to successfully recycle nutrients from wastewater while also creating significant biomass. However, further study on-site at a larger scale is required to evaluate the method's practicability and cost-effectiveness in wastewater treatment plants. Microalgae bioremediation of wastewaters could be very promising due to their rapid growth rate, ability to adapt to different wastewaters, and ability to uptake nutrients or remove pollutants from wastewater, as well as the assimilation of carbon dioxide. Despite the enormous potential of microalgal bioremediation of wastewaters, the majority of the studies were limited to indoor lab-scale microalgal cultivation studies with a small number of strains to determine the removal efficiencies of targeted elements or compounds.

Algal-microbe interactions in wastewater, as well as the use of consortia for increased nutrient recovery and biomass production, have recently become hot topics. Research into the use of algae and bacteria for wastewater or effluent treatment has progressed to the point where the focus is now on coupling algae and bacteria with the development of monetarily viable inventive bio-products, influencing the circular economy and bringing a multipurpose arrangement to fruition. Financing opportunities are currently focused on the bioeconomy and the fabrication of unusual bio-products, which will aid in the expansion of this exploration field. Individual microorganisms' roles in microalgae-bacterial or microalgae-microalgae consortia require a deeper understanding of long-term outdoor operation. For microalgal bioremediation of wastewater, the development of a strain and application-specific energy-saving biomass harvesting is required; in this case, a self-settling strain or bioflocculating strain consortia should be developed. Furthermore, to improve the economic viability and environmental sustainability of wastewater bioremediation, the harvested microalgal biomass must be valorized in a multiproduct biorefinery approach.

REFERENCES

Ahmad, Imran, Norhayati Abdullah, Koji Iwamoto, and Ali Yuzir. 2021. "The contribution of microalgae in bio-refinery and resource recovery: a sustainable approach leading to circular bioeconomy." *Chemical Engineering Transactions* 89: 391–396.

Ahmad, Imran, Norhayati Abdullah, Iwamoto Koji, Ali Yuzir, and Shaza Eva Muhammad. 2021. "Evolution of photobioreactors: a review based on microalgal perspective." IOP Conference Series: Materials Science and Engineering. 1142 012004

Ahmad, Imran, Norhayati Abdullah, I Koji, A Yuzir, and SE Mohamad. 2021. "Potential of microalgae in bioremediation of wastewater." *Bulletin of Chemical Reaction Engineering & Catalysis* 16 (2): 413–429.

Ahmad, Shamshad, Arya Pandey, and Vinayak Vandan Pathak. 2020. "Bioremediation of industrial waste for environmental safety." *Bioremediation of Industrial Waste for Environmental Safety* (May 2019). [1st ed.] XXII, 436 [447].

Ahmad, I, A Yuzir, SE Mohamad, K Iwamoto, and N Abdullah. 2021. "Role of microalgae in sustainable energy and environment." IOP Conference Series: Materials Science and Engineering. 1051. 5th International Conference on Advanced Technology and Applied Sciences 2020 ICATAS 2020 in conjunction with the 6th Malaysia-Japan Joint International Conference 2020 MJJIC 2020 (ICATAS-MJJIC 2020) 7-9 October, Kuala Lumpur, Malaysia.

Ajayan, Kayil Veedu et al. 2015. "Phycoremediation of Tannery wastewater using microalgae *Scenedesmus* species." *International Journal of Phytoremediation* 17 (10) (urriak): 907–916.

Ali, Shazia et al. 2021. "Resource recovery from industrial effluents through the cultivation of microalgae: a review." *Bioresource Technology* 337 (June): 125461.

Ali, Sameh S, Michael Kornaros, Alessandro Manni, Rania Al-Tohamy, Abd El-Raheem, R El-Shanshoury, Ibrahim M Matter, Tamer Elsamahy, Mabrouk Sobhy, and Jianzhong Sun. 2021. "Advances in microorganisms-based biofertilizers: major mechanisms and applications." In *Biofertilizers*, Woodhead Publishing, United Kingdom 371–385.

Al-Jabri, Hareb et al. 2021. "Treatment of wastewaters by microalgae and the potential applications of the produced biomass—a review." *Water (Switzerland)* 13 (1): 27.

Almomani, Fares, and Rahul R Bohsale 2021. "Bio-sorption of toxic metals from industrial wastewater by algae strains *Spirulina platensis* and *Chlorella vulgaris*: application of isotherm, kinetic models and process optimization." *Science of the Total Environment* 755: 142654.

Aly, Magda M, and Reda H Amasha. 2019. "Removal of dangerous heavy metal and some human pathogens by dried Green Algae collected from Jeddah Coast." *Pharmacophore* 10 (3): 5–13.

Anastopoulos, Ioannis, and George Z Kyzas. 2015. "Progress in batch biosorption of heavy metals onto algae." *Journal of Molecular Liquids* 209: 77–86.

Arora, Kanika et al. 2021. "Valorization of wastewater resources into biofuel and value-added products using microalgal system." *Frontiers in Energy Research* 9 (April): 1–25.

Arun, Jayaseelan, Kannappan Panchamoorthy Gopinath, PanneerSelvam SundarRajan, Vargees Felix, Marudai JoselynMonica, and Rajagopal Malolan. 2020. "A conceptual review on microalgae biorefinery through thermochemical and biological pathways: bio-circular approach on carbon capture and wastewater treatment." *Bioresource Technology Reports* 11: 100477.

Aryal, Mahendra. 2021. "A comprehensive study on the bacterial biosorption of heavy metals: materials, performances, mechanisms, and mathematical modellings." *Reviews in Chemical Engineering* 37 (6): 715–754.

Awogbemi, Omojola et al. 2021. "An overview of the classification, production and utilization of biofuels for internal combustion engine applications." *Energies* 14 (18): 1–42.

Baglieri, Andrea, Sarah Sidella, Valeria Barone, Ferdinando Fragalà, Alla Silkina, Michèle Nègre, and Mara Gennari. 2016. "Cultivating *Chlorella vulgaris* and *Scenedesmus quadricauda* microalgae to degrade inorganic compounds and pesticides in water." *Environmental Science and Pollution Research* 23 (18): 18165–18174.

Batista, Ana Paula et al. 2015. "Combining urban wastewater treatment with biohydrogen production – an integrated microalgae-based approach." *Bioresource Technology* 184: 230–235.

Bhattacharya, Munna, and Saswata Goswami. 2020. "Microalgae–a green multi-product biorefinery for future industrial prospects." *Biocatalysis and Agricultural Biotechnology* 25: 101580.

Bošnjaković, Mladen, and Nazaruddin Sinaga. 2020. "The perspective of large-scale production of algae biodiesel." *Applied Sciences (Switzerland)* 10 (22): 1–26. doi:10.3390/app10228181.

Bwapwa, JK, AT Jaiyeola, and R Chetty. 2017. "Bioremediation of acid mine drainage using algae strains: a review." *South African Journal of Chemical Engineering* 24: 62–70. doi:10.1016/j.sajce.2017.06.005.

Campbell, Iona et al. 2019. "The environmental risks associated with the development of seaweed farming in Europe – prioritizing key knowledge gaps." *Frontiers in Marine Science* 6: 107.

Chandra, Rajesh et al. 2021. "An approach for dairy wastewater remediation using mixture of microalgae and biodiesel production for sustainable transportation." *Journal of Environmental Management* 297: 113210.

Christenson, Logan, and Ronald Sims. 2011. "Production and harvesting of microalgae for wastewater treatment, biofuels, and bioproducts." *Biotechnology Advances*, 29(6): 686–702.

Craggs, Rupert J, Tryg J Lundquist, and John R Benemann. 2013. "Wastewater treatment and algal biofuel production." In Borowitzka, M., Moheimani, N. (eds.) Algae for Biofuels and Energy, *Developments in Applied Phycology*, vol 5. Springer, Dordrecht. 153–163.

Culaba, Alvin B, Aristotle T Ubando, Phoebe Mae L Ching, Wei-Hsin Chen, and Jo-Shu Chang. 2020. "Biofuel from microalgae: sustainable pathways." *Sustainability* 12 (19): 8009.

Danouche, M et al. 2020. "Heavy metals phycoremediation using tolerant green microalgae: enzymatic and non-enzymatic antioxidant systems for the management of oxidative stress." *Journal of Environmental Chemical Engineering* 8 (5) (urriak): 104460.

Danouche, Mohammed, Naïma El Ghachtouli, and Hicham El Arroussi. 2021. "Phycoremediation mechanisms of heavy metals using living green microalgae: physicochemical and molecular approaches for enhancing selectivity and removal capacity." *Heliyon* 7 (7): e07609

Dayana Priyadharshini, Stephen et al. 2021. "Phycoremediation of wastewater for pollutant removal: a green approach to environmental protection and long-term remediation." *Environmental Pollution* 290: 117989.

de Almeida, Heleno Cavalcante et al. 2020. "Phycoremediation potential of microalgae species for ethidium bromide removal from aqueous media." *International Journal of Phytoremediation* 22 (11) (irailak): 1168–1174.

Dianursanti Dianursanti, and Desya Pramadhanti. 2020. "Utilization of miroalgae Spirulina platensis as anti-bacterial compound in soap." AIP Conference Proceedings 2255, 040020.

Dineshkumar, R, R Kumaravel, J Gopalsamy, Mohammad Nurul Azim Sikder, and P Sampathkumar. 2018. "Microalgae as bio-fertilizers for rice growth and seed yield productivity." *Waste and Biomass Valorization* 9 (5): 793–800.

Do, Jeong-Mi et al. 2019. "A feasibility study of wastewater treatment using domestic microalgae and analysis of biomass for potential applications." *Water* 11 (11). https://doi.org/10.3390/w11112294. https://www.mdpi.com/2073-4441/11/11/2294.

El Hafiane, F, and B El Hamouri. 2005. "Anaerobic reactor/high rate pond combined technology for sewage treatment in the Mediterranean area." *Water Science and Technology* 51 (12): 125–132.

El Nemr, Ahmed et al. 2015. "Removal of toxic chromium from aqueous solution, wastewater and saline water by marine red alga *Pterocladia capillacea* and its activated carbon." *Arabian Journal of Chemistry* 8 (1): 105–117.

Fallahi, Alireza et al. 2021. "Interactions of microalgae-bacteria consortia for nutrient removal from wastewater: a review." *Chemosphere* 272: 129878. https://doi.org/10.1016/j.chemosphere.2021.129878.

Fan, Jie et al. 2020. "Performance of *Chlorella sorokiniana*-activated sludge consortium treating wastewater under light-limited heterotrophic condition." *Chemical Engineering Journal* 382: 122799.

Fazal, Tahir et al. 2021. "Integrating bioremediation of textile wastewater with biodiesel production using microalgae (*Chlorella vulgaris*)." *Chemosphere* 281: 130758.

Ferreira, Ana F et al. 2020. "Pyrolysis of *Scenedesmus obliquus* biomass following the treatment of different wastewaters." *BioEnergy Research* 13 (3): 896–906.

Fon Sing, Sophie, Andreas Isdepsky, Michael A Borowitzka, and Navid Reza Moheimani. 2013. "Production of biofuels from microalgae." *Mitigation and Adaptation Strategies for Global Change* 18 (1): 47–72.

Gani, Paran et al. 2016. "Effects of different culture conditions on the phycoremediation efficiency of domestic wastewater." *Journal of Environmental Chemical Engineering* 4 (4): 4744–4753.

Gani, P, NM Sunar, H Matias-Peralta, RMSR Mohamed, AAA Latiff, and UK Parjo. 2017. Extraction of hydrocarbons from freshwater green microalgae (*Botryococcus* sp.) biomass after phycoremediation of domestic wastewater. *International Journal of Phytoremediation*, 19 (7): 679–685.

Gatamaneni Loganathan, Bhalamurugan, Valerie Orsat, and Mark Lefsrud. 2020. "Phycoremediation and valorization of synthetic dairy wastewater using microalgal consortia of *Chlorella variabilis* and *Scenedesmus obliquus*." *Environmental Technology (United Kingdom)*. 42(20): 3231–3244.

Ge, Shijian et al. 2018. "Centrate wastewater treatment with *Chlorella vulgaris*: simultaneous enhancement of nutrient removal, biomass and lipid production." *Chemical Engineering Journal* 342: 310–320.

Giarikos, Dimitrios G et al. 2021. "Effects of nitrogen depletion on the biosorption capacities of *Neochloris minuta* and *Neochloris alveolaris* for five heavy metals." *Applied Water Science* 11 (2): 1–15.

Green, Franklin Bailey, TJ Lundquist, and WJ Oswald. 1995. "Energetics of advanced integrated wastewater pond systems." *Water Science and Technology* 31 (12): 9–20.

Green, FB, TJ Lundquist, NWT Quinn, MA Zarate, IX Zubieta, and WJ Oswald. 2003. "Selenium and nitrate removal from agricultural drainage using the AIWPS® technology." *Water Science and Technology* 48 (2): 299–305.

Gross, Martin, Wesley Henry, Clayton Michael, and Zhiyou Wen. 2013. "Development of a rotating algal biofilm growth system for attached microalgae growth with in situ biomass harvest." *Bioresource Technology* 150: 195–201.

Guleri, Sakshi, and Archana Tiwari. 2020. "Algae and ageing." In *Microalgae Biotechnology for Food, Health and High Value Products*, 267–293. Springer, Singapore.

Guo, Suolian, Ping Wang, Xinlei Wang, Meng Zou, Chunxue Liu, and Jihong Hao. 2020. "Microalgae as biofertilizer in modern agriculture." In *Microalgae Biotechnology for Food, Health and High Value Products*, 397–411. Springer. Singapore

Hamidian, Najmeh, and Hajar Zamani. 2021. "Potential of *Chlorella sorokiniana* cultivated in dairy wastewater for bioenergy and biodiesel production." *BioEnergy Research*, 15(1): 1–12.

Huang, Wenli et al. 2020. "Achieving partial nitrification and high lipid production in an algal-bacterial granule system when treating low COD/NH4–N wastewater." *Chemosphere* 248: 126106.

Huo, Shuhao et al. 2020. "Co-culture of *Chlorella* and wastewater-borne bacteria in vinegar production wastewater: enhancement of nutrients removal and influence of algal biomass generation." *Algal Research* 45: 101744.

Hwang, Jae-Hoon, and Woo Hyoung Lee. 2021. "Continuous photosynthetic biohydrogen production from acetate-rich wastewater: influence of light intensity." *International Journal of Hydrogen Energy* 46 (42): 21812–21821.

Jerney, Jacqueline, and Kristian Spilling. 2018. "Large scale cultivation of microalgae: open and closed systems." In: Spilling, K. (Ed) *Biofuels from Algae. Methods in Molecular Biology*, vol 1980. Humana, New York, NY.

Ji, Bin. 2021. "Towards environment-sustainable wastewater treatment and reclamation by the non-aerated microalgal-bacterial granular sludge process: recent advances and future directions." *Science of the Total Environment* 806: 150707.

Ji, Xiyan et al. 2018. "The interactions of algae-bacteria symbiotic system and its effects on nutrients removal from synthetic wastewater." *Bioresource Technology* 247: 44–50.

Ji, X, H Li, J Zhang, H Saiyin, and Z Zheng. 2019. "The collaborative effect of *Chlorella vulgaris-Bacillus licheniformis* consortia on the treatment of municipal water." *Journal of Hazardous Materials*, 365: 483–493.

Jochum, Michael, Luis P Moncayo, and Young-Ki Jo. 2018. "Microalgal cultivation for biofertilization in rice plants using a vertical semi-closed airlift photobioreactor." *PloS ONE* 13 (9): e0203456.

Johnson, Daniel B, Lance C Schideman, Thomas Canam, and Robert JM Hudson. 2018. "Pilot-scale demonstration of efficient ammonia removal from a high-strength municipal wastewater treatment side stream by algal-bacterial biofilms affixed to rotating contactors." *Algal Research* 34: 143–153.

Kang, Yeeun et al. 2021. "Potential of algae–bacteria synergistic effects on vegetable production." *Frontiers in Plant Science* 12: 556.

Kangas, Patrick, Walter Mulbry, Philip Klavon, H Dail Laughinghouse, and Walter Adey. 2017. "High diversity within the periphyton community of an algal turf scrubber on the Susquehanna River." *Ecological Engineering* 108: 564–572.

Khan, Muhammad Imran, Jin Hyuk Shin, and Jong Deog Kim. 2018. "The promising future of microalgae: current status, challenges, and optimization of a sustainable and renewable industry for biofuels, feed, and other products." *Microbial Cell Factories.* BioMed Central Ltd. doi:10.1186/s12934-018-0879-x.

Kholssi, Rajaa, Evan AN Marks, Jorge Miñ6n, Olimpio Montero, Abderrahmane Debdoubi, and Carlos Rad. 2019. "Biofertilizing effect of *Chlorella sorokiniana* suspensions on wheat growth." *Journal of Plant Growth Regulation* 38 (2): 644–649.

Kholssi, Rajaa, Priscila Vogelei Ramos, Evan AN Marks, Olimpio Montero, and Carlos Rad. 2021. "2Biotechnological uses of microalgae: a review on the state of the art and challenges for the circular economy." *Biocatalysis and Agricultural Biotechnology* 36: 102114.

Khoo, Kuan Shiong et al. 2021. "Microalgal-bacterial consortia as future prospect in wastewater bioremediation, environmental management and bioenergy production." *Indian Journal of Microbiology* 61 (3): 262–269.

Kumar, Lakhan et al. 2022. "Chapter 14 - Microbial remediation of tannery wastewater." In Shah, Maulin P et al. (eds.). *Development in Wastewater Treatment Research and Processes*, 303–328. Elsevier, Amsterdam.

Kumari, Preety et al. 2020. "Phycoremediation of wastewater by Chlorella pyrenoidosa and utilization of its biomass for biogas production." *Journal of Environmental Chemical Engineering* 9(1): 104974.

Kumar, Arun, and Jay Shankar Singh. 2020. "Biochar coupled rehabilitation of cyanobacterial soil crusts: a sustainable approach in stabilization of arid and semiarid soils." In Singh, J., Singh, C. (eds.) *Biochar Applications in Agriculture and Environment Management.* 167–191. Springer, Cham.

Lahin, FA, R Sarbatly, and E Suali. 2016. "Polishing of POME by Chlorella sp. in suspended and immobilized system." *IOP Conference Series: Earth and Environmental Science* 36:012030.

Lau, PS, NFY Tam, and YS Wong. 1998. "Effect of carrageenan immobilization on the physiological activities of *Chlorella vulgaris.*" *Bioresource Technology* 63 (2): 115–121.

Lei, Yong-Jia et al. 2018. "Microalgae cultivation and nutrients removal from sewage sludge after ozonizing in algal-bacteria system." *Ecotoxicology and Environmental Safety* 165: 107–114.

Lian, Jie et al. 2018. "The effect of the algal microbiome on industrial production of microalgae." *Microbial Biotechnology* 11 (5): 806–818.

Liao, Huan et al. 2019. "Impact of ocean acidification on the energy metabolism and antioxidant responses of the Yesso scallop (*Patinopecten yessoensis*)." *Frontiers in Physiology* 10 (JAN): 1–10.

Lin, Chao et al. 2019. "Algal-bacterial symbiosis system treating high-load printing and dyeing wastewater in continuous-flow reactors under natural light." *Water (Switzerland)* 11 (3): 469.

Lin, Zeyu et al. 2020. "Application of algae for heavy metal adsorption: a 20-year meta-analysis." *Ecotoxicology and Environmental Safety.* doi:10.1016/j.ecoenv.2019.110089.

Loganathan, Bhalamurugan Gatamaneni, Valerie Orsat, and Mark Lefsrud. 2021. "Phycoremediation and valorization of synthetic dairy wastewater using microalgal consortia of *Chlorella variabilis* and *Scenedesmus obliquus.*" *Environmental Technology* 42 (20): 3231–3244.

Malla, Fayaz A et al. 2015. "Phycoremediation potential of *Chlorella minutissima* on primary and tertiary treated wastewater for nutrient removal and biodiesel production." *Ecological Engineering* 75: 343–349. doi:10.1016/j.ecoleng.2014.11.038.

Mallick, Nirupama. 2020. "Immobilization of microalgae." In Guisan, J., Bolivar, J., López-Gallego, F., Rocha-Martín, J. (eds.) *Immobilization of Enzymes and Cells. Methods in Molecular Biology*, vol 2100. Humana, New York, NY.

Mao, Yilin et al. 2021. "Analysis of the status and improvement of microalgal phosphorus removal from municipal wastewater." *Processes* 9 (9): 1486

Marazzi, Francesca et al. 2020. "Interactions between microalgae and bacteria in the treatment of wastewater from milk whey processing." *Water* 12 (1): 297.

Marella, Thomas Kiran et al. 2020. "Wealth from waste: diatoms as tools for phycoremediation of wastewater and for obtaining value from the biomass." *Science of the Total Environment* 724: 137960. doi:10.1016/j.scitotenv.2020.137960.

Marks, Evan AN, Olimpio Montero, and Carlos Rad. 2019. "The biostimulating effects of viable microalgal cells applied to a calcareous soil: increases in bacterial biomass, phosphorus scavenging, and precipitation of carbonates." *Science of the Total Environment* 692: 784–790.

Matamoros, Víctor, Raquel Gutiérrez, Ivet Ferrer, Joan García, and Josep M Bayona. 2015. "Capability of microalgae-based wastewater treatment systems to remove emerging organic contaminants: a pilot-scale study." *Journal of Hazardous Materials* 288: 34–42.

Meng F, Xi L, Liu D, et al. 2019. Effects of light intensity on oxygen distribution, lipid production and biological community of algal-bacterial granules in photo-sequencing batch reactors. *Bioresour Technol* 272:473–481. doi: /10.1016/j.biortech.2018.10.059

Montingelli, ME, Silvia Tedesco, and AG Olabi. 2015. "Biogas production from algal biomass: a review." *Renewable and Sustainable Energy Reviews* 43: 961–972.

Morais, Etiele G et al. 2021. "Microalgal systems for wastewater treatment: technological trends and challenges towards waste recovery." *Energies* 14 (23): 1–26.

Mostafa SS, AS El-Hassanin, AS Soliman, et al. 2019. Microalgae growth in effluents from olive oil industry for biomass production and decreasing phenolics content of wastewater. *Egyptian Journal of Aquatic Biology and Fisheries*, 23(1): 359–365.

Mtaki, Kulwa, Margareth S Kyewalyanga, and Matern SP Mtolera. 2021. "Supplementing wastewater with NPK fertilizer as a cheap source of nutrients in cultivating live food (*Chlorella vulgaris*)." *Annals of Microbiology* 71 (1): 7.

Mu, Ruimin et al. 2021. "Advances in the use of microalgal-bacterial consortia for wastewater treatment: community structures, interactions, economic resource reclamation, and study techniques." *Water Environment Research : A Research Publication of the Water Environment Federation* 93 (8) (abuztuak): 1217–1230.

Mujtaba, Ghulam, Muhammad Rizwan, and Kisay Lee. 2017. "Removal of nutrients and COD from wastewater using symbiotic co-culture of bacterium *Pseudomonas putida* and immobilized microalga *Chlorella vulgaris*." *Journal of Industrial and Engineering Chemistry* 49: 145–151.

Mustafa, Shazia et al. 2021. "Microalgae biosorption, bioaccumulation and biodegradation efficiency for the remediation of wastewater and carbon dioxide mitigation: prospects, challenges and opportunities." *Journal of Water Process Engineering* 41 (March): 102009. doi:10.1016/j.jwpe.2021.102009.

Mutale-Joan, Chanda, Benhima Redouane, Elmernissi Najib, Kasmi Yassine, Karim Lyamlouli, Sbabou Laila, Youssef Zeroual, and El Arroussi Hicham. 2020. "Screening of microalgae liquid extracts for their bio stimulant properties on plant growth, nutrient uptake and metabolite profile of *Solanum lycopersicum* L." *Scientific Reports* 10 (1): 1–12.

Nagarajan, Dillirani, Duu-Jong Lee, Chun-Yen Chen, and Jo-Shu Chang. 2020. "Resource recovery from wastewaters using microalgae-based approaches: a circular bioeconomy perspective." *Bioresource Technology* 302: 122817.

Naruka, M, M Khadka, S Upadhayay, and S Kumar. 2019. "Potential applications of microalgae in bioproduct production: a review." *Octa Journal of Bioscience* 7: 01–05.

Nie, Jing, Yuqing Sun, Yaoyu Zhou, Manish Kumar, Muhammad Usman, Jiangshan Li, Jihai Shao, Lei Wang, and Daniel CW Tsang. 2020. "Bioremediation of water containing pesticides by microalgae: mechanisms, methods, and prospects for future research." *Science of the Total Environment* 707: 136080.

Özçimen, Didem, Anıl Tevfik Koçer, Benan İnan, and Tugba Özer. 2020. "Bioethanol production from microalgae." In Jacob-Lopes, E, Queiroz, MI, Maroneze, MM, Zepka, LQ, (eds.) *Handbook of Microalgae-Based Processes and Products, Fundamentals and Advances in Energy, Food, Feed, Fertilizer, and Bioactive Compounds*. Academic Press, Cambridge, MA, USA, 2020, 373–389.

Panahi, Yunes et al. 2019. "Impact of cultivation condition and media content on *Chlorella vulgaris* composition." *Advanced Pharmaceutical Bulletin* 9 (2): 182.

Park, JBK, RJ Craggs, and AN Shilton. 2011. "Wastewater treatment high rate algal ponds for biofuel production." *Bioresource Technology* 102 (1): 35–42.

Pathak, Vinayak V et al. 2015. "Experimental and kinetic studies for phycoremediation and dye removal by *Chlorella pyrenoidosa* from textile wastewater." *Journal of Environmental Management* 163: 270–277.

Pereira, Maria IB et al. 2019. "Mixotrophic cultivation of *Spirulina platensis* in dairy wastewater: effects on the production of biomass, biochemical composition and antioxidant capacity." *PLoS ONE* 14 (10): e0224294.

Plöhn, Martin et al. 2021. "Wastewater treatment by microalgae." *Physiologia Plantarum* 173 (2): 568–578.

Podder, MS, and CB Majumder 2015. "Phycoremediation of arsenic from wastewaters by *Chlorella pyrenoidosa*." *Groundwater for Sustainable Development* 1 (1–2): 78–91.

Qi, Feng et al. 2021. "Convergent community structure of algal–bacterial consortia and its effects on advanced wastewater treatment and biomass production." *Scientific Reports* 11 (1): 1–12.

Rachidi, Farid, Redouane Benhima, Laila Sbabou, and Hicham El Arroussi. 2020. "Microalgae polysaccharides bio-stimulating effect on tomato plants: growth and metabolic distribution." *Biotechnology Reports* 25:e00426.

Rashad, S et al. 2018. "Cyanobacteria cultivation using olive milling wastewater for bio-fertilization of celery plant." *Global Journal of Environmental Science and Management* 5 (2): 167–174.

Raven, John A, Christopher S Gobler, and Per Juel Hansen. 2020. "Dynamic CO_2 and pH levels in coastal, estuarine, and inland waters: theoretical and observed effects on harmful algal blooms." *Harmful Algae* 91 (April 2019): 101594.

Rebello, Sharrel et al. 2021. "Cleaner technologies to combat heavy metal toxicity." *Journal of Environmental Management* 296 (July): 113231.

Reda, Marwa M et al. 2020. "Fatty acid profiles and fuel properties of oils from castor oil plants irrigated by microalga-treated wastewater." *Egyptian Journal of Botany* 60 (3): 797–804.

Renuka, Nirmal et al. 2021. "Insights into the potential impact of algae-mediated wastewater beneficiation for the circular bioeconomy: a global perspective." *Journal of Environmental Management* 297: 113257.

Renuka, Nirmal, Abhishek Guldhe, Radha Prasanna, Poonam Singh, and Faizal Bux. 2018. "Microalgae as multi-functional options in modern agriculture: current trends, prospects and challenges." *Biotechnology Advances* 36 (4): 1255–1273.

Reyimu, Zubaidai, and Didem Özçimen. 2017. "Batch cultivation of marine microalgae *Nannochloropsis oculata* and *Tetraselmis suecica* in treated municipal wastewater toward bioethanol production." *Journal of Cleaner Production* 150 (maiatzak): 40–46.

Rodionova, Margarita V, Roshan Sharma Poudyal, Indira Tiwari, Roman A Voloshin, Sergei K Zharmukhamedov, Hong Gil Nam, Bolatkhan K Zayadan, Barry D Bruce, Harvey JM Hou, and Suleyman I Allakhverdiev. 2017. "Biofuel production: challenges and opportunities." *International Journal of Hydrogen Energy* 42 (12): 8450–8461.

Roles, John et al. 2020. "Charting a development path to deliver cost competitive microalgae-based fuels." *Algal Research* 45: 101721.

Roncero-Ramos, B, MA Muñoz-Martín, Y Cantón, S Chamizo, E Rodríguez-Caballero, and P Mateo. 2020. "Land degradation effects on composition of pioneering soil communities: an alternative successional sequence for dryland cyanobacterial biocrusts." *Soil Biology and Biochemistry* 146: 107824.

Sakarika, Myrsini, and Michael Kornaros. 2019. "*Chlorella vulgaris* as a green biofuel factory: comparison between biodiesel, biogas and combustible biomass production." *Bioresource Technology* 273: 237–243.

Sánchez-zurano, Ana et al. 2021. "Abaco: a new model of microalgae-bacteria consortia for biological treatment of wastewaters." *Applied Sciences (Switzerland)* 11 (3): 1–24.

Saravanan, A et al. 2021. "A review on algal-bacterial symbiotic system for effective treatment of wastewater." *Chemosphere* 271: 129540.

Sepehri, Arsalan, Mohammad-Hossein Sarrafzadeh, and Maryam Avateffazeli. 2020. "Interaction between *Chlorella vulgaris* and nitrifying-enriched activated sludge in the treatment of wastewater with low C/N ratio." *Journal of Cleaner Production* 247: 119164.

Serrà i Ramos, Albert, Raúl Artal López, Jaume Garcia Amorós, Elvira Gómez, and Laetitia Philippe. 2020. "Circular zero-residue process using microalgae for efficient water decontamination, biofuel production, and carbon dioxide fixation." *Chemical Engineering Journal*, 388: 124278.

Sharma, Jyoti et al. 2020. "Microalgal consortia for municipal wastewater treatment – lipid augmentation and fatty acid profiling for biodiesel production." *Journal of Photochemistry and Photobiology B: Biology* 202: 111638.

Sukačová, Kateřina, Martin Trtílek, and Tomáš Rataj. 2015. "Phosphorus removal using a microalgal biofilm in a new biofilm photobioreactor for tertiary wastewater treatment." *Water Research* 71: 55–63.

Sydney, Eduardo Bittencourt, Alessandra Cristine Novak Sydney, Júlio Cesar de Carvalho, and Carlos Ricardo Soccol. 2019. "Potential carbon fixation of industrially important microalgae." In *Biofuels from Algae*, 67–88. Elsevier.

Topal, Murat, Erdal Öbek, and E Işıl Arslan Topal. 2020. "Phycoremediation of precious metals by *Cladophora fracta* from mine gallery waters causing environmental contamination." *Bulletin of Environmental Contamination and Toxicology* 105 (1) (uztailak): 134–138. https://doi.org/10.1007/s00128-020-02879-w.

Umamaheswari, Jagannathan, and Subramainam Shanthakumar. 2021. "Paddy-soaked rice mill wastewater treatment by phycoremediation and feasibility study on use of algal biomass as biofertilizer." *Journal of Chemical Technology and Biotechnology* 96 (2): 394–403.

Whitton, Rachel et al. 2015. "Microalgae for municipal wastewater nutrient remediation: mechanisms, reactors and outlook for tertiary treatment." *Environmental Technology Reviews* 4 (1): 133–148.

Wollmann, Felix, Stefan Dietze, Jörg-Uwe Ackermann, Thomas Bley, Thomas Walther, Juliane Steingroewer, and Felix Krujatz. 2019. "Microalgae wastewater treatment: biological and technological approaches." *Engineering in Life Sciences* 19 (12): 860–871.

Wuang, Shy Chyi et al. 2016. "Use of Spirulina biomass produced from treatment of aquaculture wastewater as agricultural fertilizers." *Algal Research* 15: 59–64.

Xin, Xiaying, Gordon Huang, and Baiyu Zhang. 2021. "Review of aquatic toxicity of pharmaceuticals and personal care products to algae." *Journal of Hazardous Materials* 410 (August 2020): 124619.

Xu, Zhihui et al. 2020. "Development of integrated culture systems and harvesting methods for improved algal biomass productivity and wastewater resource recovery – a review." *Science of the Total Environment* 746: 141039.

Yilmaz, Erdem, and Mehmet Sönmez. 2017. "The role of organic/bio–fertilizer amendment on aggregate stability and organic carbon content in different aggregate scales." *Soil and Tillage Research* 168: 118–124.

Yong, Jasmine Jill Jia Yi et al. 2021. "Prospects and development of algal-bacterial biotechnology in environmental management and protection." *Biotechnology Advances* 47 (September 2020): 107684.

Zapata, Daniela et al. 2021. "Phytohormone production and morphology of *Spirulina platensis* grown in dairy wastewaters." *Algal Research* 59: 102469.

Zhang, Cui et al. 2019. "Phycoremediation of coastal waters contaminated with bisphenol A by green tidal algae *Ulva prolifera*." *Science of the Total Environment* 661 (apirilak): 55–62.

Zhang, Kan, Feifei Zhang, and Yi Rui Wu. 2021. "Emerging technologies for conversion of sustainable algal biomass into value-added products: a state-of-the-art review." *Science of the Total Environment* 784: 147024.

Zhao, Xuefei, Kuldip Kumar, Martin A Gross, Thomas E Kunetz, and Zhiyou Wen. 2018. "Evaluation of revolving algae biofilm reactors for nutrients and metals removal from sludge thickening supernatant in a municipal wastewater treatment facility." *Water Research* 143: 467–478.

6 Recycling of Fruit By-Products for Wastewater Treatment Applications

Nour F. Attia, Eman M. Abd El-Monaem, Sally E. A. Elashery, and Abdelazeem S. Eltaweil

6.1 INTRODUCTION

The rapid increase in worldwide population in conjunction with a severe shortage of drinking water is anticipated to water crisis over many places on the earth planet [1]. Therefore, in attempts to eliminate the shortage of pure water, several approaches have been executed such as the desalination process of seawater; however, this process has high energy consumption and is only useful for domestic use [2]. Interestingly, another alternative approach was introduced involving the treatment of wastewater to be suitable and used for industrial and agricultural applications [3–5]. There are various tools that have been implemented in wastewater treatments; however, the cost-effective and easier one in terms of process efficiency, industrial applicability, and energy saving is the adsorption process, and it can be directed for the removal of both organic and inorganic pollutants [6–10]. On the other hand, to facilitate easier and cost-effective industrial wastewater treatment, design and development of renewable porous adsorbents derived from biowastes is required, achieving dual benefits to environment [11–17]. On the other hand, the fruit shells and stones (fruit by-products) are considered as environmental waste and produced annually in huge quantities. Hence, recycling and converting them to porous carbons via carbonization process yields cost-effective and highly porous materials suitable for exploitation for various industrial applications, recording positive environmental and economic value. Recently, variety of fruit wastes have been extensively utilized for high-tech applications after conversion to porous carbon materials such as mandarin peel [18], orange peel (OR) [19, 20], banana peel (BP) [21], tamarind shell [22], and Indonesian snake fruit peel [23]. It is reported that some of the fruit's shells have environmental concerns about soils [24–26], and the shells constitute around 30–50% of their masses [18]. The conversion process can be done through carbonization of fruit by-products followed by activation process or via single-step yielding porous carbons [17, 18]. Recently we and others reported two-step recycling process (carbonization and activation) of different biomass and fruit shells and recently also shortened the synthesis process via one-pot

DOI: 10.1201/9781003354475-6

carbonization and/or activation process [18, 17]. Hence, in the current book chapter, recycling process of different fruit by-products was discussed. Their environmental exploitation in wastewater treatment application was also reviewed.

6.2 RECYCLING ROUTES

The recycling of fruit by-products involves the treatment of released shells with a specific treatment based on the final desired applications; however, in this chapter the recycling process will be dedicated for the synthesis of porous carbon materials. Therefore, there are two main approaches; the first one is the carbonization of the yielded fruit by-products, and the second one is activation of the obtained carbons [17, 18]. For carbonization process, after the fruit by-product is washed and dried, it is subjected to pyrolysis process at a high temperature (600–900°C) under inert atmosphere flow (nitrogen and argon), yielding carbon substance after the emission of degraded volatile compounds [12, 17, 27, 28]. The properties of the obtained carbon materials can be controlled by various parameters such as temperature, heating rate, holding time, and inert gas flow. Although the porous properties of the obtained porous carbon in this process are limited, rich surface-active sites were obtained in the yielded porous carbon, which were stemmed from naturally doping high electron density elements [14, 15, 18]. However, to incorporate more pores in the yielded porous carbon, the obtained carbon is exposed to activation process via physical and/or chemical activation process to widen the existing pores and create new ones. Basically, the physical activation process was conducted via heating the carbon at a high temperature under the flow of CO_2 to afford oxidation to a portion of carbon, creating micropores and polishing the surface chemistry of the obtained porous carbons [28, 29]. However, in chemical activation, the obtained carbon was immersed and/or mixed with chemical reagents (KOH, NaOH, $ZnCl_2$, etc.) and then heated at a high temperature under inert atmosphere flow, yielding activated carbon [27–29]. The as-obtained carbons after the activation process are denoted as activated carbons and are characterized with high surface area and considerable pore volume, which can be utilized in various industrial applications such as gas storage, energy gas storage, greenhouse gases, gas separation, electrochemical charge storage and wastewater treatments [12, 18, 27–29]. Basically, the porous materials can be divided into three categories based on the diameter of pore size: (i) microporous materials which contain pores with size less than 2 nm, (ii) mesoporous materials which feature with pore size in the range of 2–50 nm and (iii) macroporous materials which have pore size larger than 50 nm [30, 31]. However, the porous carbon materials obtained from the recycling of fruits via carbonization process, mainly microporous materials, feature [18]. Interestingly, after the activation process, it converted into hierarchical porous structures containing micropores and mesopores and was denoted as micro-mesoporous material [18]. This feature is unique and very useful for various applications specially CO_2 capture as CO_2 will accommodate in micropores at low pressure and then fill mesopores at high pressure [12, 18, 27]. Thus, the surface area and pore volume of porous carbon derived from fruit by-products reach to ~3000 m^2/g and a total pore volume of ~3 cm^3/g [12, 18, 27]. On the other hand, another unique merit of porous carbon derived from fruit by-products is the naturally doped high electron density elements (O, N, and S) which act as adsorption active sites for the removal of toxic pollutants from wastewater [11, 14, 18]. Thus, our focus in this chapter will be directed to porous carbon materials derived from fruit by-products and their application for wastewater treatment via the adsorption process.

6.3 WATER TREATMENT APPLICATIONS

6.3.1 Removal of Synthetic Dyes

In one attempt, Priyantha et al. [32] investigated the adsorption aptitude of dragon fruit skin (DFS) toward the adsorption of one of the most noxious dyes, which is methylene blue (MB). The concrete results pointed out the superb adsorption performance of DFS toward MB since the calculated maximum adsorption capacity (q_{max}) under Langmuir reached 640 mg/g. The morphology of DFS was studied by SEM before and after the adsorption of MB, clarifying a total change in the morphology (Figure 6.1A, B). SEM of DFS depicted a rough surface with much hollowness,

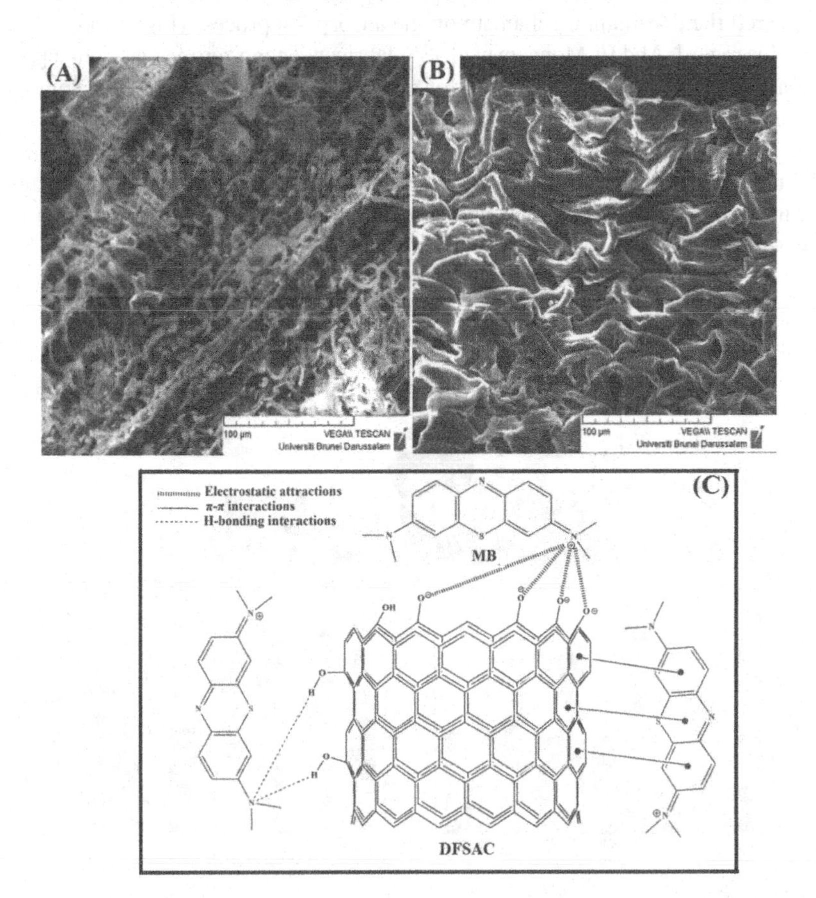

FIGURE 6.1 SEM of (A) DFS, (B) MB-loaded DFS [32], and (C) the plausible adsorption mechanism of MB onto DFSAC, reproduced with permission [34]. (Copyright 2021, Elsevier.)

clarifying the suitability of the DFS surface to adsorb the large MB molecules, while the DFS surface after the adsorption of MB became smoother since the pore was almost covered by MB molecules. It was deduced that the optimum pH was 4.5 (ambient conditions). This result was in line with the Seidmohammadi et al. study [33] since zeta potential measurements indicated that the point of zero charges (pH_{ZPC}) of raw oak fruit hulls was 3.5. So, there were strong electrostatic interactions between the negatively charged raw oak fruit hulls and the cationic MB molecules at a pH > 3.5. On the contrary, the Jawad et al. study [34] inferred that the optimum pH to adsorb MB onto DFS-derived activated carbon (DFSAC) was 10 (strong alkaline conditions). This finding may be ascribed to the strong electrostatic interaction between the cationic MB molecules and the anionic DFSAC surface (pH_{ZPC} = 9.2). Furthermore, it was proposed that the adsorption mechanism of MB onto DFSAC was controlled by π-π interaction between the aromatic ring of MB and the hexagonal DFSAC. Besides, the formed H-bonds between the N atoms in MB and the H atoms of DFSAC are one more possible mechanism. In addition, the electrostatic interaction mechanism between MB and DFSAC that was proved by pH study and ZP measurements was considered the dominant mechanism on the adsorption process (Figure 6.1C).

In this regard, Abd El-Monaem et al. [35] fabricated nano zero-valent iron-supported lemon-derived biochar (nZVI-LBC) for removing MB from an aqueous solution. It was reported that nZVI-LBC had a super high adsorption capacity toward MB since qmax was 1959.94 mg/g within 5 min. This propitious adsorption behavior may be attributed to the possibility to adsorb MB onto nZVI-LBC via several mechanisms (Figure 6.2A) involving the strong attraction forces, H-bond and n-π interaction between MB and nZVI-LBC [36, 37].

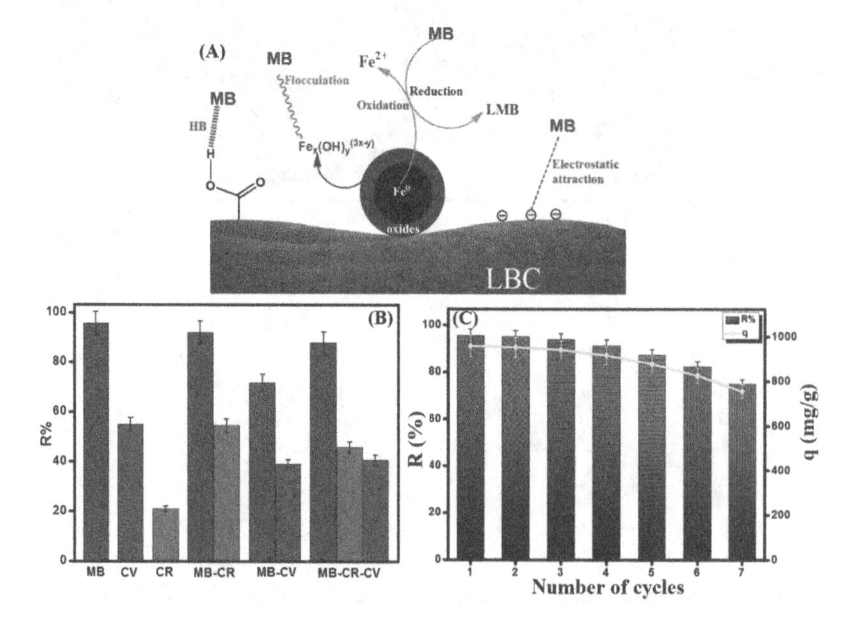

FIGURE 6.2 (A) The proposed adsorption mechanism of MB onto nZVI-LBC, (B) selectivity test, and (C) recyclability test [35].

Furthermore, the reduction of MB to the colorless leucomethylene blue (LMB) by the electrons transferred from Fe^0 is represented in the following equations [38, 39]:

$$Fe^0 \rightarrow Fe^{2+} + 2e^- \tag{6.1}$$

$$MB + ne^- \rightarrow LMB \tag{6.2}$$

In addition, the possibility of complexation and flocculation mechanisms since MB may form a complex with iron oxide/hydroxide that may flocculate in the reaction solution is clarified in the following equation:

$$MB + nFe_x(OH)_y^{(3x-y)} \rightarrow \left[MB\left[Fe_x(OH)_y^{(3x-y)}\right]n\right] \tag{6.3}$$

Moreover, the selectivity test (Figure 6.2B) implied the selectivity of nZVI-LBC toward methylene blue (MB) compared to crystal violet (CV) and Congo red (CR). It was noticed an increase in the removal percent of CR in the binary MB-CR and ternary MB-CV-CR system, which may be assigned to the interaction of CR with MB [40]. On the contrary, the removal percent of CV declined in the binary MB-CR and ternary MB-CV-CR system, which is most likely due to the competition between CR and CV for the adsorption sites of nZVI-LBC. Importantly, the recyclability test (Figure 6.2C) attested the significance of the magnetic property of the fabricated composite as it provided a perfect and facile separation. Consequently, nZVI-LBC showed high adsorption capacity (755.52 mg/g) and removal percent (75.01%) after seven sequential adsorption/desorption cycles using ethanol as an eluent.

In another study, Sajab et al. [41] developed the adsorption performance of oil palm empty fruit bunch (EFB) via citric acid (CA) and polyethylenimine (PEI) functionalization for removing MB and phenol red (PR). It was found that CA-EFB and PEI-EFB revealed excellent adsorption behaviors toward MB (103.1 mg/g) and PR (158.7 mg/g), respectively. The reusability test was executed using various eluents, including HCl, CH_3COOH, and NaOH. It was observed that HCl is the preferred eluent to desorb both MB and PR from CA-EFB and PEI-EFB as the desorption percent reached 69.8 and 64.2%, respectively, suggesting that the adsorption of both MB and PR occurred via electrostatic interaction and ion exchange [42], whereas CH_3COOH and NaOH desorbed 12.8% and 8% of MB and 8.7% and 13.5% of PR, respectively. Furthermore, CA-EFB showed an excellent removal performance after the 7th cycle since the removal percent declined by merely 10%, while the leaching of the ungrafted PEI from PEI-EFB that caused a significant diminution in the removal percent of PR reached 30%.

In this perspective, Bello [43] highlighted the development of EFB via chemical activation (EFBAC) for the efficient removal of malachite green (MG) from an aqueous medium. Brunauer, Emmett and Teller (BET) measurements pointed out that surface area, pore volume and pore diameter of EFBAC were 1254 m^2/g, 1.115 cm^3/g and 2.54 nm, respectively. This high surface area and good porosity could be explained by the dehydration of KOH, forming K_2O that reacted with the produced CO_2 from the water-shift reaction to form K_2CO_3. The metallic potassium intercalation caused an expansion of EFBAC, resulting in an increase in the specific surface area and the pore volume [44, 45]. As a result, EFBAC exhibited an advanced adsorption behavior toward MG since q_{max} was 356.27 mg/g. The pH study depicted that the adsorption of MG onto EFBAC was favorable at a wide pH range (6–10). This finding may be related to the surface charges of EFBAC since pH_{PZC} was 2.4. Subsequently, at low pH, the functional groups of EFBAC protonated, leading to electrostatic repulsion between the positively charged EFBAC and the cationic MG. Conversely, at high pH, the EFBAC function groups deprotonated and strong electrostatic interactions between MG and the negatively charged EFBAC occurred [46].

In another study, Mustapha et al. [47] inspected the adsorptive removal of MG onto the treated EFB by the alkaline method. It was deduced that the alkaline method treatment enhanced the porosity of EFB since the SEM image (Figure 6.3A) of the raw EFB showed a blockage surface, which may be due to the presence of dirt and cementing substances. Meanwhile, the SEM image (Figure 6.3B) of the treated EFB surface possessed many opened cavities since the alkaline treatment approach removes the dirt substances and activates the OH groups of EFB [48]. Thence, the treated EFB possessed a promising adsorption capacity toward MG (1250 mg/g) compared to the raw EFB (714.3 mg/g).

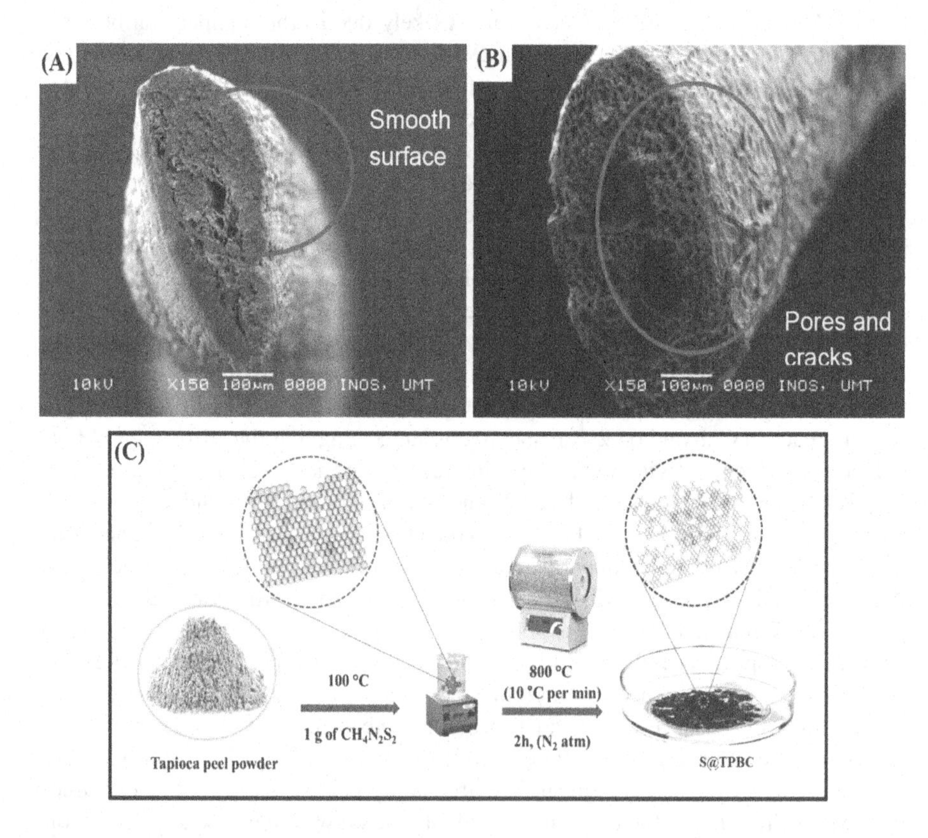

FIGURE 6.3 SEM of (A) raw EFB, (B) treated EFB [47], and (C) the fabrication approach of S@TPBC. Reproduced with permission [49]. (Copyright 2021, Elsevier.)

In another attempt, Vigneshwaran et al. [49] developed the tapioca peel by preparing sulfur-treated tapioca peel biochar (Figure 6.3C) to adsorb MG and Rhodamine B (RB) from aqueous media. It was found that the adsorption of both MG and RB onto S@TPBC was controlled by electrostatic interactions, H-bonding and surface interactions. Kinetics study pointed out that the adsorption of both MG and RB fitted the pseudo-second order model, emphasizing the presence of electrostatic interactions and complexation between S@TPBC and the dyes. Besides, the abundant OH groups onto the S@TPBC surface formed H-bonding (H-bond donor) with nitrogen and oxygen of the dyes (H-bond acceptor). Indeed, H-bonding was weak compared to complexation and electrostatic interactions; however, plenty of H-bonding affected the adsorption process [50]. Moreover, both adsorption processes were strongly affected by the surface area and porosity of S@TPBC, the surface roughness, and the solution pH.

6.3.2 Removal of Heavy Metal

In one investigation, Pap et al. [51] fabricated plum stone-derived activated carbon (PSAC) (Figure 6.4A) for the removal of lead (Pb(II)), cadmium (Cd(II)), and nickel (Ni(II)). It was found that the q_{max} of Pb(II), Cd(II), and Ni(II) onto PSAC was 172.43, 112.74, and 63.74 mg/g, respectively. More importantly, the mutual effect of metal ions on the adsorption aptitude of the metal ions on PSAC was studied in a binary and ternary system (Figure 6.4B). In the binary systems, the existence of Cd(II) declined the removal efficiency of Ni(II) by about 6% in the Ni(II)-Cd(II)

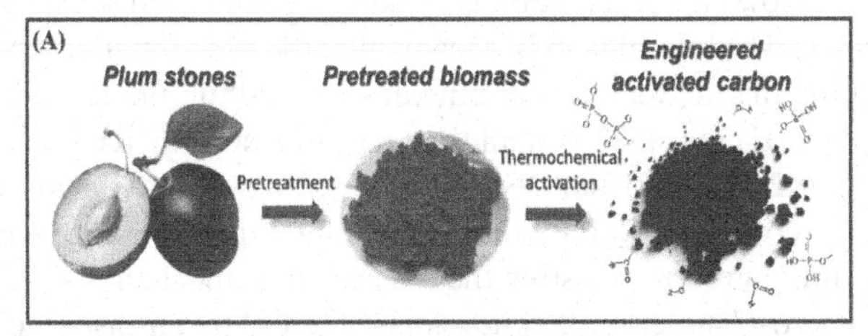

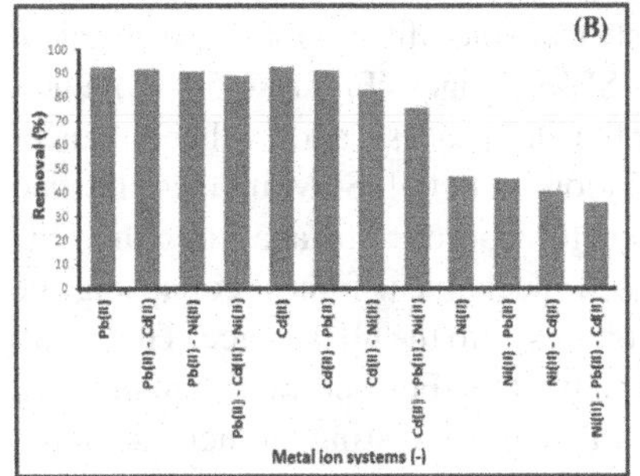

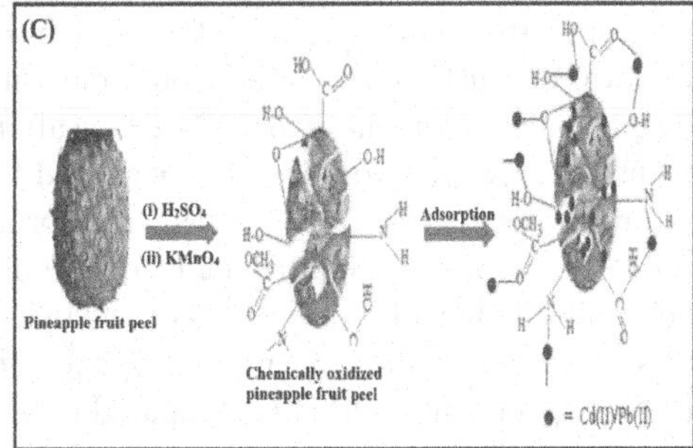

FIGURE 6.4 (A) The preparation of PSAC from plum stones and (B) interaction between metal ions in binary and ternary system, reproduced with permission [51]. (C) Adsorption mechanism of Cd (II) and Pb (II) onto pineapple fruit peel. Reproduced with permission [53]. (Copyright 2017, Elsevier; Copyright 2016, Taylor and Francis.)

system, while the removal efficiency of Ni(II) in the Ni(II)-Pb(II) system dwindled only 1%. This finding reflected the strong inhibitory effect of Ni(II) on the adsorption of Cd(II) and Pb(II). In the ternary systems, it was noticeable that the removal efficiency of Pb(II) slightly decreased in the presence of Cd(II) and Ni(II), evincing the inconsiderable inhibitory effect of both Cd(II) and Ni(II) on the adsorption of Pb(II) onto PSAC. This result may be anticipated by the large atomic radius and the high electronegativity of Pb(II), facilitating the binding of Pb(II) to the oxygen-active sites of PSAC. Moreover, it was suggested that the adsorption of Cd(II), Ni(II), and Pb(II) onto PSAC occurred via ion exchange, ion trapping into the pores, and surface adsorption; in addition to the electrostatic interactions between the phosphoric and oxygen functional groups and the cationic metal ions [52]. This proposition was in line with the Ahmad et al. [53] study that suggested the adsorptive removal of Cd(II) and Pb(II) onto the oxidized pineapple fruit peel via the electrostatic interaction between the deprotonated OH and COOH groups in the pineapple fruit peel and the metal ions (Fig. 6.4C). In addition, the metal chelation by the pineapple fruit peel with both Pb(II) and Cd(II) could form a coordination bond by accepting electrons from oxygen or nitrogen of pineapple into the vacant d-orbital [54]. Some studies proposed the possibility of the metal cations to form unidentate, bidentate, and bridge complexes with COO^- [55, 56].

In yet another attempt, Phuengphai et al. scrutinized the adsorption performances of passion fruit peel (PFP), dragon fruit peel (DFP), and rambutan peel (RP) toward the removal of Cu(II). It was reported that the optimum pH to adsorb Cu(II) onto PFP was 5, while the higher adsorption capacity of Cu(II) onto DFP and RP was obtained at pH 4. This finding may be attributed to the less ionization of the functional groups of the biosorbents at low acidic conditions, leading to an electrostatic repulsion between Cu(II) ions and biosorbents [57]. Meanwhile, the functional groups ionized with the raising in the pH medium, generating an electrostatic interaction between the biosorbents and Cu(II) ions [58]. Notably, the q_{max} of Cu(II) by RP (192.30 mg/g) was higher than PFP (92.59 mg/g) and DFP (121.95 mg/g), owing to the looser porosity of the PR surface compared to the other biosorbents. In addition, PR possesses more oxygen-containing groups, boosting the ion exchange mechanism.

In this vein, Romero-Cano et al. [59] developed the OP by an instant controlled pressure drop (DIC) process, followed by surface modification for the efficient removal of Cu(II) from an aqueous solution. SEM images (Figure 6.5A–C) elucidated the improvement of the OP surface after the DIP process, and further enhancement was observed after the chemical modification. In detail, SEM images clarified the nonporous surface of OP that transformed to a porous surface after the DIC process, which may be attributed to the induced mechanical forces, collapsing the cell walls of OP and forming micro- and macropores onto the OP surface. This result was confirmed by the increase in the peak intensity of the Fourier transform infrared (FTIR) spectrum of OP-DIC compared to the raw OP, suggesting an increase in the active sites (viz. OH and COOH) on the OP surface after the DIP process. Moreover, the experimental results showed that proceeding DIP process for three cycles at pressure 3 bar for 90 seconds boosted the number of acid and base sites by about 1.56 and 4.07 folds, respectively, compared to raw OP, whereas the chemical modification of OP-DIP declined the number of base sites and increased the acid sites amount

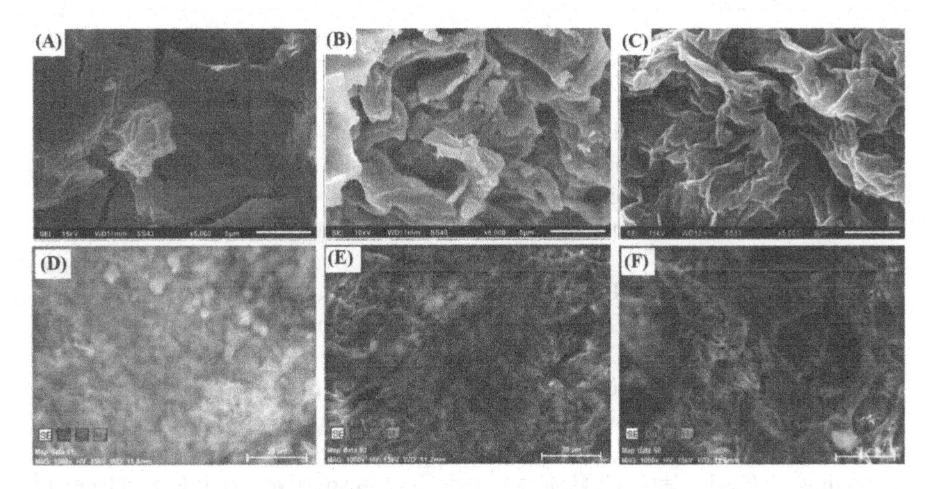

FIGURE 6.5 SEM of (A) OP, (B) OP-DIP, and (C) OP-DIP-AC and SEM-EDS mappings of (D) OP, (E) OP-DIP, and (F) OP-DIP-AC after the adsorption of Cu (II). Reproduced with permission [59]. (Copyright 2016, Elsevier.)

since the amount of acid and base sites depends on the type of chemical modifications and directly affects the pH_{PZC} of the modified material. Moreover, the SEM of OP-DIP-AC depicted an increase in the cavity width, reflecting the positive impact of the chemical modification on the porosity of the biosorbent that directly boosts its adsorption capacity. It was found that the adsorption capacities of Cu(II) onto OP, OP-DIP, and OP-DIP-AC were 31.79, 67.32, and 106.91 mg/g, respectively. Such enhancement in the adsorption aptitude of Cu(II) may be due to the higher amount of acid sites on OP-DIP and OP-DIP-AC than raw OP. SEM-EDS mappings (Figure 6.5D–F) confirmed the increase in the adsorbed Cu(II) ions onto OP-DIP-AC and OP-DIP than raw OP.

In one attempt, Chen et al. [60] studied the adsorptive removal of Cd(II) onto four fruit wastes including OP, litchi peel (LP), BP, and pomegranate peel (PP). It was found that q_{max} of Cd(II) onto LP, OP, PP, and BP was 230.5, 170.3 132.5, and 98.4 mg/g, respectively, at pH 5 and 25°C. The higher q_{max} of Cd(II) onto LP is most likely due to plenty of negative charges onto the LP surface as clarified from ZP measurements since pH_{PZC} of LP reached -18 mV. This finding suggested the domination of the electrostatic interaction on the Cd(II) adsorption onto the biosorbents. In addition, fruit peels contain abundant functional groups, including OH, COOH, OCH_3, $CONH_2$, NH_2, SH, amino acids, and polyphenols, increasing the possibility of various interactions between Cd(II) and the biosorbents to occur [61]. Thence, it was proposed that the adsorption of Cd(II) onto LP, OP, PP, and BP occurred via electrostatic interaction, ion exchange, surface adsorption, chelation, complexation, and precipitation [62]. Feng et al.'s study proposed that these mechanisms often coexist together [63]. Moreover, the desorption study reflected the excellent reusability of the biosorbents since the removal percent of Cd(II) by LP, OP, PP, and BP declined only to 6.51, 7.57%, 8.38, and 11.08%, respectively, after the 10th cycle. Furthermore, Mallampati et al. [64] investigated the reusability of Hami melon peel at diverse pH, showing low desorption percent of the metal cations (Ni(II) and Pb(II)) at pH 10. Nonetheless, the desorption percent of Ni(II) and Pb(II) reached 96% at pH 4 within 10 min. In addition, the weight loss of the biosorbent at a pH range of 4–10 was negligible without any change in the morphology or the chemical composition of the Hami melon peel. It was observed degradation in the peel when the pH was raised to 12, so regeneration of the Hami melon peel was executed at acidic medium pH to maintain the efficiency of the peel.

In another attempt, Patil et al. [65] scrutinized the adsorb ability of mahogany fruit shell activated carbon (MFSAC) toward Pb(II) ions. XRD pattern of MFSAC revealed two peaks at $2\theta = 21.58°$ and 43.36°, which are ascribed to [002] and [100] planes, respectively [66]. Moreover, the FTIR spectrum of MFSAC clarified that the absorption peaks situated the peaks at 3787.97, 3429.03, 2932.84, 2321.40, 1614.95, 1225.21, 1730.20, and 1029.91 cm^{-1} are assigned to free OH, bonded OH, aromatic stretching of C–H, C≡C, C=C, C=O, and C–O, respectively. FTIR spectrum of MFSAC after the adsorption of Pb(II) elucidated the disappearance of the peaks at 1730.20 and 1029.91 cm^{-1}, which may be attributed to the interaction between Pb(II), C=O, and C–O. Besides, the XPS wide spectra implied the adsorption of Pb(II) onto MFSAC since the distinguishing peak of Pb(II) appeared at 139 eV. The Pb 4f spectrum showed the peaks at 140.96 and 136.24 eV, which are assigned to Pb 4f5/2 and 4f7/2, reflecting that the adsorbed Pb is in the oxidation state +2 [67]. It was reported that the calculated q_{max} of Pb(II) under Langmuir was 322.58 mg/g at pH 4.5 and 25°C. Furthermore, the adsorption of Pb(II) onto MFSAC obeyed Langmuir isotherm model and pseudo-second order kinetic model. Besides, the thermodynamics study pointed out that the Pb(II) adsorption onto MFSAC was a spontaneous and endothermic process.

6.3.3 Removal of Pharmaceutical Pollutants

In one study, Cheng et al. [68] fabricated pomelo peel-derived biochar (PPBC) for the adsorptive removal of tetracycline. It was reported that PPBC possessed an exceptional adsorption capacity toward tetracycline (TC), oxytetracycline (OTC), and chlortetracycline (CTC), which reached 476.19, 407.50, and 555.56 mg/g, respectively. Indeed, for such polyprotic pollutants (TCs exist as a cation at $pH < 3.3$, a zwitterion at pH 3.3 to 7.8 and an anion at $pH > 7.8$), the solution pH has a vast effect on the adsorption efficacy [69]. Meanwhile, ZP measurements revealed that pH_{PZC} was 3.2, so the PPBC was positively charged at $pH < 3.2$ and negatively charged at $pH > 3.2$. Consequently, the adsorption of TCs onto PPBC was studied at a wide pH scale, clarifying the poor adsorption capacity of TCs at $pH < 3.2$. This finding may be due to the strong repulsion forces between the cationic TCs and the positively charged PPBC, while the adsorption capacity of TCs onto PPBC increased with the increase in pH, owing to the electrostatic interactions between the zwitterionic TCs and the negatively charged PPBC. Notably, it was observed that the optimum pH to adsorb TCs onto PPBC was 8.5, although repulsion forces were between the negatively charged PPBC and the anionic TCs. These results inferred that the electrostatic interaction was not the dominant mechanism of the TC adsorption, and there were other mechanisms that controlled the processes [70, 71]. In the light of SEM and BET, PPBC possessed high surface area and a porous structure. Hence, it was proposed that the pore filling was the controlled mechanism on the TC adsorption processes [72]. In addition, Raman spectrum confirmed the presence of graphitic layers in PPBC, suggesting the occurrence of the π-π interaction mechanism between the aromatic ring of TCs and graphene layers of PPBC [73]. This plausible mechanism was consistent with the Yu et al. study [74] that scrutinized the adsorptive performance of the grapefruit peel-derived biochar (GPBC) toward TC. BET measurements revealed that GPBC has a high S_{BET} (130.83) and the pore diameter was 2.35 nm, suggesting the possibility of the TC molecules (<1.27 nm) to diffuse into the GPBC pores which evinced pore filling effect on the TC adsorption process. Moreover, the FTIR spectrum of GPBC after the TC adsorption process showed a shift in the belonging peak to OH, suggesting the formation of H-bonding between TC and GPBC. Besides, the shift occurred in the related peak to C=O, indicating the presence of the π-π interaction or the electron acceptor-donor interaction between TC and GPBC [75]. The GPCB fabrication and the proposed TC adsorption mechanism are represented in Figure 6.6.

In another investigation, Sathish Kumar et al. [76] highlighted the removing of diclofenac (DC) by terminalia catappa peel (TCP). It was deduced that the optimum pH of the DC adsorption onto TCP was 5 since the removal percent of DC reached 89%; however, the raising in pH caused a significant decrease in the DC adsorption (51%). This result was consistent with the Attia et al. study [77] as the DC molecules become neutral at pH< pKa and they interact with the adsorbent via non-electrostatic

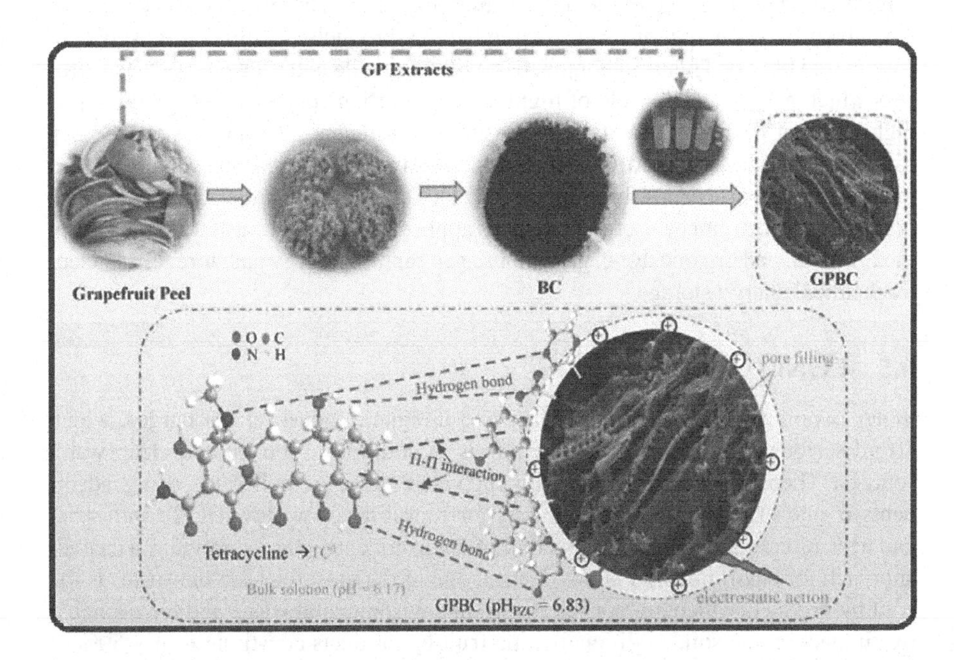

FIGURE 6.6 Represented the fabrication of GPBC and the plausible TC adsorption mechanism. (Reproduced with permission [74]; Copyright 2020, Elsevier).

interactions. While at pH > pKa, the DC molecules exist in an anionic form, resulting in electrostatic repulsion forces between DC and the adsorbent. Since pKa of DC is 4 ± 0.2, the adsorption aptitude of DC onto TCP decreased in the basic conditions [78]. Moreover, the impact of the system temperature on the adsorption efficacy of DC was studied at a temperature range of 20–50°C. It was found a diminution in the removal percent of DC at low temperature, which may be ascribed to the functional groups of TCP that are not active. Also, the removal percent of DC declined with the raising in the system temperature, which is most likely due to the increase in the DC solubility, resulting in a stronger interaction between the DC molecules and the solute than the interaction between TCP and DC [79]. In addition, the H-bonding between DC and TCP is broken at a high temperature. As a result, there was an increase in the Brownian movement of the DC molecules in solution [80]. Therefore, a reusability test was implemented at 60°C, revealing that the desorption percent of DC reached 92% after 6 h.

Thence, the removal percent of DC by TCP was still over 85% after the 8th cycle and gradually declined to 50% after the 12th cycle.

6.4 ECONOMIC FEASIBILITY FOR INDUSTRIAL APPLICATIONS

Importantly, to afford porous carbons for industrial application (produce in ton scale), the carbon precursors should be available at a cost-effective price and in a sustainable and abundant form. Hence, in the usage of fruit by-product waste as porous carbon precursors, the abundant, renewable and cost-effective process is achieved [18]. This is due to the huge availability of fruit by-product waste, as huge annual world production of fruits and the fruit by-products constitute almost half of fruit mass [18, 24]. This is in conjunction with the facile and efficient synthesis process which affords ~30% yield of highly porous carbon containing naturally doped high electron density elements which afford an efficient adsorption site [18, 24]. Interestingly, the developed fruit by-product-derived porous carbons characterized by high textural properties (high specific surface area, pore volume, and hierarchical porous structures) can be useful for various applications such as wastewater treatment (heavy metal and organic dyes), greenhouse gas capture, energy gas storage, and electrochemical energy storage.

6.5 CONCLUSION

Water is polluted with various organic and inorganic industrial discharges, which afford serious environmental issues; this is in addition to the limited clean water sources. Therefore, wastewater treatment via removal of pollutants using adsorbents is one of the promising solutions for recycling of wastewater for industrial and agricultural applications. Recycling of environmental wastes provides a cleaner approach for environmental issues. Hence, one of the most released biowastes is the fruit by-products which are considered as an environmental waste and have a negative impact on the soil. Additionally, the fruit by-products constitute nearly 50% of the fruit masses. Therefore, recycling of this undesirable waste to porous carbons is a beneficial approach from environment and economy point of view. Variety of fruit by-products were recycled via carbonization and activation processes, yielding highly porous carbon materials naturally doped with unique adsorption sites. The developed porous carbon-based adsorbents derived from recycling of fruit by-products afford a cost-effective and efficient approach for the production of an efficient adsorbent for industrial wastewater treatments.

REFERENCES

1. T. Sreeprasad, S. M. Maliyekkal, T. Pradeep, Reduced graphene oxide-metal/metal oxide composites: Facile synthesis and application in water purification. J. Hazard. Mater. 186 (2011) 921–931.
2. W. Patrick, Egypt looks to the sea to meet its need for water, The National. URL https://www.thenational.ae/business/egypt-looks-to-the-sea-to-meet-its-need-for-water-1.58966 2017 (Accessed 15.05.2020).
3. W. H. Sun, H. M. Ang, M. O. Tadé, Adsorptive remediation of environmental pollutants using novel graphene-based nanomaterials. Chem. Eng. J. 226 (2013) 336–347.
4. A. S. Eltaweil, A. M. Omer, H. G. El-Aqapa, N. M. Gaber, N. F. Attia, G. M. El-Subruiti, M. S. Mohy-Eldin, E. M. Abd El-Monaem, Chitosan based adsorbents for the removal of phosphate and nitrate: A critical review. Carbohyd. Polym. 274 (2021) 118671.
5. A. Galal, M. M. Zaki, N. F. Atta, S. H. Samaha, H. E. Nasr, N. F. Attia, Electroremoval of copper ions from aqueous solutions using chemically synthesized polypyrrole on polyester fabrics. J. Water Proc. Eng. 43 (2021) 102287.
6. C. Gao, X.-Y. Yu, R.-X. Xu, J.-H. Liu, X.-J. Huang, AlOOH-reduced graphene oxide nanocomposites: One-pot hydrothermal synthesis and their enhanced electrochemical activity for heavy metal ions. ACS Appl. Mater. Inter. 4 (2012) 4672–4682.
7. L. Zhang, L. Yan, W. Xu, X. Guo, L. Cui, L. Gao, Q. Wei, B. Du, Adsorption of Pb (II) and Hg(II) from aqueous solution using magnetic $CoFe_2O_4$-reduced graphene oxide. J. Mol. Liq. 191 (2014) 177–182.
8. E. M. Abd El-Monaem, A. S. Eltaweil, H. M. Elshishini, M. Hosny, M. M. Abou Alsoaud, N. F. Attia, G. M. El-Subruiti, A. M. Omer. Sustainable adsorptive removal of antibiotic residues by chitosan composites: An insight into current developments and future recommendations. Arab. J. Chem. 15 (2022) 103743.
9. N. F. Attia, M. A. Diab, A. S. Attia, M. F. El-Shahat, Greener approach for fabrication of antibacterial graphene-polypyrrole nanoparticle adsorbent for removal of Mn^{2+} from aqueous solution. Synth. Metals. 282 (2021) 116951.
10. M. A. Diab, N. F. Attia, A. S. Attia, M. F. El-Shahat, Green synthesis of cost-effective and efficient nanoadsorbents based on zero- and two-dimensional nanomaterials for Zn^{2+} and Cr^{3+} removal from aqueous solutions. Synth. Metals. 265 (2020) 116411.
11. N. F. Attia, J. Park, H. Oh, Facile tool for green synthesis of graphene sheets and their smart free-standing UV protective film. Appl. Surf. Sci. 458 (2018) 425–430.
12. M. Sevilla, R. Mokaya, Energy storage applications of activated carbons: Supercapacitors and hydrogen storage. Energy Environ. Sci. 7 (2014) 1250–1280.
13. N. F. Attia, S. M. Lee, H. J. Kim, K. E. Geckeler, Nanoporous carbon-templated silica nanoparticles: Preparation, effect of different carbon precursors, and their hydrogen storage adsorption. Micro. Meso. Mater. 173 (2013) 139–146.
14. J. Park, N. F. Attia, M. Jung, M. Lee, K. Lee, J. Chung, H. OH, Sustainable nanoporous carbon for CO_2, CH_4, N_2, H_2 adsorption and CO_2/CH_4 and CO_2/N_2 separation. Energy. 158 (2018) 9–16.
15. J. Park, M. Jung, H. Jang, K. Lee, N. F. Attia, H. Oh, A facile synthesis tool of nanoporous carbon for promising H_2, CO_2, and CH_4 sorption capacity and selective gas separation. J. Mater. Chem. A. 6 (2018) 23087–23100.

16. S. E. A. Elashery, N. F. Attia, M. M. Omar, H. M. I. Tayea, Cost-effective and green synthesized electroactive nanocomposite for high selective potentiometric determination of clomipramine hydrochloride. Microchem. J. 151 (2019) 104222.
17. N. F. Attia, M. Jung, J. Park, S. Y. Cho, H. Oh, Facile synthesis of hybrid porous composites and its porous carbon for enhanced H_2 and CH_4 storage. Int. J. Hydrogen Energy. 45 (2020) 32797–32807.
18. M. Jung, J. Park, K. Lee, N. F. Attia, H. Oh, Effective synthesis route of renewable nanoporous carbon adsorbent for high energy gas storage and CO_2/N_2 selectivity. Renew. Energy. 161 (2020) 30–42.
19. A. El Nemr, O. Abdelwahab, A. El-Sikaily, A. Khaled, Removal of direct N blue-106 from artificial textile dye effluent using activated carbon from orange peel: Adsorption isotherm and kinetic studies. J. Hazard. Mater. 161 (2009) 102–110.
20. K. Y. Foo, B. H. Hameed, Preparation, characterization and evaluation of adsorptive properties of orange peel based activated carbon via microwave induced K_2CO_3 activation. Bioresour. Technol. 104 (2012) 679–686.
21. E. M. Lotfabad, J. Ding, K. Cui, A. Kohandehghan, W. P. Kalisvaart, M. Hazelton, D. Mitlin, High-density sodium and lithium ion battery anodes from banana peels. ACS Nano. 8 (2014) 7115–7129.
22. S. T. Senthilkumar, R. K. Selvan, J. S. Melo, C. Sanjeeviraja, High performance solid-state electric double layer capacitor from redox mediated gel polymer electrolyte and renewable tamarind fruit shell derived porous carbon. ACS Appl. Mater. Interfaces. 5 (2013) 10541–10550.
23. A. A. Arie, H. Kristianto, E. Demir, R. D. Cakan, Activated porous carbons derived from the Indonesian snake fruit peel as anode materials for sodium ion batteries. Mater. Chem. Phys. 217 (2018) 254–261.
24. N. R. Kim, Y. S. Yun, M. Y. Song, S. J. Hong, M. Kang, C. Leal, Y. W. Park, H.-J. Jin, Citrus- peel-derived, nanoporous carbon nanosheets containing redox-active heteroatoms for sodium-ion storage. ACS Appl. Mater. Interfaces. 8 (2016) 3175–3181.
25. H. A. El-aal, F. T. Halaweish, Food preservative activity of phenolic compounds in orange peel extracts (*Citrus sinensis* L.). Lucrări Stiinţifice. 53 (2010) 233–240.
26. P. Nowicki, J. Kazmierczak-Razna, R. Pietrzak, Physicochemical and adsorption properties of carbonaceous sorbentsprepared by activation of tropical fruit skins with potassium carbonate. Mater. Des. 90 (2016) 579–585.
27. J. Park, S. Y. Cho, M. Jung, K. Lee, Y. C. Nah, N. F. Attia, H. Oh, Efficient synthetic approach for nanoporous adsorbents capable of pre-and post-combustion CO_2 capture and selective gas separation. J. CO_2 Util. 45 (2021) 101404.
28. M. Jung, J. Park, S. Y. Cho, S. E. A. Elashery, N. F. Attia, H. Oh, Flexible carbon sieve based on nanoporous carbon cloth for efficient CO_2/CH_4 separation. Surf. Interfaces. 23 (2021) 100960.
29. N. F. Attia, M. Jung, J. Park, H. Jang, K. Lee, H. Oh, Flexible nanoporous activated carbon cloth for achieving high H_2, CH_4, and CO_2 storage capacities and selective CO_2/CH_4 separation. Chem. Eng. J. 379 (2020) 122367.
30. K. S. W. Sing, D. H. Everett, R. A. W. Haul, L. Moscou, R. A. Pierotti, J. Rouquerol, T. Siemieniewska, Reporting physisorption data for gas/solid systems with special reference to the determination of surface area and porosity (Recommendations 1984). Pure Appl. Chem. 57 (1985) 603–612.
31. N. F. Attia, S. M. Lee, H. J. Kim, K. E. Geckeler, Nanoporous polypyrrole: Preparation and hydrogen storage properties. Int. J. Energy Res. 38 (2014) 466–476.
32. N. Priyantha, L. Lim, M. Dahri, Dragon fruit skin as a potential biosorbent for the removal of methylene blue dye from aqueous solution. Int. Food Res. J. 22 (2015) 2141–2148.

33. A. Seidmohammadi, G. Asgari, A. Dargahi, M. Leili, Y. Vaziri, B. Hayati, A. Shekarchi, A. Mobarakian, A. Bagheri, S. Nazari Khanghah, A comparative study for the removal of methylene blue dye from aqueous solution by novel activated carbon based adsorbents, progress in color. Prog. Color. 12 (2019) 133–144.
34. A. H. Jawad, A. S. Abdulhameed, L. D. Wilson, S. S. A. Syed-Hassan, Z. A. Alothman, M. R. Khan, High surface area and mesoporous activated carbon from KOH-activated dragon fruit peels for methylene blue dye adsorption: Optimization and mechanism study. Chin. J. Chem. Eng. 32 (2021) 281–290.
35. A. El-Monaem, M. Eman, A. M. Omer, G. M. El-Subruiti, M. S. Mohy-Eldin, A. S. Eltaweil, Zero-valent iron supported-lemon derived biochar for ultra-fast adsorption of methylene blue. Biomass Convers. Biorefin. (2022) 1–13. https://doi.org/10.1007/s13399-022-02362-y
36. H. Li, L. Liu, J. Cui, J. Cui, F. Wang, F. Zhang, High-efficiency adsorption and regeneration of methylene blue and aniline onto activated carbon from waste edible fungus residue and its possible mechanism. RSC Adv. 10 (2020) 14262–14273.
37. S. Fan, Y. Wang, Z. Wang, J. Tang, J. Tang, X. Li, Removal of methylene blue from aqueous solution by sewage sludge-derived biochar: Adsorption kinetics, equilibrium, thermodynamics and mechanism. J. Environ. Chem. Eng. 5 (2017) 601–611.
38. M. Elkady, H. Shokry, A. El-Sharkawy, G. El-Subruiti, H. Hamad, New insights into the activity of green supported nanoscale zero-valent iron composites for enhanced acid blue-25 dye synergistic decolorization from aqueous medium. J. Mol. Liq. 294 (2019) 111628.
39. B.-S. Trinh, P. T. Le, D. Werner, N. H. Phuong, T. L. Luu, Rice husk biochars modified with magnetized iron oxides and nano zero valent iron for decolorization of dyeing wastewater. Processes. 7 (2019) 660.
40. S. Bentahar, A. Dbik, M. El Khomri, N. El Messaoudi, A. Lacherai, Adsorption of methylene blue, crystal violet and Congo red from binary and ternary systems with natural clay: Kinetic, isotherm, and thermodynamic. J. Environ. Chem. Eng. 5 (2017) 5921–5932.
41. M. S. Sajab, C. H. Chia, S. Zakaria, P. S. Khiew, Cationic and anionic modifications of oil palm empty fruit bunch fibers for the removal of dyes from aqueous solutions. Bioresour. Technol. 128 (2013) 571–577.
42. I. Mall, V. Srivastava, G. Kumar, I. Mishra, Characterization and utilization of mesoporous fertilizer plant waste carbon for adsorptive removal of dyes from aqueous solution. Colloids Surf. A Physicochem. Eng. Asp. 278 (2006) 175–187.
43. O. S. Bello, Adsorptive removal of malachite green with activated carbon prepared from oil palm fruit fibre by KOH activation and CO_2 gasification. S. Afr. J. Chem. 66 (2013) 32–41.
44. G. Stavropoulos, A. Zabaniotou, Production and characterization of activated carbons from olive-seed waste residue. Microporous Mesoporous Mater. 82 (2005) 79–85.
45. R.-L. Tseng, S.-K. Tseng, F.-C. Wu, Preparation of high surface area carbons from corncob with KOH etching plus CO_2 gasification for the adsorption of dyes and phenols from water. Colloids Surf. A Physicochem. Eng. Asp. 279 (2006) 69–78.
46. M.-H. Baek, C. O. Ijagbemi, O. Se-Jin, D.-S. Kim, Removal of malachite green from aqueous solution using degreased coffee bean. J. Hazard. Mater. 176 (2010) 820–828.
47. R. Mustapha, A. Ali, G. Subramaniam, A. A. A. Zuki, M. Awang, M. H. C. Harun, S. Hamzah, Removal of malachite green dye using oil palm empty fruit bunch as a low-cost adsorbent. Biointerface Res. Appl. Chem. 11 (2021) 14998–15008.
48. R. Naseer, N. Afzal, S. Saeed, H. Mujhahid, S. Faryal, S. Aslam, Effect of Bronsted base on topological alteration of rice husk as an efficient adsorbent comparative to rice husk ash for azo dyes. Pol. J. Environ. Stud. 29 (2020) 2795–2802.

49. S. Vigneshwaran, P. Sirajudheen, P. Karthikeyan, S. Meenakshi, Fabrication of sulfur-doped biochar derived from tapioca peel waste with superior adsorption performance for the removal of malachite green and Rhodamine B dyes. Surf. Interfaces. 23 (2021) 100920.
50. P. Sirajudheen, P. Karthikeyan, S. Vigneshwaran, M. Nikitha, C. A. Hassan, S. Meenakshi, Ce (III) networked chitosan/β-cyclodextrin beads for the selective removal of toxic dye molecules: Adsorption performance and mechanism. Carbohydr. Polym. Tech. App. 1 (2020) 100018.
51. S. Pap, T. Š. Knudsen, J. Radonić, S. Maletić, S. M. Igić, M. T. Sekulić, Utilization of fruit processing industry waste as green activated carbon for the treatment of heavy metals and chlorophenols contaminated water. J. Clean. Prod. 162 (2017) 958–972.
52. M. Özacar, İ. A. Şengil, H. Türkmenler, Equilibrium and kinetic data, and adsorption mechanism for adsorption of lead onto valonia tannin resin. Chem. Eng. J. 143 (2008) 32–42.
53. A. Ahmad, A. Khatoon, S.-H. Mohd-Setapar, R. Kumar, M. Rafatullah, Chemically oxidized pineapple fruit peel for the biosorption of heavy metals from aqueous solutions. Desalination Water Treat. 57 (2016) 6432–6442.
54. P. Chand, A. K. Shil, M. Sharma, Y. B. Pakade, Improved adsorption of cadmium ions from aqueous solution using chemically modified apple pomace: Mechanism, kinetics, and thermodynamics. Int. Biodeterior. Biodegradation. 90 (2014) 8–16.
55. J. P. Chen, L. Yang, Study of a heavy metal biosorption onto raw and chemically modified *Sargassum* sp. via spectroscopic and modeling analysis. Langmuir. 22 (2006) 8906–8914.
56. L. H. Velazquez-Jimenez, A. Pavlick, J. R. Rangel-Mendez, Chemical characterization of raw and treated agave bagasse and its potential as adsorbent of metal cations from water. Ind. Crops. Prod. 43 (2013) 200–206.
57. K. Keawkim, A. Khamthip, Removal of Pb^{2+} ion from industrial wastewater by new efficient biosorbents of Oyster plant (*Tradescantia spathacea* steam) and Negkassar leaf (*Mammea siamensis* T. Anderson). Chiang. Mai. J. Sci. 45 (2018) 369–379.
58. C. L. Sanders, K. H. Eginger, Optimizing care of the mechanically ventilated patient in the emergency department through the utilization of validated sedation scoring scales. J. Emerg. Nurs. 43 (2017) 84–86.
59. L. A. Romero-Cano, L. V. Gonzalez-Gutierrez, L. A. Baldenegro-Perez, Biosorbents prepared from orange peels using instant controlled pressure drop for Cu (II) and phenol removal. Ind. Crops Prod. 84 (2016) 344–349.
60. Y. Chen, H. Wang, W. Zhao, S. Huang, Four different kinds of peels as adsorbents for the removal of Cd (II) from aqueous solution: Kinetics, isotherm and mechanism. J. Taiwan. Inst. Chem. Eng. 88 (2018) 146–151.
61. H.-P. Chao, C.-C. Chang, A. Nieva, Biosorption of heavy metals on citrus maxima peel, passion fruit shell, and sugarcane bagasse in a fixed-bed column. J. Ind. Eng. Chem. 20 (2014) 3408–3414.
62. G. S. Simate, S. Ndlovu, The removal of heavy metals in a packed bed column using immobilized cassava peel waste biomass. J. Ind. Eng. Chem. 21 (2015) 635–643.
63. N. Feng, X. Guo, S. Liang, Y. Zhu, J. Liu, Biosorption of heavy metals from aqueous solutions by chemically modified orange peel. J. Hazard. Mater. 185 (2011) 49–54.
64. R. Mallampati, L. Xuanjun, A. Adin, S. Valiyaveettil, Fruit peels as efficient renewable adsorbents for removal of dissolved heavy metals and dyes from water. ACS Sustain. Chem. Eng. 3 (2015) 1117–1124.
65. S. A. Patil, U. P. Suryawanshi, N. S. Harale, S. K. Patil, M. M. Vadiyar, M. N. Luwang, M. A. Anuse, J. H. Kim, S. S. Kolekar, Adsorption of toxic Pb (II) on activated carbon derived from agriculture waste (Mahogany fruit shell): Isotherm, kinetic and thermodynamic study. J. Environ. Anal. Chem. 102 (2020) 8270–8286.

66. H. Chen, P. Xia, W. Lei, Y. Pan, Y. Zou, Z. Ma, Preparation of activated carbon derived from biomass and its application in lithium–sulfur batteries. J. Porous Mater. 26 (2019) 1325–1333.
67. A. Hesham, R. Philip, An XPS study of the adsorption of lead on goethite (α-FeOOH). Appl. Surf. Sci. 136 (1998) 46–54.
68. D. Cheng, H. H. Ngo, W. Guo, S. W. Chang, D. D. Nguyen, X. Zhang, S. Varjani, Y. Liu, Feasibility study on a new pomelo peel derived biochar for tetracycline antibiotics removal in swine wastewater. Sci. Total Environ. 720 (2020) 137662.
69. N. F. Attia, S. E. A. Elashery, A. M. Zakria, A. S. Eltaweil, H. Oh, Recent advances in graphene sheets as new generation of flame retardant materials. Mater. Sci. Eng. B. 274 (2021) 115460.
70. A. M. Omer, A. El-Monaem, M. Eman, G. M. El-Subruiti, A. El-Latif, M. Mona, A. S. Eltaweil, Fabrication of easy separable and reusable MIL-125 (Ti)/MIL-53 (Fe) binary MOF/CNT/alginate composite microbeads for tetracycline removal from water bodies. Sci. Rep. 11 (2021) 1–14.
71. W. Xiong, G. Zeng, Z. Yang, Y. Zhou, C. Zhang, M. Cheng, Y. Liu, L. Hu, J. Wan, C. Zhou, Adsorption of tetracycline antibiotics from aqueous solutions on nanocomposite multi-walled carbon nanotube functionalized MIL-53 (Fe) as new adsorbent. Sci. Total Environ. 627 (2018) 235–244.
72. K. Premarathna, A. U. Rajapaksha, B. Sarkar, E. E. Kwon, A. Bhatnagar, Y. S. Ok, M. Vithanage, Biochar-based engineered composites for sorptive decontamination of water: A review. J. Chem. Eng. 372 (2019) 536–550.
73. H. Jin, S. Capareda, Z. Chang, J. Gao, Y. Xu, J. Zhang, Biochar pyrolytically produced from municipal solid wastes for aqueous As (V) removal: Adsorption property and its improvement with KOH activation. Bioresour. Technol. 169 (2014) 622–629.
74. H. Yu, L. Gu, L. Chen, H. Wen, D. Zhang, H. Tao, Activation of grapefruit derived biochar by its peel extracts and its performance for tetracycline removal. Bioresour. Technol. 316 (2020) 123971.
75. B. Wang, Y.-S. Jiang, F.-Y. Li, D.-Y. Yang, Preparation of biochar by simultaneous carbonization, magnetization and activation for norfloxacin removal in water. Bioresour. Technol. 233 (2017) 159–165.
76. P. Sathishkumar, M. Arulkumar, V. Ashokkumar, A. R. M. Yusoff, K. Murugesan, T. Palvannan, Z. Salam, F. N. Ani, T. Hadibarata, Modified phyto-waste terminalia catappa fruit shells: A reusable adsorbent for the removal of micropollutant diclofenac. RSC Adv. 5 (2015) 30950–30962.
77. T. M. S. Attia, X. L. Hu, Synthesized magnetic nanoparticles coated zeolite for the adsorption of pharmaceutical compounds from aqueous solution using batch and column studies. Chemosphere. 93 (2013) 2076–2085.
78. E. M. Cuerda-Correa, J. R. Domínguez-Vargas, F. J. Olivares-Marín, J. B. de Heredia, On the use of carbon blacks as potential low-cost adsorbents for the removal of non-steroidal anti-inflammatory drugs from river water. J. Hazard. Mater. 177 (2010) 1046–1053.
79. Z. Zhou, S. Lin, T. Yue, T.-C. Lee, Adsorption of food dyes from aqueous solution by glutaraldehyde cross-linked magnetic chitosan nanoparticles. J. Food Eng. 126 (2014) 133–141.
80. H. Li, G. Huang, C. An, J. Hu, S. Yang, Removal of tannin from aqueous solution by adsorption onto treated coal fly ash: Kinetic, equilibrium, and thermodynamic studies. Ind. Eng. Chem. Res. 52 (2013) 15923–15931.

7 Metal Oxide-Based Antibacterial Nano-Agents for Wastewater Treatment

Mohamed S. Selim, Nesreen A. Fatthallah, Shimaa A. Higazy, Hekmat R. Madian, Zhifeng Hao, and Xiang Chen

7.1 INTRODUCTION

All living organisms require clean water for survival. However, the contamination of current water resources has intensified as a result of fast industrialization and the massive population explosion [1]. The demand for clean water in agriculture has greatly increased. Clean and fresh water with a wide range of pollutants is caused by industrial, domestic, and agricultural activities [2]. Heavy metal ions and dyes are the two primary types of contaminants. The polluted water shouldn't be consumed unless it has been purified first. After the ions of heavy metals have entered the water, it is very complicated to purify them totally [3]. These contaminants are hazardous to all living organisms and have a large influence on ecosystems. To avoid their detrimental effects on humans and the environment, these pollutants must be removed from contaminated water. Water supply entrances are currently confronted with a variety of issues. Millions of people around the world do not have access to clean drinking water [4]. Immediate action is essential in developing nations with insufficient wastewater treatment. On the other hand, the existing wastewater management technologies can provide safe drinking water that meets both environmental and human requirements.

Recent advancements in nanoscience and nanotechnology point to possibilities for bettering water resources [5, 6]. Nanoscience and nanotechnology are expected to promote highly effective and economical solutions for wastewater treatment [7]. Environmental friendly and efficient strategies for removing these contaminants are critical [8]. Ultrafiltration, reverse osmosis, solvent extraction, and evaporation are examples of wastewater treatment strategies that have been used in the literature. These methods, on the other hand, eliminate pollutants from water without producing innocuous end products [9].

Oxidation can quickly produce complete breakdown, either chemically or photochemically [10]. To decrease the impact of pollutants, each oxidative process generates free radicals of OH for using as an efficient oxidizing agent [11]. The primary

DOI: 10.1201/9781003354475-7

application of nanomaterials is to overcome the problems of major water and wastewater [12]. Nanomaterials are employed in a variety of industries including biomedicine, environmental detection, electronics, and cosmetics. The improvement of surface area to volume ratio and quantum characteristics are two physical aspects of these small nanomaterials.

Nanoparticles (NPs) have features that are significantly different from those of conventional materials, including biological, electrical, optical, and magnetic features [13]. Nanomaterials possess characteristics like high adsorption, catalytic activity, and reactivity [14]. NPs have received much attention in recent decades, and they've been used successfully in a variety of industrial sectors [15, 16]. They are widely employed in wastewater treatment [17–19]. These nanomaterials have high adsorption reactivity and capacity because of their huge surface area and small diameters [20]. Bacterial strains, contaminants, and inorganic and organic pollutants are eliminated by diverse types of nanomaterials around the world [21, 22]. Carbon nanotubes, nano-metal oxide particles, and nanostructured composites [23, 24] are all potential technologies for using in various wastewater ecosystems.

Nanomaterials offer novel water supply and water-resource solutions. Several wastewater treatment methods have been developed in the last several years [9, 25]. Solvent extraction, gravity separation, reverse osmosis, precipitation, microfiltration, electrodialysis, flotation coagulation, ultrafiltration, electrolysis, ion exchange, distillation, and adsorption are the most important methods [26]. Nano-metal oxide particles for wastewater treatment have a scarcity of review work in the literature. Vilardi et al. [19] talked about how metal oxide NPs are currently being used and how they affect biological wastewater treatment processes. They also summarized the various ways for measuring metal oxide NPs' suppression of nitrification, as well as the results achieved utilizing these methods. The potential impacts of four NP kinds (Ag, ZnO, TiO_2, and zero-valent iron) on the treatment of wastewater were investigated by Yang et al. [27]. They reported about metal oxide NPs that affect anaerobic sludge digesting and wastewater treatment.

Sing and coworkers [28] looked at how NPs could be employed in wastewater treatment. Different nano-metal oxides were studied in-depth in this review. MgO, TiO_2, MnO_2, Fe_3O_4/Fe_2O_3, MnO_2, Al_2O_3, and CeO_2 were among the metal oxide NPs studied by Junbai et al. [29] and their uses in water treatment. While some metal oxides are discussed in several publications for the treatment of wastewater, there is no enough data about the mechanism of these nano-metal oxides' antibacterial action [30, 31]. Abdelbasir and Shalan [32] reported the applications of different nanomaterials for wastewater treatment. Different metal oxides such as TiO_2, Fe_3O_4/Fe_2O_3, Al_2O_3, MnO_2, Mn_2O_3, and ZnO are discussed in terms of industrial uses in this chapter. To give a comprehensive analysis, it focuses on the applications of nano-metal oxides in wastewater treatment and the probable procedures employed to solve various issues confronting present wastewater treatment systems. CuO, Cu_2O, ZnO, iron oxide, Ag, and TiO_2 NPs are all extensively discussed. These nanomaterials are excellent for wastewater managements because they are dissolved and/or oxidized in aqueous solutions and metal ions are released, resulting in toxicity of metal ions. The nano-metal oxide particles are stable chemically and can be used for many applications including coatings, adsorption, and bacterial resistance activity.

7.2 WASTEWATER POLLUTION AND IMPACTS

Organic and inorganic components and microbes, as well as dangerous heavy metals, are all found in wastewater. These pollutants alter the physical, biological, and chemical features of pure water [33]. Industrial wastewater from waste sources contains urine, residential components, and organic and inorganic compounds [34]. Figure 7.1 depicts several wastewater sources. In wastewater, dangerous compounds such as radionuclides, heavy metals, and large microorganisms, including bacteria, viruses, and protozoa, can be detected. Wastewater is also a major source of waterborne illnesses such as typhoid and cholera, which can be fatal [35]. The contaminants found in water are illustrated in Figure 7.2 [36]. Pathogen toxicity and the risks of wastewater pollution to crops, humans, and animals should be studied during the wastewater treatment. To preserve the environment from contamination, the treatment of wastewater at the governmental and individual levels should be considered. For water purification from various impurities, wastewater treatment might include physical, chemical, and biological methods [37, 38]. The wastewater's physical features include dyes, total solids, and others (volatile, fixed, suspended, and dissolved) [39].

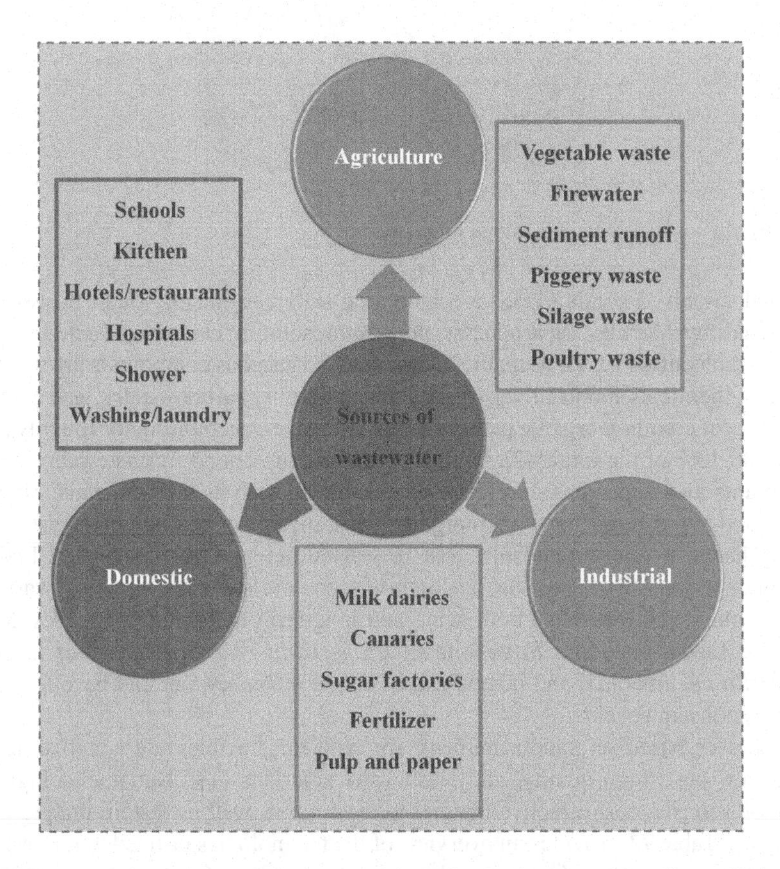

FIGURE 7.1 Different sources of wastewater including agriculture, domestic, and industrial applications.

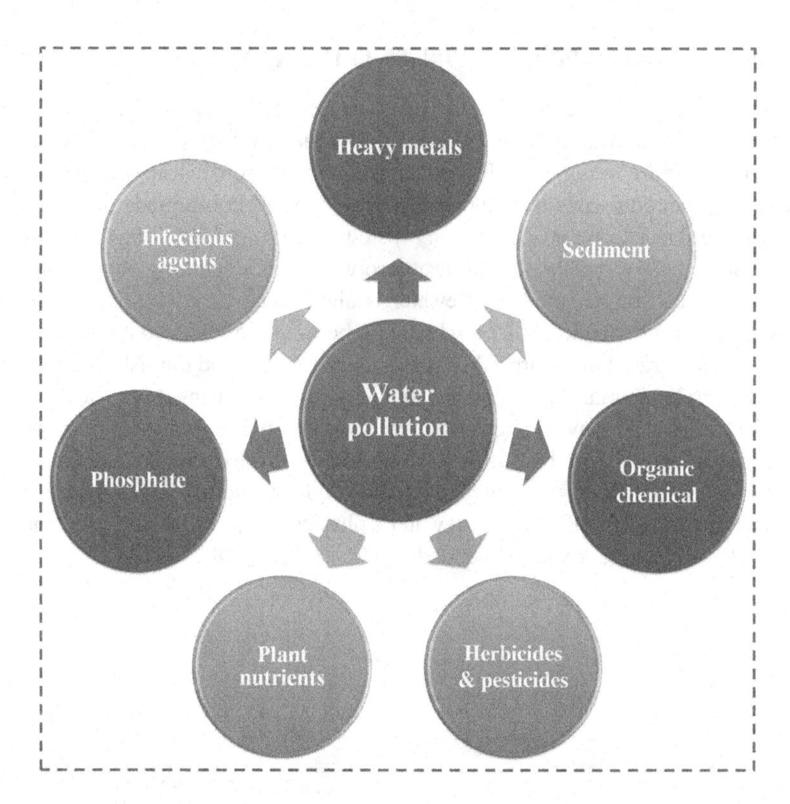

FIGURE 7.2 Major pollutants found in water.

Total dissolved solids (TDS) are inorganic salts and metals found in wastewater, including chlorides, bicarbonates, potassium, sodium, calcium, magnesium, and organic molecules [40, 41]. Organic, inorganic, and gaseous compounds are the three types of chemicals found in wastewater. Fats and oils, carbohydrates, and proteins are the most common organic pollutants in wastewater, accounting for roughly 50%, 40%, and 10% of the total [42]. Organic contaminants found in wastewater include surfactants and impurities. Inorganic contaminants such as heavy metals, nitrogen compounds, and other harmful inorganic substances can be found in wastewater. Living harmful microorganisms exist in wastewater and are considered biological contaminants. Bacteria, viruses, and protozoa are the most common wastewater microbes, and they have both acute and long-term health consequences. Many bacterial kinds, including *Enterobacter*, *Escherichia coli*, *Streptococcus faecalis*, *Klebsiella pneumoniae*, and others, can be found in wastewater and have less catastrophic consequences.

Processes based on nanotechnology are efficient, flexible, and adaptive, resulting in low-cost, high-quality, and wastewater solutions [43]. Nanotechnology can be employed in a cost-effective manner to clean up as well as restore unique water resources. Table 7.1 provides an overview of the technologies utilized to remove pollutants from wastewater. It also lists the most significant limitations of the traditional water filtration systems.

TABLE 7.1
Main Characteristics and Limitations of the Traditional Methods Employed to Treat the Wastewaters

Conventional Methods	Main Characteristics	Limitations
Flotation	Separation process	The initial capital cost is high, as is the pH-dependent selectivity, as well as the ongoing maintenance and operational expenditures
Chemical precipitation		It's crucial to use a lot of reagents Its product could be a low-quality mixture, which would limit its use
Extraction of liquid-liquid solvents		High-tech apparatus
Evaporation		For large amounts of wastewater, it is expensive Small and medium-sized businesses have high investment costs
Flocculation/coagulation		This process is complicated and inefficient since it requires alkaline additions to attain an optimal pH
Flotation		Initial investment costs are substantial, as is pH-dependent selectivity, as well as operational and maintenance costs
Membrane filtration	Nondestructive separation	Capital costs are expensive for small and medium-sized businesses A lot of energy is required Membrane filtration systems come in a wide range of designs
Thermal oxidation/ incineration	Combustion-based destruction	The initial and running costs are both substantial
Ion exchange	Nondestructive process	The initial cost is substantial, as is the running cost
Electrochemistry	Electrolysis	Equipment has a high original cost, as well as a high maintenance cost
Advanced oxidation process	Emerging process and destructive techniques	Technical constraints and limited throughput make it economically unviable for medium and small-sized industries
Nanofiltration	Use of solid materials and nondestructive techniques	This method necessitates a lot of water cleaning energy as well as pre-treatment The retention of salt and univalent ions was low
Ultrafiltration		This process uses a lot of energy and doesn't get rid of dissolved inorganics
Microfiltration		Metals, fluoride, salt, nitrates, volatile organics, pigments, and other contaminants are not removed with this procedure Cleaning is required on a regular basis, and membrane fouling might arise
Carbon filter		Fluoride, nitrates, salt, metals, and other contaminants are not removed with this procedure Mold can grow here, and undissolved particles can cause blockage

7.3 NANOSTRUCTURED MATERIALS FOR WASTEWATER MANAGEMENT

NPs' form and size, as well as shape, are the most important characteristics. Surface and interfacial properties can be altered when chemical substances are present. These compounds can prevent aggregation by altering the particle's outermost layer while keeping the charges. A fairly intricate structure is designed based on the NP's growth path as well as lifespan. During the conventional production process of NPs, various distinct agents condense over their surface in the cooling, which is subject to various ambient atmospheres. As a result, sophisticated chemical reactions near the surface are required, and just few particle model systems have been documented. At the liquid-NP interface, polyelectrolytes were employed for changing their surface features as well as their reactions with the surroundings. In various sectors, they are used for colloidal dispersion lubrication, cohesion, and organized flocculation [44]. NPs are divided into groups based on their size, shape, and physical and chemical properties. The atomic and molecular origins of NPs give them diverse chemical and physical capabilities. Smaller clusters have radically different electrical and optical properties, as well as reactivity, than larger or extended surfaces. Stronger polar and electrostatic contacts, weaker van der Waals forces, and covalent interactions can affect particle interactions at the nanoscale scale. According to the viscosity and polarity of the fluid, NPs' interaction can cause agglomerations. Surface modification can increase or decrease a coagulating colloid's ability to aggregate. Details of NP-NP interactions, as well as NP-fluid interactions, are crucial for developing advanced materials.

The properties of the active surface layer are difficult to be quantified due to the small number of molecules present. Surface charge, solvation, and energy are all significant factors to be considered. Individual and collective NPs must be regulated by attractive or repulsive forces of interaction. Aggregates and/or agglomerates are formed as a result of the interaction between NPs, which cause a change in their behavior. The NPs are intended to attract water and are very porous, which permits their water absorption in a sponge-like manner while resisting dissolved salts and other contaminants. Organic chemicals and microorganisms are repelled by hydrophilic NPs placed in the microorganisms' membrane [45]. Various kinds of NPs are utilized in the treatment of wastewater.

7.3.1 NANO-METAL OXIDE PARTICLES

Metal oxide NPs are produced entirely from metal precursors and are used in a wide range of chemistry, physics, and material sciences applications. They are produced in a variety of structural geometries and electronic structures that can be metallic, semiconductors, and insulators. Such NPs possess unusual opto-electrical features due to their famous surface plasmon resonance with specific location features. Noble metals including Cu, Au, and Ag as well as alkali NPs have a high visible light absorption of electromagnetism. The morphology, diameter size, and crystal facets of the designed nano-metal particles play a significant role in the bacterial resistance application [46]. Metal NPs offer a wide range of uses due to their enhanced optical characteristics. In SEM analysis, gold NP coating is widely employed to boost the electronic beam, allowing for efficient SEM captures. Metal oxides with nanoscale dimensions exhibit a number of unique features, such as the selectivity and removal capacity of heavy metals. They have tremendous potential as heavy metal adsorbents.

Examples of metal oxide nanoparticles include ZnO, CuO, TiO_2, Al_2O_3, Fe_2O_3, MnO_2, Mn_2O_3, and ZrO_2. Metal oxide NPs come in a variety of shapes and sizes and are applied for different applications (Figure 7.3). Metal oxide NP's antibacterial activity and dye removal from wastewater are dependent on their morphology and size arrangement. As a result, synthesis approaches are primarily concerned with size and shape control and distribution of NPs. Herein, we reported the antibacterial performance of various metal oxides against different bacterial organisms.

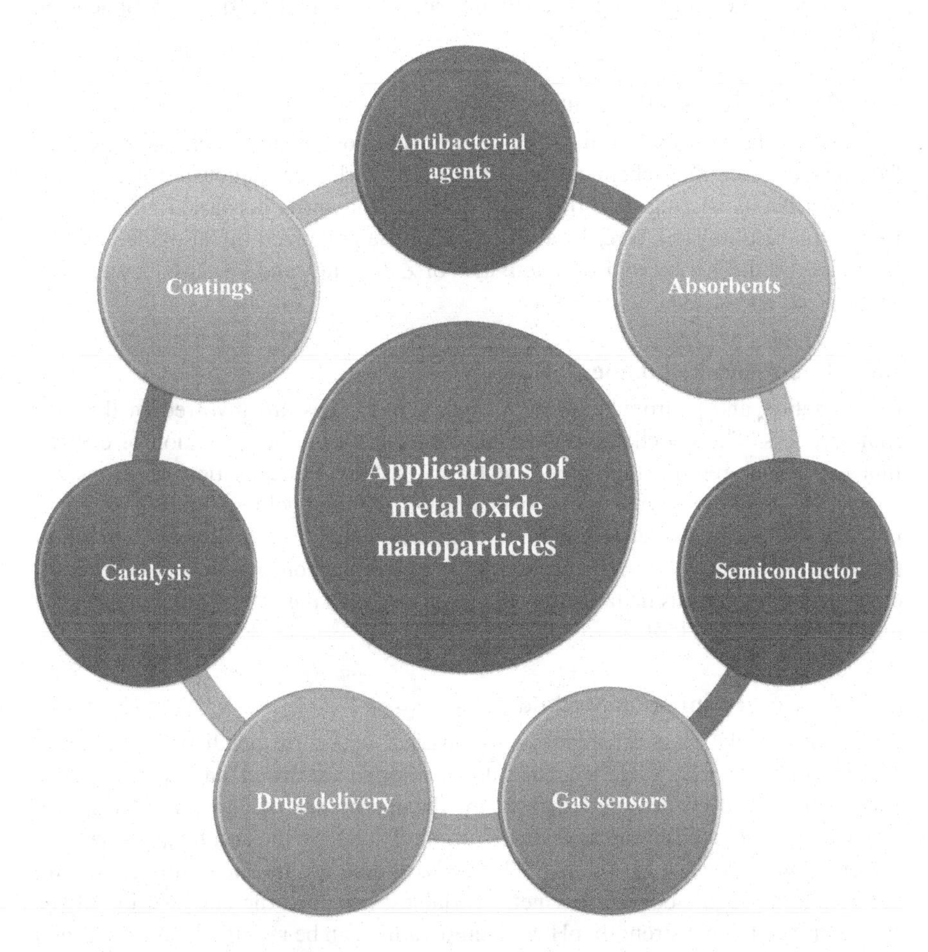

FIGURE 7.3 Common applications of metal oxide NPs.

7.3.2 Preparation of Nano-Metal Oxide Particles

The synthesis procedure influences the properties and applications of metal oxide NPs. It is inevitable to control the preparation process of NPs to produce effective antibacterial active agent. This chapter has concisely reviewed the most often used approaches for preparing nano-metal oxide particles.

7.3.2.1 Chemical Methodology

During the synthesis process, the reaction mixture (surfactants, precursors, and solvents) is gradually brought to a high temperature, which is the most favored chemical approach [47]. The primary phases in manufacturing extremely monodisperse metal oxide NPs in this method are monomer synthesis, nucleation, and growth. The growth process and boiling point of the solvent affect the NP's size and diameter. Chemically synthesized Ce_2O, CdO, MgO, Fe_2O_3, Fe_3O_4, ZnO, TiO_2, V_2O_5, and graphene-related nanomaterials were tested for antibacterial potential [16].

7.3.2.2 Biosynthesis Procedure

To facilitate the synthesis of metal oxide NPs, microorganism secretions or extracts from plants containing chemical precursors are swirled continuously for 24 hours with continuous stirring at 120°C [48]. The supernatant is discarded after centrifuging the resulting mixture. In a watch glass, the pelletized metal oxide NPs are naturally dried. The microbicidal activities of ZnO, CuO, and Fe_3O_4 NPs were biosynthesized and tested.

7.3.2.3 Sol-Gel Technique

Condensation and hydroxylation of precursor molecules are involved in this wet route process. The gel characteristics have an impact on the formation of crystalline nanostructures. Furthermore, the hydrolysis and condensation reactions are controlled by agitation, precursor concentration, pH, and gel temperature [49]. It is possible to obtain a predefined nanostructure that includes monodisperse and amorphous phases and to manage the homogeneity of reaction products. The bactericidal activity of various metal oxide NPs generated with the sol-gel method has been tested and evaluated [50].

7.3.2.4 Co-Precipitation Method

In an aqueous solution, a salt precursor is converted to its corresponding metal hydroxide by adding NaOH or NH_4OH. The salts of chloride are rinsed, and the hydroxides can be eliminated via heating for preparing the appropriate metal oxides. Nuclei are formed on the crystalline surface via diffusion of solute in the co-precipitation after a short-term nucleation period [51, 52]. The NPs' distribution and diameter size are not able to be regulated due to kinetic variables; however, the composition of the salts employed, ionic strength, pH, and temperature can be modified. Antimicrobial activity of Fe_3O_4 and ZnO NPs prepared by co-precipitation method has been evaluated.

7.3.2.5 Electrochemical Method

The bulk metal is maintained in the anode and converted into metal clusters using electrolyte solutions and a two-electrode arrangement [53]. Tetraalkylammonium salts are utilized as support electrolytes to keep the metal clusters stable. During the oxidation of the bulk metal at the anode, the cations of metal transfer to the cathode. To remove the dissolved oxygen, electrolysis is performed under nitrogen. The metal is oxidized by residual oxygen (in the electrolytic bath) into equivalent metal oxides. Metallic powders do not clump together when ammonium stabilizers are used. The size of a cluster is determined by its current density. The harmful effects of Cu_2O, CuO, and Cr_2O_3 NPs on microorganisms have been confirmed.

7.3.2.6 Wet Chemical Approach

This approach represents a straightforward and economic route which can be applied to any surface, regardless of its shape [54]. The precursors are blended and boiled for 45 minutes in ultrapure water before being centrifuged. ZnO, Fe_3O_4, and CuO NPs were developed and applied as antibacterial active agents.

7.3.2.7 Pyrolytic

The solution is atomized, mixed with soluble polymeric matrix, and integrated into the porous resin before heating the precursor material. Sintering, aggregation, and agglomeration are all prevented by these methods. Metal oxide NPs act as catalysts when combined with the polymer. However, this method is ineffective to create metal oxide nanocomposites. Away from the forgoing data, cobalt oxide (Co_3O_4) was reported to be prepared by this approach and approved a high bactericidal activity [55].

7.3.2.8 Microwave-Assisted Technique

The precursor materials can be dissolved separately in distilled H_2O, and the prepared solution is then stirred continuously at room temperature for 10 minutes until a white gel is formed [56]. Microwave energy is used to irradiate the mixture, which is then allowed to cool naturally to the RT. The precipitate is filtered under vacuum, rinsed with ethanol and distilled H_2O, and then dried at 80°C for 1 h. MgO and CaO NPs were then prepared and applied as antimicrobial agents.

7.3.2.9 Hydrothermal/Solvothermal Method

This method can control the metal oxide NPs' size and morphology and binary structure which are performed in an aqueous/solvent solution in an autoclave at ~200°C under pressure. In normal conditions, aqueous solvents/mineralizers are difficult to be dissolved, but they are dispersed and recrystallized in a heterogeneous reaction [57, 58]. A surfactant-assisted solvothermal technique using $AlCl_3$ and NaOH was performed for 24 h at 200°C, yielding γ-Al_2O_3 nanorods (NRs) with 10 nm diameter size (Figure 7.4) [59, 60]. Controlling the hydrothermal/solvothermal temperature allows the production of various types of NPs. ZnO NRs, ZnO nanosheets, and MgO NPs were developed and applied as antibacterial nanostructured materials.

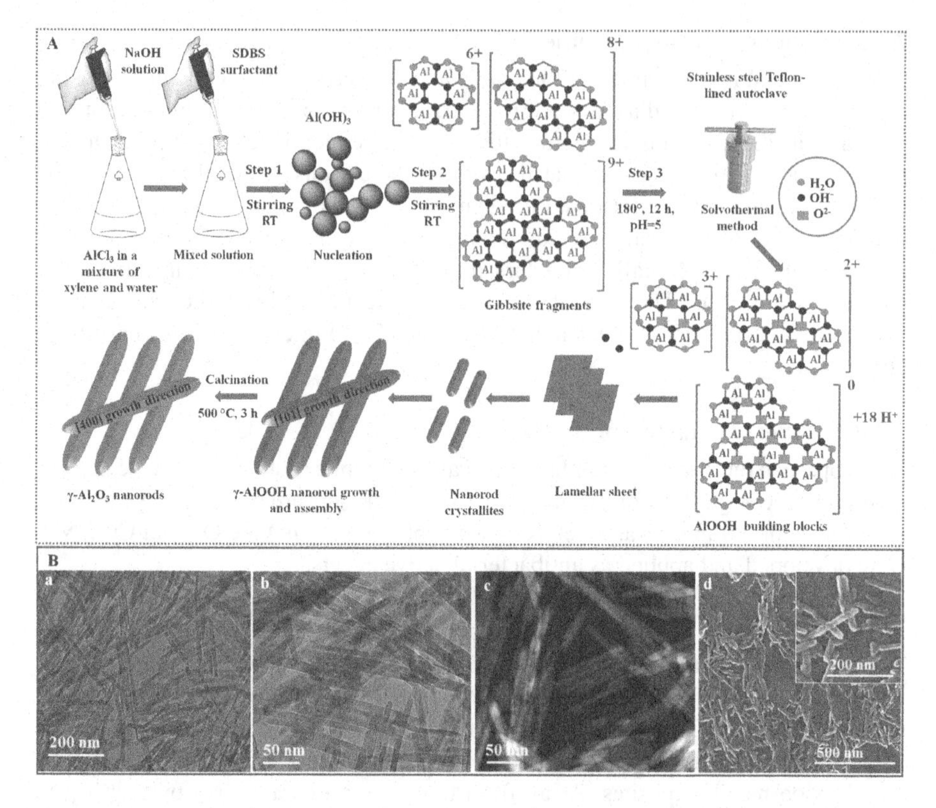

FIGURE 7.4 (A) A solvothermal method used to prepare γ-AlOOH NRs which are then transferred to γ-Al_2O_3 through a calcination method. It was demonstrated how planar gibbsite fragments can be transformed into γ-AlOOH building units. (B) (a–d) are the γ–Al_2O_3 NRs' HRTEM images at different magnifications, while (c and d) represent the γ–Al_2O_3 NRs' TEM in dark-field and SEM images [59]. (Copyright 2020, reproduced with permission from Elsevier.)

7.4 ANTIBACTERIAL MECHANISMS OF METAL OXIDE NPs

Scientists are fascinated by metal oxide NPs as antibacterial agents because of their unique chemical, physical, biological, magnetic, and optical features. Modern research are directed to study the antibacterial mechanisms of metal oxide NPs [16, 17]. We go over many mechanisms of metal oxide NPs' antibacterial activity that have been discovered in recent investigations. Recent research reported that bacterial growth inhibition or killing is caused by a combination of one or more mechanisms, which changes based on the metal oxide NPs' chemistry. The formation of reactive oxygen species (ROS) is a main method for antibacterial performance. Other mechanisms of action include adsorption method, electrostatic interaction damage to cell membranes, disruption of metal ions' balance, cytotoxicity, photo-killing, and enzyme and protein malfunction.

7.4.1 Cell Walls' Biochemical Nature and Adsorption Methods

The cell wall is a polymeric structure that displays the main army of microorganisms. The walls represent the defense barriers, survival, and disease resistivity [61]. Gram-positive and gram-negative organisms' adsorption methods are produced via different cell membrane components. Murein is a peptidoglycan found in the cell walls of gram-positive and gram-negative bacteria, and it is the abundant barrier to overcome extraneous materials. It is thicker in gram-positive bacterial organisms compared to gram-negative ones. However, cell walls of both gram-positive and gram-negative bacteria contain additional complicated ingredients compared to those found in eukaryotic microbes [62]. Lipopolysaccharide is commonly termed endotoxin and is the dominant lipid constituent of the outer membrane of gram-negative bacteria in addition to other protein types (lipoprotein and porin) [62]. Because of the increased protection provided by the outer membrane, gram-negative bacterial organisms are more resistant to antibacterial materials than gram-positive bacterial strains [63, 64]. Lipopolysaccharide is a unique material in the cell walls of the gram-negative bacteria that offer a heavily burdened area which protects against the NPs' attack [65]. Gram-positive bacterial organisms are more susceptible to antibacterial nano-agents than gram-negative strains. This is because the gram-negative cell wall contains lipopolysaccharides, lipoproteins, and phospholipids, which produce a binding barrier and resist the NPs' penetration [66, 67]. In contrast, gram-positive bacteria have a crucial slender cell wall with the anionic polymer teichoic acid together with the thick outer cell walls. This enables the NPs to disperse along the phosphate molecular chain, preventing aggregation and allowing NPs' penetration (adsorption). Cell permeability, membrane destruction, and death are then the final syndromes [61].

7.4.2 Cell Membranes' Electrostatic Contact Damages

Metal cations are drawn to the polymers' electronegative chemical units in bacterial membranes. The negative charge is borne on the bacterial surfaces at a biological pH, owing to the protein's COOH units. Electrostatic interaction takes place because of the charges' difference between metal oxide NPs and the membranes of bacteria. Therefore, the nano-metal oxide particles accumulate onto the cellular surface allowing bacterial killing. Microorganisms are harmed by the combination of membrane polymers and cationic metal oxide NPs. The negative charge of gram-negative bacterial strains is higher than that of gram-positive organisms; thus, the gram-negative strains' electrostatic interaction is stronger [68]. Gram-negative bacteria have more charge per unit surface of lipopolysaccharide in the lipid bilayer's outer leaflet than other phospholipids, making them extremely negatively charged [69]. The accessible surface area for interaction is proportional to the electrostatic interaction forces, which metal oxide NPs supply in abundance compared to their natural particles of a large size, resulting in toxicity. The greater the metal oxide NPs' efficacy in suppressing bacterial growth, the smaller the surface area and particle size. When nano-metal oxide particles are bound to a cell membrane, it changes its permeability and structure. The disruption of cell walls is caused by positively charged metal oxide NPs forming a strong connection with membranes. As a result of the membrane breakdown, permeability increases, allowing more metal oxide NPs to enter the microbial system while cellular contents leak out, including membrane proteins and lipopolysaccharide cellular features. The membrane's pores exhibit nanometer sizes, while the size of bacteria is in the micrometer range, allowing metal oxide NPs to enter bacterial cells. Protein and lipid intensity increased when exposed to nano-ZnO particles, but not to nucleic acid intensity, demonstrating C = C and $C–H_2$ deformation in the lipids.

The degradation of cellular membranes and lipids, resulting in intracellular content leaking, was assumed to constitute the antibacterial process [70]. The metal oxide NPs' zeta potential causes membrane rupture and intracellular content loss when they contact with the membranes of bacteria. The proton motive force is dissipated when nano-metal oxide particles accumulate within the cellular surface, altering the membrane's chemiosmotic potential and inducing proton leakage. Metal oxide NPs' electrostatic contact with the cell surface limits bacterial growth.

Metal oxide NPs bind to mesosomes, affecting cell division and respiration as well as DNA replication. The antibacterial and immunomodulatory effects of hierarchical ZnO NRs and SiC nanowires (NWs) were studied by Askar et al. [68] (Figure 7.5). An advanced hexamethylenetetramine-supported hydrothermal procedure was employed to create wurtzite ZnO NRs with growth orientations of [0001], lengths of 1 μm, and average widths of 40 nm. ZnO NRs were compared to SiC NWs with a direction of [111] and width of 50–80 nm, studying the antibacterial action. The antibacterial efficiency of the hierarchical NRs and NWs was tested using different microbial strains, such as gram-positive and gram-negative bacterial organisms, and fungus. SEM analysis demonstrated that ZnO NRs have stronger killing activity than SiC NWs. *Acinetobacter baumannii*, *Pseudomonas aeruginosa*, *Salmonella typhi* ATCC 6539, and *S. aureus* ATCC 29213, MIC values for NRs of ZnO were 7.9, 3.9, 7.91, and 31.25 g/mL, respectively. The high activity of ZnO NRs is attributed to their huge surface area, [0001] polar facets, hexagonal wurtzite structure, tiny diameter, and length. Despite the fact that different investigations [69, 70] have identified multiple mechanisms of microbial cell membrane destruction, more research are still needed.

7.4.3 Disturbance of the Metal/Metal Ion Hometal Oxidestasis

By assisting catalysts, cofactors, and coenzymes, these components influence metabolic activity [71], which is critical for microbial survival. Metabolic activities will be disrupted when bacteria have a metal ion overabundance. Metal ions bind to DNA by cross-linking interaction which damages the helical structure of the molecule. Electrostatic interactions are triggered by metal ions released from the metal oxide NPs, which have a positive charge. The metal ions reduce the permeability of the outer membrane by neutralizing the charges on the lipopolysaccharide. As the membranes become disorganized, bacterial growth is reduced, and the permeability contributes to metal oxide NP accumulation in the cells is increased. Ag and ZnO NPs are disrupted outer membrane's lipopolysaccharide and proteins. The use of metal oxide NPs such as TiO_2 NPs in the treatment of *Escherichia coli* revealed significant metal oxide NP binding to the cellular membrane, which affects active transport and

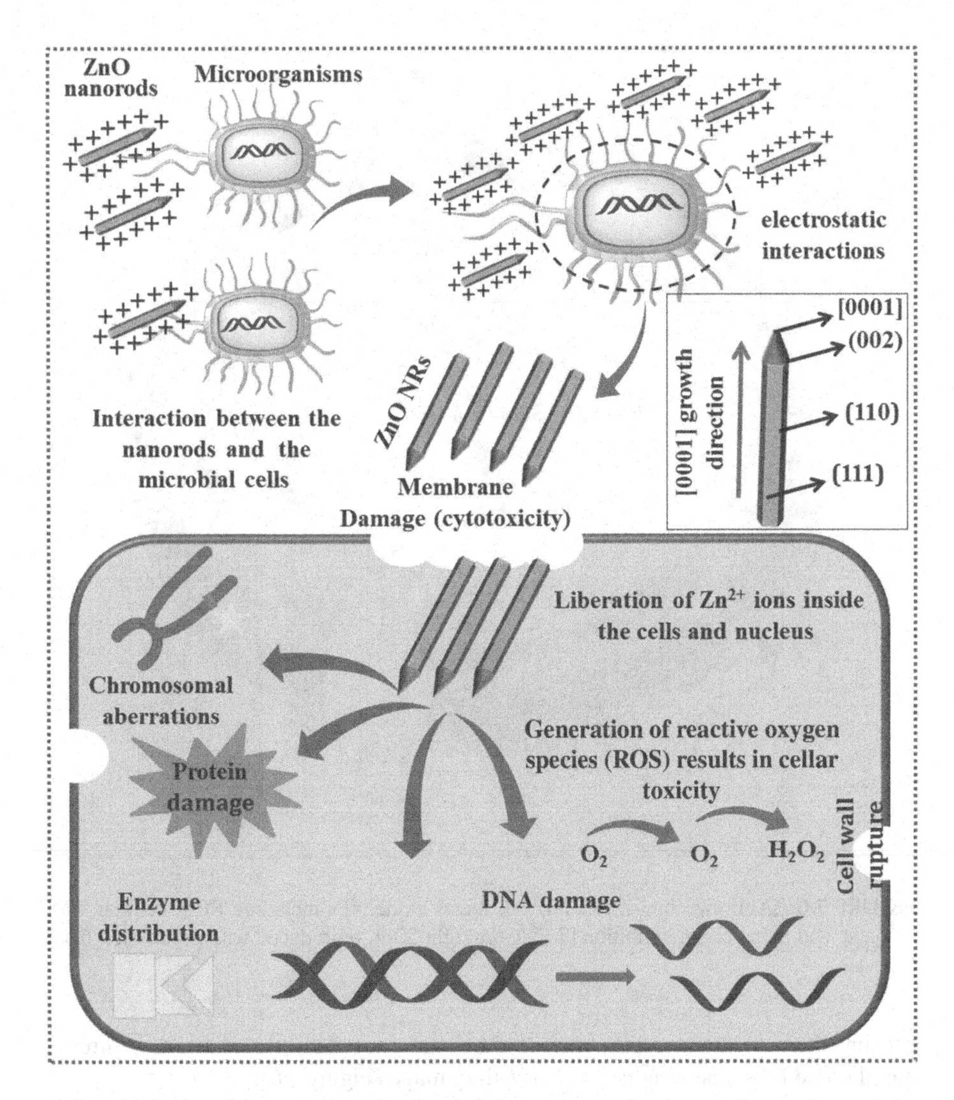

FIGURE 7.5 Representation of ZnO NRs' antibacterial mechanism against various microbial strains. It involves nanorods' interactions with bacterial cells, cellular wall breakage, oxidative stress manifested in DNA damage, and bacterial toxicity that leads to death [68]. (Copyright 2021, reproduced with permission from Elsevier.)

dehydrogenase enzymes' activities [72]. Gram-negative and gram-positive bacterial strains have both been found to be successfully killed by long-chain polycations deposited on cellular surfaces [73]. Metal ions produced by nano-metal oxide particles in cells are decreased, via thiol –SH units in proteins and enzymes, to metal atoms. Crucial metabolic proteins become inactive, and cells die as a result [74]. By reacting with the peptidoglycan layer's –SH units, metal ions produce cellular wall disintegration [75]. Adsorption, hydrolysis, and dissolution slowly release metal ions from metal oxide NPs [76], and they are abrasive and toxic to microbial organisms,

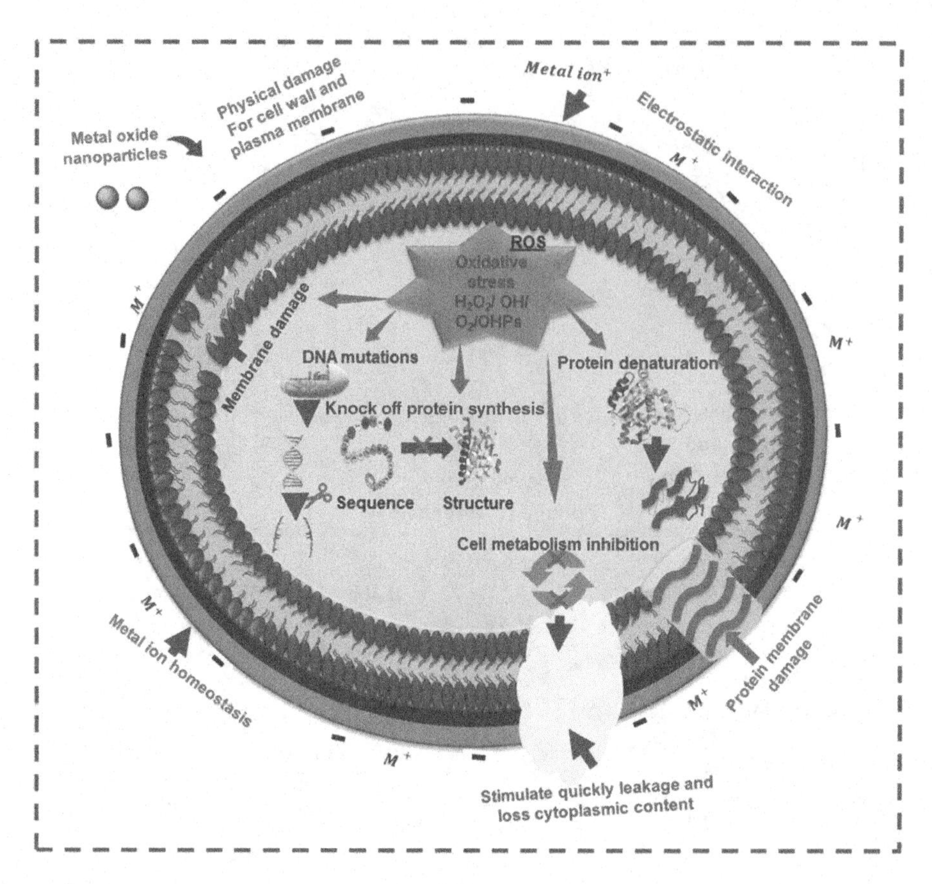

FIGURE 7.6 Antimicrobial mechanism of metal oxide NPs including ROS, cellular wall damage, and metabolism inhibition [76]. (Copyright 2014, reproduced with permission from Elsevier.)

causing them to disintegrate. Surface flaws are caused by the abrasive nature of metal oxide NPs, resulting in cellular wall damage (Figure 7.6).

7.4.4 Oxidative Stress and ROS Production

ROS, such as H_2O_2, hydroxyl radicals ($OH^{\bullet}$), superoxide anion ($O_2^{\bullet}$), and organic hydroperoxides, are produced when nano-metal oxides penetrate bacterial cells. This causes damage to nearly all organic macromolecules, such as nucleic acids, carbohydrates, amino acids, proteins, and lipids, and causes bacterial death. The ROS generation is influenced by pro-oxidant units on the nano-metal oxide's surface, cell–particle interactions, and active redox cycling [77]. Due to particle size reduction and changes in electronic properties, reactive units arise on the surface of NPs. At these reactive sites, electron donor/acceptor active sites interact with molecular oxygen, resulting in the production of $O_2^{\bullet}$. Through Fenton-type reactions, more ROS can be produced by this $O_2^{\bullet}$. ROS quantity produced in bacterial strains is determined by the metal oxide NPs' physicochemical properties, including electrophilicity, diffusibility, and surface area.

Metal oxide NPs largely block respiratory enzymes by producing ROS [78, 79]. In the electron transport chain, $O_2^{\bullet}$ damages the iron–sulfur (Fe–S) clusters, decreases ATP synthesis, and liberates more ferrous ions. When bacteria come into contact with metal oxide NPs, electrostatic interactions produce cell morphological alterations, membrane deformation, and damage. The Fenton reaction oxidizes these ferrous ions, producing additional $OH^{\bullet}$ and causing damage to DNA, proteins, and lipids [80]. Microorganisms' DNA and proteins are harmed by H_2O_2, a powerful oxidant that kills cells [81]. When fungi were exposed to ZnO NPs, ROS showed anticandidal activity via cytotoxicity and apoptosis. ZnO NPs have been found to stifle fungal development by interfering with cell activity and hypha formation [82]. Increased free radicals will cause unsaturated phospholipids in membranes to peroxide, resulting in additional peroxyl radical intermediates and severe damage. Lipid peroxidation damages membrane architecture and generates changes in membrane fluidity, integrity, and lateral structure. This can influence the membrane's characteristics and functions. Due to membrane failure, more ions leak and cause ion imbalance [83]. ROS-induced lipid peroxidation increases the malondialdehyde (MDA), which is a marker of cell membrane damage [84]. MDA generation in bacterial cells was boosted by TiO_2 NPs. MDA changes proteins by carbonylation or forming protein–MDA adducts since it is reactive [85, 86].

ROS generation destroys membrane phospholipids, nucleic acids, and lipoproteins, resulting in ROS and bacteria′s final death. Bacteria are protected from oxidative stress by glutathione, a nonenzymatic antioxidant. Excessive ROS are produced when metal ions are released from metal oxide NPs, which oxidize cellular glutathione. Furthermore, peroxidation of lipids in the bacterial membrane is caused by oxidized glutathione, resulting in cell rupture. ROS formation in the microbial cells disrupts the antioxidant enzyme activity. Oxidative stress in microorganisms is caused by an imbalance of oxidants and antioxidants. ROS production is a main strategy used by nano-metal oxide particles to kill bacteria.

7.4.5 Dysfunction of Proteins and Enzymes

Metal oxide NPs' antibacterial action can also be attributed to protein malfunctioning [87]. Protein-bound carbonyls are formed after the oxidation of the side chains of amino acid and are catalyzed by metal ions. Carbonylation levels inside proteins act as a signal for oxidative stress and protein destruction. In the case of enzymes, the carbonylation of protein can cause loss of their catalytic function, resulting in protein breakdown. Furthermore, these ions also bind to the –SH units in many enzymes and proteins, rendering them inactive. Many metals have a proclivity for reacting with soft bases, including phosphorus and sulfur, which are the fundamental components of DNA and proteins. These soft bases can interact with the metals released, causing DNA damage and cell death [88]. Furthermore, metal oxide NPs are capable of inactivating bacterial Fe–S dehydratases. Cu-depleted Fe–S dehydratases in *E. coli* cause bacteriostasis to occur [89]. Tellurium (IV), a metalloid oxyanion, oxidized Fe–S clusters and produced ROS [90]. Metal oxide NPs can bind to noncatalytic sites and restrict enzyme action in addition to blocking the catalytic site.

7.4.6 Inhibition of Signal Transduction and Toxicity

Because of their electrical characteristics, metal oxide NPs interact with microorganism's nucleic acids, especially plasmid and genomic DNA [91]. Microbes' cell division is inhibited by metal oxide NPs, which disrupt chromosomal and plasmid DNA replication processes. In aerobic microorganisms, DNA damage was mediated by Fe_2O_3 NPs [92]. The ions of metal can attach to the 30S ribosomal unit in rare cases, causing the ribosome complex to stall and prevent proteins from being translated from mRNA. Metal oxide NPs have been shown to disrupt bacterial signal transmission. Phosphotyrosine residues are dephosphorylated in metal oxide NPs, which prevent signal transmission and, as a result, bacterial growth [93]. In bacteria, phosphotyrosine is required for signal transduction.

7.4.7 Photokilling

When metal oxide NPs come into contact with bacteria under solar-light irradiation, photokilling occurs. Photosensitization of transition metal oxides is possible [94]. Photochemical changes in the cellular membranes, the permeability of Ca^{2+} ions, caused protein/DNA destruction as well as aberrant division for cells under light. The fundamental reason for electron release through light is the generation of O_2. Metal oxide NPs capture these electrons, resulting in increased ROS production. Photochemical reactions demonstrated by Fe_3O_4 NPs were found to limit the growth of nosocomial bacteria [95]. In the presence of TiO_2, microorganisms were completely killed by UV radiation after 50 minutes [96].

7.4.8 Other Strategies

When Ag-TiO_2 is exposed to sunlight, it acts as a photocatalytic material, which is demonstrated as actual toward *E. coli.* The plasmon resonance surface effect supports the generation of photo-induced electrons under sunlight irradiation, which produce ROS within bacterial cells and their destruction [97]. Antibiotics have the most difficulty in destroying biofilms. Metal oxide NPs have a significant advantage over biofilm growth. SPIONs and SPIONs in combination with Zn and Fe exhibited antibacterial activity toward methicillin-resistant Staphylococcus aureus when supplemented with fructose metabolites. Fructose may help biofilms in absorbing SPIONs and reduce NP treatment toxicity [98]. SPIONs in combination with ZnO and ultrasonic stimulation could effectively reduce *S. aureus* populations [99].

Selim et al. [100] developed γ–MnOOH, α-Mn_2O_3, and γ–AlOOH NRs with advanced antibacterial action toward yeasts and gram-negative and gram-positive bacterial organisms (Figure 7.7). These NRs' antibacterial action increases following the order of γ-MnOOH < γ-AlOOH < α-Mn_2O_3 NRs, according to biological research. Because of the α-Mn_2O_3 NRs' increased surface area and smaller diameter, they have a better antibacterial action. These NRs clung tightly onto the cells

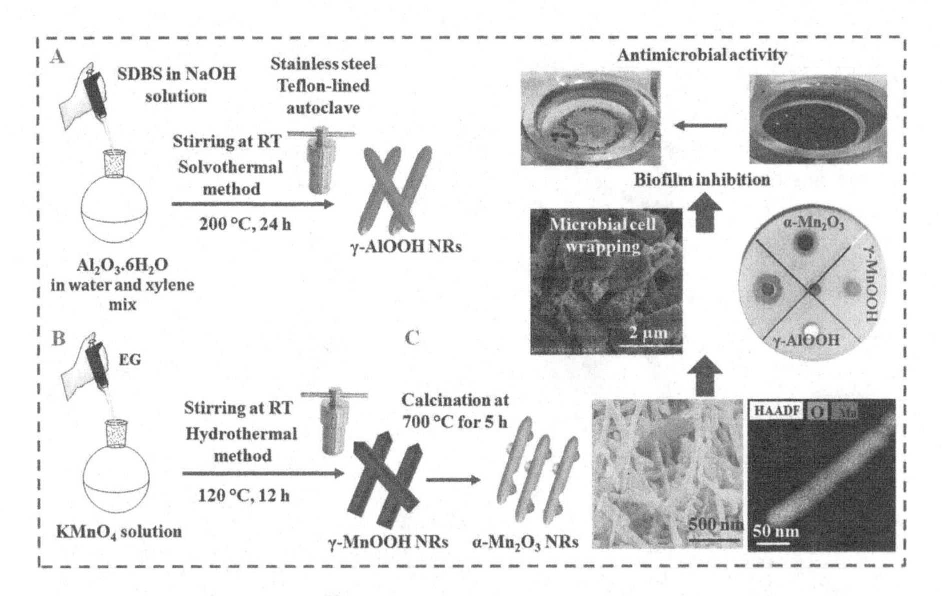

FIGURE 7.7 Preparation strategies of (A) γ-AlOOH NRs using a solvothermal method, (B) β–MnOOH using a hydrothermal method, and (C) α-Mn_2O_3 via a calcination method. The highest antibacterial mechanism for α-Mn_2O_3 was demonstrated. SEM and dark-field TEM images of α-Mn_2O_3 NRs as well as SEM image of the NRs in the inhibition zone against *S. aureus* pathogenic bacteria and agar well diffusion method were elucidated [100]. (Copyright 2020, reproduced with permission from the Royal Society of Chemistry.).

of bacteria, resulting in morphological disturbance and cellular wrapping as well as microbiological killing. The fabricated NRs make an excellent stand for inhibiting bacterial development.

Fatthallah et al. [101] investigated the antibacterial effectiveness in an aqueous solution of hierarchical γ-Al_2O_3, ZnO, and β-MnO_2 (Figure 7.8). These antimicrobial inhibitors are made in a variety of hierarchical shapes and sizes with good surface orientation. H-NRs' antibacterial efficacy was tested by different gram-negative bacteria including *P. aeruginosa and E. coli* as well as gram-positive bacterial strains in an experimental setting. The minimum inhibitory concentration and the minimum bactericidal concentration (MIC and MBC) were calculated. γ-Al_2O_3 H-NRs with [400] dominating surfaces exhibited the highest antibacterial performance against gram-negative and gram-positive bacterial strains with MIC values of 1.146 and 0.250 μg/mL, MBC values of 1.146 and 0.313 μg/mL, and MIC/MFC evaluations of 0.375 and 0.375 μg/mL.

When nano-metal oxide particles come in contact with microorganisms, aggregation is significant because ultrasound stimulation can cause agitation for the clumped bacteria, NP aggregation was prevented, and particle-microbe interaction is increased. When ultrasound was combined with ZnO NPs, the production of H_2O_2 was increased. Despite the fact that this ultrasound stimulation is successful, the stimulus will help it enter the systemic circulation. Using Van-Fe_3O_4@Au nanoeggs, photo-thermal killing for (vancomycin-resistant Enterococcus) VRE and

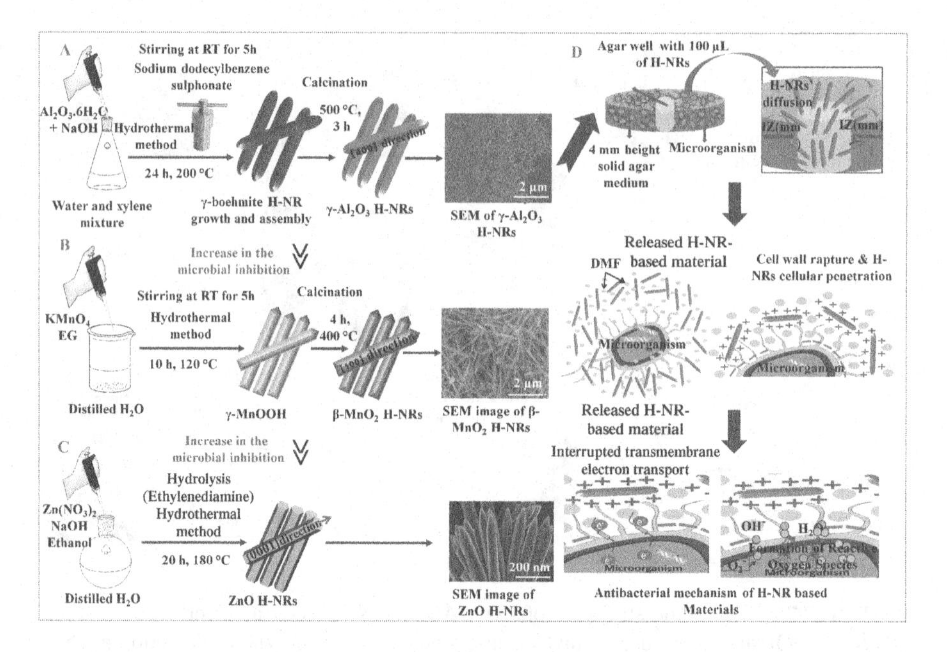

FIGURE 7.8 Preparation methods for the created (A) H-NRs of γ-Al_2O_3 using a surfactant-based solvothermal technique followed by annealing thermally; (B) β-MnO_2 H-NRs were synthesized using a hydrothermal procedure and then calcined; (C) wet chemical method, including an ethylenediamine-based hydrothermal technique, was used to make ZnO H-NRs. The antibacterial effect of H-NRs was assessed using an agar well diffusion scenario; (D) the antibacterial mechanisms for γ-Al_2O_3 H-NRs (the most effective nanomaterial) toward distinct microbiological organisms are also depicted [101]. (Copyright 2021, reproduced with permission from Elsevier.)

methicillin-resistant Staphylococcus aureus was demonstrated [102]. Van-Fe_3O_4@Au nanomaterial exhibited magnetic properties and increased photothermal death for gram-positive and gram-negative bacterial organisms when exposed to near-IR irradiation at 808 nm.

Metal ions released from the nano-metal oxides exhibited antibacterial properties [103]. According to this study, the antibacterial activities were induced by Zn^{2+} ions released from nano-ZnO particles, Cu^{2+} ions are released from CuO NPs, and metals are also released from Fe_2O_3, Co_3O_4, Cr_2O_3, and NiO metal oxides [103]. Distinct metal oxide NPs have different physicochemical features, as well as different dosages to trigger antibacterial effect. In the case of antibiotics, chronic infections with bacteria necessitate a high dose, as side effects and resistance spread are issues to solve [104]. NPs have maximum efficacy against microorganisms even at low doses [105]. The various processes by which the various metal oxide NPs exert their antibacterial effect are not totally known. It is inevitable to study how metal oxide NPs interact with microorganisms. The best dose of metal oxide NPs is once again a hot topic of debate. As proven by several research, the effective concentration of a specific metal oxide NP will differ for different metal oxide NPs (Table 7.2) [106–126].

TABLE 7.2
Nano-Metal Oxide Particles and Their Antimicrobial Activity

Metal Oxide NPs	Size	Dose and Exposure Time	Microorganisms Used	Antimicrobial Activity	Ref.
Sb_2O_3 NPs	90–210 nm	50, 100, 150, 200, 250, 300, and 350 mg/L 1 day	*E. coli*, *Bacillus subtilis*, and *S. aureus*	All the examined microorganisms were toxic	[36]
Co_3O_4 NPs	10–25 nm	2, 4, 8, 16, 32, 64, and 128 μg/mL 24 h	*S. aureus* and *E. coli*	Demonstrated antimicrobial activity against *E. coli* and *S. aureus*	[55]
CdO NPs	60 nm	0.01%, 0.5%, and 1% 2 days	*E. coli*	Antibacterial activity against *E. coli*	[72]
γ-Al_2O_3 NRs	γ-Al_2O_3 (20 nm (W) <0.5 μm (L))	0.1875, 0.375, 0.750, 1.25, and 2.5 μg/mL 1 day	*E. coli*, *S. aureus*, *B. subtilis*, *P. aeruginosa*, and *B. thuriginesis*	All the strains tested showed antibacterial activity	[101]
β–MnO_2 NRs	β–MnO_2 (30 nm (W) >1 μm (L))				
ZnO NRs	ZnO (40 nm (W) 1 μm (L)).				
Bi_2O_3 NPs	75.87–89.04 nm	2.0, 9.3, and 11.29 ppm 1 day	*Acinetobacter baumannii*, *E. coli*, and *Pseudomonas aeruginosa*	All bacteria examined showed no antimicrobial action	[106]
CaO NPs	15–180 nm	100 ppm 6 h and 1 day	*Lactobacillus plantarum*	Bactericidal action that is satisfactory	[107]

(Continued)

TABLE 7.2 ***(Continued)***
Nano-Metal Oxide Particles and Their Antimicrobial Activity

Metal Oxide NPs	Size	Dose and Exposure Time	Microorganisms Used	Antimicrobial Activity	Ref.
NPs of CaO	16 nm	0.125–32 mM 1 day	*Candida tropicalis, P. aeruginosa,* and *S. epidermidis*	Gram-positive bacteria have a stronger antibacterial activity than gram-negative strains	[108]
NPs of CeO_2	6–40 nm	0.5% 18 h	*S. oneidensis, E. coli*, and *B. subtilis*	Antibacterial action was demonstrated against *E. coli* and *B. subtilis*, but not against *S. oneidensis*	[109]
Cr_2O_3 NPs	65, 79, and 41 nm	50 mg/mL 2 days	*P. aeruginosa* and *E. coli*	Both of *P. aeruginosa* and *E. coli were inhibited*	[110]
CuO NPs	50–100 nm	1, 50, 100, and 200 μg/disk Overnight	*E. coli, S. aureus, S. dysenteriae, S. pneumoniae*, and *Vibrio cholerae*	*V. cholerae* and *S. dysenteriae* have strong antibacterial action	[111]
CuO NRs	100 nm (D)	1 mg/mL 2 h	*E. coli* and *B. anthracis*	Maximum antibacterial performance (100%) against *E. coli*	[112]
Fe_3O_4 NPs	8 nm	5, 100, 150, and 200 μg/mL 8 h	*E. coli*	On *E. coli*, it has a microbiostatic impact	[113]
Fe_3O_4 NPs	10–120 nm	50 mg/mL 1 day	*S. epidermis, E. coli, Shigella flexneri, S. aureus, Brevibacillus brevis, P. aeruginosa, B. subtilis, V. cholerae*, and *S. aureus*	Antibacterial activity was moderate against six gram-positive bacterial organisms and two gram-negative strains	[114]
α-Fe_2O_3 NPs	2–540 nm	10 μg/mL 10 h	*P. aeruginosa*	Biofilm formation and increased *P. aeruginosa* growth	[115]

(Continued)

TABLE 7.2 *(Continued)*
Nano-Metal Oxide Particles and Their Antimicrobial Activity

Metal Oxide NPs	Size	Dose and Exposure Time	Microorganisms Used	Antimicrobial Activity	Ref.
Nanowires of MgO	6 nm (D) as well as 10 μm (L)	100, 300, 500, 700, 900, and 1100 mg/mL 45 min	*Bacillus* sp. and *E. coli*	Increased MgO nanowire concentration resulted in lower bacteriostatic efficacy against *E. coli* and *Bacillus* sp.	[116]
TiO_2 NPs	12 nm for ethyl alcohol and 7 nm for water	1%, 1.5%, and 2.0% 1 day	*K. pneumoniae* and *S. aureus*	Antimicrobial activity of TiO_2 NPs (7 nm) was superior than TiO_2 NPs (7 nm) (12 nm)	[117]
V_2O_5 nanowires with H_2O_2 and Br^-	300 nm (L) 20 nm (W)	0.075 mg/mL 180 min	*E. coli and S. aureus*	Gram-positive and gram-negative bacteria have high antibacterial action	[118]
ZnO and AgO NP	ZnO (50 nm) AgO (20 nm)	1% 12 h	*Lactobacillus* sp. and *Streptococcus mutans*	Antibacterial activity was approved against the two microorganisms	[119]
Chitosan/ZnO composite	99.3 nm and 603 nm	5 mg/mL 1 day	*M. luteus, S. aureus,* and *C. albicans*	*S. aureus* and *M. luteus* both showed antibacterial activities	[120]
Graphene oxide (GO) decorated with Cu, Mn, Ag, and Zn	80–200 nm	0, 10, 25, 50, 75, 150, 225, and 300 μm 1 day	*E. coli, S. aureus*, and methicillin-resistant Staphylococcus aureus	Antibacterial activity against the three pathogenic microorganisms investigated	[121]
Fe_3O_4-encapsulated silica sulfonic acid NPs	22 nm	0.02/30 mL 1 day	*S. aureus and E. coli*	Both organisms are inactivated to a greater extent (*S. aureus* and *E. coli*) under light than dark conditions	[122]

(Continued)

TABLE 7.2 *(Continued)*
Nano-Metal Oxide Particles and Their Antimicrobial Activity

Metal Oxide NPs	Size	Dose and Exposure Time	Microorganisms Used	Antimicrobial Activity	Ref.
α-Fe_2-xAg_xO_3 nanocrystals	37.13–41.87 nm	450, 600, and 700 µg 1 day	*E. coli, P. fluorescens, B. subtilis,* and *S. aureus*	Exhibited more antibacterial performance toward *B. subtilis* and *P. fluorescens* than *S. aureus*	[123]
Graphene sheets / ZnO composite	40 nm	0.01 and 0.05 g 1 day	*S. Typhi* and *E. coli*	*S. Typhi* showed greater antibacterial activity than *E. coli*	[124]
TiO_2/Ag NPs	TiO_2/Ag with 112 nm	15.62–1000 mg/L 2 days	*S. aureus* and *Candida* spp.	*Candida* sp. biofilm-forming cells and methicillin-resistant *Staphylococcus aureus* have antibacterial properties	[125]
NiO NPs	40–60 nm	100, 50, 25, 12.5, 6.25 3.12, 1.56, 0.78, 0.39, and 0.195 mg/mL 24 h	*S. marcescens, S. epidermidis, M. luteus* and *B. subtilis, E. coli, P. aeruginosa, K. pneumoniae,* and *S. aureus*	NiO NPs act as antibacterial agents; the maximum sensitivity was shown in *S. epidermidis*	[126]

(W): Width; (L): Length; (D): Diameter

More research is needed to identify whether simple, functional, hybrid, or surface modifications of metal oxide NPs represent the most effective agents as antibacterial material, and these forms have a similar mechanism whether alterations can achieve different results.

7.5 CONCERNS ABOUT THE APPLICATIONS OF NANO-METAL OXIDE PARTICLES FOR BACTERIAL RESISTANCE

The nano-metal oxide particles' interaction with organisms is being studied by researchers all over the world [127]. Metal oxide NPs have higher toxicity degrees than their bulk counterparts [128]. NPs' morphology, size, oxidant-generation capability, dose, exposure time, capacity for entering various barriers, solubility, material's aggregation, and other characteristics all pose a threat to metal oxide NPs being used for environmental antibacterial applications. Evaluation of the side effects and risks is obligatory before antimicrobial agents can be put on the market. The nano-metal oxide's application varies depending on the cytotoxicity, dosage, size and shape, exposure time, and cell and tissue types. Several dominating parameters for the antibacterial applications of these NPs are discussed as follows.

7.5.1 NPs' Cytotoxicity

Toxicity is a major concern when using nanomaterials. Despite this, human toxicity pathways are still mostly unknown. Because of its capacity to promote hemolysis, NPs can disrupt blood circulation pathways. Organ malfunction and damage can also be caused by NPs. Smaller sizes of NPs are more harmful to the biological system than those with larger sizes [129]. The nano-metal oxide's toxicity varies depending on the dosage, exposure time, and cell and tissue types. The capacity to destroy biomolecules and cell membranes is the most serious worry, followed by their pro-oxidant character. Despite the fact that there are a few reports on the metal oxide NPs in vivo and in vitro cytotoxicity, more research is needed. CuO NPs may cause hepatotoxicity and nephrotoxicity when they interact with cell components [130]. Although some studies have found that NPs have no significant in vivo life-threatening toxicity, their accumulation could be harmful to body cells [131, 132].

It is obvious that different investigations have produced contradictory results; probably, in vivo and in vitro effects are not comparable. In fact, some data suggests that Au NPs are harmful in certain taxa, while other research suggests that they are not [133]. Toxicity research isn't often done in a systematic method, which explains the disparities in results [134]. Furthermore, most research has focused on toxicity assessments rather than toxicity mechanisms [135]. The toxicity of metallic NPs varies based on its oxidation state, solubility, shape, and ligands, as well as the surrounding environment and health conditions [136]. The toxicity effects of several metal/metal oxide NPs at different concentrations are summarized in Table 7.3 [137–146].

TABLE 7.3
The Toxic Effects of Metal/Metal Oxide NPs

Metal/Metal Oxide NPs	Model (Cell Lines, Animals)	Particle Size (nm)	Dose and Incubation Time	Summary of Effects	Ref.
ZnO, TiO_2, and SiO_2 NPs	Human embryonic stem cell (hESC)-derived fibroblasts (hESC-Fib)	TiO_2; and ZnO (< 100 nm) SiO_2 (10–20 nm)	1250, 250, and 50 _g/mL SiO_2, TiO_2, and ZnO NPs, respectively 24 h	• Evidence of NPs' antimigratory effects on hESC-Fib was presented • The cell death caused by these NPs was mostly due to membrane damage and apoptosis	[137]
Au, Ag, CuO, ZnO, and TiO_2 NPs	Human colorectal adenocarcinoma cells (HT29)		2 and 10 µg/mL 24 h	• The toxic effects of the studied NPs on cell viability, oxidative stress, DNA damage, and programmed cell death vary significantly and are mostly dependent on the kind and size of the NPs	[138]
SeNPs	Sprague-Dawley (SD) rats	79.88 nm	0, 2, 4, and 8 mg Se/kg BW for two weeks	• At a dose of 8.0 mg Se/kg BW, SeNPs dramatically reduced the antioxidant capacity of rats' serum, liver, kidneys, and heart • Pharmacological SeNPs have the greatest toxicity in the liver	[139]
ZnO, CuO, CeO_2, Fe_2O_3, WO_3, V_2O_5, TiO_2, Al_2O_3, and MgO	Female and male mice (C57Bl/6J)		0.5 mg/mL, at 4 and 24 h ZnO (0.3 mg/mL) CuO and V_2O_5 (1 mg/mL) WO_3 (0.4 mg/mL)	• CuO and ZnO NPs had the highest toxicity in both in vitro and in vivo acute conditions • At 4 and 24 hours, there was a significant rise in bronchoalveolar lavage neutrophils and proinflammatory cytokines, as well as a loss of macrophage viability for ZnO and CuO, but not for V_2O_5 or WO_3 • Both female and male mice exhibited these effects	[140]

(Continued)

TABLE 7.3 *(Continued)*
The Toxic Effects of Metal/Metal Oxide NPs

Metal/Metal Oxide NPs	Model (Cell Lines, Animals)	Particle Size (nm)	Dose and Incubation Time	Summary of Effects	Ref.
ZnO	**Cancer cells** (human lung adenocarcinoma A549, human hepatocellular carcinoma HepG2, and human bronchial epithelial BEAS-2B) **Rat cells** (hepatocytes and astrocytes)	21.34 ± 7.67 nm	0, 5, 10, and 15 μg/mL 24 h	• ZnO NPs destroy all three types of cancer cells in mammalian cells, but have no effect on normal rat hepatocytes or astrocytes	[141]
Au	Mouse	4.4, 22.5, 29.3, and 36.1 nm	Weighed and assessed every two days for 28 days	• There was inflammation, as well as damage to the liver and kidneys • There is no discernible toxicological response in the reproductive system	[142]
Al_2O_3	Chinook salmon cells (CHSE-214)	20–80 nm	10, 25, 50, and 100 μg/mL	• Cell viability in CHSE-214 cells treated to Al_2O_3 NPs was shown to be dose dependent • When CHSE-214 cells were exposed to Al_2O_3 NPs for 6, 12, and 24 hours, significant morphological alterations were detected	[143]
Au	Fungi	0.7–400 nm	19.697 mg/L 48 h	• Toxicities were higher in larger and nonspherical Au NPs	[144]

(Continued)

TABLE 7.3 *(Continued)*
The Toxic Effects of Metal/Metal Oxide NPs

Metal/Metal Oxide NPs	Model (Cell Lines, Animals)	Particle Size (nm)	Dose and Incubation Time	Summary of Effects	Ref.
TiO_2	Human bronchial epithelium cell line (BEAS-2B)	50 ± 9 nm (bipyramids), 108 ± 47 nm (rods), 75 ± 25 *nm* (platelets), 20 ± 5 nm (P25), and 150 ± 50 nm (food grade)	5, 10, 20, 50, and 80 µg/mL 20, 50, 80, 120, and 160 µg/mL with light	• Rods showed the most cytotoxicity after light exposure, commencing at a dose of 10 g/ml • Bipyramids, P25, and platelets showed a similar cytotoxic impact (doses of 50 g/ml) • Rods and bipyramids did not cause any genotoxic or oxidative damage • Biological impacts were lower in the dark than they were after light exposure	[145]
ZnO	Chinese hamster lung fibroblast cell line (V-79)	30–50 nm	1, 5, 10, 15, and 20 µg/mL	• Oxidative stress and gene mutation were caused by ZnO NPs • After 6 hours of ZnO NP exposure, all genotoxicity endpoints such as chromosomal break, DNA damage, and mutagenicity were detected • Cell cycle arrest, ultrastructural changes, and cell death were all observed	[146]

7.5.2 NPs' Dosage and Clearance

In NP applications, dosage is a critical concern. Currently, the dose of NPs required to disturb cells in vitro is quite large, making their usage in people nearly impossible. Only a few clinical research on NP doses are available at this time. Dosage adjustment and evaluation are critical for achieving important therapeutic goals and reducing toxicity [147, 148]. Also, NPs have a low rate of removal from biological systems. This can cause them to build up in the system for a long time. The charge and size of NPs have a significant impact on their removal from the biological system. Some NPs can be eliminated by the kidney, but others that have not been destroyed will be maintained in the body for a long time [149].

7.5.3 Size, Morphology, and Stability of the NPs

The antibacterial potential of NPs is strongly influenced by their size, morphology, and stability. These parameters are influenced by the synthesis technique and reducing/stabilizing agents. Controlling these factors is thus a fundamental difficulty for producing efficient and effective NPs. The size of NPs utilized as antibacterial agents typically ranges from 1 to 100 nanometers. Particles of a diameter of 10 μm to 10 nm, on the other hand, are often more effective because they can quickly infiltrate and interact with cells. However, manufacturing NPs with controlled size and morphology is frequently a challenge [150]. Different shapes of NPs are frequently manufactured, including nanorods, nanowires, nanoprism, and spherical, flower-shaped, pyramidal, octahedral, and cubic structures. The antibacterial activity of NPs can be greatly influenced by their morphologies [150].

7.5.4 NPs' Interactions with the Cells

The quick aggregation of NPs when used as an antibacterial agent is a disadvantage. The creation of these aggregates causes an increase in size, which reduces cell penetration and increases toxicity. For example, ZnO NPs can interact strongly with organic acids in biological systems, resulting in the production of bioconjugates. Furthermore, the presence of amino acids has a significant impact on the antibacterial action of NPs. This protein is abundant in biological systems and is frequently a barrier to achieving a safer product design as well as increased performance. Additionally, CuO NPs can interact with amino acids, affecting its bacterial activity significantly [151].

7.5.5 Scale-Up/Optimization

There is a lack of effective guideline formulation for manufacture, scale-up, physiochemical property evaluation, standardization, biocompatibility, and protocols for comparing results from in vivo and in vitro research. During large-scale production, inconsistency in size, shape, morphology, and other features may be visible [150]. Producing NPs with controlled size, quality, and lack of aggregates is also a significant difficulty. Furthermore, nanomaterial effectiveness and potency are notoriously difficult to anticipate.

7.5.6 Instrumentation and Variation in Microbes and Human Diseases

NP manufacturing also necessitates high-throughput equipment and technologies. This makes it challenging to produce high-quality NPs in a continuous and uniform manner. Variations in strains and infections caused by different microbes can affect NP activity and make treatment more difficult [150]. Metal oxide NPs' microbicidal properties are influenced by their physicochemical qualities as well as the type of bacteria used. As a result, NPs have great potential as future antibacterial agents and have great potential in nanomedicine, but some concerns should be considered before application.

7.6 CONCLUSION

Many studies have been conducted to date for improving the drinking water's quality, which can greatly affect the environmental safety and human health. Microorganisms can be successfully killed by the nanostructured metal oxides. The notion of nanomaterial production has been elevated in order to place a greater emphasis on the potential for its implementation. Heavy metals and organic contaminants are best absorbed by metal oxide NPs, which have demonstrated promising results in a variety of applications. Immobilization carriers are NPs that can be employed as support carriers for biosensors and biosorbents. Their physical and chemical features have contributed to their effectiveness, but their use in wastewater treatment remains limited. The most utilized nano-metal oxide particles to treat wastewater includes CuO, Cu_2O, ZnO, and TiO_2. They have been employed for many fields, including photocatalytic activity, adsorption, and antimicrobial actions, according to the literature. ZnO is the most utilized metal oxide NP among these five. In adsorption, iron oxide is commonly utilized. Despite the fact that nano-metal oxide particles are effective antibacterial materials for a wide range of applications, their uncontrolled use and release into the natural environment raises some health concerns. To achieve more effective and eco-friendly use of metal oxide NPs, these challenges must be addressed.

7.7 CHALLENGES AND FUTURE PERSPECTIVES

Metal oxide NPs have antibacterial properties, indicating a strong desire to use them as antimicrobials in medicine as soon as possible. The discovery of new antibiotics materials that can kill microorganisms with high rate and efficiency is of great interest. Because globalization has risen considerably, different microorganisms have spread quickly over the world. In the recent several years, different metal oxide NPs with antibacterial properties against diverse bacterial and fungal strains have been widely described. Coating the surface of medical instruments, textiles, and prosthetic could be influenced by the antimicrobial activities of metal oxide NPs. Research is in progress to develop new antibiotics, so researchers are constantly looking for new antimicrobial materials to counteract the growth of antibiotic resistance. Metal oxide NPs hold a lot of potential in terms of overcoming microbial resistance. The possible toxicity of metal oxide NPs as antimicrobials is the next hurdle to overcome. Although metal oxide NPs may be a feasible alternative to traditional antibiotic agents, concerns about systemic cytotoxicity and beneficial microbial disturbance should be addressed and a suitable delivery route must be explored.

Knowledge of metal oxide NPs' interactions with the organs and cells of human, including their capability to penetrate the blood–testis and blood–brain barriers, is critical. Metal oxide NPs will be studied in the environmental sciences, medicine, agriculture, and food science as a result of the knowledge gained via studies in the coming years. Microbes use ATP-binding cassette (ABC) transporters to expel conventional drugs and gain resistance. The cell's entry of divalent metallic ions via ATP-binding cassette (ABC) transporters has yet to be investigated by scientists. Because they contribute in the translocation of essential substrate materials through membranes, as well as RNA translation and DNA repair, these ABC transporters are required for pathogenicity and cell survival. These alterations in the protein structure will prevent the cell from surviving undesirable changes. More investigations about the interaction of metal oxide NPs with ABC transporters are inevitable. Nanotechnologists, microbiologists, and toxicologists must figure out how to employ the proper metal oxide NPs and how much of them to use against specific microorganisms. Modification of nano-metal oxide particles will be aided by a better understanding of molecular routes and their safe use. This can develop advanced antibacterial nano-agents with lower cytotoxicity in people and limit or prevent unintended environmental consequences. The use of metal oxide NPs to tackle MDR microorganisms isn't far off, and it looks promising.

ACKNOWLEDGMENTS

This work was supported by the financial support from the National Natural Science Foundation of China (Grant No.: 52150410401). Also, this work was supported by the 65th Batch of General Aid from China Postdoctoral Science Foundation (No. 2019M652822).

REFERENCES

1. Bhat, S.A., Sher, F., Hameed, M., Bashir, O., Kumar, R., Vo, D-V.N., Ahmad, P., Lima, E.C. 2022. Sustainable nanotechnology based wastewater treatment strategies: achievements, challenges and future perspectives. *Chemosphere*. 288, Part 3, 132606.
2. Shah, B.A., Oluyinka, O.A., Patel, A.V., Bagia, M.I. 2018. Impacts of industrial and agricultural chemical wastes on freshwater resources. *JSM Chem*. 6 (1), 1052.
3. Malik, L.A., Bashir, A., Qureashi, A., Pandith, A.H. 2019. Detection and removal of heavy metal ions: a review. *Environ. Chem. Lett.* 17, 1495–1521.
4. Herschy, R.W. 2012. Water quality for drinking: WHO guidelines. *Encycl. Earth Sci. Ser.* 876–883. https://doi.org/10.1007/978-1-4020-4410-6_184.
5. Bavasso, I., Vilardi, G., Stoller, M., Chianese, A., Palma, L.D. 2016. Perspectives in nanotechnology based innovative applications for the environment. *Chem. Eng. Trans.* 47, 55–60.
6. (a) Samak, N.A., Selim, M.S., Hao, Z., Xing, J. 2020. Controlled-synthesis of alumina-graphene oxide nanocomposite coupled with DNA/sulfide fluorophore for eco-friendly "Turn off/on" H_2S nanobiosensor. *Talanta* 211, 120655; (b) Selim, M.S., El-Safty, S.A.,

Abbas, M.A., Shenashen, M.A. 2021. Facile design of graphene oxide-ZnO nanorod-based ternary nanocomposite as a superhydrophobic and corrosion-barrier coating. *Colloid. Surf. A Physicochem. Engin. Asp.* 611, 125793.

7. Qu, X., Brame, J., Li, Q., Alvarez, P.J.J. 2013. Nanotechnology for a safe and sustainable water supply: enabling integrated water treatment and reuse. *Acc. Chem. Res.* 46 (3), 834–843.
8. deMendonça, V.R., Mourão, H.A.J.L., Malagutti, A.R., Ribeiro, C. 2019. The role of the relative dye/photocatalyst concentration in TiO_2 assisted photodegradation process. *Photochem. Photobiol.* 90 (1), 66–72.
9. Anjaneyulu, R.B., Mohan, B.S., Naidu, G.B., Muralikrishna, R. 2018. Visible light enhanced photocatalytic degradation of methylene blue by ternary nanocomposite, MoO_3/Fe_2O_3/rGO. *J. Asian Ceram. Soc.* 6 (3), 183–195.
10. Zangeneh, H., Zinatizadeh, A.A.L., Habibi, M., Akia, M., Isa, M.H. 2015. Photocatalytic oxidation of organic dyes and pollutants in wastewater using different modified titanium dioxides: a comparative review. *J. Indust. Engin. Chem.* 26, 1–36.
11. Rial, J.B., Ferreira, M.L. 2021. Challenges of dye removal treatments based on IONzymes: beyond heterogeneous Fenton. *J. Water Process Eng.* 41, 102065.
12. Gautam, P.K., Singh, A., Misra, K., Sahoo, A.K., Samanta, S.K. 2019. Synthesis and applications of biogenic nanomaterials in drinking and wastewater treatment. *J. Environ. Manag.* 231, 734–748.
13. (a) Selim, M.S., Yang, H., Li, Y., Wang, F.Q., Li, X., Huang, Y. 2018. Ceramic hyperbranched alkyd/γ-Al_2O_3 nanorods composite as a surface coating. *Prog. Org. Coat.* 120, 217–227; (b) Selim, M.S., El-Safty, S.A., El-Sockary, M.A., et al. 2015. Modeling of spherical silver nanoparticles in silicone-based nanocomposites for marine antifouling. *RSC Adv.* 5 (78), 63175; (c) Selim, M.S., Yang, H., Wang, F.Q., Fatthallah, N.A., Li, X., Li, Y., Huang, Y. 2019. Superhydrophobic silicone/SiC nanowire composite as a fouling release coating material. *J. Coat. Technol. Res.* 16, 1165–1180; (d) Selim, M.S., El-Safty, S.A., Fatthallah N.A., Shenashen, M.A. 2018. Silicone/graphene oxide sheet-alumina nanorod ternary super hydrophobic antifouling coating. *Prog. Org. Coat.* 121, 160–172.
14. (a) Selim, M.S., Shenashen, M.A., El-Safty, S.A., et al. 2017. Recent progress in marine foul-release polymeric nanocomposite coatings. *Prog. Mater. Sci.* 87, 1–17; (b) Mostafa, M.S., Chen, L., Selim, M.S., Betiha, M.A., Zhang, R., Gao, Y., Zhang, S., Ge, G. 2021. Novel cyanate intercalated CoBi layered double hydroxide for ultimate charge separation and superior water splitting. *J. Clean. Prod.* 313, 127868.
15. (a) Biju, V. 2014. Chemical modifications and bioconjugate reactions of nanomaterials for sensing, imaging, drug delivery and therapy. *Chem. Soc. Rev.* 43 (3), 744–764; (b) Selim, M.S., Elmarakbi, A., Azzam, A.M., Shenashen, M.A., EL-Saeed, A.M., El-Safty, S.A. 2018. Ecofriendly design of superhydrophobic nano-magnetite/silicone composites for marine foul-release paints. *Prog. Org. Coat.* 116, 21–34; (c) Selim, M.S., Yang, H., Wang, F.Q., Li, X., Huang, Y., Fatthallah, N.A. 2018. Silicone/Ag@SiO_2 core–shell nanocomposite as a self-cleaning antifouling coating material. *RSC Adv.* 8, 9910–9921; (d) Selim, M.S., El-Safty, S.A., Azzam, A.M., Shenashen, M.A., El-Sockary M.A., Abo Elenien, O.M. 2019. Superhydrophobic silicone/TiO_2–SiO_2 nanorod-like composites for marine fouling release coatings. *Chem. Select.* 4, 3395.
16. (a) Selim, M.S., Samak, N.A., Hao, Z., Xing, J. 2020. Facile design of reduced graphene oxide decorated with Cu_2O nanocube composite as antibiofilm active material. *Mater. Chem. Phys.* 239, 122300; (b) Selim, M.S., El-Safty S.A., Shenashen, M.A. 2019. Chapter 8 - Superhydrophobic foul resistant and self-cleaning polymer coating. In *Superhydrophobic Polymer Coatings*, Ed. S.K. Samal, S. Mohanty, S.K. Nayak, 181–203, Elsevier, Amsterdam, Netherlands.

17. (a) Chiavola, A., D'Amato, E., Stoller, M., Chianese, A., Boni, M.R. 2016. Application of iron based nanoparticles as adsorbents for arsenic removal from water. *Chem. Eng. Trans.* 47, 325–330; (b) Selim, M.S., Mo, P.J., Hao, Z., Fatthallah, N.A., Chen, X. 2020. Blade-like structure of graphene oxide sheets decorated with cuprous oxide and silicon carbide nanocomposites as bactericidal materials. *J. Colloid. Inter. Sci.* 578, 698–709.
18. Chiavola, A., Stoller, M., Di Palma, L., Boni, M.R. 2017. Magnetic core nanoparticles coated by titania and alumina for water and wastewater remediation from metal contaminants. *Chem. Eng. Trans.* 60, 205–210.
19. Vilardi, G., Ochando-Pulido, J.M., Stoller, M., Verdone, N., Di Palma, L. 2018. Fenton oxidation and chromium recovery from tannery wastewater by means of iron-based coated biomass as heterogeneous catalyst in fixed-bed columns. *Chem. Eng. J.* 351, 1–11.
20. Mauter, M., Zucker, I., Perreault, F., Werber, J., Kim, J., Elimelech, M. 2018. The role of nanotechnology in tackling global water challenges. *Nat. Sustain.* 1, 166–175.
21. Varjani, S.J., Gnansounou, E., Pandey, A. 2017. Comprehensive review on toxicity of persistent organic pollutants from petroleum refinery waste and their degradation by microorganisms. *Chemosphere.* 188, 280–291.
22. Du, C., Song, Y., Shi, S., Jiang, B., Yang, J., Xiao, S. 2020. Preparation and characterization of a novel Fe_3O_4-graphene-biochar composite for crystal violet adsorption. *Sci. Total Environ.* 711, 134662.
23. Usmani, M., Khan, I., Bhat, A., et al. 2016. Current trend in the application of nanoparticles for waste water treatment and purification: a review. *Curr. Org. Synth.* 13, 1.
24. (a) Selim, M.S., Fatthallah, N.A., Higazy, S.A., Hao, Z., Mo, P.J. 2022. A comparative study between two novel silicone/graphene-based nanostructured surfaces for maritime antifouling, *J. Colloid Interface Sci.* 606, Part 1, 367–383; (b) Samak, N.A., Selim, M.S., Hao, Z., Xing, J. 2022. Immobilized arginine/tryptophan-rich cyclic dodecapeptide on reduced graphene oxide anchored with manganese dioxide for microbial biofilm eradication. *J. Hazard. Mater.* 426, 128035; (c) Selim, M.S., Azzam, A.M., Higazy, S.A., El-Safty, S.A., Shenashen, M.A. 2022. Novel graphene-based ternary nanocomposite coatings as ecofriendly antifouling brush surfaces. *Prog. Org. Coat.* 167, 106803.
25. Stoller, M., Sacco, O., Vilardi, G., Pulido, J.M.O., Di Palma, L. 2018. Technical–economic evaluation of chromium recovery from tannery wastewater streams by means of membrane processes. *Desalin. Water Treat.* 127, 57–63.
26. Saleh, T.A., Gupta, V.K. 2012. Column with CNT/magnesium oxide composite for lead (II) removal from water. *Environ. Sci. Pollut. Res.* 19 (4), 1224–1228.
27. Yang, Y., Zhang, C., Hu, Z. 2013. Impact of metallic and metal oxide nanoparticles on wastewater treatment and anaerobic digestion. *Environ. Sci. Process Impacts.* 15 (1), 39–48.
28. Singh, S., Kumar, V., Romero, R., Sharma, K., Singh, J. 2019. Applications of nanoparticles in wastewater treatment. In *Nanobiotechnology in Bioformulations. Nanotechnology in the Life Sciences*, Ed. R. Prasad, V. Kumar, M. Kumar, D. Choudhary, 395–418, Springer, Switzerland.
29. Fei, J., Li, J. 2010. Metal oxide nanomaterials for water treatment. In *Nanotechnologies for the Life Sciences,* Ed. C.S. Kumar, Wiley, Weinheim, https://doi.org/10.1002/9783527610419.ntls0145.
30. Yang, J., Hou, B., Wang, J., Tian, B., Bi, J., Wang, N., Li, X., Huang, X. 2019. Nanomaterials for the removal of heavy metals from wastewater. *Nanomaterials (Basel, Switzerland).* 9 (3), 424.
31. Lu, H., Wang, J., Stoller, M., Wang, T., Bao, Y., Hao, H. 2016. An overview of nanomaterials for water and wastewater treatment. *Adv. Mater. Sci. Engin.* 2016, 4964828.
32. Abdelbasir, S.M., Shalan, A.E. 2019. An overview of nanomaterials for industrial wastewater treatment, Korean. *J. Chem. Eng.* 36 (8), 1209–1225.

33. Sharma, S., Bhattacharya, A. 2017. Drinking water contamination and treatment techniques. *Appl. Water Sci.* 7, 1043–1067.
34. Raychaudhuri, A., Gurjar, R., Bagchi, S., Behera, M. 2022. Chapter 12 - Application of microbial electrochemical system for industrial wastewater treatment. In *Advances in Green and Sustainable Chemistry, Scaling Up of Microbial Electrochemical System*, Ed. D.A. Jadhav, S. Pandit, S. Gajalakshmi, M.P. Shah, 195–215, Elsevier, Amsterdam, Netherlands.
35. Forstinus, N., Ikechukwu, N., Emenike, M., Christiana, A. 2015. Water and waterborne diseases: a review. *IJTDH.* 12 (4), 1–14.
36. Shi, C., Wang, X., Zhou, S., Zuo, X., Wang, C. 2022. Mechanism, application, influencing factors and environmental benefit assessment of steel slag in removing pollutants from water: a review. *J. Water Process Engin.* 47, 102666.
37. Saleh, T.A., Mustaqeem, M., Khaled, M. 2022. Water treatment technologies in removing heavy metal ions from wastewater: a review. *Environ. Nanotechnol. Monit. Manag.* 17, 100617.
38. Paul, A.K., Mukherjee, S.K., Hossain, S.T. 2022. Chapter 23 - Application of nanomaterial in wastewater treatment: recent advances and future perspective. In *Development in Wastewater Treatment Research and Processes*, Ed. M. Shah, S. Rodriguez-Couto, J. Biswas, 515–542, Elsevier, Amsterdam, Netherlands.
39. Peces, M., Astals, S., Mata-Alvarez, J. 2014. Assessing total and volatile solids in municipal solid waste samples. *Environ. Technol.* 35 (24), 3041–3046.
40. Khanam, T., Wan Nur, W.R.S., Rashedi, A. 2016. Particle size measurement in waste water influent and effluent using particle size analyzer and quantitative image analysis technique. *Adv. Mater. Res.* 1133, 571–575.
41. Chowdhary, P., Raj, A., Bharagava, R.N. 2018. Environmental pollution and health hazards from distillery wastewater and treatment approaches to combat the environmental threats: a review. *Chemosphere.* 194, 229–246.
42. Alisawi, H.A.O. 2020. Performance of wastewater treatment during variable temperature. *Appl. Water Sci.* 10, 89.
43. Khan, S., Naushad, M., Al-Gheethi, A., Iqbal, J. 2021. Engineered nanoparticles for removal of pollutants from wastewater: current status and future prospects of nanotechnology for remediation strategies. *J. Environ. Chem. Eng.* 9 (5), 106160.
44. Mazilu, M., Musat, V., Innocenzi, P., Kidchob, T., Marongiu, D. 2012. Liquid-phase preparation and characterization of zinc oxide nanoparticles. *Part. Sci. Technol.* 30 (1), 32–42.
45. Gehrke, I., Somborn-Schulz, A.G.A. 2015. Innovations in nanotechnology for water treatment. *Nanotech. Sci. Appl.* 8, 1–17.
46. Dreaden, E.C., Alkilany, A.M., Huang, X., Murphy, C.J., El-Sayed, M.A. 2012. The golden age: gold nanoparticles for biomedicine. *Chem. Soc. Rev.* 41 (7), 2740–2779.
47. (a) Selim, M.S., El-Safty, S.A., El-Sockary, M.A., et al. 2015. Tailored design of Cu_2O nanocube/silicone composites as efficient foul-release coatings. *RSC Adv.* 5, 19933; (b) Selim, M.S., El-Safty, S.A., Shenashen, M.A., Higazy, S.A., Elmarakbi, A. 2020. Progress in biomimetic leverages for marine antifouling using nanocomposite coatings. *J. Mater. Chem. B.* 8, 3701–3732; (c) Selim, M.S., Shenashen, M.A., Elmarakbi, A., Fatthallah, N.A., Hasegawa, S., El-Safty, S.A. 2017. Synthesis of ultrahydrophobic and thermally stable inorganic–organic nanocomposites for self-cleaning foul release coatings. *Chem. Engin. J.* 320, 653–666.
48. Abboud, Y., Saffaj, T., Chagraoui, A., et al. 2014. Biosynthesis, characterization and antimicrobial activity of copper oxide nanoparticles (CO NPs) produced using brown alga extract (*Bifurcaria bifurcata*). *Appl. Nanosci.* 4, 571–576.
49. Danks, A.E., Hall, S.R., Schnepp, Z. 2016. The evolution of 'sol–gel' chemistry as a technique for materials synthesis. *Mater. Horiz.* 3, 91–112.

50. Parashar, M., Shukla, V.K., Singh, R. 2020. Metal oxides nanoparticles via sol–gel method: a review on synthesis, characterization and applications. *J. Mater. Sci. Mate. Electron.* 31, 3729–3749.
51. Stankic, S., Suman, S., Haque, F., Vidic, J. 2016. Pure and multi metal oxide nanoparticles: synthesis, antibacterial and cytotoxic properties. *J. Nanobiotechnol.* 14, 73.
52. Kaur, N., Sharma, R., Sharma, V. 2022. Chapter 15 - Advances in the synthesis and antimicrobial applications of metal oxide nanostructures, In *Elsevier Series in Advanced Ceramic Materials, Advanced Ceramics for Versatile Interdisciplinary Applications*, Ed. S. Singh, P. Kumar, D.P. Mondal, 339–369, Elsevier, Amsterdam, Netherlands.
53. Lawrence, M.J., Kolodziej, A., Rodriguez, P. 2018. Controllable synthesis of nanostructured metal oxide and oxyhydroxide materials via electrochemical methods. *Curr. Opin. Electrochem.* 10, 7–15.
54. (a) Wu, X., Zheng, L., Wu, D. 2005. Fabrication of superhydrophobic surfaces from microstructured ZnO-based surfaces via a wet-chemical route. *Langmuir* 21, 2665–2667; (b) Selim, M.S., Yang, H., Wang, F.Q., Fatthallah, N.A., Huang, Y., Kuga, S. 2019. Silicone/ZnO nanorod composite coating as a marine antifouling surface. *Appl. Surf. Sci.* 466, 40–50.
55. Ghosh, T., Dash, S.K., Chakraborty, P., et al. 2014. Preparation of antiferromagnetic CO_3O_4 nanoparticles from two different precursors by pyrolytic method: in vitro antimicrobial activity. *RSC Adv.* 4, 15022–15029.
56. Mirzaei, A., Neri, G. 2016. Microwave-assisted synthesis of metal oxide nanostructures for gas sensing application: a review. *Sens. Actuat. B Chem.* 237, 749–775.
57. Selim, M.S., Hao, Z., Jiang, Y., Yi, M., Zhang, Y. 2019. Controlled-synthesis of β-MnO_2 nanorods through a γ-manganite precursor route. *Mater. Chem. Phys.* 235, 121733.
58. Mamaghani, A.H., Haghighat, F., Lee, C.S. 2019. Hydrothermal/solvothermal synthesis and treatment of TiO_2 for photocatalytic degradation of air pollutants: preparation, characterization, properties, and performance. *Chemosphere.* 219, 804–825.
59. Selim, M.S., Mo, P.J., Zhang, Y.P., Hao, Z., Wen, H. 2020. Controlled-surfactant-directed solvothermal synthesis of γ-Al_2O_3 nanorods through a boehmite precursor route. *Ceram. Inter.* 46 (7), 9289–9296.
60. Selim, M.S., Shenashen, M.A., Fatthallah, N.A., Elmarakbi, A., El-Safty, S.A. 2017. In Situ fabrication of one-dimensional based lotus-like silicone/γ–Al_2O_3 nanocomposites for marine fouling release coatings. *Chem. Select.* 2 (30), 9691–9700.
61. Caveney, N.A., Li, F.K., Strynadka, N.C. 2018. Enzyme structures of the bacterial peptidoglycan and wall teichoic acid biogenesis pathways. *Curr. Opin. Struct. Biol.* 53, 45–58.
62. Shan, L.I., Wenling, Q., Mauro, P., Stefano, B. 2020. Antibacterial agents targeting the bacterial cell wall. *Curr. Med. Chem.* 27 (17), 2902–2926.
63. Epand, R.M., Walker, C., Epand, R.F., Magarvey, N.A. 2016. Molecular mechanisms of membrane targeting antibiotics. *Biochim. Biophys. Acta.* 1858 (5), 980–987.
64. Wang, L.L., Hu, C., Shao, L.Q. 2017. The antimicrobial activity of nanoparticles: present situation and prospects for the future. *Int. J. Nanomed.* 12, 1227–1249.
65. Ebbensgaard, A., Mordhorst, H., Aarestrup, F.M., Hansen, E.B. 2018. The role of outer membrane proteinsand lipopolysaccharides for the sensitivity of *Escherichia coli* to antimicrobial peptides. *Front. Microbiol.* 9, 2153.
66. Domínguez, A.V., Algaba, R.A., Canturri, A.M., Villodres Á.R., Smani, Y. 2020. Antibacterial activity of colloidal silver against gram-negative and gram-positive bacteria. *Antibiotics (Basel).* 9 (1), 36.
67. Shaikh, S., Nazam, N., Rizvi, S., Ahmad, K., Baig, M.H., Lee, E.J., Choi, I. 2019. Mechanistic insights into the antimicrobial actions of metallic nanoparticles and their implications for multidrug resistance. *Inter. J. Mol. Sci.* 20 (10), 2468.

68. Askar, A.A., Selim, M.S., El-Safty, S.A., Hashem, A.I., Selim, M.M., Shenashen, M.A. 2021. Antimicrobial and immunomodulatory potential of nanoscale hierarchical one-dimensional zinc oxide and silicon carbide materials. *Mater. Chem. Phys.* 263, 124376.
69. Paracini, N., Schneck, E., Imberty, A., Micciulla, S. 2022. Lipopolysaccharides at solid and liquid interfaces: models for biophysical studies of the gram-negative bacterial outer membrane. *Adv. Colloid Inter. Sci.* 301, 102603.
70. Chen, Y., Duan, X., Zhou, X., Wang, R., Wang, S., Ren, N.Q., Ho, S.H. 2021. Advanced oxidation processes for water disinfection: features, mechanisms and prospects. *Chem. Engin. J.* 409, 128207.
71. Robinson, P.K. 2015. Enzymes: principles and biotechnological applications. *Essays Biochem.* 59, 1–41.
72. Chai, H., Yao, J., Sun, J., Zhang, C., Liu, W., Zhu, M., Ceccant, B. 2015. The effect of metal oxide nanoparticles on functional bacteria and metabolic profiles in agricultural soil. *Bull. Environ. Contam. Toxicol.* 94, 490–495.
73. Ali, I.M., Ibrahim, I.M., Ahmed, E.F., Abbas, Q.A. 2016. Structural and characteristics of manganese doped zinc sulfide nanoparticles and its antibacterial effect against gram-positive and gram-negative bacteria. *Open J. Biophys.* 6, 1–9.
74. Kulikov, S.N., Khairullin, R.Z., Varlamov, V.P. 2015. Influence of polycations on antibacterial activity of lysostaphin. *Appl. Biochem. Microbiol.* 51 (6), 683–687.
75. Miranda, R.R., Gorshkov, V., Korzeniowska, B., Kempf, S.J., Neto F.F., Kjeldsen F. 2018. Co-exposure to silver nanoparticles and cadmium induce metabolic adaptation in HepG2 cells, *Nanotoxicology* 12 (7) 781–795.
76. Dizaj, S.M., Lotfipour, F., Barzegar-Jalali, M., Zarrintan, M.H., Adibkia, K. 2014. Antimicrobial activity of the metals and metal oxide nanoparticles. *Mater. Sci. Engin. C.* 44, 278–284.
77. Sims, C.M., Hanna, S.K., Heller, D.A., et al. 2017. Redox-active nanomaterials for nanomedicine applications. *Nanoscale.* 9, 15226–15251.
78. Padil, T.V.V., Černik, M. 2013. Green synthesis of copper oxide nanoparticles using gum karaya as a biotemplate and their antibacterial application. *Int. J. Nanomed.* 8, 889–898.
79. Hajipour, M.J., Fromm, K.M., Ashkarran, A.A., et al. 2012. Antibacterial properties of nanoparticles. *Trends Biotechnol.* 30, 499–511.
80. Juan, C.A., Pérez de la Lastra, J.M., Plou, F.J., Pérez-Lebeña, E. 2021. The chemistry of reactive oxygen species (ROS) revisited: outlining their role in biological macromolecules (DNA, Lipids and Proteins) and induced pathologies. *Int. J. Mol. Sci.* 22, 4642.
81. Kumar, S.R., Imlay, J.A. 2013. How *Escherichia coli* tolerates profuse hydrogen peroxide formation by a catabolic pathway. *J. Bacteriol.* 195, 4569–4579.
82. Rosenberg, M., Visnapuu, M., Vija, H., Kisand, V., Kasemets, K., Kahru, A., Ivask, A. 2020. Selective antibiofilm properties and biocompatibility of nano-ZnO and nano-ZnO/Ag coated surfaces. *Sci. Rep.* 10, 13478.
83. Raghunath, A., Perumal, E. 2017. Metal oxide nanoparticles as antimicrobial agents: a promise for the future. *Inter. J. Antimicrob. Agents.* 49 (2), 137–152.
84. Ayala, A., Munoz, M.F., Arguelles, S. 2014. Lipid peroxidation: production, metabolism, and signaling mechanisms of malondialdehyde and 4-hydroxy-2-nonenal. *Oxid. Med. Cell. Longev.* 2014, 360438.
85. Stanić, V., Tanasković, S.B. 2020. Chapter 11 - Antibacterial activity of metal oxide nanoparticles. In *Micro and Nano Technologies, Nanotoxicity*, Ed. S. Rajendran, A. Mukherjee, T.A. Nguyen, C. Godugu, R.K. Shukla, 241–274, Elsevier, Amsterdam, Netherlands.
86. Estévez, M., Padilla, P., Carvalho, L., Martín, L., Carrapiso, A., Delgado, J. 2019. Malondialdehyde interferes with the formation and detection of primary carbonyls in oxidized proteins. *Redox Biol.* 26, 101277.

87. Ezealigo, U.S., Ezealigo, B.N., Aisida, S.O., Ezema, F.I. 2021. Iron oxide nanoparticles in biological systems: antibacterial and toxicology perspective. *JCIS Open*. 4, 100027.
88. Godoy-Gallardo, M., Eckhard, U., Delgado, L.M., de Roo Puente, Y.J.D., Hoyos-Nogués, M., Gil, F.J., Perez, R.A. 2021. Antibacterial approaches in tissue engineering using metal ions and nanoparticles: from mechanisms to applications. *Bioactive Mater.* 6 (12) 4470–4490.
89. Lemire, J., Harrison, J., Turner, R. 2013. Antimicrobial activity of metals: mechanisms, molecular targets and applications. *Nat. Rev. Microbiol.* 11, 371–384.
90. Vrionis, H.A., Wang, S., Haslam, B., Turner, R.J. 2015. Selenite protection of Tellurite toxicity toward *Escherichia coli*. *Front. Mol. Biosci.* 2, 69.
91. Giannousi, K., Lafazanis, K., Arvanitidis, J., Pantazaki, A., Dendrinou-Samara, C. 2014. Hydrothermal synthesis of copper based nanoparticles: antimicrobial screening and interaction with DNA. *J. Inorg. Biochem.* 133, 24–32.
92. Arakha, M., Pal, S., Samantarrai, D., et al. 2015. Antimicrobial activity of iron oxide nanoparticle upon modulation of nanoparticle–bacteria interface. *Sci. Rep.* 5, 14813.
93. Raghunath, A., Perumal, E. 2017. Metal oxide nanoparticles as antimicrobial agents: a promise for the future. *Int. J. Antimicrob. Agents.* 49 (2), 137–152.
94. Chen, H., Wang, L. 2014. Nanostructures sensitization of transition metal oxides for visible-light photocatalysis. *Beilstein. J. Nanotechnol.* 5, 696–710.
95. Fasiku, V.O., Owonubi, S.J., Malima, N.M., Hassan, D., Revaprasadu, N. 2020. Chapter 15 - Metal oxide nanoparticles: a welcome development for targeting bacteria. In *Antibiotic Materials in Healthcare*, Ed. V. Kokkarachedu, V. Kanikireddy, R. Sadiku, 261–286, Academic Press, United States.
96. Prakash, J., Cho, J., Mishra, Y.K. 2022. Photocatalytic TiO_2 nanomaterials as potential antimicrobial and antiviral agents: scope against blocking the SARS-COV-2 spread. *Micro Nano Engin.* 14, 100100.
97. Devi, L.G., Nagaraj, B. 2014. Disinfection of *Escherichia coli* gram negative bacteria using surface modified TiO_2: optimization of Ag metallization and depiction of charge transfer mechanism. *Photochem. Photobiol.* 5, 1089–1098.
98. Durmus, N.G., Taylor, E.N., Kummer, K.M., Webster, T.J. 2013. Enhanced efficacy of superparamagnetic iron oxide nanoparticles against antibiotic-resistant biofilms in the presence of metabolites. *Adv. Mater.* 25, 5706–5713.
99. Seil, J.T., Webster, T.J. 2012. Antibacterial effect of zinc oxide nanoparticles combined with ultrasound. *Nanotechnology.* 23, 495101.
100. Selim, M.S., Hamouda, H., Hao, Z., Shabana, S., Chen, X. 2020. Design of γ–AlOOH, γ-MnOOH, and α-Mn_2O_3 nanorods as advanced antibacterial active agents. *Dalton Trans.* 49, 8601–8613.
101. Fatthallah, N.A., Selim, M.S., El Safty, S.A., Selim, M.M., Shenashen, M.A. 2021. Engineering nanoscale hierarchical morphologies and geometrical shapes for microbial inactivation in aqueous solution. *Mater. Sci. Engin. C.* 122, 111844.
102. Huang, W.C., Tsai, P.J., Chen, Y.C. 2009. Multifunctional Fe_3O_4@Au nanoeggs as photothermal agents for selective killing of nosocomial and antibiotic-resistant bacteria. *Small.* 5, 51–56.
103. Wang, D., Lin, Z., Wang, T., Yao, Z., Zheng, S., Lu, W., et al. 2016. Where does the toxicity of metal oxide nanoparticles come from: the nanoparticles, the ions, or a combination of both? *J. Hazar. Mater.* 308, 328–334.
104. Belete, T.M. 2019. Novel targets to develop new antibacterial agents and novel alternatives to antibacterial agents. *Human Microb. J.* 11, 100052.
105. Beyth, N., Houri-Hadded, Y., Domb, A., Khan, W., Hazan, R. 2015. Alternative antimicrobial approach: nano-antimicrobial materials, Evid based complement. *Alternat. Med.* 2015, 246012.

106. Jassim, A.M., Farhan, S.A., Salman, J.A., Khalaf, K.J., Al Marjani, M.F., Mohammed, M.T. 2015. Study the antibacterial effect of bismuth oxide and tellurium nanoparticles. *Int. J. Chem. Biol. Sci.* 1, 81–84.
107. Tang, Z.X., Yu, Z., Zhang, Z.L., Zhang, X.Y., Pan, Q.Q., Shi, L.E. 2013. Sonication-assisted preparation of CaO nanoparticles for antibacterial agents. *Quim. Nova.* 36, 933–936.
108. Roy, A., Gauri, S.S., Bhattacharya, M., Bhattacharya, J. 2013. Antimicrobial activity of CaO nanoparticles. *J. Biomed. Nanotechnol.* 9, 1570–1578.
109. Pelletier, D.A., Suresh, A.K., Holton, G.A., McKeown, C.K., Wang, W., Gu, B., et al. 2010. Effects of engineered cerium oxide nanoparticles on bacterial growth and viability. *Appl. Environ. Microbiol.* 76, 7981–7989.
110. Rakesh, Ananda, S., Gowda, N.M.M. 2013. Synthesis of chromium(III) oxide nanoparticles by electrochemical method and *Mukia maderaspatana* plant extract, characterization, $KMnO_4$ decomposition and antibacterial study. *MRC.* 2, 127–135, http://dx.doi.org/10.4236/mrc.2013.24018.
111. Sutradhar, P., Saha, M., Maiti, D.J. 2014. Microwave synthesis of copper oxide nanoparticles using tea leaf and coffee powder extracts and its antibacterial activity. *Nanostruct. Chem.* 4, 86, http://dx.doi.org/10.4236/mrc.2013.24018
112. Pandey, P., Merwyn, S., Agarwal, G.S., Tripathi, B.K., Pant, S.C. 2012. Electrochemical synthesis of multi-armed CuO nanoparticles and their remarkable bactericidal potential against waterborne bacteria, *J. Nanopart. Res.* 14, 709–721.
113. Chatterjee, S., Bandyopadhyay, A., Sarkar, K. 2011. Effect of iron oxide and gold nanoparticles on bacterial growth leading towards biological application. *J. Nanobiotechnol.* 9, 34.
114. Behera, S.S., Patra, J.K., Pramanik, K., Panda, N., Thatoi, H. 2012. Characterization and evaluation of antibacterial activities of chemically synthesized iron oxide nanoparticles. *WJNSE.* 2, 196–200.
115. Borcherding, J., Baltrusaitis, J., Chen, H., et al. 2014. Iron oxide nanoparticles induce *Pseudomonas aeruginosa* growth, induce biofilm formation, and inhibit antimicrobial peptide function. *Environ. Sci. Nano.* 1, 123–132.
116. Al-Hazmi, F., Alnowaiser, F., Al-Ghamdi, A.A., et al. 2012. A new large-scale synthesis of magnesium oxide nanowires: structural and antibacterial properties. *Superlattices Microstruct.* 52, 200–209.
117. Sundaresan, K., Sivakumar, A., Vigneswaran, C., Ramachandran, T. 2011. Influence of nano titanium dioxide finish, prepared by sol-gel technique, on the ultraviolet protection, antimicrobial, and self-cleaning characteristics of cotton fabrics. *J. Ind. Text.* 41, 259–277.
118. Natalio, F., Andre, R., Hartog, A.F., Stoll, B., Jochum, K.P., Wever, R., et al. 2012. Vanadium pentoxide nanoparticles mimic vanadium haloperoxidases and thwart biofilm formation. *Nat. Nanotechnol.* 7, 530–535.
119. Kasraei, S., Sami, L., Hendi, S., Alikhani, M.Y., Rezaei-Soufi, L., Khamverdi, Z. 2014. Antibacterial properties of composite resins incorporating silver and zinc oxide nanoparticles on *Streptococcus mutans* and *Lactobacillus*. *Restor. Dent. Endod.* 39, 109–114.
120. Dhillon, G.S., Kaur, S., Brar, S.K. 2014. Facile fabrication and characterization of chitosan based zinc oxide nanoparticles and evaluation of their antimicrobial and antibiofilm activity. *Int. Nano Lett.* 4, 107.
121. Richtera, L., Chudobova, D., Cihalov, K., Kremplova, M., Milosavljevic, V., Kopel, P., et al. 2018. The composites of graphene oxide with metal or semimetal nanoparticles and their effect on pathogenic microorganisms. *Materials.* 8, 2994–3011.
122. Naemimi, H., Nazifi, Z.S., Amininezhad, S.M. 2015. Preparation of Fe_3O_4 encapsulated silica sulfonic acid nanoparticles and study of their in vitro antimicrobial activity. *J. Photochem. Photobiol. B.* 149, 180–188.

123. Bhushan, M., Muthukamalam, S., Sudharani, S., Viswanath, A.K. 2015. Synthesis of α-Fe_2–xAgxO_3 nanocrystals and study of their optical, magnetic and antibacterial properties. *RSC Adv.* 5, 32006–320014.
124. Bykkam, S., Narsingam, S., Ahmadipour, M., et al. 2015. Few layered graphene sheet decorated by ZnO nanoparticles for anti-bacterial application. *Superlattices Microstruct.* 83, 776–784.
125. Andre, R.S., Zamperini, C.A., Mima, E.G., Longo, V.M., Albuquerque, A.R., Sambrano, J.R., et al. 2015. Antimicrobial activity of TiO_2: Ag nanocrystalline heterostructures: experimental and theoretical insights. *Chem. Phys.* 459, 87–95.
126. Ilbeigi, G., Kariminik, A., Moshafi, M.H. 2019. The antibacterial activities of NiO nanoparticles against some gram-positive and gram-negative bacterial strains. *Int. J. Basic. Sci. Med.* 4 (2), 69–74.
127. Cervantes-Avilés, P., Barriga-Castro, E., Palma-Tirado, L., Cuevas-Rodríguez, G. 2017. Interactions and effects of metal oxide nanoparticles on microorganisms involved in biological wastewater treatment. *Microsc. Res. Tech.* 80 (10), 1103–1112.
128. Murthy, S., Effiong, P., Fei, C.C. 2020. Chapter 11 - Metal oxide nanoparticles in biomedical applications. In *Metal Oxides, Metal Oxide Powder Technologies*, Ed. Y. Al-Douri, 233–251, Elsevier, Amsterdam, Netherlands.
129. Dos-Santos, C.A., Seckler, M.M., Ingle, A.P., et al. 2014. Silver nanoparticles: therapeutical uses, toxicity, and safety issues. *J. Pharm. Sci.* 103, 1931–1944.
130. Baptista, P.V., Mccusker, M.P., Carvalho, A., et al. 2018. Nano-strategies to fight multidrug resistant bacteria—a battle of the titans. *Front. Microbiol.* 9, 1441.
131. Wei, L., Lu, J., Xu, H., Patel, A., Chen, Z.S., Chen, G. 2015. Silver nanoparticles: synthesis, properties, and therapeutic applications. *Drug. Discov. Today.* 20, 595–601.
132. Zaidi, S., Misba, L., Khan, A.U. 2017. Nano-therapeutics: a revolution in infection control in post antibiotic era. *Nanomedicine.* 13, 2281–2301.
133. Rambanapasi, C., Zeevaart, J.R., Buntting, H., et al. 2016. Bioaccumulation and subchronic toxicity of 14 nm gold nanoparticles in rats. *Molecules.* 21, E763.
134. Minetto, D., Libralato, G., Marcomini, A., Ghirardini, A.V. 2017. Potential effects of TiO_2 nanoparticles and $TiCl_4$ in saltwater to *Phaeodactylum tricornutum* and *Artemia franciscana. Sci. Total Environ.* 579, 1379–1386.
135. Borase, H.P., Patil, C.D., Suryawanshi, R.K., et al. 2017. Mechanistic approach for fabrication of gold nanoparticles by *Nitzschia diatom* and their antibacterial activity. *Bioprocess Biosyst. Engin.* 40, 1437–1446.
136. Sengul, A.B., Asmatulu, E. 2020. Toxicity of metal and metal oxide nanoparticles: a review, *Environ. Chem. Lett.* 18, 1659–1683.
137. Handral, H.K., Ashajyothi, C., Sriram, G., Kelmani, C.R., Dubey, N., Cao, T. 2021. Cytotoxicity and genotoxicity of metal oxide nanoparticles in human pluripotent stem cell-derived fibroblasts. *Coatings.* 11, 107.
138. Schneider, T., Westermann, M., Glei, M. 2017. In vitro uptake and toxicity studies of metal nanoparticles and metal oxide nanoparticles in human HT29 cells. *Arch. Toxicol.* 91 (11), 3517–3527.
139. Haidong, W., Yudan, H., Lujie, L., et al. 2020. Prooxidation and cytotoxicity of selenium nanoparticles at nonlethal level in Sprague-Dawley rats and buffalo rat liver cells. *Oxid. Med. Cell Longev.* 2020, 7680276.
140. Areecheewakul, S., Adamcakova-Dodd, A., Givens, B.E., et al. 2020. Toxicity assessment of metal oxide nanomaterials using *in vitro* screening and murine acute inhalation studies. *NanoImpact.* 18, 100214.
141. Akhtar, M.J., Ahamed, M., Kumar, S., Khan, M.M., Ahmad, J., Alrokayan, S.A. 2012. Zinc oxide nanoparticles selectively induce apoptosis in human cancer cells through reactive oxygen species. *Int. J. Nanomed.* 7, 845–857.

142. Chen, H., Dorrigan, A., Saad, S., Hare, D.J., Cortie, M.B., Valenzuela, S.M. 2013. In vivo study of spherical gold nanoparticles: inflammatory effects and distribution in mice, *PLoS ONE*. 8, e58208.
143. Srikanth, K., Mahajan, A., Pereira, E., Duarte, A.C., Venkateswara, R.J. 2015. Aluminium oxide nanoparticles induced morphological changes, cytotoxicity and oxidative stress in Chinook salmon (Chse-214) cells. *J. Appl. Toxicol.* 35 (10), 1133–1140.
144. Liu, K., He, Z., Byrne, H.J., Curtin, J.F., Tian, F. 2018. Investigating the role of gold nanoparticle shape and size in their toxicities to fungi. *Int. J. Environ. Res. Public Health.* 15, E998.
145. Gea, M., Bonetta, S., Iannarelli, L., et al. 2019. Shape-engineered titanium dioxide nanoparticles (TiO-NPs): cytotoxicity and g enotoxicity in bronchial epithelial cells. *Food Chem. Toxicol.* 127, 89–100.
146. Jain, A.K., Singh, D., Dubey, K., Maurya, R., Pandey, A.K. 2019. Zinc oxide nanoparticles induced gene mutation at the HGPRT locus and cell cycle arrest associated with apoptosis in V-79 cells. *J. Appl. Toxicol.* 39 (5), 735–750.
147. Hua, S., de Matos, M.B.C., Metselaar, J.M., Storm, G. 2018. Current trends and challenges in the clinical translation of nanoparticulate nanomedicines: pathways for translational development and commercialization. *Front. Pharmacol.* 9, 790.
148. Grumezescu, A.M. 2018. *Nanoscale Fabrication, Optimization, Scale-up and Biological Aspects of Pharmaceutical Nanotechnology*. Elsevier Inc., London.
149. Lin, Z., Monteiro-Riviere, N.A., Riviere, J.E. 2015. Pharmacokinetics of metallic nanoparticles. *Wiley Interdiscip. Rev. Nanomed. Nanobiotechnol.* 7, 189–217.
150. Mba, I.E., Nweze, E.I. 2021. Nanoparticles as therapeutic options for treating multidrug-resistant bacteria: research progress, challenges, and prospects. *World J. Microbiol. Biotechnol.* 37, 108.
151. Badetti, E., Calgaro, L., Falchi, L., et al. 2019. Interaction between copper oxide nanoparticles and amino acids: influence on the bacterial activity. *Nanomaterials.* 9, 792.

8 Electrochemical Treatment as a Promising Advanced Technique for Industrial Wastewater Treatment

Mohamed S. Hellal and Enas M. Abou-Taleb

8.1 INTRODUCTION

Water is the most abundant substance on Earth. Oceans, seas, lakes, rivers, and ice represent about 71% of the Earth's composition. However, only 2.5% of this amount is freshwater and 87.5% is salt water (UN 2020). The water supply demands increased with the growth of cities, intensification of agriculture, expansion of irrigation, and rapid development of industry, besides climate change and other factors (Abou-Elela et al. 2018). One in six people in the world has no access to safe drinking water (WHO/UNICEF 2021). Alongside the scarcity of fresh water, anthropogenic contamination is the major problem of significant reduction of existing freshwater resources. This has resulted from discharges from industrial firms and fertilizer runoffs from the fields, besides brine water incursion into freshwater aquifers near to coastal areas because of over pumping of groundwater (Abou-Taleb et al. 2020). Natural water resources should be utilized rationally through expanding freshwater resources and the development of sustainable solutions for pollution prevention of water resources. The change in physical, chemical, or biological properties of natural water is defined as pollution or contamination. This may be caused due to liquid, solid, or gaseous compounds added to the water, which could be hazardous to health.

8.1.1 Water Pollution Sources and Types

Clean water is ingested by different disciplines of human life, such as industrial plants, mining, energy production, agriculture, municipal utilities, health, ecosystems, and other services (D'Odorico et al. 2018). Water resulting from industries and municipalities cannot be directly discharged back to the environment or returned to be reused in the technological cycle as it is polluted with various impurities such as organic and inorganic pollutants (Aboutaleb et al. 2018). Types of water pollutants can be categorized by origin, nature, size, toxicity, etc. Source classification divides pollutants into human-made and natural pollutants. Pollution conducted by human activities is generated from industry, agriculture, mining, municipal and other.

DOI: 10.1201/9781003354475-8

Pollution from natural sources includes volcanic activity, biological decay of dead works and animals, epithermal deposits, and leaching of pollutants from rocks and lands (Tarzia et al. 2002). According to the pollution source, contaminants can be classed as chemical compounds (organic and inorganic), biological (pathogens and viruses), thermal and radioactive pollution (Ahamad et al. 2020). The common and largest sources of water pollution are coming from the industrial process which is responsible for the following changes in natural water quality:

- Increased organic matter content from wastewater in terms of biological oxygen demand (BOD) and chemical oxygen demand (COD) within water bodies.
- Decrease of dissolved oxygen content due to the presence of organic hydrophobic compounds, thermal pollution and other pollution that could prevent oxygen diffusion from the air.
- Eutrophication of water bodies due to the discharge of water polluted with phosphates, nitrates, nitrites, and ammonium.
- Nitric oxide, nitrogen oxide, and sulfur dioxide in the ambient air cause acid precipitation which results in a decrease in the pH of freshwater.

Inorganic pollutants present in industrial wastewaters, including alkalis, acids, heavy metals, etc., result from industrial operations such as lead, Zn, and nickel ore factories and phosphate, sulfate, and nitrogen fertilizer factories. Industrial effluents including petroleum compounds, ammonia, phenols, aldehydes, resins and other harmful substances are generated from petrochemical, refinery, organic, and synthetic power plant and coking plant discharge. Paper, tannery, food, textile and dying industries are among the resources of organic contamination of surface water.

8.1.2 Treatment Techniques of Wastewater

Types of wastewater treatment are classified into either chemical, mechanical, physicochemical, biological, or combination of these types where few methods are employed in combination. The use of any of these methods in a specific case depends on the type of wastewater, its origin, and its hazardous content. Mechanical treatment is a pretreatment step for contaminant elimination by gravity separation or filtration (Moussavi et al. 2021). In this process, suspended solids are entrapped, according to their size and density through several types of equipment such as grit chambers, sieves, septic tanks, grids, clarifiers, oil traps, etc. This process can achieve 60–80% removal efficiency of insoluble contaminates and suspended solids in both municipal and industrial wastewater (Nowobilska-Majewska and Bugajski 2019). Chemical treatment involves the addition of chemical reagents that react with contaminants and, in most cases, precipitate as insoluble particles. This process can be utilized as a pretreatment, primary treatment, or secondary treatment according to the wastewater type and may include neutralization, oxidation, and reduction processes. Physicochemical treatment of wastewater involves the elimination of dissolved inorganic impurities, degradation of organic compounds, metal recovery, etc. Coagulation, flotation, crystallization, electrochemical treatment, ion exchange, sorption, and extraction are most of the common physicochemical treatment methods. Biological treatment is the use of microorganisms for the elimination of organic pollutants and dissolved nutrients such as nitrogen and phosphorus. It may include both aerobic and anaerobic microorganisms which could decrease the COD and BOD content of water (Hellal et al. 2021). There are several methods for biological treatment such as digestion tanks, conventional activated sludge, biofilters, membrane bioreactors, and constructed wetlands. However, there are some pollutants that couldn't be removed by the previous traditional treatment techniques have highly soluble organic matter. These pollutants could be removed using advanced oxidation processes (AOPs). These processes are novel techniques using reactive chemical groups for the degradation of organic compound into simple form of inorganic compound such as carbon dioxide. AOP is a comprehensive set of treatment methods including, Fenton, photocatalysis, ozonation, photodegradation, persulphate oxidation and recently electro-oxidation (EOX).

Recently, due to water and energy crises, the treatment of wastewater for reuse with minimal cost is the main target (Abou-Taleb et al. 2020). Selection of the most appropriate treatment technology specially in the industrial sector which is cost-effective and produces high-quality treated wastewater amenable for reuse is the current challenge (Abou-Elela et al. 2018). The advantages and disadvantages of common treatment techniques are summarized in Table 9.1 (Sillanpää and Shestakova 2017c). As indicated in Table 8.1, electrochemical treatment has various benefits, including environmental compatibility, versatility, energy efficiency, safety, selectivity, amenability to automation, and cost-effectiveness. Recently, electrochemical treatment methods such as electro-coagulation (EC), EOX, and electro-flotation have attracted more and more attention in the treatment of different types of wastewater (Aboutaleb et al. 2018).

8.1.3 Categories of Electrochemical Wastewater Treatment Methods

The electrochemical wastewater treatment is one of the basic types of electrolysis processes. To carry on the electrolysis, an electrical power source is utilized to supply a proper potential at the anode and cathode, which are put in an electrochemical cell, resulting in an electrochemical reaction. Electrochemical treatment is characterized by several phases of physical and chemical phenomena and by relative complexity. There are many factors that determine the mechanism and rate of reaction, optimal reactor design and operational conditions. According to the physicochemical characteristics, electrochemical wastewater treatment methods could be divided into two broad categories: EC and EOX.

8.1.3.1 Electro-Oxidation of Wastewater

EOX is a chemical reaction, involving the loss of one or more electrons by an atom or a molecule at the anode surface made of a catalyst material during the passage of direct electric current through the electrochemical systems (anode, cathode, and an electrolyte solution) (Qiao and Xiong 2021). The main application of EOX is to wastewater containing highly soluble organic compounds. Direct and indirect mechanisms are the main two mechanisms for EOX of organic compounds in wastewater.

TABLE 8.1
Advantages and Disadvantages of Different Treatment Techniques

Treatment Technique		Advantages	Disadvantages
Physicochemical treatment	Adsorption	• Could achieve 100% pollutant removal • Simplicity of application • Cost-effective operating	• Regeneration of saturated adsorbents • Utilization of spent adsorbents • pH and temperatures control the efficiency • Demanding the utilization of retentate
	Membrane filtration	• Immediate water softening • Different flow rate systems could be utilized	• Membrane fouling • Concentrate separation • Membranes' lifetime is short • Utilization of retentate and spent membranes is required • High capital and operation costs
Biological treatment	Aerobic	• High level of pollutant removal efficiency • Common and more spread method	• Require large land areas • The demand of disposal of waste-activated sludge • Nutrient addition is required • Disinfection is mandatory
	Anaerobic	• Possibility to recycle anaerobic sludge, which could produce biogas	• Toxic compounds are dangerous for microorganisms • Efficiency depends on temperature and pH • Not efficient for nitrogen removal
AOPs	Fenton	• Achievement of complete mineralization • Rapid decomposition reactions • Minimum energy requirements	• Chemical addition is required • Degradation is pH dependent • Sludge with ferric hydroxide content
	Photocatalysis and degradation	• Complete mineralization • Chemicals are not required	• Not suitable for turbid and colored wastewaters • pH dependent
	Ozonation	• Could achieve high degradation rate • Chemicals are not required • pH is not changed during the treatment • Disinfection	• High energy required process • Toxicity of ozone • Formation of toxic by-products
Electrochemical treatment		• Organic matter could be completely decomposed and mineralized • Easy to implement and operate • No requirement of chemical addition in industrial wastewater applications	• Electrode polarization and contamination • Electrode corrosion • Prohibitive cost of electrodes

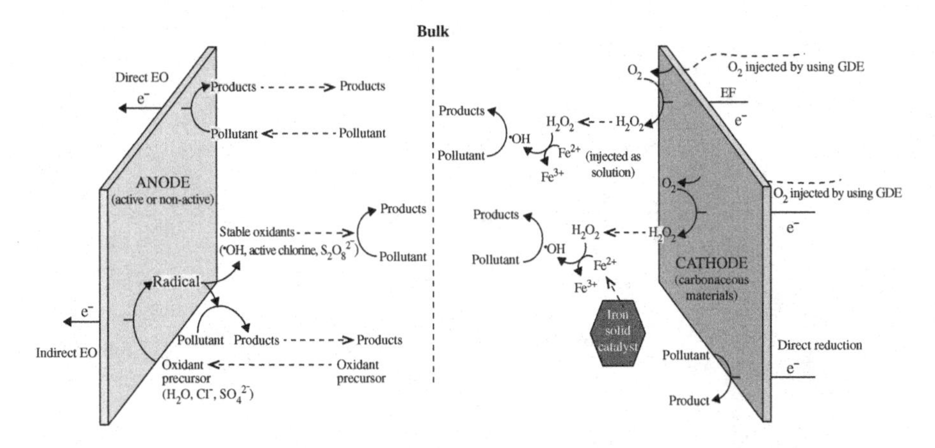

FIGURE 8.1 Mechanism of direct and indirect EOX of wastewater. (Reprinted with permission from Martínez-Huitle and Panizza 2018.)

Figure 8.1 shows the schematic diagram for the mechanism of EOX of wastewater (Martínez-Huitle and Panizza 2018). In the direct EOX process, electrons are transferred to the surface of the electrode without the involvement of other substances. The method involves a direct interaction of the active oxygen and the organic contaminants with their subsequent oxidation. This is only probable in the case of using electrodes manufactured from catalytic substances. The indirect oxidation includes generation of different electroactive species during electrolysis, which play a part in the oxidation process (Barrios et al. 2015). The most common method of indirect oxidation of organics is the production of highly reactive hypochlorite ions. For instance, if the treatment solution includes NaCl, chlorine gas is generated at the anode (Eqn. (8.1)) followed by reaction with H_2O molecules, resulting in the production of hypochlorous and hydrochloric acids (Eqn. (8.2)) (Qiao and Xiong 2021).

$$2Cl^- \rightarrow Cl_2 + 2e^- \tag{8.1}$$

$$Cl_2 + H_2O \rightarrow HOCl + H^+ + Cl^- \tag{8.2}$$

$$HOCl \rightarrow H^+ + OCl^- \tag{8.3}$$

8.1.3.2 Electro-Coagulation (EC)

EC involves the physicochemical process of colloidal particles' coagulation under the effect of a direct electrical charge. In the electro-treatment of wastewater with iron or aluminum electrodes, anodic metal dissolution occurs electrochemically (Potrich et al. 2020). Dissolved cations of a metal are hydrolyzed and transformed into a coagulant, which initiates the adherence and gathering of the flakes. Coagulation generally means a phase separation through the loss of aggregate stability in dispersed systems (Kuzin and Kruchinina 2021). A wide range of pollutants can be disposed of out of water through EC. These pollutants, most of which are negatively

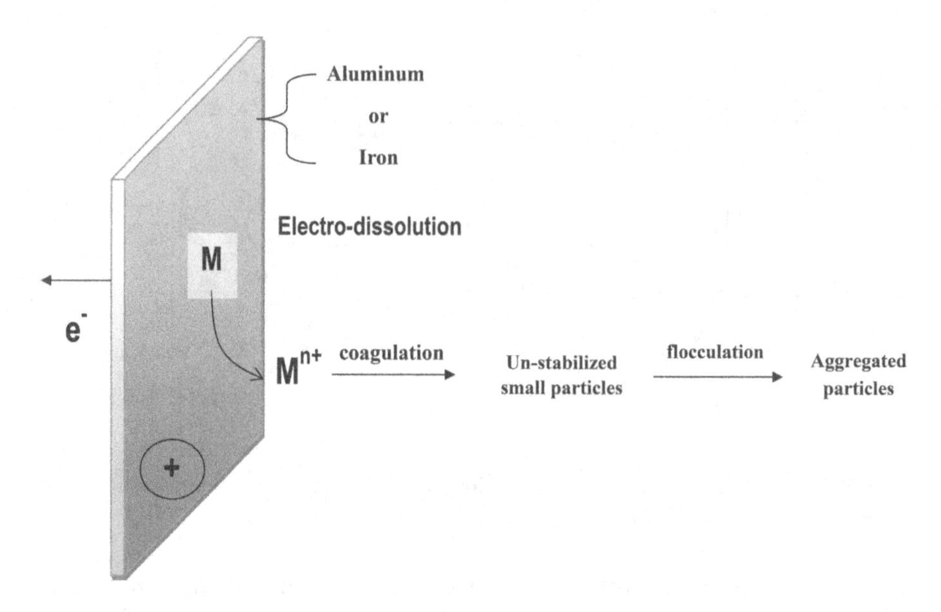

FIGURE 8.2 Mechanism of electro-coagulation.

charged, may be organic pollutants, inorganic colloids, pathogenic microorganisms and others (Hakizimana et al. 2017). In this regard, adding positively charged cations can destabilize and neutralize colloids, requiring them to coagulate. The simplified EC procedure is illustrated in Figure 8.2.

Aluminum (Al) and iron (Fe) electrodes are the most commonly used metals in EC cells, being more readily available, reliable and non-toxic (Potrich et al. 2020). The quantity of metal dissolved and deposited depends on the quantity of electricity that passes through the reaction solution. Current density influences not only the response time of the system but also the dominant mode of separating pollutants (Garcia-Segura et al. 2017). EC with aluminum electrodes is usually used to treat concentrated wastewater containing oil and organic pollutants. Inorganic wastes as well as non-ferrous metals (Zn, Cu, Ni, Cd, Cr(III), etc.) could be removed from wastewater through EC process with iron electrodes. The EC reactions of aluminum and iron electrode are presented in Eq. (4.7) as following:

For iron electrodes:

$$Fe_{(s)} \rightarrow Fe^{n+}_{(aq)} + ne^- \tag{8.4}$$

$$4Fe^{2+}_{(aq)} + 10H_2O + O_2 \rightarrow 4Fe(OH)_3 + 8H^+ \tag{8.5}$$

For aluminum electrodes:

$$Al_{(s)} \rightarrow Al^{3+}_{(aq)} + 3e^- \tag{8.6}$$

$$Al^{3+}_{(aq)} + nH_2O \rightarrow Al(OH)_n^{3-n} + nH^+ \tag{8.7}$$

8.2 DESIGN PARAMETERS AND FACTORS AFFECTING THE ELECTROCHEMICAL TREATMENT EFFICIENCY

Designing electrochemical treatment reactors includes several controls and parameters for specific treatment processes. Depending on the application, designing of the reactor for a certain treatment process must be optimized for process evaluation through energy consumption and efficiency, current efficiency, cell voltage, and treatment efficiency. The final design should be in accordance with a number of key features which include the following (Walsh and Ponce de León 2018):

1. Capital and operating cost reduction through simple reactor engineering design, affordable components, and low cell potential difference.
2. Practical and reliable design, installation, operating, maintenance, and surveillance procedures.
3. Suitable facilities for controlling and monitoring the temperature, potential, current density, and mass transport regime adequate for the removal of pollutants, by appropriate flow distributions.
4. Simplicity and versatility are perhaps the least quantified and most neglected factors, but perhaps the most important, to achieve at an elegant and durable design to attract users.
5. Planning for future developments by designing a modular form that facilitates scale-up by adding unit cells or by increasing the size of each unit.

There is no one-size-fits-all design; rather, reactors have been designed for every pollutant assessed by different studies (Sillanpää and Shestakova 2017b). In addition, over the past few years, the new research trend has been to integrate this new technology with other traditional treatment technologies to increase treatment efficiency or address different pollutants' removal (Muddemann et al. 2019; Chen 2004; Butler et al. 2011).

Designing electrochemical treatment reactor is the critical point of the treatment process as the process operation and the cost are dependent on the performance of the reactor since it affects many of the other units of the process, such as settlers and filters (Mann 2009). There is no standard design for the electrochemical reactors' construction. Commonly, electrochemical reactors are compact and simple to operate. These reactors may be rectangular tanks or cylindrical vessels in which a pack of electrodes is positioned. Most of the reactors are of the horizontal or vertical flow type as indicated by Figure 8.3 A, B. The electrochemical reactor could be designed in combination with other treatment processes such as filtration or sedimentation in one compact unit, leading to the reduction of required working areas and obtain treated wastewater in one treatment step (Sillanpää and Shestakova 2017b). Figure 8.4 shows an example of a combination of EC cells inside a vertical clarifier for the treatment of wastewater containing oil and grease. Suitable reactor design based on choosing a suitable geometry and the placement of the electrodes is necessary to optimize the procedure in terms of energy consumption and increase the oxidation rate. Thus, the combination of design parameters, phenomena, and operating conditions affects the treatment efficiency of the reactor (Rajkumar and Palanivelu 2004). Operational

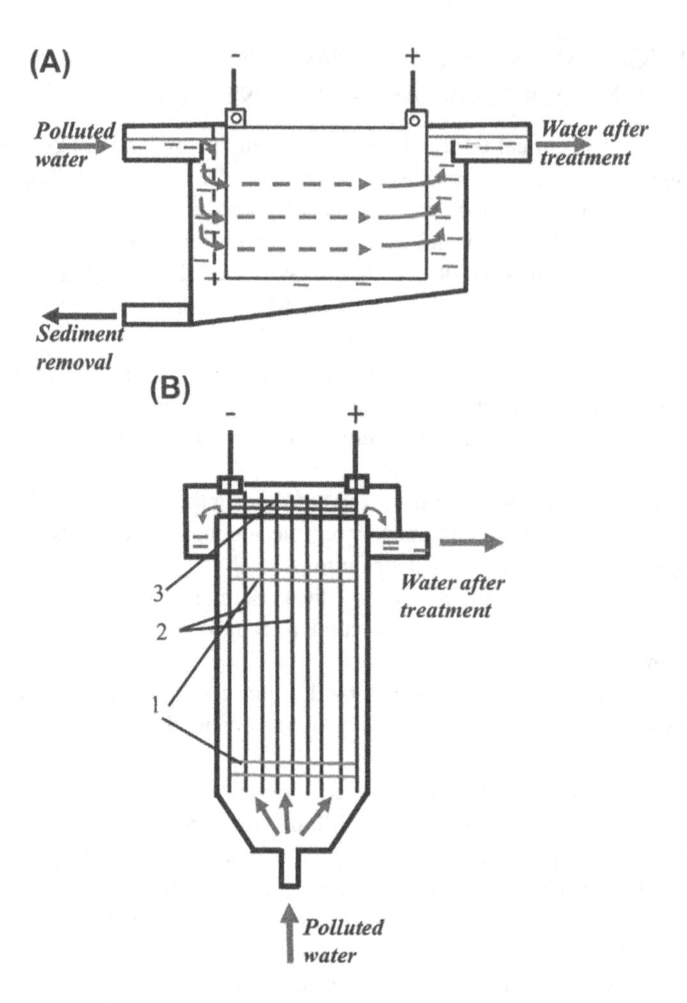

FIGURE 8.3 Design examples of (A) horizontal and (B) vertical model of the EC unit. (1) Insulating inserts; (2) electrodes; (3) current-conducting inserts. (Reprinted with permission from Sillanpää and Shestakova 2017b.)

parameters which influence the treatment process include treatment time, current density, pH, electrode material, number, and arrangement, the distance between them, as well as the conductivity of the water to be treated (Muddemann et al. 2019; Rajasekhar et al. 2020; Bracher et al. 2021).

8.2.1 Electrode Material and Shape

Electrode material selection is one of the main factors that could boost the process treatment efficiency. There are several electrode materials that have been investigated in electrochemical treatment studies. These include but are not limited to aluminum, silver, cadmium, chromium, cesium, iron, magnesium, and zinc for EC

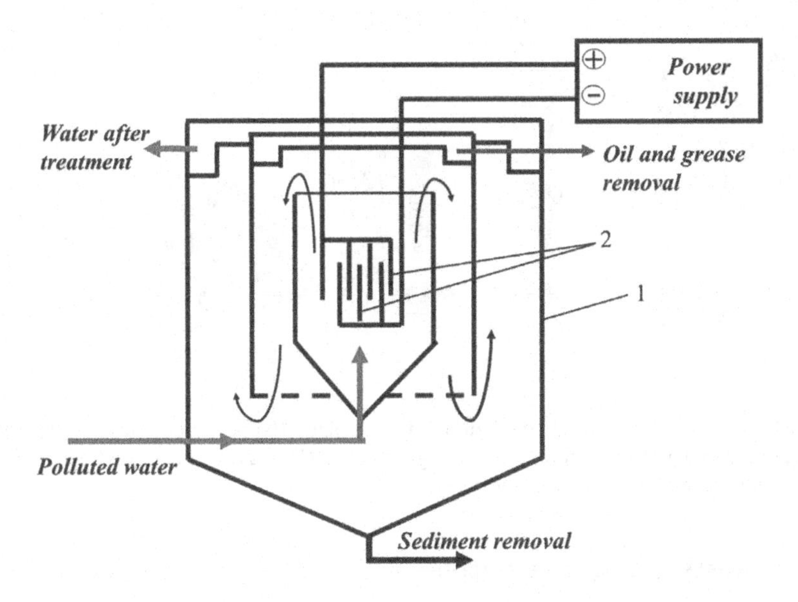

FIGURE 8.4 Combination of EC unit with other treatment steps for oil and grease wastewater treatment. (1) Vertical settler; (2) electrodes. (Reprinted with permission from Sillanpää and Shestakova 2017b.)

process (Butler et al. 2011), while stainless steel (SS), graphite, platinum, gold, lead, and other noble materials are used for EOX process (Barrios et al. 2015). Multivalent metal electrodes like iron or aluminum are used for EC due to their coagulation properties of multivalent ions (Abu Ghalwa and Saqer 2015). Iron, aluminum, SS, and graphite electrodes can be used in combination with EC and EOX as they are anodically soluble (Sharma and Simsek 2019). Besides the traditional materials, there are chemically resistant electrodes such as graphite, nickel, boron-doped diamond, SnO_2, and PbO_2. They offer greater removal efficiency of high organic matter content of wastewater. The most common anodes used in the EOX reactors are SS and graphite electrodes (Kolpakova, Oskina, and Sabitova 2018; Abou-Taleb et al. 2021). These anodes are of low cost and readily available, while other alloy electrodes such as iridium oxide and titanium/iridium oxide electrodes are very expensive. The shape of the electrodes utilized in electrochemical treatment is a key parameter while designing the reactor. The common traditional shape of the electrodes utilized in the electro-treatment cell is plate shape. However, several studies developed new and novel designs of electrode shapes (Magnisali, Yan, and Vayenas 2022). These shapes included, but were not limited to, cylindrical rode (Abou-Taleb, Hellal, and Kamal 2021), perforated cylindrical (Al-Shannag et al. 2013), cylindrical in helical wire (Hamdan and El-Naas 2014), parallel rotating discs (Xu et al. 2018), mesh wire (Abdel-Salam et al. 2018), and perforated discoid (Hashim et al. 2015). Figure 8.5 illustrates some of the new developed electrode shapes (Magnisali, Yan, and Vayenas 2022).

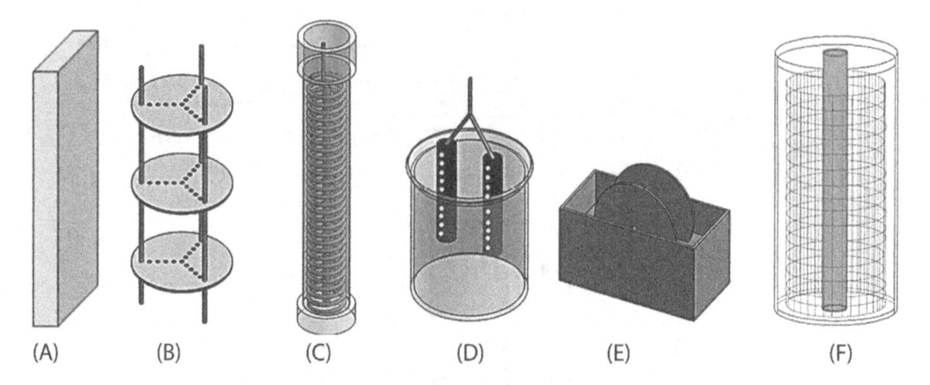

FIGURE 8.5 Some electrode configurations: (A) plate, (B) perforated discoid, (C) rod in helical wire, (D) perforated rods, (E) rotating discs, and (F) rod in mesh wire. (Reprinted with permission from Magnisali, Yan, and Vayenas 2022.)

8.2.2 Distance between Electrodes

The binary distance between electrodes has an important effect on the reactions in the electrochemical reactor. This parameter is a main factor when determining the reactor size and the power consumption and has a considerable effect on the total cost treatment process cost (Ghosh et al. 2008). Reduction of energy consumption could be obtained because of small inter-electrode space applied in the electrochemical treatment reactor. Determining the electrodes' inter-space depends on the practicality and field conditions. All literature studies stated that the electrode spacing ranged between 2 and 10 mm depending on other design parameters and waste characteristics (Rajkumar and Palanivelu 2004; Sillanpää and Shestakova 2017b; Muddemann et al. 2019; Qiao and Xiong 2021). The more increased electrode binary spacing, the more increase in cell voltage, and consequently energy consumption increases (Shahedi et al. 2020). Close arrangement of less than 10 mm results in minimal energy consumption. A 10-mm spacing between electrodes was observed to be appropriate for electrochemical treatment of wastewater from pulp and paper mill industry (Uğurlu et al. 2008). Spacing less than 10 mm reduces the rate of liquid medium movement between the electrodes and affects the removal efficiency of pollutants.

8.2.3 Current Density

Current density is considered as a measure of electrochemical reaction rate. According to Faraday's law, the rate of an electrochemical reaction is directly proportional to the amount of substance that undergoes a reaction at the electrode within a given time and the current that flows through the electrode. In other words, the rate of reaction is proportional to the quantity of electricity that passes through the system. (Sillanpää and Shestakova 2017a). As a result, the rate of the electrochemical reaction could be quantified as a function of electric current value. Since electrodes can be of different sizes, the same potential can produce different currents. Consequently, the reaction rate typically refers to an electrode surface unit. As per Eq. (8.8), the current (I) applied to electrode area (A), this relation is the current density (CD).

$$CD = \frac{I}{A} \tag{8.8}$$

The operating current density is very important for the design of an electrochemical reactor, given that this is the only operating parameter that can be directly examined. In such processing techniques, the gap between electrodes is fixed and the current could be variable. The current density directly determines the coagulant dose in EC and the redox generation rates in EOX and has a strong influence on the solution mixture and mass transfer to the electrodes. At high current densities, the extent of release of ion at anode increases, and in sequence the number of hydroxy-cationic complexes also increases, resulting in an increase in pollutant removal. However, the maximum permissible current density may not be the most effective method of operating the reactor.

8.2.4 pH

The solution pH is an important factor in electrochemical treatment processes (Moussa et al. 2017). It directly affects electrodes' dissolution, solution conductivity, the hydroxide speciation, and colloidal particles' zeta potential. For example, bulk acid solution inhibits electrochemical processing due to the hydroxyl ions that are generated and consumed by the acid in the solution. At acidic pH, aluminum and iron electrodes could be dissolved without electricity, whereas oxidation of Fe(II) to Fe(III) occurs only at pH above 5 (Sasson et al. 2009). The pollutants' origin determines the efficiency of the electro-treatment process; however, the best pollutant removal was found near pH 7. According to several studies, the optimum pH of wastewater conducted to treatment via electrochemical means ranges between 6.5 and 7.5 (Rajkumar and Palanivelu 2004; Garcia-Segura et al. 2017; Moussa et al. 2017; Rajasekhar et al. 2020).

8.2.5 Conductivity

The high concentration of dissolved solids increases the ionic strength in the bulk solution, resulting in an increase in current density. It is therefore necessary to monitor the solution conductivity during the electrochemical process. The conductivity of the solution depends both on the type and on the concentration of electrolyte. Different types of electrolyte are available like sodium chloride, barium chloride, potassium chloride, and sodium sulfate. Sodium chloride is the common additive to increase conductivity because of its availability and cost-effectiveness. The presence of chlorides in the solutions leads to the formation of Cl_2 and OCl^- at the anode. The OCl^- is a strong oxidizing agent, which can oxidize organic matter present in the solution. It is therefore recommended to study the effect of conductivity values of electrolysis solution to ensure an efficient water treatment.

8.2.6 Temperature

There are few studies on the effect of temperature on the removal of pollutants through electrochemical treatment process. Temperature may positively or negatively affect the process of electrochemical treatment. The increase in temperature reduces metal hydroxide formation, resulting in a decrease in removal efficiency (Chen 2004). At low temperature, anodic dissolution rate decreased and, consequently, pollutant removal efficiency decreased (Vasudevan et al. 2010). Temperature affects the treatment process by shifting the reaction rate, solubility of metal hydroxides, solution conductivity, and kinetics of gas bubbles, or small colloidal particles.

8.2.7 Reaction Time

Reaction time also has a significant impact on the elimination of pollutants and the effectiveness of electrochemical processing. It determines the size of the reactor, the amount of flocs produced, and the cost of the process. Increasing the treatment time to the optimum increases the pollutant removal efficiency but does not exceed the optimum level, considering that the above optimal reaction time and increased coagulant dose do not increase the removal of the pollutant due to the presence of a enough number of flocs (Abou-Taleb et al. 2021). Electrolysis time adversely affects the treatment cost as energy and electrode consumption increased at a longer electrolysis time.

8.3 THE REMOVAL EFFICIENCY OF POLLUTANTS IN SEVERAL TYPES OF INDUSTRIAL WASTEWATER

The application of electrochemical technology specially EC and EOX for industrial wastewater treatment gained high attraction worldwide. The performance of a two-step process combining EC and EOX appears to have proven effective in the treatment of industrial wastewater. Industrial applications of EC units are more spread worldwide in commercial designs than EOX. Figure 8.6 shows the percentage distribution of EC manufacturers worldwide (Magnisali et al. 2022). America holds 33% of the world production share of EC systems and is considered the leader in innovative EC designs. However, in the United States, just over 29% of producers still use traditional EC design of the typical tank with plate electrodes generally in aluminum or iron. Europe shares 22% of the world market for European industrial systems. Traditional structures occupy about half of the European commercial units share, representing more than 56% of the total models available. The most active sector in terms of EC manufacturers is Asia, which occupies 43% of the worldwide production market for EC units. About 65% of Asia's share comes from India, while China, Singapore, UAE, Israel, and Thailand represent other producer countries, where the typical design dominates at a share of almost 80%. Unfortunately, Africa shares only 1% of each of the world manufacturing EC systems, and traditional design monopolizes these regions, though there are companies from other continents that keep subsidiary offices there. In Egypt, electrochemical treatment technologies considered a new technology for industrial wastewater treatment market.

This section presents a summary of recent applications of electrochemical treatment with different types of industrial wastewater in terms of removal efficiencies,

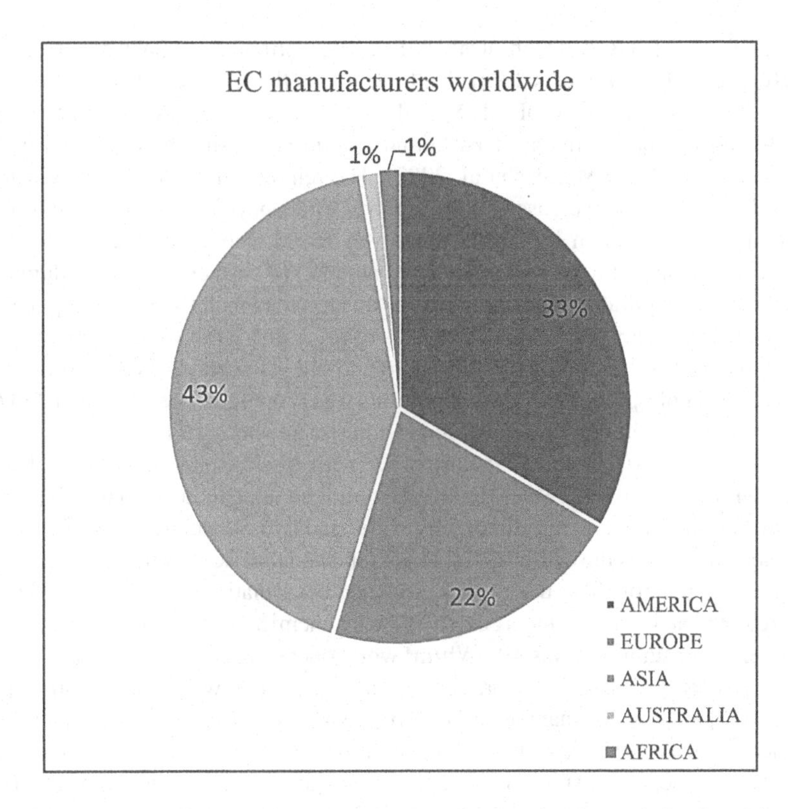

FIGURE 8.6 Manufacturing of EC unit worldwide. (Reprinted with permission from Magnisali et al. 2022.)

economic values, and essential operational parameters. The common industrial categories presented are food, textile, paper, slaughterhouse, petroleum, etc. The following are some of these industries.

8.3.1 Food Industry

Wastewaters from food industries come from different industrial categories such as dairy, juice and beverages, potato, pasta and cookies, yeast, and sugar industry. These types of industries produce wastewater containing high concentrations of COD and BOD due to the presence of soluble organic content in high level (Hellal,et al. 2023). Electrochemical treatment of these types of wastewater should be integrated with pretreatment or polishing treatment steps. EC and EOX can be used either separate or combined with each other. There are a lot of studies that utilized electro-treatment in food industry. Dairy wastewater treatment via EC combined with other treatments was investigated in many studies (Bazrafshan et al. 2013; de Freitas Silva Valente, Mendonça, and Pereira 2015; Benaissa, Kermet-Said, and Moulai-Mostefa 2016; Chakchouk et al. 2017; Akansha et al. 2020; Chezeau et al. 2020;). A real dairy effluent was treated by EC system with aluminum electrodes where raw wastewater contains high concentrations of COD (18300 mg/l), oil and grease (4570 mg/l), and

TSS (10200 mg/l) (Bazrafshan et al. 2013). The optimal treatment occurred very quickly, with only 60 minutes deemed sufficient, and the removal efficiency was high (98%, 97%, and 97.7% of COD, TSS, and O&G, respectively). Another EC reactor was utilized for the treatment of pasta and cookie processing wastewater with the addition of H_2O_2 (Roa-Morales et al. 2007). The removal efficiencies are presented. The addition of H_2O_2 increased COD removal efficiency from 80% to 90%. High variation in the optimum initial pH range was found to be 3–8. Wastewater from potato chips manufacturing was treated by batch EC (Kobya et al. 2006). Aluminum proved to be considerably superior to iron as an electrode material in this application. The COD and turbidity removal efficiencies of (68% and 98%) were acceptable with respect to initial concentrations of 2800 mg/l and 610 NTU, respectively. However, no clear single set of optimum process conditions was proposed. The pH was 6.2 to 6.5, which was near optimum; therefore, no pH adjustment was required.

El-Ezaby et al. (2021) studied the use of EC with Al electrodes and chemical coagulation with alum for wastewater treatment from fruit juice industry. The EC process was carried out using two aluminum electrodes and two SS electrodes where Al was the anode and SS was the cathode. The results showed that EC was more efficient than chemical coagulation. The use of this electrode combination resulted in an 83.78% COD removal at 40 min, compared to 57.57% by chemical coagulation. Furthermore, total energy consumption was 40 kWh/m^3 while operating cost was 1.46 $\$/m^3$.

In a laboratory-scale EC reactor, baker's yeast production wastewater treatment has been investigated by Al-Shannag et al. (2014) using iron electrodes. An EC unit was designed with six iron electrodes, achieving a maximum COD removal efficiency of 85%. Visual observation showed a complete removal of dark brown wastewater color through EC process. The corresponding electrical energy and electrode consumption per kg COD removed were around 0.493 kWh and 2.956 kg iron, respectively.

Recently, there are several factories in Egypt that tend to use electrochemical treatment techniques to obtain a quality of treated wastewater amenable for reuse. One of these factories in the field of food industry installed a readymade combined EC-EOX reactor for the treatment of a daily flow of industrial wastewater about 240 m^3/d. Figure 8.7 shows the component of the installed electrochemical cell. The challenge was the available area for the treatment as it was about 50 m^2. The combined EC-EOX units were installed within an integrated treatment system vertically in a building that consists of three stages where an equalization tank was set underground, the first stage contains the EC treatment cell with its supplies and the second stage contains an EOX cell with the polishing multimedia filter. The influent COD was about 16500 mg/L, and pH was 4. The combined treatment system achieved removal efficiency of COD about 98%, and the residual concentration discharged to public sewerage was about 350 mg/l. The treatment cells were automatically controlled through the control panel system.

8.3.2 Textile Industry

Textile wastewaters represent a major environmental problem because of day colors and their ability to obstruct the passage of light into water, which is harmful to living organisms in water bodies. Treatment of such wastewater is very difficult due to the residual color resulting from dying process. Electrochemical treatment methods were

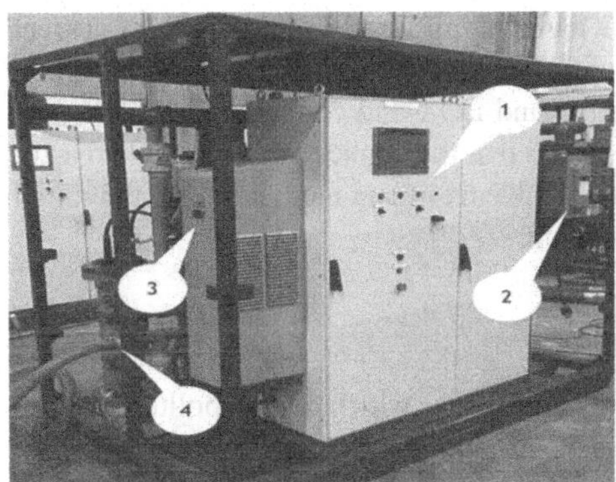

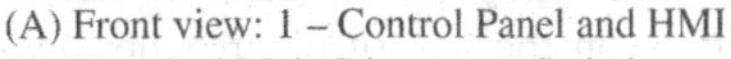

(A) Front view: 1 – Control Panel and HMI
2 – Electrical Main Disconnect Switch
3 – Air Conditioner
4 – Effluent Connection Point

(B) Rear view: 1 – Cell Influent Isolation Valve
2 – Flow Meter 3 – Mixed Oxidant Cell
4 – Influent Connection Point
5 – Cell Over-temperature Switch

FIGURE 8.7 Compact electrochemical cells (10 m^3/h) installed in food industry factory in Egypt: (A) Front view and (B) Rear view.

recently utilized for this type of wastewater. A combined EC and EOX treatment is investigated by a study carried out by Gil Pavas et al., 2017. Aluminum and iron anodes were used as sacrificial anodes in EC. The results achieved were 70% removal of COD with an operating cost of USD 1.47/m^3 at pH 4, conductivity of 3.7 mS/cm, and a current density of 4.1 mA/cm^2. This study also showed the applicability of a combined EC/EOX treatment process of such complex industrial wastewater. A combined EOX reactor with a pilot-scale fluidized biofilm process and chemical coagulation for textile wastewater treatment was studied by Kim et al. (2002). SS and titanium electrodes, 2.1 mA/cm^2 of current density, and 0.7 L/min of flow rate were chosen for the most efficient electrochemical oxidation at the pilot-scale treatment. COD and color removals of 95.4% and 98.5% were achieved by overall combined process. It was concluded that this combined process was successfully employed and much effectively decreased pollutant loading post-treatment for textile wastewater treatment at pilot scale.

8.3.3 Paper and Pulp Industry

The paper and pulp industry are one of the main water-demanding industrial processes, contributing significantly to environmental pollution, e.g. in the form of black liquor which is more difficult to handle. Wastewater generated from this industry is characterized by the presence of black color, high organic content, and suspended solids (mainly fibers). EC reactor with perforated iron electrodes was used for the treatment of recycled fiber-based paper mill wastewater (Pandey and Thakur 2020). EC process shows 95% and 67% reduction achieved for COD and color, while the TDS value was about 1340 mg/L. The operating conditions were HRT = 20 min, pH = 7.15, current density = 20 mA/cm^2, inter-spacing = 2 cm, and conductivity = 3.15 mS/cm. Another work by Özyurt et al. (2017) was executed for the pulp and paper mill wastewater separate and combined EC/EOX treatment methods. The operating conditions were conductivity at 2.10 mS/cm, pH of 5.82, temperature at 20°C and constant current intensity of 1.22 A. The highest removal of turbidity, color, and COD was 99%, 100%, and 90%, respectively, for both separate EC and EOX and combined EC/EOX processes with Al electrodes. The study concluded that the application of a single reactor for EC/EOX has been considered as a more viable treatment method due to the initial cost of investment and simplicity of application.

8.3.4 Metal Plating Industry

Metal plating industry is one of the chemical industries which release polluted wastewater with a variety of types of hazardous substances such as degreasing solvents, cyanided, alkaline cleaning agents, oil and grease, and heavy metals (such as nickel, chromium, copper, and zinc). There are few studies on the use of electrochemical treatment on the removal of pollutants from wastewater. In a study investigated by Oden and Sari-Erkan (2018), EC reactor with iron electrodes was used to remove heavy metals, color, and COD from metal plating wastewater. The optimum operating conditions were treatment time: 30 min, pH: 5, and current density: 30 mA/cm2. The COD, color, total chromium, nickel, and zinc removal efficiencies were found to be 76.2%, 99.9%, 98.9%, 96.3%, and 99.8% at these optimum operating conditions. The operating cost for electrical energy consumption and electrode consumption was found to be 2.92 and 3.64 $\$/m^3$.

A continuous flow EC reactor was investigated by Abdel-Shafy et al. (2022) for the treatment of wastewater from electroplating industry. The reactor design consisted of a rectangular plastic tank reactor with dimensions of 18 cm width, 35 cm length, and 22 cm height, with a maximum capacity of 10 L (Figure 8.8). The reactor divided into six sets as follows: The first five sets consisted of two SS electrodes

FIGURE 8.8 EC cell unit for electroplating wastewater consisted of stainless-steel plate electrodes, a DC power supply, voltmeter, and electric pump. (Reprinted with permission from Abdel-Shafy et al. 2022.)

assembled in two parallel ones. The last set was designed to collect the clear treated water through a hole connected with a hose to eject the clear treated water. The study revealed that at a potential of 10 volts, contact time of 30 min, and pH 9, the removal rates of metals were 98.9%, 97.4%, and 96.6% for Ni, Zn, and Cu, respectively. The study showed that the advantage of EC technique is achieving high treatment efficiency over using chemical reagents and high construction cost of other conventional processes. However, the authors didn't present the economic study and possibility application of this reactor, especially the electrical power consumption and the electrodes' material consumption.

8.3.5 Petroleum and Oil Industry

Polluted water from oil refining industry contains a variety of chemical compounds, like aliphatic and aromatic hydrocarbons, organic salts, phenols, oils and greases, and metals. In recent years, treatment of refinery wastewater via electrochemical methods such as EOX and EC has recently great attention. Abou-Taleb et al. (2021) designed a pilot-scale EOX unit with cylindrical-shaped graphite and SS electrodes for the removal of phenol from petroleum wastewater. The reactor was operated at a current density of 3 mA/cm^2 and reaction time of 15 min. A complete removal of phenol was obtained at these conditions with a phenol starting value of about 6.8 mg/L. Also, 50–60% of COD and BOD were removed. Abdel-Salam et al. (2018) investigated the use of flow-by porous graphite electrode in EOX reactor for the treatment of produced water from gas and oil industry. A bench-EOX reactor with flow-by porous graphite electrodes was used for the oxidation of organic matter in produced water which was collected from natural gas processing field (Figure 8.9). The maximum removal efficiency of COD was 66.52% at a current density of 1.41 mA/cm^2, a flow rate of 50 mL/min, and pH of 7.3 with an influent COD value of 2845 mg O_2/L and a residual concentration of 832 mgO_2/L. The energy consumption at these conditions was 2.12 kWh/kg_{COD}.

Also, Abdelwahab et al. (2009) studied the use of EC to remove phenol from oil refinery waste effluent using a cell with horizontally oriented aluminum cathode and a horizontal aluminum screen anode. The results showed that, at solution pH 7 and high current density, remarkable removal of 97% of phenol after 2 h can be achieved starting from 30 mg/L phenol concentration.

8.4 TECHNO-ECONOMIC EVALUATION FOR DIFFERENT TYPES OF ELECTROCHEMICAL TREATMENT

The use of the electrochemical treatment process in industry as a wastewater treatment process is dependent on its economic viability. An economic analysis of such a process would consider the capital costs (Capex) and operating costs (Opex) imposed. The fixed capital cost consists mainly of the total cost of the purchasing equipment (electrolyte, reaction cell, and pumps) and the cost of setting it up. Operating cost can be the costs of electrodes and power consumptions; maintenance, sludge dewatering, and disposal should be equally considered (Kobya et al. 2016). However, the major contribution to operational cost is electrode material (kg/m^3) and electrical

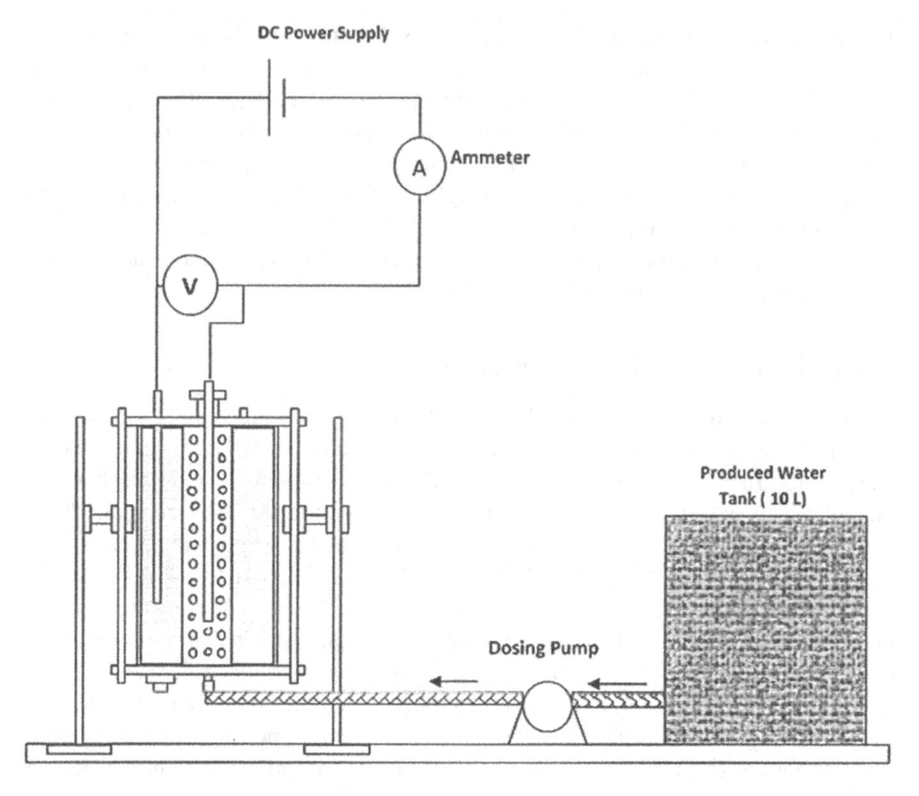

FIGURE 8.9 EOX cell with flow-by porous graphite electrode. (Reprinted with permission from Abdel-Salam et al. 2018.)

power consumption (kWh/m^3). The cost of maintenance can be decreased if reversing of current between anode and cathode is used. In EC process, the operating cost is a function of electrode material consumption and electrical power consumption, while in EOX the operating cost is only the electrical power consumption besides maintenance. Operating cost of EC using aluminum electrodes is higher than that of using iron electrodes, and this is due to their respective prices of each metal (about 0.5–0.8 US$/kg for iron and 1.5–3 US$/kg for aluminum) (Hakizimana et al. 2017). Compared to traditional chemical coagulation, EC offers some advantages such as the buffer effect that allows to avoid final acidic pH, the disinfection effect, the high removal efficiency of dissolved pollutants, low sludge production, and the avoidance of adding some chemical anions (SO_4). EC was found to be an eco-friendly and cost-effective process when it is operated at a current density below 20 mA/cm^2. Although sludge disposal may become an issue in EC process, there is an option for sludge or coagulant recovery from sludge as there is no chemical additive to the process (Kurniawan et al. 2020). Abou-Taleb et al. (2021) investigated the use of novel pilot-scale EOX unit with graphite and SS electrodes for the oxidation of phenol from a real petroleum wastewater. The design shown in Figure 8.10 was applied for a treatment of continuous mode with a flow rate of 1 m^3/d. The optimum

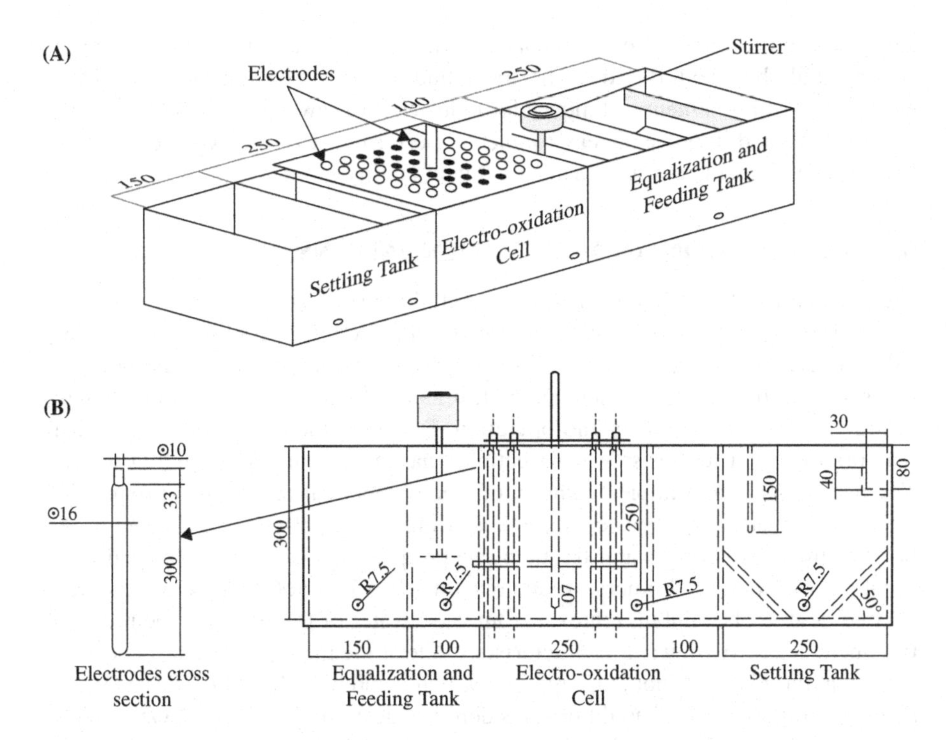

FIGURE 8.10 (A) Schematic diagram of the EOX unit and (B) cross-section of EOX and electrode. (Reprinted with permission from Abou-Taleb et al. 2021.)

conditions required for the complete removal of phenol were 3 mA/cm^2 current density and time, 15 min. Also, the energy consumption was 0.79 kWh/m^3, and the total running cost was 0.051 \$/m^3, which is very cost-effective compared to other treatment methods.

Another study by Abdel-Shafy et al. (2019) on the use of EC reactor with magnesium-rod electrode for cooling tower blowdown water treatment was investigated. The operating conditions were 60 min of treatment time, a current density of 14.29 mA/cm^2, 1 cm electrode distance, and 300 rpm stirring speed. The maximum removal efficiency obtained was 51.80% for hardness and 93.70% for silica ions. The energy and electrode consumption and total operating cost were found to be 0.82 kg/m^3, 1.81 kWh/m^3, and 0.88\$/m^3, respectively. Ramesh et al. (2021) performed a complete cost analysis to evaluate the feasibility of using EOX process for the treatment of biologically pretreated textile wastewater. The study was a comparison of an EOX with an already existing chemical treatment plant with a capacity of 1000 m^3/d in a textile factory. Based on the total operation cost, an overall cost reduction was assessed by comparison of the EOX with the current conventional treatment system. In this study, the electrical power consumptions were 3.06 kW h/m^3 at 64% of COD, and 90% of color removal rate was achieved. The calculated Capex and Opex were 0.11\$/m^3 and 0.45\$/m^3, respectively. By considering the savings from the replacing of the chemical treatment system with EOX, the direct payback period was calculated as

1.90 years and around 39% cost associated with wastewater treatment and fabric processing could be saved. Based on the above findings, it can be concluded that using electrochemical treatment techniques for industrial wastewater is cost-effective and technically could improve the environmental performance of the existing treatment facility, as it is clean, and sustainable treatment facility.

8.5 CONCLUSION AND RECOMMENDATIONS

Electrochemical treatment is an efficient technique which could provide sufficient removal rates of pollutants, which cannot be achieved by the traditional, mechanical, biological, chemical, and physicochemical water treatment methods. They are environmentally friendly, which could eliminate soluble and toxic pollutants with minimal costs for chemicals and operation. Recently, the use of electrochemical reactors for water and wastewater treatment has spread and is expanding widely. The development and implementation of electrochemical reactor is a progressive trend for industrial wastewater treatment. Moreover, electrochemical reactor can be combined with other water treatment techniques, such as biological treatment, flotation, filtration, and others, in one reactor when it is not possible to achieve a desired degree of pollutant removal from wastewater or to minimize the required treatment areas. Wastewater characteristics determine the type of electrochemical treatment method that should be applied, which could be EC or EOX. Several variables are important and should be considered in designing an EC or EOX reactor, such as current density, electrode material and shape, treatment time, pH, conductivity, and temperature of the solution. To date, meaningful advancement has been made in the development of cathode and anode materials, as well as reactor geometry, to obtain sufficient treatment level. Aluminum and iron electrodes are widely used in industrial-scale EC reactors because of their availability and cost. Graphite and metal electrodes are effective for EOX reactor due to their robustness. Since the electrochemical treatment methods are easy to operate, significant operator control is not required. However, further studies are necessary to relate the new developments in experimental scientific studies with industrial application through scaling-up, commercialization, and marketing of this scientific experience, which could reach the same level of understanding and robustness rather than those alternative processes.

REFERENCES

Abdel-Salam, Omar E., Enas M. Abou Taleb, and Ahmed A. Afify. 2018. "Electrochemical Treatment of Chemical Oxygen Demand in Produced Water Using Flow-by Porous Graphite Electrode." *Water and Environment Journal* 32 (3): 404–11. doi:10.1111/wej.12336.

Abdel-Shafy, Hussein, Mahmoud A.I. Hewehy, Taha Razek, Maamoun Hamid, and Rehan Morsy. 2019. "Treatment of Industrial Electroplating Wastewater by Electrochemical Coagulation Using Carbon and Aluminum Electrodes." *Egyptian Journal of Chemistry* 62 (0): 383–92. doi:10.21608/ejchem.2019.11667.1742.

Abdel-Shafy, Hussein I., Rehan M. M. Morsy, Mahmoud A. I. Hewehy, Taha M. A. Razek, and Maamoun M. A. Hamid. 2022. "Treatment of Industrial Electroplating Wastewater for Metals Removal via Electrocoagulation Continuous Flow Reactors." *Water Practice and Technology* 17 (2): 555–66. doi:10.2166/wpt.2022.001.

Abdelwahab, O., N. K. Amin, and E. S Z El-Ashtoukhy. 2009. "Electrochemical Removal of Phenol from Oil Refinery Wastewater." *Journal of Hazardous Materials* 163 (2–3): 711–16. doi:10.1016/j.jhazmat.2008.07.016.

Abou-Elela, S. I., S. A. El-Shafai, M. E. Fawzy, M. S. Hellal, and O. Kamal. 2018. "Management of Shock Loads Wastewater Produced from Water Heaters Industry." *International Journal of Environmental Science and Technology* 15 (4): 743–54. doi:10.1007/s13762-017-1433-9.

Aboutaleb, Enas, Gamal Kamel, and Mohamed Hellal. 2018. "Investigation of Effective Treatment Techniques for Olive Mill Wastewater." *Egyptian Journal of Chemistry* 61(3): 320–30. doi:10.21608/ejchem.2018.2655.1224.

Abou-Taleb, Enas M., Mohammed Eid M. Ali, Mohamed S. Hellal, Kholod H. Kamal, Shimaa M. Abdel Moniem, Nabila S. Ammar, and Hanan S. Ibrahim. 2020. "Sustainable Solutions for Urban Wastewater Management and Remediation." *Egyptian Journal of Chemistry* 63 (2): 405–15. doi:10.21608/ejchem.2019.13605.1840.

Abou-Taleb, Enas M., Mohamed S. Hellal, and Kholod H. Kamal. 2021. "Electro-oxidation of Phenol in Petroleum Wastewater Using a Novel Pilot-scale Electrochemical Cell with Graphite and Stainless-steel Electrodes." *Water and Environment Journal* 35 (1). Blackwell Publishing Ltd: 259–68. doi:10.1111/wej.12624.

Abu Ghalwa, Nasser M, and Alaa M Saqer. 2015. "Removal of Reactive Red 24 Dye by Clean Electrocoagulation Process Using Iron and Aluminum Electrodes." *Journal of Chemical Engineering & Process Technology* 07 (01): 1–7. doi:10.4172/2157-7048.1000269.

Ahamad, Arif, Sughosh Madhav, Amit K. Singh, Ashutosh Kumar, and Pardeep Singh. 2020. "Types of Water Pollutants: Conventional and Emerging." In, 21–41. doi:10.1007/978-981-15-0671-0_3.

Akansha, J., P. V. Nidheesh, Ashitha Gopinath, K. V. Anupama, and M. Suresh Kumar. 2020. "Treatment of Dairy Industry Wastewater by Combined Aerated Electrocoagulation and Phytoremediation Process." *Chemosphere* 253, 126652. doi:10.1016/j.chemosphere.2020.126652.

Al-Shannag, Mohammad, Khalid Bani-Melhem, Zaid Al-Anber, and Zakaria Al-Qodah. 2013. "Enhancement of COD-Nutrients Removals and Filterability of Secondary Clarifier Municipal Wastewater Influent Using Electrocoagulation Technique." *Separation Science and Technology (Philadelphia)* 48 (4): 673–80. doi:10.1080/01496395.2012.707729.

Al-Shannag, Mohammad, Zakaria Al-Qodah, Kholoud Alananbeh, Nahla Bouqellah, Eman Assirey, and Khalid Bani-Melhem. 2014. "Cod Reduction of Baker's Yeast Wastewater Using Batch Electrocoagulation." *Environmental Engineering and Management Journal* 13 (12): 3153–60. doi:10.30638/eemj.2014.354.

Barrios, J. A., E. Becerril, C. De León, C. Barrera-Díaz, and B. Jiménez. 2015. "Electrooxidation Treatment for Removal of Emerging Pollutants in Wastewater Sludge." *Fuel* 149, 26–33. Elsevier Ltd: 26–33. doi:10.1016/j.fuel.2014.10.055.

Bazrafshan, Edris, Hossein Moein, Ferdos Kord Mostafapour, and Shima Nakhaie. 2013. "Application of Electrocoagulation Process for Dairy Wastewater Treatment." *Journal of Chemistry*. 2013, 1–8. doi:10.1155/2013/640139.

Benaissa, Faiza, Hadjira Kermet-Said, and Nadji Moulai-Mostefa. 2016. "Optimization and Kinetic Modeling of Electrocoagulation Treatment of Dairy Wastewater." *Desalination and Water Treatment* 57 (13): 5988–94. doi:10.1080/19443994.2014.985722.

Bracher, Gustavo Holz, Elvis Carissimi, Delmira Beatriz Wolff, Cristiane Graepin, and Andressa Paola Hubner. 2021. "Optimization of an Electrocoagulation-Flotation System for Domestic Wastewater Treatment and Reuse." *Environmental Technology (United Kingdom)* 42 (17): 2669–79. doi:10.1080/09593330.2019.1709905.

Butler, Erick, Yung Tse Hung, Ruth Yu Li Yeh, and Mohammed Suleiman Al Ahmad. 2011. "Electrocoagulation in Wastewater Treatment." *Water (Switzerland)* 3 (2), 495–525 doi:10.3390/w3020495.

Chakchouk, I., N. Elloumi, C. Belaid, S. Mseddi, L. Chaari, and M. Kallel. 2017. "A Combined Electrocoagulation-Electrooxidation Treatment for Dairy Wastewater." *Brazilian Journal of Chemical Engineering* 34 (1): 109–17. doi: 10.1590/0104-6632.20170341s20150040.

Chen, Guohua. 2004. "Electrochemical Technologies in Wastewater Treatment." *Separation and Purification Technology* 38 (1): 11–41. doi:10.1016/j.seppur.2003.10.006.

Chezeau, Benoit, Lilya Boudriche, Christophe Vial, and Amel Boudjemaa. 2020. "Treatment of Dairy Wastewater by Electrocoagulation Process: Advantages of Combined Iron/aluminum Electrodes." *Separation Science and Technology (Philadelphia)* 55 (14): 2510–27. doi:10.1080/01496395.2019.1638935.

D'Odorico, Paolo, Kyle Frankel Davis, Lorenzo Rosa, Joel A. Carr, Davide Chiarelli, Jampel Dell'Angelo, Jessica Gephart, Graham K. MacDonald, David A. Seekell, Samir Suweis, Maria Cristina Rulli. 2018. "The Global Food-Energy-Water Nexus." *Reviews of Geophysics*. 56(3), 456-531 doi:10.1029/2017RG000591.

El-Ezaby, Khaled H., Maie I. El-Gammal, and Youmna A. Shaaban. 2021. "Using Electro- and Alum Coagulation Technologies for Treatment of Wastewater from Fruit Juice Industry in New Damietta City, Egypt." *Environmental Monitoring and Assessment* 193 (6) 370. Springer International Publishing. doi:10.1007/s10661-021-09149-0.

Freitas Silva Valente, Gerson de, Regina Célia Santos Mendonça, and José Antônio Marques Pereira. 2015. "The Efficiency of Electrocoagulation Using Aluminum Electrodes in Treating Wastewater from a Dairy Industry." *Ciencia Rural* 45 (9): 1713–19. doi:10.1590/0103-8478cr20141172.

Garcia-Segura, Sergi, Maria Maesia S.G. Eiband, Jailson Vieira de Melo, and Carlos Alberto Martínez-Huitle. 2017. "Electrocoagulation and Advanced Electrocoagulation Processes: A General Review about the Fundamentals, Emerging Applications and Its Association with Other Technologies." *Journal of Electroanalytical Chemistry* 801, 267–299. doi:10.1016/j.jelechem.2017.07.047.

Ghosh, D., C. R. Medhi, and M. K. Purkait. 2008. "Treatment of Fluoride Containing Drinking Water by Electrocoagulation Using Monopolar and Bipolar Electrode Connections." *Chemosphere* 73 (9): 1393–1400. doi:10.1016/j.chemosphere.2008.08.041.

GilPavas, Edison, Paula Arbeláez-Castaño, José Medina, and Diego A. Acosta. 2017. "Combined Electrocoagulation and Electro-Oxidation of Industrial Textile Wastewater Treatment in a Continuous Multi-Stage Reactor." *Water Science and Technology* 76 (9): 2515–25. doi:10.2166/wst.2017.415.

Hakizimana, Jean Nepo, Bouchaib Gourich, Mohammed Chafi, Youssef Stiriba, Christophe Vial, Patrick Drogui, and Jamal Naja. 2017. "Electrocoagulation Process in Water Treatment: A Review of Electrocoagulation Modeling Approaches." *Desalination* 404, 1–21. doi:10.1016/j.desal.2016.10.011.

Hamdan, Shaima S., and Muftah H. El-Naas. 2014. "An Electrocoagulation Column (ECC) for Groundwater Purification." *Journal of Water Process Engineering* 4 (C): 25–30. doi:10.1016/j.jwpe.2014.08.004.

Hashim, Khalid S, Andy Shaw, Rafid Alkhaddar, and Montserrat Ortoneda Pedrola. 2015. "Controlling Water Temperature during the Electrocoagulation Process Using an Innovative Flow Column-Electrocoagulation Reactor." *International Journal of Environmental and Ecological Engineering*. 9 (8), 964–967.

Hellal, Mohamed Saad., Aly Al-Sayed, Mohamed Azab El-Liethy, and Gamal K. Hassan. 2021. "Technologies for Wastewater Treatment and Reuse in Egypt: Prospectives and Future Challenges." In *Handbook of Advanced Approaches Towards Pollution Prevention and Control*, 275–310. Elsevier, Oxford, UK. doi:10.1016/B978-0-12-822134-1.00010-5.

Hellal, Mohamed Saad, Hala Salah Doma, and Enas Mohamed Abou-Taleb. 2023. "Techno-Economic Evaluation of Electrocoagulation for Cattle Slaughterhouse Wastewater Treatment Using Aluminum Electrodes in Batch and Continuous Experiment." *Sustainable Environment Research* 33 (1), 2. doi:10.1186/s42834-023-00163-0.

Kim, Tak Hyun, Chulhwan Park, Jinwon Lee, Eung Bai Shin, and Sangyong Kim. 2002. "Pilot Scale Treatment of Textile Wastewater by Combined Process (Fluidized Biofilm Process-Chemical Coagulation-Electrochemical Oxidation)." *Water Research* 36 (16): 3979–88. doi:10.1016/S0043-1354(02)00113-6.

Kobya, M., E. Gengec, and E. Demirbas. 2016. "Operating Parameters and Costs Assessments of a Real Dyehouse Wastewater Effluent Treated by a Continuous Electrocoagulation Process." *Chemical Engineering and Processing - Process Intensification* 101: 87–100. doi:10.1016/j.cep.2015.11.012.

Kobya, M., H. Hiz, E. Senturk, C. Aydiner, and Erhan Demirbas. 2006. "Treatment of Potato Chips Manufacturing Wastewater by Electrocoagulation." *Desalination* 190 (1–3): 201–11. doi:10.1016/j.desal.2005.10.006.

Kolpakova, Nina A., Yulia A. Oskina, and Zhamilya K. Sabitova. 2018. "Determination of Rh(III) by Stripping Voltammetry on a Graphite Electrode Modified with Lead." *Journal of Solid State Electrochemistry* 22 (6): 1933–39. doi:10.1007/s10008-018-3898-y.

Kurniawan, Setyo Budi, Siti Rozaimah Sheikh Abdullah, Muhammad Fauzul Imron, Nor Sakinah Mohd Said, Nur 'Izzati Ismail, Hassimi Abu Hasan, Ahmad Razi Othman, and Ipung Fitri Purwanti. 2020. "Challenges and Opportunities of Biocoagulant/bioflocculant Application for Drinking Water and Wastewater Treatment and Its Potential for Sludge Recovery." *International Journal of Environmental Research and Public Health* 17 (24), 9312. doi:10.3390/ijerph17249312.

Kuzin, E. N., and N. E. Kruchinina. 2021. "Complex Coagulants Produced from Bulk Waste and Industrial Products." *Tsvetnye Metally* 2021 (1): 13–18. doi:10.17580/tsm.2021.01.01.

Magnisali, Eirini, Qun Yan, and Dimitris V. Vayenas. 2022. "Electrocoagulation as a Revived Wastewater Treatment Method-Practical Approaches: A Review." *Journal of Chemical Technology and Biotechnology* 97 (1), 9–25. doi:10.1002/jctb.6880.

Mann, Uzi. 2009. *Principles of Chemical Reactor Analysis and Design*. John Wiley & Sons, New Jersy, United States. doi:10.1002/9780470385821.

Martínez-Huitle, Carlos Alberto, and Marco Panizza. 2018. "Electrochemical Oxidation of Organic Pollutants for Wastewater Treatment." *Current Opinion in Electrochemistry.* 11, 62-71 doi:10.1016/j.coelec.2018.07.010.

Moussa, Dina T., Muftah H. El-Naas, Mustafa Nasser, and Mohammed J. Al-Marri. 2017. "A Comprehensive Review of Electrocoagulation for Water Treatment: Potentials and Challenges." *Journal of Environmental Management* 186, 24–41. doi:10.1016/j.jenvman.2016.10.032.

Moussavi, Sussan, Matthew Thompson, Shaobin Li, and Bruce Dvorak. 2021. "Assessment of Small Mechanical Wastewater Treatment Plants: Relative Life Cycle Environmental Impacts of Construction and Operations." *Journal of Environmental Management* 292. 112802 doi:10.1016/j.jenvman.2021.112802.

Muddemann, Thorben, Dennis Haupt, Michael Sievers, and Ulrich Kunz. 2019. "Electrochemical Reactors for Wastewater Treatment." *ChemBioEng Reviews.* 6 (5), 142–156. doi:10.1002/cben.201900021.

Nowobilska-Majewska, Elwira, and Piotr Bugajski. 2019. "The Analysis of the Amount of Pollutants in Wastewater after Mechanical Treatment in the Aspect of Their Susceptibility to Biodegradation in the Treatment Plant in Nowy Targ." *Journal of Ecological Engineering* 20 (8): 135–43. doi:10.12911/22998993/110393.

Oden, Muhammed Kamil, and Hanife Sari-Erkan. 2018. "Treatment of Metal Plating Wastewater Using Iron Electrode by Electrocoagulation Process: Optimization and

Process Performance." *Process Safety and Environmental Protection* 119: 207–17. doi:10.1016/j.psep.2018.08.001.

Özyurt, Baran, Şule Camcıoğlu, and Hale Hapoglu. 2017. "A Consecutive Electrocoagulation and Electro-Oxidation Treatment for Pulp and Paper Mill Wastewater." *Desalination and Water Treatment* 93: 214–28. doi:10.5004/dwt.2017.21257.

Pandey, Neha, and Chandrakant Thakur. 2020. "Study on Treatment of Paper Mill Wastewater by Electrocoagulation and Its Sludge Analysis." *Chemical Data Collections* 27: 100390. doi:10.1016/j.cdc.2020.100390.

Potrich, Mateus Cescon, Ellen de Souza Almeida Duarte, Mariana de Souza Sikora, and Raquel Dalla Costa da Rocha. 2020. "Electrocoagulation for Nutrients Removal in the Slaughterhouse Wastewater: Comparison between Iron and Aluminum Electrodes Treatment." *Environmental Technology (United Kingdom)* 43 (5), 751–765. doi: 10.1080/09593330.2020.1804464.

Qiao, Jing, and Yuzhu Xiong. 2021. "Electrochemical Oxidation Technology: A Review of Its Application in High-Efficiency Treatment of Wastewater Containing Persistent Organic Pollutants." *Journal of Water Process Engineering* 44, 102308. doi:10.1016/j.jwpe.2021.102308.

Rajasekhar, Bokam, Urmika Venkateshwaran, Nivetha Durairaj, Govindaraj Divyapriya, Indumathi M. Nambi, and Angel Joseph. 2020. "Comprehensive Treatment of Urban Wastewaters Using Electrochemical Advanced Oxidation Process." *Journal of Environmental Management* 266: 110469. doi:10.1016/j.jenvman.2020.110469.

Rajkumar, D., and K. Palanivelu. 2004. "Electrochemical Treatment of Industrial Wastewater." *Journal of Hazardous Materials* 113 (1–3): 123–29. doi:10.1016/j.jhazmat.2004.05.039.

Ramesh, Kanthasamy, Balasubramanian Mythili Gnanamangai, and Rajamanickam Mohanraj. 2021. "Investigating Techno-Economic Feasibility of Biologically Pretreated Textile Wastewater Treatment by Electrochemical Oxidation Process towards Zero Sludge Concept." *Journal of Environmental Chemical Engineering* 9 (5): 106289. doi:10.1016/j.jece.2021.106289.

Roa-Morales, Gabriela, Eduardo Campos-Medina, Juan Aguilera-Cotero, Bryan Bilyeu, and Carlos Barrera-Díaz. 2007. "Aluminum Electrocoagulation with Peroxide Applied to Wastewater from Pasta and Cookie Processing." *Separation and Purification Technology* 54 (1): 124–29. doi:10.1016/j.seppur.2006.08.025.

Sasson, Moshe Ben, Wolfgang Calmano, and Avner Adin. 2009. "Iron-Oxidation Processes in an Electroflocculation (Electrocoagulation) Cell." *Journal of Hazardous Materials* 171 (1–3): 704–9. doi:10.1016/j.jhazmat.2009.06.057.

Shahedi, A., A. K. Darban, F. Taghipour, and A. Jamshidi-Zanjani. 2020. "A Review on Industrial Wastewater Treatment via Electrocoagulation Processes." *Current Opinion in Electrochemistry* 22, 154–169. doi:10.1016/j.coelec.2020.05.009.

Sharma, Swati, and Halis Simsek. 2019. "Treatment of Canola-Oil Refinery Effluent Using Electrochemical Methods: A Comparison between Combined Electrocoagulation + Electrooxidation and Electrochemical Peroxidation Methods." *Chemosphere* 221: 630–39. doi:10.1016/j.chemosphere.2019.01.066.

Sillanpää, Mika, and Marina Shestakova. 2017a. "Electrochemical Water Treatment Methods." In *Electrochemical Water Treatment Methods*, 47–130. Elsevier, Oxford, United Kingdom. doi:10.1016/B978-0-12-811462-9.00002-5.

Sillanpää, Mika, and Marina Shestakova. 2017b. "Equipment for Electrochemical Water Treatment." In *Electrochemical Water Treatment Methods*, 227–63. Elsevier, Oxford United Kingdom. doi:10.1016/B978-0-12-811462-9.00004-9.

Sillanpää, Mika, and Marina Shestakova. 2017c. "Introduction." In *Electrochemical Water Treatment Methods*, 1–46. Elsevier, Oxford United Kingdom. doi:10.1016/B978-0-12-811462-9.00001-3.

Tarzia, M., B. De Vivo, R. Somma, R. A. Ayuso, R. A.R. McGill, and R. R. Parrish. 2002. "Anthropogenic vs. Natural Pollution: An Environmental Study of an Industrial Site under Remediation (Naples, Italy)." *Geochemistry: Exploration, Environment, Analysis* 2 (1): 45–56. doi:10.1144/1467-787302-006.

Uğurlu, M., A. Gürses, Ç Doğar, and M. Yalçin. 2008. "The Removal of Lignin and Phenol from Paper Mill Effluents by Electrocoagulation." *Journal of Environmental Management* 87 (3): 420–28. doi:10.1016/j.jenvman.2007.01.007.

UN. 2020. "Quality and Wastewater | UN-Water." *United Nations - UN Water.*

Vasudevan, Subramanyan, Sagayaraj Margrat Sheela, Jothinathan Lakshmi, and Ganapathy Sozhan. 2010. "Optimization of the Process Parameters for the Removal of Boron from Drinking Water by Electrocoagulation - A Clean Technology." *Journal of Chemical Technology and Biotechnology* 85 (7): 926–33. doi:10.1002/jctb.2382.

Walsh, Frank C., and Carlos Ponce de León. 2018. "Progress in Electrochemical Flow Reactors for Laboratory and Pilot Scale Processing." *Electrochimica Acta.* 280, 121–148. doi: 10.1016/j.electacta.2018.05.027.

WHO/UNICEF, IMI-SDG6 SDG 6 PROGRESS REPORTS. 2021. "WHO/UNICEF Joint Monitoring Program for Water Supply, Sanitation and Hygiene (JMP) – Progress on Household Drinking Water, Sanitation and Hygiene 2000 – 2020." *Imi-Sdg6 Sdg 6 Progress Reports.*

Xu, Longqian, Guangzhu Cao, Xiaojun Xu, Changhua He, Yao Wang, Qihua Huang, and Maiphone Yang. 2018. "Sulfite Assisted Rotating Disc Electrocoagulation on Cadmium Removal: Parameter Optimization and Response Surface Methodology." *Separation and Purification Technology* 195: 121–29. doi:10.1016/j.seppur.2017.12.010.

9 Lignocellulosic-Based Sorbents

A Sustainable Framework for the Adsorption of Pharmaceutical and Heavy Metal Pollutants in Wastewater

Ashish Kumar Nayak, Kalyani Rajashree Naik, and Anjali Pal

9.1 INTRODUCTION

Rapid growth in human population and thus induced needs has resulted in acclivitous infrastructure development, blooming industrialization and increased resource consumption, all of which have a massive negative impact on the environment. These human behaviors not only deplete natural resources but also release harmful contaminants into the environment, such as pharmaceutical and personal care products, heavy metals, dyes, pesticides and so on. Because of their non-degradability and continuous release, emerging micropollutants, specifically pharmaceuticals and heavy metals, have become a global concern in water systems (Valdés et al., 2014; Maszkowska et al., 2014a; Hu et al., 2017). Unusually, the aquatic environment is more vulnerable to heavy metal pollution due to unconcerned, illegal direct effluent disposal of various contaminants (Herojeet et al., 2015). Pharmaceutical product consumption has increased significantly and has become a part of daily life around the world, making it the most emerging contaminant. Global spending on medicines was estimated to reach \$1.2 trillion in 2018 and is expected to exceed \$1.5 trillion by 2023, owing primarily to an ageing population and improvements in health standards, particularly in developing countries (Aitken et al., 2019). The United States Food and Drug Administration (FDA) approved 59 new formulations for therapeutic use in 2018, and 48 new drugs in 2019 (US Food and Drug Administration, 2018; 2019). These compounds enter the aquatic environment through a variety of pathways, including human excretion (sewage), animal excretion, landfill leachate, improper disposal, hospital wastewater and industry. Traces of pharmaceutical

DOI: 10.1201/9781003354475-9

products have been found at significant concentrations (ng/L–μg/L) in rivers, dams, marine water bodies, wastewater effluent, drinking water and many other places (Camacho-Muñoz et al., 2014; Mezzelani et al., 2018). The presence of these compounds in the water cycle has heightened concerns about the efficacy of wastewater treatment processes in removing these pollutants, as it has been identified in numerous studies as a major source of pharmaceutical pollutants in drinking water (Balakrishna et al., 2017). Unintended consequences may occur in non-target organisms after exposure, such as increased resistance of microorganisms (Christou et al., 2017) and different consequences throughout the trophic web (Brodin et al., 2014), eliciting physiological changes leading to negative population levels, which depend on factors such as bioaccumulation potential, lipophilicity, biodegradability, concentration and exposure time (de Jesus Gaffney et al., 2015; Wee and Aris, 2017).

Heavy metals are released through natural processes such as volcanic eruptions, weathering and rock erosion, as well as through a variety of anthropogenic processes such as mining, ferrous and non-ferrous metal production, glass manufacturing, tanneries, energy and fuel production and use, and corrosion of pipes, agro-industrial processes, municipal waste and wastewater treatment (Ma et al., 2016; Rosa et al., 2017; Zhang et al., 2017). Heavy metals found in wastewater include zinc, arsenic, copper, lead, chromium, nickel and mercury. Because they are highly soluble in wastewater, living organisms can easily absorb them. As a result, these heavy metals may accumulate in their bodies and thus be concentrated throughout the food chain (Maigari et al., 2016; Bashir et al., 2019). These heavy metals can form harmful complexes with biological matters that contain oxygen, nitrogen and sulfur (Jan et al., 2015; Gupta et al., 2020). Some of these heavy metals, such as copper, zinc, iron and cobalt, are necessary for living beings to survive, but above a certain concentration level, they are harmful. There are also many heavy metals, such as lead, cadmium, mercury and uranium, that can cause disasters even at trace levels (Goyer, 1997). Consumption of these heavy metals can cause chronic illness as well as non-carcinogenic health hazards such as kidney, liver and neurological problems (Ezemonye et al., 2019). Several epidemics have been reported in recent decades as a result of heavy metal contamination in aquatic environments, such as the Minamata tragedy in Japan in the 1950s due to methyl mercury contamination and the Itai-Itai disease following World War II due to cadmium contamination.

For the removal of pharmaceutical and heavy metal pollutants from wastewater, a variety of water treatment methods have been used. Some are ineffective (conventional methods) (Rivera-Jaimes et al., 2018; Palli et al., 2019; Singh et al., 2019), while others have drawbacks such as high maintenance costs and energy consumption (such as membrane filtration processes and advanced oxidation processes (AOPs)) (Garcia-Ivars et al., 2017a; 2017b) and the formation of by-products that are even more toxic than the original product (AOPs, ozonation, etc.), as well as high electricity and electrode costs, pH sensitivity and toxic and biorecalcitrant by-products (electrochemical separation technologies) (Luo et al., 2015; Tahmasebi et al., 2015; Qu et al., 2017; Tahir et al., 2019; Gong et al., 2020). Adsorption has many advantages over these treatment methods, such as low energy consumption, mild operating conditions and the absence of by-products added to the system; thus, this technology can be used to benefit pharmaceutical removal. However, the cost of commercial activated carbon is high, which is a disadvantage of this system. As a result, instead of commercial activated carbon, various bioadsorbents based on lignocellulosic materials can be used effectively because they are non-toxic, highly efficient with a high removal percentage, simple to synthesize and prepare and simple to use (Fang et al., 2018; Khadir et al., 2020). Despite extensive research, the large-scale application of commercial activated carbon is hampered by higher costs. As a result, some low-cost materials such as biopolymers, metal oxides, clay, zeolite, dried plant parts, agricultural waste, microorganisms, macrophytes, sewage sludge and fly ash are being studied for pharmaceutical and heavy metal adsorption. These novel materials not only remove toxic micropollutants from water, but they also reduce the waste load that would otherwise be generated. In the last few decades, lignocellulosic materials derived from dried plant mass and agricultural waste biomass have been extensively researched for the adsorption of these micropollutants because the use of dead biomass improves the process by making it nutrient-unrestricted, faster and elevated metal and pharmaceutical uptake with excellent adsorption properties, easy handling and availability (Castro et al., 2017).

The purpose of this review article is to provide up-to-date information on the adsorption of heavy metals and pharmaceuticals from aqueous solutions, focusing on the parameters that influence these processes. To the best of our knowledge, there is a niche in articles that summarize this aspect from the last 20 years. The primary goal of this paper is to review the effects of pH, temperature, initial adsorbate concentration, contact time, bioadsorbent dose and coexisting ions using examples from recent research. Furthermore, different types of lignocellulosic-based material compositions are presented at the beginning of the chapter, along with a brief overview of the occurrence of pharmaceutical and heavy metal pollutants around the world, their fate in the environment and the effectiveness of operational parameters for removing these recalcitrant pollutants using lignocellulosic-based materials in their natural and modified form as a potential bioadsorbent. The various analytical techniques used in adsorption research are also described. The kinetics, isotherms and thermodynamics are then used to understand bioadsorbent behavior and the adsorption mechanism. Finally, eco-friendly methods for recycling the exhausted lignocellulosic-based materials are presented.

9.1.1 Lignocellulosic Materials

Lignocellulosic material, also known as biomass, is the most abundantly available plant dry matter and has enormous potential to be used as a sustainable bioadsorbent in water and wastewater treatment. These materials are initially produced through the photosynthesis process, which uses atmospheric CO_2, sunlight and water, and are thus referred to as photo mass. Lignocellulose is a natural polymer composed of lignin, cellulose, hemicellulose, tannins, inorganic components and extractives such as proteins, sugars, terpenes, gums, alkaloids, resins, fats, saponins and so on, with more than two hydroxyl groups per molecule. Figure 9.1 depicts the plant cell wall and its three components (Bamdad et al., 2018). Cellulose is a polysaccharide that is found in plant cell walls and has the molecular formula $(C_6H_{10}O_5)_n$.

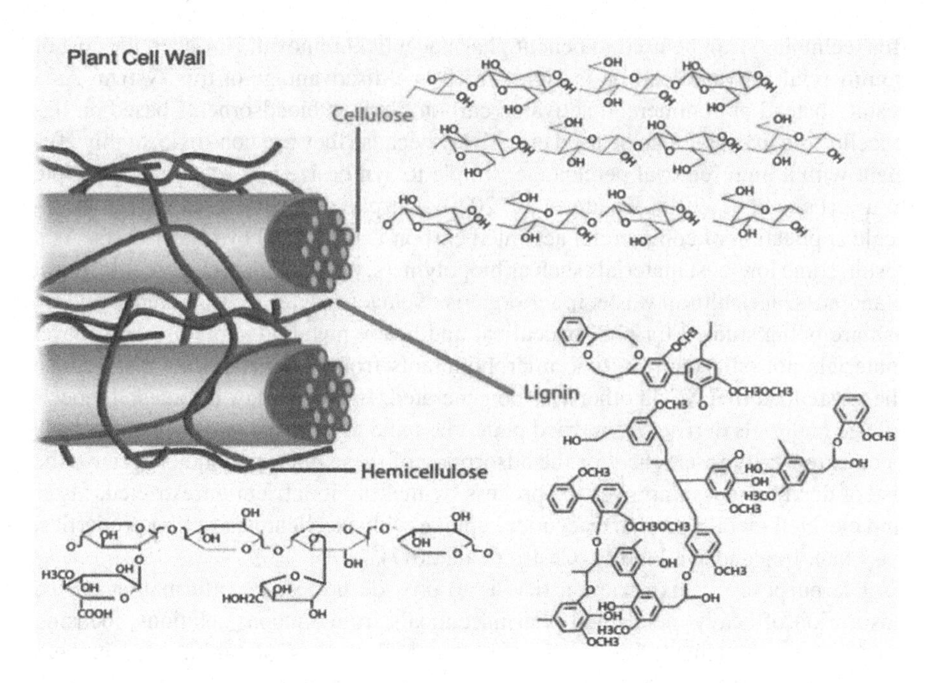

FIGURE 9.1 Schematic diagram of lignocellulosic materials and their structure. (Reprinted with permission from Bamdad et al., 2018.)

It is made up of a linear chain of many β(1→4) linked D-glucose units (Suhas et al., 2007). Hemicelluloses are branched polysaccharides composed of monomers such as glucose, xylose, mannose, arabinose, galactose, deoxy sugar rhamnose and their acidified forms that surround celluloses and act as an intermediate link between cellulose and lignin. Lignin is an aromatic, branched, amorphous, heterogeneous, three-dimensional cross-linked polymer that is closely linked with cellulose and hemicellulose polymers. It has a high surface area of 180 m^2/g and a molecular weight ranging from 2000 to 15000 g/mol (Suhas et al., 2007). Table 9.1 lists the chemical compositions of commonly used bioadsorbents. Wood, algae, agricultural, or forestry wastes are examples of lignocellulosic materials that are produced in large quantities and appear as solid waste to the environment, posing a huge disposal problem. These can be used in their natural state, as well as after being physically or chemically modified. These materials have been used to remove pharmaceuticals and heavy metals by producing activated carbon, biochar, carbon nano- or micro-tubes, as well as by direct application. A variety of materials, including sugarcane bagasse (Gupta et al., 2018; Abo El Naga et al., 2019), walnut shell (Nazari et al., 2016; Zbair et al., 2019), plant sludge (Yan et al., 2020), coconut shell (Shan et al., 2016; Yap et al., 2017), pomegranate wood (Moussavi et al., 2013), plum kernel (Paunovic et al., 2019), paper pulp (Oliveira et al., 2018), grape pulps (Güzel and Sayğılı, 2016), tomato wastes (Sayğılı and Güzel, 2016), guava seeds (Pezoti et al., 2016) and rice husk (Nche et al., 2017), have been successfully used to remove heavy metals and pharmaceutical pollutants.

TABLE 9.1
Chemical Composition of Some Common Lignocellulosic Materials

Biomass	Cellulose (%)	Hemicellulose (%)	Lignin (%)	Ash (%)	References
Rice straw	30–40	25–35	15–20	10–17	Abdolali et al. (2014); Goodman (2020)
Rice husk	30–40	20–30	20–25	17–20	Paredes-Laverde et al. (2018); Saldarriaga et al. (2020)
Sugarcane bagasse	30–40	30–35	20–30	<5	Halysh et al. (2020); Katakojwala and Mohan (2020)
Leaves	15–25	70–80	5–10	<1	Abdolali et al. (2014)
Cotton waste	30–40	20–30	20–30	5–10	Tunç et al. (2009)
Hardwood	40–55	25–40	20–25	<1	Abdolali et al. (2014)
Softwood	30–50	25–35	25–40	<5	Tabak et al. (2019)
Olive stone	20–40	15–37	10–43	<1	Guo et al. (2016); Saleem et al. (2019); Saldarriaga et al. (2020)
Nutshell and stone	25–35	25–30	30–40		Abdolali et al. (2014)
Grape marc	10	11	50		Guo et al. (2016)
Luffa cylindrica	60	30	10		Khadir et al. (2020)
Apricot nut shells	30–40	15–20	20–35	<5	Marzbali et al. (2016); Šoštarić et al. (2018)
Pomegranate seeds	27	26	40	0–5	Uçar et al. (2009)
Balanites aegyptiaca seeds	41	34	15	10	N'diaye and Kankou (2019)
Walnut shell	42	20	27	<10	Nazari et al. (2016)
Guava seeds	60	10	10	5–10	Elizalde-González and Hernández-Montoya (2009)
Coconut shell	30–40	10–25	25–50	5–10	Phan et al. (2006); Israel et al. (2011); Okon et al. (2012)

9.1.2 Preparation of Lignocellulosic-Based Bioadsorbents

Numerous studies have been conducted to investigate the bioadsorption of corn stalk, sugarcane bagasse, rice husk, coffee waste, wheat bran, nutshell, oil waste, leaves, barks, vegetable and fruit peel and other lignocellulosic wastes. Because of their ability to bind contaminants, lignocellulosic materials have proven to be excellent bioadsorbents. The method of developing lignocellulosic waste bioadsorbents is simple and inexpensive. These novel products can be used in their natural or modified forms to improve and build desired properties. The basic preparation procedure begins with precleaning the surface to remove dust, debris and any foreign particles, followed by drying to lower the moisture content and finally size reduction to obtain the required size. In some cases, additional physical or chemical modifications are

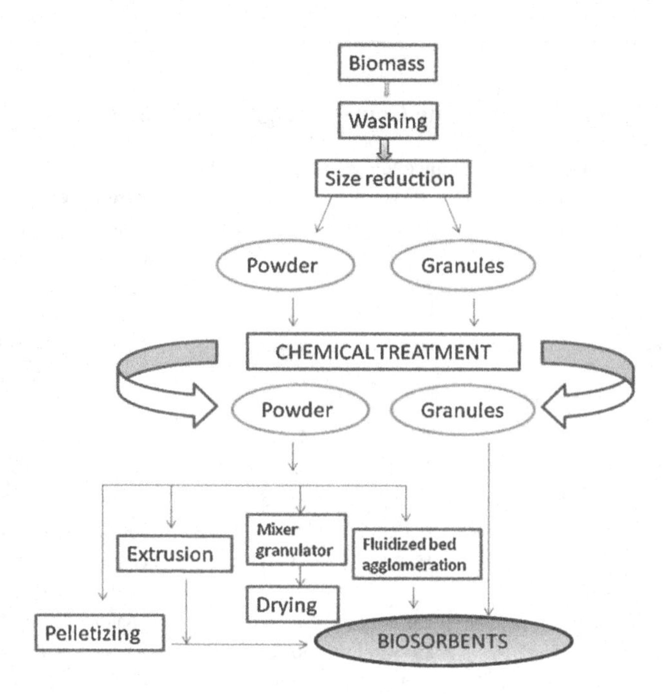

FIGURE 9.2 Schematic diagram illustrating the process of converting several forms of native lignocellulosic biomass into bioadsorbents. (Reprinted with permission from Gautam et al., 2014.)

performed as needed. Figure 9.2 depicts alternative process pathways proposed by Gautam et al. (2014) for producing lignocellulosic bioadsorbent materials that are effective and durable in repeated long-term applications. After collection, the raw materials of the bioadsorbent are washed either with tap water or with distilled water, double distilled water, de-ionized water (Richards et al., 2019) and ultra-pure water (Yahya et al., 2020). Cases of rewashing with different water were also revealed in various literatures (Zbair et al., 2019; Yu et al., 2020). Washing is sometimes performed after size reduction to remove tannin, some redundant water-soluble substances and color.

The next step is drying, which is a critical step that can be accomplished in a variety of ways, including sun drying, room temperature drying, air drying and oven drying. The first two steps take longer than the subsequent ones. Furthermore, the unmodified bioadsorbent can be physically or chemically modified to improve properties such as abundant pores and functional groups. Many modifications have been made to improve performance, such as pretreatment of bioadsorbents with acids or bases such as H_3PO_4, $KMnO_4$ and NaOH, bioadsorbent immobilization, heat treatments such as activated or pyrolyzed to produce activated carbon or biochar, addition of magnetic particles to bioadsorbent and grafting functional groups (Daneshfozoun et al., 2017; Richards et al., 2019; Verma et al., 2019; Halysh et al., 2020). These pretreatments have the potential to alter the surface characteristics/groups by eliminating or masking the groups or by exposing more adsorbate binding sites.

Yap et al. (2017) prepared microwave-synthesized magnetic biochar from coconut shell, which was pyrolyzed at 800W reaction power and 20 min reaction time and impregnated with $FeCl_3$ at a ratio of 0.5. Sun et al. (2019) successfully synthesized $KMnO_4$-treated magnetic biochar with $KMnO_4$ treatment followed by the addition of $Fe(NO_3)_3$ during carbonization for Pb(II) and Cd(II) adsorption. Fernandes et al. (2019) produced biochar from 12 different vegetative biomasses by pyrolyzing at 500°C for fluoxetine removal.

9.1.3 Characterization Techniques

Agricultural waste contains cellulose, hemicellulose, lignin, lipids, proteins, sugars, starch, polysaccharides and pigments as well as other basic chemical constituents. These constituents include carboxyl, hydroxyl and amino functional groups. These groups are found on the surfaces of the sorbate ions and molecules and can bind to them. For the development of adsorption and extraction procedures, it is essential to characterize the surface chemistry and structure of the bioadsorbent. The surface morphology and physicochemical properties of lignocellulosic bioadsorbents were studied using sophisticated analytical tools such as scanning electron microscopy (SEM), Fourier transformed infrared spectroscopy (FTIR), X-ray diffraction (XRD), energy dispersive X-ray spectroscopy (EDS), X-ray fluorescence (XRF), Brunauer-Emmett-Teller (BET) surface area, nuclear magnetic resonance (NMR), transmission electron microscopy (TEM) and vibrating sample magnetometer (VSM). To learn more about the porous structure of activated carbon, a simple and quick test for iodine number is performed to determine the microporosity of bioadsorbents with pore diameters less than 2 nm. The thermogravimetric analysis (TGA) instrument is used to test the thermal stability of bioadsorbents. Proximate and ultimate analyses are also carried out to determine the elemental and mass content of the bioadsorbent. Table 9.2 summarizes the various analytical techniques used to characterize the bioadsorbent.

9.1.3.1 Pore Volume, Size and Surface Area

The N_2-physisorption isotherm with the Brunauer-Emmett-Teller (BET) and Barrett-Joyner-Halenda (BJH) equations is often used to compute surface area, pore volume and pore size distribution. Using the N_2-physisorption isotherm, Zbair et al. (2019) found the specific surface area, total pore volume and average pore size of a walnut shell-based adsorbent to be 892.6 m^2/g, 0.510 cm^3/g and 4.62 nm, respectively. Yahya et al. (2020) developed a cashew nutshell adsorbent with specific surface area and total pore volume of 608.2 m^2/g and 0.221 cm^3/g, respectively. Lima et al. (2019) discovered activated carbons derived from agroindustrial residue, known as *Bertholletia excelsa*, which is activated with $ZnCl_2$ solutions before being pyrolyzed at 600 and 700°C temperatures, with surface area values of 1457.0 and 1419.0 m^2/g and total pore volumes of 0.275 and 0.285 cm^3/g, respectively. Nche et al. (2017) found that the specific surface areas of rice husk activated carbons were 178.1 m^2/g and 104.8 m^2/g, respectively. Nazari et al. (2016) demonstrated a remarkable yield of 41.30% on $ZnCl_2$ activated walnut shell with a high surface area of 1452.0 m^2/g. On the other hand, the total surface area and total pore volume of the corn straw

TABLE 9.2
Different Analytical Techniques Used in Bioadsorption Research

Analytical Techniques	Remarks
Scanning electron microscope (SEM)	Analysis of surface morphology of the bioadsorbent
Transmission electron microscope (TEM)	Analysis of inner morphology of the bioadsorbent
Energy dispersive X-ray spectroscopy (EDS)	Elemental analysis and chemical characterization of metal bound on the bioadsorbent
X-ray diffraction (XRD) analysis	Analysis of the crystallographic structure and chemical composition of metal bound on the bioadsorbent
Electron spin resonance spectroscopy (ESR)	Examine the active sites on the bioadsorbent
Nuclear magnetic resonance (NMR)	Examine the active sites on the bioadsorbent
Fourier transformed infrared spectroscopy (FTIR)	Examine the active sites on the bioadsorbent
X-ray photoelectron spectroscopy (XPS)	Examine the oxidation state of the metal bound on the bioadsorbent, as well as the effects of the ligand
X-ray absorption spectroscopy (XAS)	Examine the oxidation state of the metal bound on the bioadsorbent, as well as its coordination environment
Thermogravimetric analysis (TGA)	Determine the thermal stability of the bioadsorbent
Differential scanning calorimetry (DSC)	Determine the thermal stability of the bioadsorbent
X-ray fluorescence (XRF)	Determine the elemental analysis of the bioadsorbent
CHNS	Determine the percentages of carbon, hydrogen, nitrogen and sulfur in the bioadsorbent
BET surface area and pore volume	Determine the surface area, porosity and pore volume of the bioadsorbent
Proximate analysis	Determine the moisture, ash, dry matter and fix carbon content of the bioadsorbent
Vibrating sample magnetometer (VSM)	Determine the magnetic properties of the bioadsorbent
Potentiometric titration	Determine the active sites and quantities of the bioadsorbent
Point of zero charge (pH_{ZPC})	Determine the surface charges of the bioadsorbent

adsorbent were 2131.2 m^2/g and 1.128 cm^3/g, respectively, with percentages of micropore surface area and micropore volume of 91.93% and 80.43%, respectively, of the total surface area and volume (Ma et al., 2019).

9.1.3.2 FTIR Analysis

FTIR spectroscopy is one of the most common, simple and useful characterization techniques for detecting various functional groups (e.g., –OH, $–NH_2$, $–SO_3H$, –COOH and $–CONH_2$). In the study done by Verma et al. (2019), the FTIR spectra of *citrus limetta* peel magnetic biochar, as shown in Figure 9.3, the peaks found at 3435–3451 cm^{-1} represented O–H stretching of the OH group and N–H stretching of the NH_2 group in aromatic amines, primary amines and amides, whereas 2876–2933 cm^{-1} represented the $–CH_3$ and $–CH_2–$ stretching of aliphatic compounds,

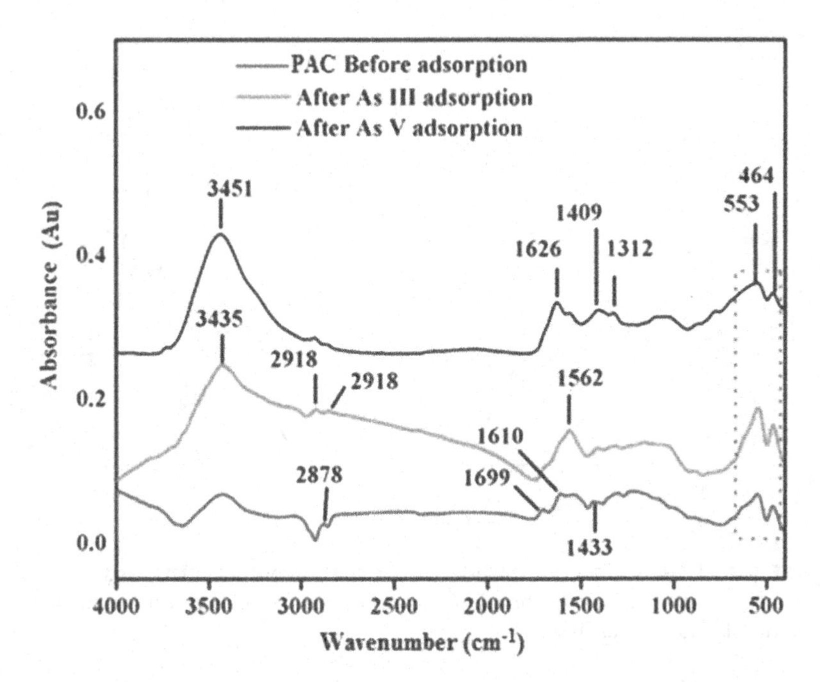

FIGURE 9.3 FTIR spectra of magnetic biochar synthesized from *Citrus limetta* peels before and after adsorption. (Reprinted with permission from Verma et al., 2019.)

the NH_2 group in primary amine, 1600 cm^{-1} represented the NH_3^+ group in amino acids and 1699 cm^{-1} represented the hydroxyl group as well as the C=O bond in carboxylic acids. The peaks at 1433 and 1359 cm^{-1} corresponded to the stretching of the –OH group in carboxylic acid and the stretching of the $–NO_2$ group in aromatic nitro compounds, respectively. The peak at 1235 cm^{-1} was attributed to the existence of the C–N bond in aromatic amines, whereas the peaks at 1162, 800, 546 and 470 cm^{-1} were attributed to the presence of C–OH, di-bust benzene, C–C=O group of carboxylic acid and C–N–C stretches of amines, respectively. The presence of similar functional groups was confirmed by Shahrokhi-Shahraki et al. (2021), as shown in Figure 9.4, and by other authors. The FTIR spectrum of grape stalk by Villaescusa et al. (2011) showed a broad absorption peak around 3350 cm^{-1}, indicating the existence of bonded hydroxyl groups, and a peak at 1620 cm^{-1} corresponding to the C–C stretching of the lignin aromatic bonds.

9.1.3.3 Zeta Potential and pH_{ZPC} Analyses

The stability of bioadsorbents is determined by the zeta potential or electro kinetic potential. The pH_{ZPC} (the pH at which the surface charge of the adsorbent is zero), which is determined by Boehm's titration, is dependent on the chemical and electronic properties of the functional groups and is used to determine the net charge and concentration of some functional groups, such as carboxyl, phenols and total basic groups. Verma et al. (2019) found the pH_{ZPC} of magnetic biochar produced from *Citrus limetta* peel and pulp to be 4.61 and 3.06, respectively.

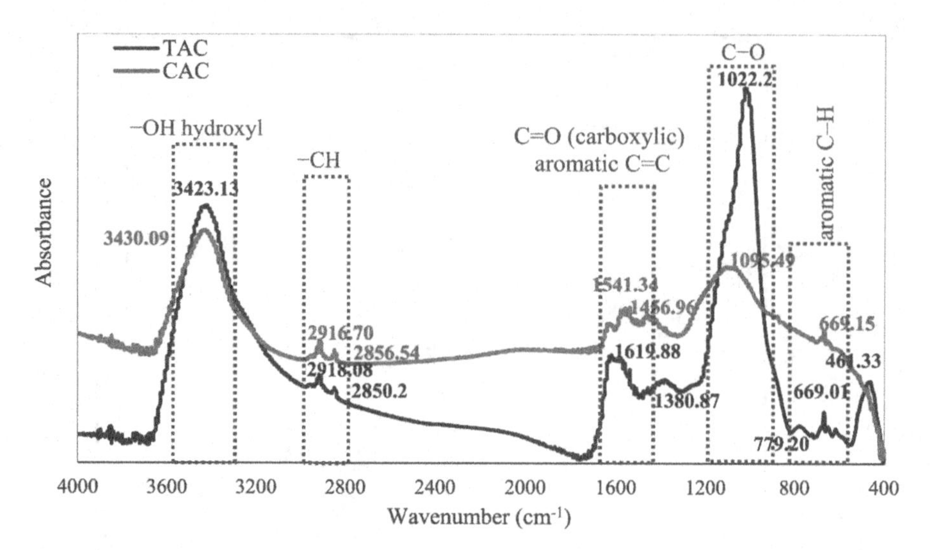

FIGURE 9.4 FTIR spectra of synthesized activated carbons (i.e., TAC and CAC) along with the positions of various functional groups associated. (Reprinted with permission from Shahrokhi-Shahraki et al., 2021.)

9.1.3.4 Optical and SEM/TEM Microscopy

Scanning electron microscopy (SEM) and transmission electron microscopy (TEM) are typically used to characterize the morphology of nanoadsorbents. As illustrated in Figure 9.5, SEM analysis of walnut shell by Zbair et al. (2019) revealed micropores on the surface microspheres with a relatively uniform diameter of approximately 4.5 mm,

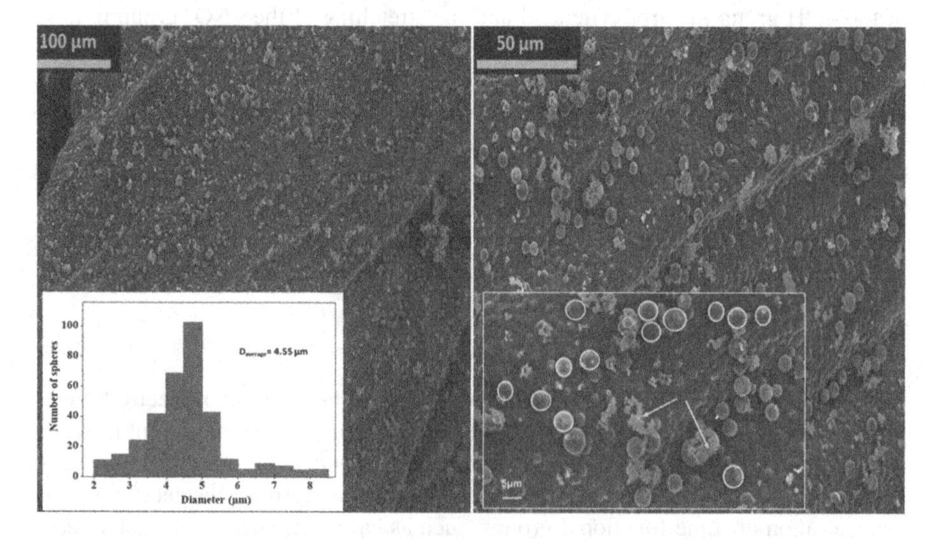

FIGURE 9.5 SEM photographs of activated carbon manufactured from raw walnut shell at a high temperature of 600°C. (Reprinted with permission from Zbair et al., 2019.)

smooth outer surfaces and regular spherical shapes, implying a potential application in the removal of heavy metal contaminants. Here activated carbon has been developed by treating raw walnut shell with hydrogen peroxide in 1:5 ratio. Further structure study on a nanoscale was accomplished solely through the use of a TEM equipment (Smith, 2007).

9.1.3.5 XRD Analysis

To know the crystallinity of the adsorbents involved, X-ray diffraction (XRD) studies are carried out, which is a powerful and non-destructive tool. Daneshfozoun et al. (2017) analyzed adsorbents at 2θ scan at a wide-angle range of 2–80°, with a step width of 2° per second at 25°C to validate the loading of Fe_2O_3 on it. Similarly, Verma et al. (2019) confirmed the existence of Fe_3O_4 by scanning XRD patterns from 10° to 80° at the rate of 2° per min. The broad peaks at 23° and 44° on the XRD spectrum indicated that the synthesized corn straw adsorbent had a degree of crystalline graphitization (Ma et al., 2019), as shown in Figure 9.6.

9.1.3.6 XPS Analysis

The X-ray photoelectron spectroscopic (XPS) technique helps in the detection of species and chemical elements on the surface of metals, as well as their oxidation states. In a study conducted by Zbair et al. (2019), the deconvolution of the C1s XPS spectrum revealed three Gaussian curves centered at 285.6, 286.8 and 288.7 eV, showing graphitic carbon (C–C and CHn), alcohol or ether (C–OH and C–O–C) and carbonyl groups (C=O), respectively. In an XPS study conducted by Ma et al. (2019), functional groups such as COOH, OH, COC and others were found on the porous

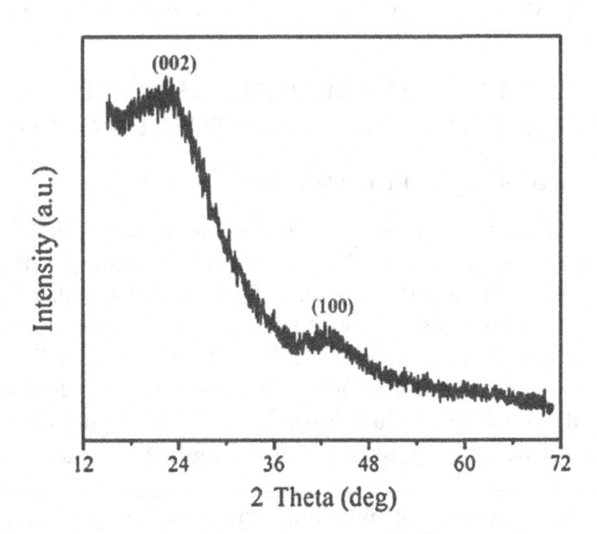

FIGURE 9.6 XRD pattern of adsorbent made of corn straw. (Reprinted with permission from Ma et al., 2019.)

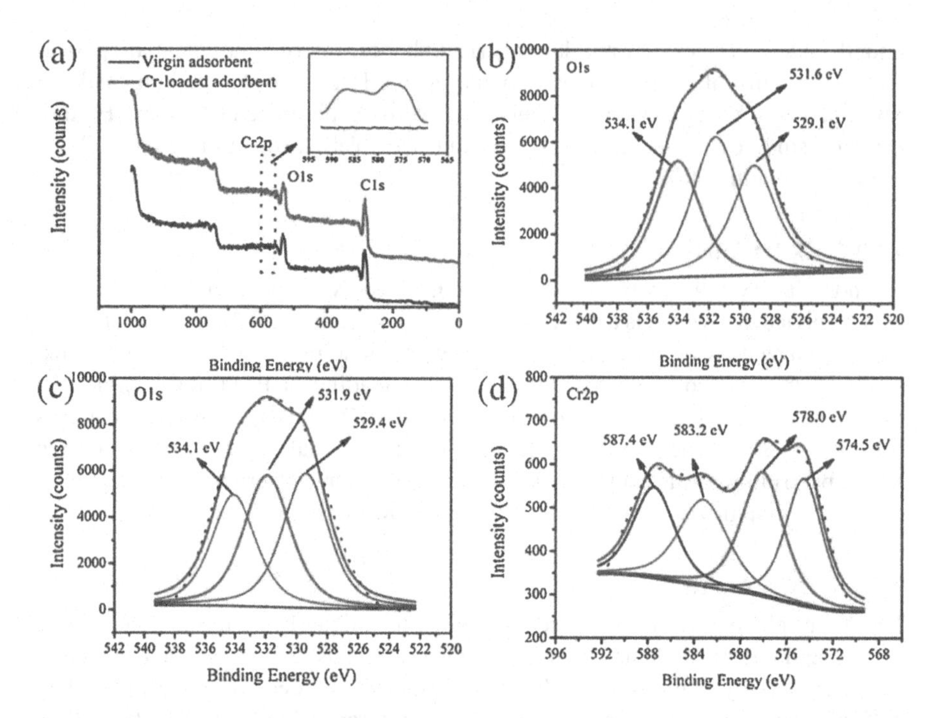

FIGURE 9.7 (A) Survey XPS spectra of porous carbon adsorbents before and after Cr adsorption, high resolution XPS spectra of (B) Ols before Cr adsorption, (C) Ols and (D) Cr2p after Cr adsorption. (Reprinted with permission from Ma et al., 2019.)

corn straw adsorbent surface, which could act as active sites during chromium adsorption, and XPS spectra between 545 eV and 595 eV confirmed the Cr2p signal, indicating that Cr(VI) was adsorbed on the carbon surface, as shown in Figure 9.7.

9.2 REMOVAL OF PHARMACEUTICALS THROUGH BIOADSORPTION USING LIGNOCELLULOSIC MATERIALS

9.2.1 Pharmaceuticals in the Environment

The environmental occurrence and ecotoxicological impacts of many pharmaceutical compounds are still unknown. Pharmaceutical wastes are generated and reach into the water sources through the following channels: manufacturing process units, hospital wastes, household wastes, sewage treatment plants, industrial services, aquaculture facilities, runoff, animal shelters, runoff of veterinary medicines from hard surfaces in farmyards, landfill leachates, sewer discharge and disposal of the carcasses of treated animals (Rodríguez-Navas et al., 2013; Aus der Beek et al., 2016). The environmental pathways through which pharmaceuticals travel are depicted in Figure 9.8.

Pharmaceuticals include anti-inflammatory drugs, broad-spectrum antibiotics, non-steroidal anti-inflammatory drugs, blood lipid regulators, antiepileptics, β-blockers, antibiotics, hormones, analgesics and cytostatic drugs, as listed in Table 9.3,

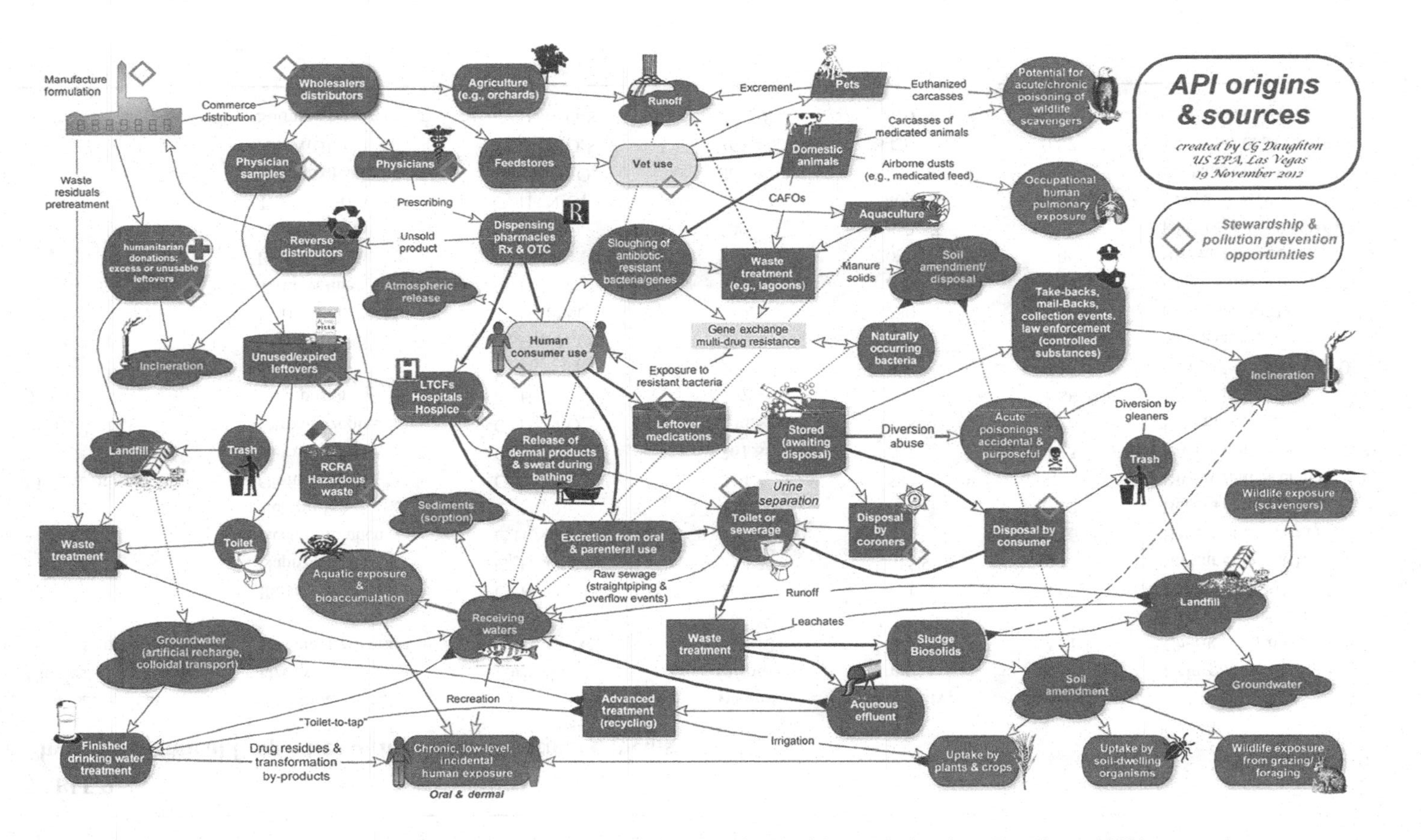

FIGURE 9.8 Pharmaceutical distribution pathways in the environment. (Reprinted with permission from Daughton, 2013.)

TABLE 9.3
Physical-Chemical Properties of the Target Pharmaceuticals

Pharmaceutical Classes	Pharmaceutical Drugs	Formula	Weight (g/mol)	Solubility (mg/L)	Log K_{ow}	References
Anti-inflammatory	Diclofenac sodium	$C_{14}H_{11}Cl_2NNaO_2$	318.13	4.82	4.51	Cherik and Louhab (2018); Abo El Naga et al. (2019)
	Ibuprofen	$C_{13}H_{18}O_2$	206.2	21	3.97	Jiang et al. (2015)
	Naproxen	$C_{14}H_{14}O_3$	230.25	15.9	3.18	Baccar et al. (2012)
	Acetaminophen	$C_8H_9NO_2$	151.17	14000	0.46	Behera et al. (2011)
	Ketoprofen	$C_{16}H_{14}O_3$	254.29	51	3.12	Behera et al. (2011)
Lipid regulators	Clofibric acid	$C_{10}H_{11}ClO_3$	214.7	582.5	2.57	Rivera-Jaimes et al. (2018)
	Bezafibrate	$C_{19}H_{20}ClNO_4$	361.8	1.55	4.25	Rivera-Jaimes et al. (2018)
Antiepileptics	Carbamazepine	$C_{15}H_{12}N_2O$	236.3	18	2.25	Behera et al. (2011)
β–blockers	Metoprolol	$C_{15}H_{25}NO_3$	267.4	60	1.88	Rivera-Jaimes et al. (2018)
	Atenolol	$C_{14}H_{22}N_2O_3$	266.0	13300	0.16	Rivera-Jaimes et al. (2018)
Antibiotics	Ciprofloxacin	$C_{17}H_{18}FN_3O_3$	331.3	150	0.28	Archana et al. (2016)
	Norfloxacin	$C_{16}H_{18}FN_3O_3$	319.3	280	0.46	Liu et al. (2011)
	Tetracycline	$C_{22}H_{24}N_2O_8$	444.4	231	−1.37	Boudrahem et al. (2019)
	Erythromycin	$C_{37}H_{67}NO_{13}$	733.9	2000	3.06	Rivera-Jaimes et al. (2018)
	Azithromycin	$C_{38}H_{72}N_2O_{12}$	749	2.37	4.02	Kleywegt et al. (2016)
	Trimethoprim	$C_{14}H_{18}N_4O_3$	290.12	400	0.9	Li et al. (2019)
	Levofloxacin	$C_{18}H_{20}FN_3O_4$	361.4		−0.39	Archana et al. (2016)
	Amoxicillin	$C_{16}H_{19}N_3O_5S$	365.4	3430	0.87	Moussavi et al. (2013)
	Sulfamethoxazole	$C_{10}H_{11}N_3O_3S$	253.28	610	0.97	Ganesan et al. (2020)

which are most commonly used to reduce inflammation, relieve pain associated with headaches, arthritis and rheumatism, and to prevent cardiovascular accidents (Imbimbo et al., 2010) as inhibitors and biocides of pathogenic microorganisms (Maszkowska et al., 2014b; Sayğılı and Güzel, 2016). Researchers discovered a concentration of 107.9 μg/L of levofloxacin, 41.4 μg/L of atenolol and 120.9 μg/L of ciprofloxacin, among other pharmaceutically active compounds, in waste water treatment plants across India (Archana et al., 2016; Mohapatra et al., 2016). Table 9.4 shows the concentrations of various pharmaceuticals in wastewater treatment plant and treated effluent.

9.2.2 Treatment Methods for Pharmaceutical-Laden Wastewater

Several wastewater treatment techniques have been used, including conventional biological treatments such as constructed wetlands, activated sludge process (ASP), trickling filter, fluidized bed reactors, up-flow anaerobic sludge blanket reactors (UASB), extended aeration and membrane bioreactors (Kleywegt et al., 2016; Mohapatra et al., 2016; Subedi et al., 2017; Rivera-Jaimes et al., 2018; Kibuye et al., 2019; Afsa et al., 2020); physicochemical treatments such as coagulation, flotation, lime softening, sedimentation and filtration; and electrochemical processes such as electrocoagulation, internal microelectrolysis, anodic oxidation, electrooxidation, electro-Fenton reactions, photo-electro-Fenton and photo-electrocatalysis, while advanced technologies such as membrane filtration, activated carbon adsorption and AOPs have become increasingly popular in recent years (Majumdar and Pal, 2020). Biological treatment methods have conventionally been employed to deal with pharmaceutical wastewater, though easily bridgeable drugs such as ibuprofen, naproxen and bezafibrate showed a high degree of removal efficiency while drugs such as sulfamethoxazole, carbamazepine and diclofenac showed limited removal (Rivera-Jaimes et al., 2018; Singh et al., 2019). Membrane microfiltration has various disadvantages, including severe membrane fouling, competition with cations and natural organic matter in solution, and the process being affected by the polarity and hydrophobicity of the compounds (Garcia-Ivars et al., 2017a; 2017b). Although AOPs such as H_2O_2/UV, Fe^{2+}/H_2O_2, Fe^{2+}/H_2O_2/UV, O_3/H_2O_2 and O_3/UV have been extensively used in removing pharmaceuticals very effectively with less demand for additional chemicals, they have the disadvantages of forming oxidation intermediates that are more toxic than the original pollutant, as well as being energy-consuming, complex and operationally expensive processes. Similarly, electrochemical degradation techniques might generate more harmful by-products than the parent pharmaceutical.

Adsorption has numerous advantages over conventional approaches, minimal energy consumption, lack of by-products added to the system, easy regeneration and reusability, and mild operating conditions; thus, it has the potential to be employed for pharmaceutical removal. Adsorption of pharmaceuticals has been accomplished by the use of activated carbons, biochar, carbon nanotubes, graphene, raw waste materials and soils such as kaolinite. Rather than relying on commercial activated carbon, synthetic activated carbon can be used successfully as a low-cost bioadsorbent.

TABLE 9.4
Concentration of Various Pharmaceuticals in Water

Pharmaceutical Name	Treatment System	Concentration in ng/L		Location	Reference
		Raw Wastewater	Treated Wastewater		
Ibuprofen	ASP-Chlorination	1200	980	India	Subedi et al. (2017)
	UASB-ASP-Chlorination	1400	630	India	Subedi et al. (2017)
	Facultative aerated lagoon	16328	749	USA	Mohapatra et al. (2016)
Acetaminophen	ASP-Chlorination	900	690	India	Subedi et al. (2017)
	UASB-ASP-Chlorination	4500	340	India	Subedi et al. (2017)
	CASP	134322	58	USA	Mohapatra et al. (2016)
	CAS-UV	11600	Not detected	Mexico	Rivera-Jaimes et al. (2018)
Ketoprofen	Denitrification-Oxidation tank- Disinfection	961 ± 1003	213 ± 393	Italy	Palli et al. (2019)
	Aerated lagoon	3300	790	Tunisia	Afsa et al. (2020)
Naproxen	ASP-TF-Anoxic tank	452.17	3765.07	USA	Kibuye et al. (2019)
	ASP-Chlorination	5500	4520	Canada	Kleywegt et al. (2016)
Diclofenac	Denitrification-Oxidation tank-Final disinfection	2065 ± 739	2364 ± 539	Italy	Palli et al. (2019)
	Wetland system	750	695	Spain	Fernández-López et al. (2016)
Mefenamic acid	ASP-UV	328	392	Korea	Behera et al. (2011)
Sulfamethoxazole	WWWT2	1941	394.2	USA	Singh et al. (2019)
	WWWT2	360.4	1301	Philippines	Singh et al. (2019)
	WWWT1	177.6	65.4	Switzerland	Singh et al. (2019)
	Aerated lagoon	240	140	Tunisia	Afsa et al. (2020)
Trimethoprim	ASP-Chlorination	223	234	Canada	Kleywegt et al. (2016)

(Continued)

TABLE 9.4 *(Continued)*
Concentration of Various Pharmaceuticals in Water

Pharmaceutical Name	Treatment System	Concentration in ng/L		Location	Reference
		Raw Wastewater	Treated Wastewater		
Ciprofloxacin	WWWT1	48,103	180.1	China	Singh et al. (2019)
	Aerated lagoon	580	134	Tunisia	Afsa et al. (2020)
Azithromycin	WWWT1	4585.1	613.2	China	Singh et al. (2019)
	ASP-Chlorination	1160	1100	Canada	Kleywegt et al. (2016)
Tetracycline	WWWT2	241		India	Singh et al. (2019)
Paracetamol	ASP-Sand-filtration-Chlorination	12300 ± 180	157 ± 20	Saudi Arabia	Al-Qarni et al. (2016)
	ASP-Nitrification- Denitrification	1208	409	Tunisia	Moslah et al. (2018)
Carbamazepine	ASP-UV	127	74	Korea	Behera et al. (2011)
	Aerated lagoon	1100	920	Tunisia	Afsa et al. (2020)
Atenolol	ASP-UV	11239	5911	Korea	Behera et al. (2011)
	Aerated lagoon	2400	2100	Tunisia	Afsa et al. (2020)
	ASP-Chlorination	2290	2080	Canada	Kleywegt et al. (2016)
Caffeine	Aerated lagoon	439400	19800	Tunisia	Afsa et al. (2020)
Metformin	ASP-Chlorination	147000	45700	Canada	Kleywegt et al. (2016)

All of the treatment approaches listed in Table 9.4 are preceded by preliminary treatments.
ASP: Activated sludge process; CAS: Conventional activated sludge treatment; CASP: Cyclic activated sludge process; TF: Trickling filter; UASB: Up-flow anaerobic sludge blanket reactor; UV: UV disinfection; WWWT1: Waste water treatment plant 1 (Singh et al., 2019), WWWT2: Waste water treatment plant 2 (Singh et al., 2019).

9.2.3 Adsorption

Adsorption refers to the physical and/or chemical interactions that result in the formation, transport and accumulation of adsorptive molecules on the interfacial layer. Bioadsorption is the process of adsorption by biomass through a variety of mechanisms such as electrostatic attraction, complexation, ion exchange, covalent binding, van der Waals attraction, adsorption and microprecipitation. It has the ability to remove dissolved substances in an aqueous environment, and its cheap cost, security and energy efficiency make it an ideal choice for systems requiring large volumes of removal at low operating costs.

9.2.3.1 Utilization of Lignocellulosic Biomass

Lignocellulosic biomass has been used in its natural or physically or chemically modified state to produce activated carbon, biochar and micro- and nanocarbon tubes, as well as through direct application (Table 9.5). Sugarcane bagasse, *Albizia lebbeck* seed pods, peach stones, rice husk, olive waste, *Eucalyptus grandis*, grape pulps and stalk, Jatropha shell, paper mill sludge, babassu coconut mesocarp, *Luffa cylindrica*, *Bertholletia excelsa* (Brazil nutshell), lotus stalk, macadamia nutshell, pomegranate wood, walnut shell, tomato waste, *Moringa oleifera* Lam., seed husk and guava seeds have all been used successfully in the removal of pharmaceutical pollutants.

The efficiency of adsorption is dependent on a number of parameters, including the adsorbent, absorbate and solvent qualities, as well as pH, temperature, agitation speed, porosity and surface area, operational circumstances, chemical properties and superficial net charge. Additionally, very common analyses such as total organic carbon (TOC), thermo-gravimetric and proximate analyses, point of zero charge (pH_{ZPC}), zeta potential, iodine number, scanning electron microscopy (SEM), nitrogen physisorption, toxicity testing using bioassays, Fourier transformed infrared spectroscopy (FTIR) and X-ray diffraction (XRD) have been used for the characterization of bioadsorbents (Kumari et al., 2011; Baccar et al., 2012; Mazzeo et al., 2015; Moro et al., 2017; Jaria et al., 2019).

Cabrita et al. (2010) demonstrated comparable removal efficiencies of acetaminophen by activated carbons prepared from cork powder and peach stones, as well as a fast rate of adsorption like commercial carbons. Czech et al. (2020) investigated the adsorption of naproxen, caffeine and triclosan on textile waste, i.e., cotton-based carbon microtubes, in which tests of binary solute systems revealed that the mechanism of adsorption and the amount of pharmaceutical adsorbed were affected. According to Jaria et al. (2019), the adsorbent made with the highest precursor/activating agent ratio (1:1) and the highest temperature (800°C) resulted in percentages of average adsorption of 78% and specific surface area between 1380 and 1630 m^2/g for the removal of sulfamethoxazole, carbamazepine and paroxetine at activated carbon doses of 0.015 mg/L only. Khadir et al. (2020) investigated the removal of ibuprofen from aquatic media using *Luffa cylindrica* and discovered that when the solution was not agitated, the removal percentage reduced dramatically. Lima et al. (2019) found a high removal percentage (up to 98.83%) of acetaminophen for the treatment of synthetic hospital effluents comprising different pharmaceuticals as well as organic

TABLE 9.5
Adsorptive Behavior of Lignocellulosic Materials in Pharmaceutical Removal

Adsorbent	Experimental Condition	Adsorption Isotherm, Kinetics	Adsorption Capacity	Thermodynamic Elucidation	Pharmaceutical	Reference
Sugar cane bagasse activated with $ZnCl_2$	Dose: 0.4 g/L pH: 2 Temp: 25°C Contact time: 15 min Initial concentration: 250 mg/L	Pseudo-second order, Langmuir isotherm	315 mg/g		Diclofenac sodium	Abo El Naga et al. (2019)
Albizia lebbeck seed pods activated with KOH	Dose: 1.2 g/L 200 rpm pH: 8.7 Temp: 30°C Contact time: 90 min (for ciprofloxacin) 60 min (for norfloxacin) Initial concentration: 20–100 mg/L	Pseudo-second order, Langmuir isotherm	131.14 mg/g 166.99 mg/g	Endothermic Exothermic, Spontaneous for both the cases	Ciprofloxacin Norfloxacin	Ahmed and Theydan (2014)
Peach stones activated with H_3PO_4	Dose: 2.4 g/L 250 rpm Temp: 30°C Contact time: 72 h Initial concentration: 100 mg/L	Pseudo-second order, Langmuir isotherm	1106.1 mg/g	Endothermic	Tetracycline	Álvarez-Torrellas et al. (2016)
Rice husk activated with H_3PO_4	Dose: 2.4 g/L 250 rpm Temp: 30°C Contact time: 72 h Initial concentration: 100 mg/L	Pseudo-second order, Langmuir isotherm	5600.0 mg/g	Endothermic	Tetracycline	Álvarez-Torrellas et al. (2016)

(Continued)

TABLE 9.5 *(Continued)*
Adsorptive Behavior of Lignocellulosic Materials in Pharmaceutical Removal

Adsorbent	Experimental Condition	Adsorption Isotherm, Kinetics	Adsorption Capacity	Thermodynamic Elucidation	Pharmaceutical	Reference
Rice husk activated with H_3PO_4	Dose: 2.4 g/L 250 rpm Temp: 30°C Contact time: 72 h Initial concentration: 100 mg/L	Pseudo-second order, Guggenheim-Anderson-de Boer model	110.4 mg/g	Endothermic	Ibuprofen	Álvarez-Torrellas et al. (2016)
Olive waste	Dose: 0.3 g/L 200 rpm pH: 4.12 Temp: 25°C	Pseudo-second order, Langmuir isotherm	39.5 mg/g 24.7 mg/g 56.2 mg/g 12.6 mg/g	Athermic	Naproxen Ketoprofen Diclofenac Ibuprofen	Baccar et al. (2012)
Olive stones activated with H_3PO_4	Dose: 0.4 g/L pH: 6 Temp: 25°C Equilibrium time: 1 h Initial concentration: 100 mg/L	Pseudo-first order, Langmuir isotherm	186 mg/g		Tetracycline	Boudrahem et al. (2019)
Peach stones activated with K_2CO_3	Dose: 0.66 g/L pH: 5.8 700 rpm Temp: 30°C Contact time: 48 h Initial concentration: 120 mg/L	Pseudo-second order, Langmuir isotherm	204 mg/g		Acetaminophen	Cabrita et al. (2010)

(Continued)

TABLE 9.5 *(Continued)*
Adsorptive Behavior of Lignocellulosic Materials in Pharmaceutical Removal

Adsorbent	Experimental Condition	Adsorption Isotherm, Kinetics	Adsorption Capacity	Thermodynamic Elucidation	Pharmaceutical	Reference
Olive stone activated with $ZnCl_2$	Dose: 1.5 g/L pH: 6 ± 0.3 300 rpm Temp: 20°C Contact time: 3 h Initial concentration: 31.81 mg/L	Pseudo-second order, Langmuir isotherm	30.67 mg/g	Exothermic Spontaneous	Diclofenac	Cherik and Louhab (2018)
Grape slurry waste	Dose: 1 g/l pH: 4–9 150 rpm Temp: 23 ± 2 °C Equilibrium time: 90–150 min Initial concentration: 30 mg/L	Pseudo-second order, Langmuir isotherm, Freundlich isotherm	2.971 mg/g 2.838 mg/g 3.104 mg/g	Exothermic Non-spontaneous	Amoxicillin Ampicillin Chloramphenicol	Chitongo et al. (2018)
Eucalyptus biochar	Dose: 3 g/L pH: 6–7 450 rpm Temp: 25°C Contact time: 1 h Initial concentration: 20 mg/L	Langmuir isotherm	6.41 mg/g		Fluoxetine	Fernandes et al. (2019)
Jatropha shell	Dose: 0.5 g/L pH: 2 180 rpm Temp: ambient Equilibrium time: 60 min Initial concentration: 50 mg/L	Pseudo-first order, Freundlich isotherm	206.2 mg/g		Sulfamethoxazole	Ganesan et al. (2020)

(Continued)

TABLE 9.5 *(Continued)*
Adsorptive Behavior of Lignocellulosic Materials in Pharmaceutical Removal

Adsorbent	Experimental Condition	Adsorption Isotherm, Kinetics	Adsorption Capacity	Thermodynamic Elucidation	Pharmaceutical	Reference
Grape pulps activated with $ZnCl_2$	Dose: 0.2 g/L pH: 5.7 Contact time: 8 h Temp: 35°C Initial concentration: 150–500 mg/L	Pseudo-second order, Langmuir isotherm	625 mg/g		Tetracycline	Güzel and Sayğılı (2016)
Activated carbon of babassu coconut mesocarp	Dose: 0.111 g/L pH: 6.12 150 rpm Temp: 25.15°C Equilibrium time: 200 min Initial concentration: 100mg/L	Pseudo-second order, Sips isotherm	50.4 mg/g	Exothermic Spontaneous	Acetylsalicylic acid	Hoppen et al. (2019)
Paper mill sludge activated with KOH	Temperature: 25°C Contact time: 550 min	Pseudo-second order, Zhu-Gu isotherm	191.6 ± 4.8 mg/g		Fluoxetine	Jaria et al. (2015)
Luffa cylindrica	Equilibrium time: 90 min 100 rpm pH: 5 Temp: 25°C	Pseudo-second order, Langmuir isotherm	19.157 mg/g	Exothermic Spontaneous	Ibuprofen	Khadir et al. (2020)
Olive stones	Dose: 5 g/L pH: 4.2 500 rpm Temp: 23±2°C Contact time: 3 h Initial concentration: 20 mg/L	Pseudo-second order, BET isotherm	0.26347 mg/g		Diclofenac	Larous and Meniai (2016)

(Continued)

TABLE 9.5 *(Continued)*
Adsorptive Behavior of Lignocellulosic Materials in Pharmaceutical Removal

Adsorbent	Experimental Condition	Adsorption Isotherm, Kinetics	Adsorption Capacity	Thermodynamic Elucidation	Pharmaceutical	Reference
Biochar	Dose: 1 g/L pH-independent Contact time: 120 min Initial concentration: 5 mg/L	Pseudo-second order, Langmuir isotherm	7.25 mg/g		Diclofenac	Li et al. (2019)
Brazil nutshells activated with $ZnCl_2$	Dose: 1.5 g/L pH: 7 adsorbent Temperature: 25°C Contact time: 30 min	Avrami kinetic adsorption model, Liu model	411.0 mg/g 309.7 mg/g	Exothermic Spontaneous	Acetaminophen	Lima et al. (2019)
Bertholletia excelsa activated with $ZnCl_2$	Dose: 1.5 g/L pH: 7 Temp: 45°C Contact time: 30 min Initial concentration: 400 mg/L	Avrami kinetic adsorption model, Liu model	451 mg/g 454.7 mg/g	Exothermic Spontaneous	Amoxicillin	Lima et al. (2019)
Lotus stalk activated with H_3PO_4	Dose: 0.1 g/L pH: 5.5 200 rpm Temp: 21±1°C Contact time: 24 h Initial concentration: 31.93 mg/L	Pseudo-second order, Langmuir isotherm	1428.57 μmol/g		Norfloxacin	Liu et al. (2011)
Lotus stalk activated with H_3PO_4	Dose: 0.2 g/L pH: 3–7 200 rpm Temp: 25°C Contact time: 72 h Initial concentration: 58.06 mg/L	Langmuir isotherm	333.23 mg/g		Trimethoprim	Liu et al. (2012)

(Continued)

TABLE 9.5 *(Continued)*
Adsorptive Behavior of Lignocellulosic Materials in Pharmaceutical Removal

Adsorbent	Experimental Condition	Adsorption Isotherm, Kinetics	Adsorption Capacity	Thermodynamic Elucidation	Pharmaceutical	Reference
Lotus stalk activated with $H_4P_2O_7$	Dose: 0.2 g/L pH: 3–7 200 rpm Temp: 25°C Contact time: 24 h Initial concentration: 29.03 mg/L	Langmuir isotherm	345.36 mg/g		Trimethoprim	Liu et al. (2012)
Biochar from spent olive stones	Dose: 0.2 g/L pH: 6.5 400 rpm Temp: 30°C Contact time: 72 h Initial concentration: 0.5 mg/L	Pseudo-first order	150 mg/g		Sulfamethoxazole	Magioglou et al. (2019)
NaOH- activated macadamia shells	Dose: 1 g/L pH: 3–10 Contact time: 3 h Initial concentration: 250–800 mg/L	Pseudo-second order, Temkin isotherm	455.8 mg/g		Tetracycline	Martins et al. (2015)
Apricot nut shells activated with H_3PO_4	Dose: 0.6 g/L pH: 6.5 250 rpm Temp: 30°C Contact time: 72 h Initial concentration: 150 mg/L	Pseudo-second order, Freundlich isotherm	308.33 mg/g	Endothermic Spontaneous	Tetracycline	Marzbali et al. (2016)

(Continued)

TABLE 9.5 *(Continued)*
Adsorptive Behavior of Lignocellulosic Materials in Pharmaceutical Removal

Adsorbent	Experimental Condition	Adsorption Isotherm, Kinetics	Adsorption Capacity	Thermodynamic Elucidation	Pharmaceutical	Reference
Pomegranate wood activated with NH_4Cl	Dose: 1 g/L pH: 7.1 150 rpm Temp: 10°C Contact time: 120 min Initial concentration: 200 mg/L	Pseudo-second order, Langmuir isotherm	233 mg/g	Exothermic Spontaneous	Acetaminophen	Mashayekh-Salehi and Moussavi (2016)
Cork activated with KOH	Dose: 0.66 g/L pH: 5 Temp: 30°C Contact time: 24 h Initial concentration: 100 mg/L	Pseudo-second order, Langmuir isotherm	174.4 mg/g		Ibuprofen	Mestre et al. (2014)
Sucrose-derived activated carbon	Dose: 0.66 g/L pH: 5 Temp: 30°C	Pseudo-second order, Langmuir isotherm	514 mg/g		Paracetamol	Mestre et al. (2015)
Argania spinosa tree nutshells	AC/TiO_2: 9% Dose 0.1 g/L Temp: 25°C Initial concentration: 50 mg/L Equilibrium time: 60 min	Pseudo-second order, Langmuir isotherm	153.8 mg/g 105.3 mg/g 125.0 mg/g		Diclofenac Carbamazepine Sulfamethoxazole	Mouchtari et al. (2020)
NH_4Cl- activated carbon pomegranate wood waste	Dose: 0.8 g/L pH: 6 100 rpm Temp: 25°C Contact time: 6 h Initial concentration: 50–500 mg/L	Pseudo-second order, Langmuir isotherm	437 mg/g	Spontaneous Endothermic	Amoxicillin	Moussavi et al. (2013)

(Continued)

TABLE 9.5 *(Continued)*
Adsorptive Behavior of Lignocellulosic Materials in Pharmaceutical Removal

Adsorbent	Experimental Condition	Adsorption Isotherm, Kinetics	Adsorption Capacity	Thermodynamic Elucidation	Pharmaceutical	Reference
Balanites aegyptiaca seeds	Dose: 1 g/L pH: 6.12 70 rpm Temp: 25°C Contact time: 6 h Initial concentration: 100 mg/L	Pseudo-second order, Langmuir isotherm	4.28 mg/g		Caffeine	N'diaye and Kankou (2019)
$ZnCl_2$- activated walnut shell	Dose: 0.6 g/L pH: 6.5 200 rpm Temp: 30°C Contact time: 20 h Initial concentration: 100 mg/L	Pseudo-second order, Langmuir isotherm	233.1 mg/g		Cephalexin	Nazari et al. (2016)
Rice husk activated carbon	pH: 2–3.5 Equilibrium time: 100 min Initial concentration: 100 mg/L	Pseudo-second order, Langmuir isotherm	20.964 mg/g		Paracetamol	Nche et al. (2017)
Paper pulp non-activated and activated	pH independent for CBZ pH: 5.5 for SMX 80 rpm Temp: 25°C Conductivity: 0.26 mS/cm Initial concentration: 5 mg/L Dissolved organic carbon: Dose: 0.07 g/L for CBZ Dose: 0.3 g/L for SMX	Pseudo-second order, Langmuir isotherm	92 ± 19 mg/g (bleached pulp) 13.0 ± 0.6 mg/g (H_3PO_4 activated)		Carbamazepine (CBZ) Sulfamethoxazole (SMX)	Oliveira et al. (2018)

(Continued)

TABLE 9.5 *(Continued)*
Adsorptive Behavior of Lignocellulosic Materials in Pharmaceutical Removal

Adsorbent	Experimental Condition	Adsorption Isotherm, Kinetics	Adsorption Capacity	Thermodynamic Elucidation	Pharmaceutical	Reference
Wild kernel biochar activated with KOH	Dose: 0.4 g/L pH: 6 Contact time: 3 h Concentration: 125.3 mg/L 150 rpm Temp: 22 ± 1°C	Electrostatic attraction, pseudo-second order, Langmuir isotherm	73.14 mg/g		Naproxen	Paunovic et al. (2019)
NaOH- activated guava seeds	Dose:1 g/L pH: 4 Contact time: 4 h Temp: 25°C Initial concentration: 50–800 mg/L	Pseudo-second order, Langmuir isotherm, endothermic	570.4 mg/g		Amoxicillin	Pezoti et al. (2016)
Moringa oleifera Lam. seed husks	pH: 3–9 Temp: 45°C Equilibrium time: 18 h	Pseudo-first order, Langmuir isotherm	17.48 mg/g	Endothermic Spontaneous	Acetaminophen	Quesada et al. (2019)
Tomato wastes activated with $ZnCl_2$	Dose: 0.2 g/L pH: 5.7 Temp: 25°C, Contact time: 8 h Initial concentration: 200–400 mg/L	Pseudo-second order, Langmuir isotherm	500 mg/g		Tetracycline	Sayğılı and Güzel (2016)
Phosphorous-doped microporous carbonous material	Dose: 2.0 g/L pH: 6–7 Contact time: 120 min Tempe: 22 ± 1°C Initial concentration: 20 mg/L	Pseudo-second order, Freundlich isotherm	17.193 mg/g 17.685 mg/g 19.265 mg/g 17.657 mg/g 21.116 mg/g 23.332 mg/g		Sulfamethoxazole Carbamazepine Ketoprofen Naproxen Diclofenac Ibuprofen	Sekulic et al. (2019)

(Continued)

TABLE 9.5 *(Continued)*
Adsorptive Behavior of Lignocellulosic Materials in Pharmaceutical Removal

Adsorbent	Experimental Condition	Adsorption Isotherm, Kinetics	Adsorption Capacity	Thermodynamic Elucidation	Pharmaceutical	Reference
Ultrafine magnetic coconut shell biochar/Fe_3O_4	Dose: 0.2 g/L pH: 2 170 rpm Temp: 25°C Contact time: 8 h Initial concentration: 50 mg/L	Langmuir isotherm	62.7 mg/g 94.2 mg/g		Carbamazepine Tetracycline	Shan et al. (2016)
Paper mill sludge	Dose: 0.02 g/L pH: 6.5 80 rpm Temp: 25 ± 0.1°C Equilibrium time: 240 min Initial concentration: 50 mg/L	Pseudo-second order, Langmuir isotherm	209 ± 27 mg/g 407 ± 14 mg/g		Sulfamethoxazole Paracetamol	Silva et al. (2019)
Walnut shell	pH: 6–8 Temp: 30°C Initial concentration: 40 mg/L	Pseudo-second order, Langmuir isotherm	107.4 mg/g 93.5 mg/g		Metronidazole Sulfamethoxazole	Teixeira et al. (2019)
Grape stalk	pH: 6.0 Temp: 20°C 30 rpm Initial concentration: 20 mg/L	Pseudo-first order, Langmuir isotherm	1.692 mg/g		Paracetamol	Villaescusa et al. (2011)
Aerobic granular sludge modified by $ZnCl_2$	Dose: 0.75 g/L pH: 4–9 160 rpm Temp: 25°C Contact time: 48 h Initial concentration: 80 mg/L	Pseudo-second order, Langmuir isotherm	93.44 mg/g	Spontaneous Endothermic	Tetracycline	Yan et al. (2020)

and inorganic salts. Mouchtari et al. (2020) discovered that the photocatalytic activities of activated carbon/TiO_2 composite in the removal of the three pharmaceuticals (50–100%) from solution by combining adsorption and photodegradation processes were extremely high.

9.2.4 Adsorption Modeling

9.2.4.1 Adsorption Kinetics

The kinetics of adsorption describes the rate of adsorbate uptake on adsorbent, controls the equilibrium time and is important for process control. Adsorption rate is determined by kinetic analysis, which can be limited by different mass transfer resistances, primarily resistance to external diffusion and resistance to intraparticle diffusion, depending on temperature and pressure conditions, as well as nature of the adsorbent and adsorbate. Adsorption begins with external mass transfer of the adsorbate onto the adsorbent, followed by internal diffusion and finally sorption of the adsorbate onto the adsorbent. The adsorption kinetic model can be based on sorption or diffusion. Most studies have discovered that the adsorption kinetics of pharmaceutical is pseudo-first order and pseudo-second order. However, the Elovich (Largitte and Pasquier, 2016), Avrami (Avrami, 1939) and intraparticle diffusion (Weber Jr. and Morris, 1963) models have also been widely applied. Li et al. (2019) discovered that both the film and intraparticle diffusion steps of analytes to macroalgae and wood chippings occurred.

The pseudo-first-order kinetic model is one of the most important models that uses the first-order mechanism based on adsorption capacity to describe the kinetic rate of liquid-solid phase adsorption. The most commonly used kinetic model is the pseudo-second-order kinetic model, which assumes that the rate of adsorption of adsorbate is proportional to the available sites on the adsorbent and the driving force is proportional to the available fraction of active sites and supports the chemisorption mechanism. The Elovich kinetic model was initially used to investigate the kinetics of chemisorption of gases onto heterogeneous solids, but its applicability in wastewater processes has been observed to be significant (Nayak and Pal, 2019). The Avrami kinetic model elucidates the kinetics of phase transformation under the assumption of spatially random nucleation (Avrami, 1939).

9.2.4.2 Adsorption Isotherm

Adsorption isotherms describe the interaction of an adsorbate and an adsorbent in any system. When adsorption and desorption rates are equal, dynamic adsorption equilibrium is achieved. Adsorption isotherms describe the equilibrium relationship between the concentration of a solute absorbed in solid phase and the concentration in solution. Table 2.3 enumerates several commonly used adsorption isotherms such as Langmuir, Sips, Temkin, Freundlich, Redlich-Peterson and Dubinin-Radushkevich.

The Langmuir model estimates the maximum adsorption capacity based on the assumption that sorption occurs at a finite number of sites within the adsorbent at a constant temperature and is valid for homogeneous and monolayer coverage of adsorbate onto an identical set of well-defined localized adsorption sites without lateral interaction and steric hindrance between adsorbed molecules (Langmuir, 1916).

The Freundlich isotherm is an empirical model developed for non-linear adsorption that takes into account surface heterogeneity and encompasses multi-layer coverage through the exponential distribution of active sites, in which infinite surface coverage is predicted without any saturation onto the bioadsorbent surface (Freundlich, 1906). Sips isotherm is an empirical model developed by combining the Langmuir and Freundlich isotherms. It is based on the theory that the active sites of adsorbent have heterogeneous energies and can be applied in a multi-element system by taking into account the equal competition of the ions for the active sites (Sips, 1948). When $n = 1$ and the initial adsorbate concentration is high, the Sips model predicts monolayer Langmuir adsorption; however, at low adsorbate concentrations, this model does not obey Henry's law and reduces to the Freundlich isotherm (Eqn. 9.10). In the Temkin isotherm, adsorption is assumed to be multi-layered and is characterized by a uniform distribution of binding energies (Temkin and Pyzhev, 1940). As the coverage of the adsorbent surface increases, the heat of adsorption for all molecules decreases linearly (Yang, 1993). The Redlich and Peterson model is an empirical hybrid of the Langmuir and Freundlich models, which has been widely used in homogeneous or heterogeneous adsorption processes (Redlich and Peterson, 1959). When $\beta = 1$, the isotherm reduces to the Langmuir equation; when $\beta = 0$ or $C_e \sim 0$, the isotherm approaches to Henry's equation; and when $C_e \sim \infty$, i.e., at high concentration, the isotherm reduces to the Freundlich equation (Eqn. 9.12). The numerous adsorption kinetic and isotherm models that have been utilized in most research is listed in Table 9.6.

9.2.4.3 Adsorption Thermodynamics

Information on the inherent energetic changes associated with the adsorption process may be obtained from several thermodynamic parameters such as Gibbs free energy (ΔG, kJ/mol), enthalpy (ΔH, kJ/mol) and entropy (ΔS, J/mol K), which can be computed using the following equations:

$$\ln K = -\frac{\Delta G}{RT} \tag{9.13}$$

$$\Delta G = \Delta H - T\Delta S \tag{9.14}$$

where K is the equilibrium thermodynamic constant, R is the gas constant (8.314 J/mol K) and T (K) is the absolute temperature.

Adsorption is favorable and spontaneous at a particular temperature in the case that $\Delta G < 0$. The presence of a $\Delta H < 0$ value indicates that the process is exothermic and involves physical adsorption, chemical adsorption, or a combination of the two. In contrast, $\Delta H > 0$ shows an endothermic process and suggests the occurrence of chemisorption. $\Delta S > 0$ denotes that the solid-liquid interface has more randomness, which can be attributed to the fact that the translational energy obtained by the displaced solvent molecules (e.g., water) is more than that of the translational energy lost by the adsorbate molecules. Thermodynamic analysis by Marzbali et al. (2016) revealed that the adsorption of tetracycline onto apricot nut shells activated with H_3PO_4 was endothermic and spontaneous. The negative $\Delta H = -23.45$ kJ/mol as studied by Hoppen et al. (2019) indicates exothermic nature of adsorption and weak bonding between adsorbate and adsorbent, $\Delta S = 69.04$ J/mol K suggests an

TABLE 9.6
Adsorption Models

Adsorption Kinetic Models			
Model	**Equation**		**Reference**
Pseudo-first order	$q_t = q_e\left(1 - e^{-k_1 t}\right)$	(9.1)	Largitte and Pasquier (2016)
Pseudo-second order	$q_t = \frac{k_2 q_e^2 t}{1 + k_2 q_e t}$	(9.2)	Nayak et al. (2019)
Elovich	$q_t = \frac{1}{\beta}\ln\left(1 + \alpha\beta t\right)$	(9.3)	Roginsky and Zeldovich (1934)
Avrami	$q_t = q_e\left(1 - e^{k_{AV} t^{n_{AV}}}\right)$	(9.4)	Avrami (1939)
Intraparticle diffusion	$q_t = k_{id}\sqrt{t} + C_i$	(9.5)	Weber Jr. and Morris (1963)
Adsorption Isotherm Models			
Langmuir	$q_e = q_m \frac{K_L C_e}{1 + K_L C_e}$	(9.6)	Langmuir (1916)
Freundlich	$q_e = K_f C_e^{\frac{1}{n}}$	(9.7)	Freundlich (1906)
Dubinin-Radushkevich	$q_e = q_{mDR}\exp\left(-K_{DR}\varepsilon^2\right)$	(9.8)	Dubinin and Radushkevich (1947)
	$\varepsilon = RT\ln\left(1 + \frac{1}{C_e}\right)$	(9.9)	Dubinin (1960)
Sips	$q_e = \frac{q_{mS} K_S C_e^{\frac{1}{n}}}{1 + K_S C_e}$	(9.10)	Sips (1948)
Temkin	$q_e = \frac{RT}{b_{TE}}\ln\left(K_{TE} C_e\right)$	(9.11)	Temkin and Pyzhev (1940)
Redlich and Peterson	$q_e = \frac{K_{TE} C_e}{1 + \alpha_{RP} C_e^{\beta_{RP}}}$	(9.12)	Redlich and Peterson (1959)

q_t (mg/g) is the amount of adsorbate adsorbed at time t, q_e (mg/g) is the adsorption capacity at equilibrium, k_1 (1/min) is the pseudo-first-order rate constant, t (min) is the contact time, k_2 is the rate constant of pseudo-second-order model, α and β are Elovich constants, k_{AV} and n_{AV} are Avrami constants, k_{id} and C_i are intraparticle diffusion constants, q_m (mg/g) is the maximum adsorbed concentration of adsorbate at the equilibrium, Ce (mg/L) refers to the concentration of adsorbate in the liquid phase at equilibrium, K_L (L/mg) is the Langmuir equilibrium constant, K_f (mg/g (L/mg)$^{1/n}$) is the Freundlich equilibrium constant, n is the degree of non-linearity, q_{mDR} (mg/g) is the adsorption capacity of the D-R isotherm, ε is the Polanyi potential, K_{DR} (mol^2/J^2) is the constant related to mean free energy of adsorption, R (8.314 J/mol K) is the universal gas constant, T (K) is the absolute temperature, q_{mS} (mg/g) is the Sips maximum adsorption capacity, K_S (mg/L)$^{-1/n}$ is the affinity coefficient of Sips model, b_{TE} (J/mol) is the Temkin constant related to heat of adsorption, K_{TE} (L/mg) is the equilibrium binding constant related to maximum binding energy, K_{RP} (L/g) is the Redlich-Peterson model isotherm constant, α_{RP} (mg/L)$^{-\beta}$ is the Redlich-Peterson model constant, and β is the Redlich-Peterson model exponent which lies between 0 and 1.

increase in random interaction between the solid-liquid interface with the concentration of pharmaceutical, and $\Delta G < 0$ for all temperatures indicates that the process was spontaneous. The thermodynamic parameters in the study done by Quesada et al. (2019) indicated that the process is spontaneous ($\Delta G = -21.659$ kJ/mol), endothermic ($\Delta H = 23.110$ kJ/mol) and reversible ($\Delta S = 0.141$ J/mol K), and imply that the adsorption is controlled by physical forces such as hydrogen bonds and π-stacking. Adsorption of cephalexin on walnut shell activated carbon was discovered by Nazari et al. (2016) to be spontaneous ($\Delta G < 0$ for increase in temperature), endothermic ($\Delta H = 50.114$ kJ/mol) and with random orientation at the solid-liquid interface ($\Delta S = 0.1817$ J/mol K).

9.2.5 Adsorption Mechanisms Involved During the Adsorption of Pharmaceutical Wastewater

There are a number of mechanisms that can describe the nature of the adsorption of pharmaceutical contaminants from aqueous solutions on the surface of carbon-based adsorbents, including acid-base, π-π, electrostatic, van der Waals, hydrophobic and H-bond interactions (Baccar et al., 2012; Liu et al., 2012). FTIR peaks (shifts or new appearances) can suggest potential adsorption mechanisms such as electrostatic interaction, H-bonds, n-π and π-π electron donor-acceptor (EDA) interactions and so on (Tran et al., 2017). Baccar et al. (2012) investigated the adsorption capability of ibuprofen, ketoprofen, naproxen and diclofenac onto olive waste activated carbon in both single and mixed solutions. They discovered that their adsorption capacities were quite different and were mainly governed by their pK_a and octanol/water coefficient (K_{ow}). The adsorption mechanism was related to pH, which indicated π-π interactions as well as others, i.e., for pH < pH_{ZPC} (pH = 2.01), the adsorbent was positively charged, while the acid drugs were neutral, indicating the involvement of hydrogen bonding and/or van der Waals interaction, whereas at pH > 8.65, electrostatic repulsion between adsorbent and drugs was observed. Similarly, in another study done by Álvarez-Torrellas et al. (2016), FTIR spectra of the adsorbents before and after adsorption revealed a peak at 3400 cm^{-1}, indicating the presence of hydroxylic groups in the π–π EDA interactions, and a clear shift of the peak from 1710 to 1729 cm^{-1}, indicating the development of H-bonding between the amide groups in tetracycline. According to Giles et al. (1960), the adsorption isotherms were classified as S-type, indicating multi-layer adsorption. Liu et al. (2012) identified the major mechanisms of trimethoprim adsorption onto the four carbons as π-π EDA interaction, cation-π bonding, hydrogen bonding, Lewis acid-base interaction and cation exchange. Marzbali et al. (2016) discovered that adsorption is substantially dependent on the initial pH of the solution, indicating the participation of numerous intermolecular interactions such as H-bonds and EDA. The FTIR spectrum of the adsorbent revealed a broad signal in the region of 3201–3163 cm^{-1} with peaks at 1440 cm^{-1} and 1710 cm^{-1}, confirming acetaminophen adsorption through hydrogen bonding and π-stacking (Quesada et al., 2019). According to Sekulic et al. (2019), the principal mechanism for physical adsorption of pharmaceuticals onto activated carbon is H-bonding and n-π and π-π EDA interaction. Villaescusa et al. (2011) indicated the π-stacking interactions between the aromatic ring of paracetamol and the lignin syringyl and guaiacyl moieties of grape stalk, as well as hydrogen bonding, as the mechanisms responsible for paracetamol sorption with a favorable hydrophobic effect. According to the classification used by Abo El Naga et al. (2019), the equilibrium adsorption isotherm of diclofenac on the adsorbent surface corresponded to the L-type. In summary, Giles et al. (1960) demonstrated a strong interaction between bioadsorbent and pharmaceutical pollutant species. Lima et al. (2019) postulated that hydrogen bonds, EDA interactions, π-π interactions and dispersion interactions would be involved in the adsorption mechanism.

9.2.6 Regeneration Capabilities of Exhausted Bioadsorbents Loaded with Pharmaceutical Pollutants

Regeneration capability denotes the ability to desorb adsorbates from the adsorbent surface. Different regeneration techniques such as steam desorption, thermal swing, ball milling, solvent extraction, pH shift, microbiological, electrochemical and ultrasonic may be used; the choice depends on the type of adsorbent, the quality of the water or effluent, regeneration costs and future treatment goals. Lima et al. (2019) demonstrated a regeneration efficiency of up to 74% utilizing a mixture of 0.1 M NaOH and 20% CH_3CH_2OH solutions and that the mixture could be reused up to four times, ensuring continuous usage of the prepared bioadsorbent for acetaminophen removal from wastewater. In a study conducted by Abo El Naga et al. (2019), the regenerated sugarcane bagasse-activated carbon sample illustrated a record adsorption efficiency of 92.4% even after five cycles of repeated usage, and the FTIR spectrum revealed that the regenerated adsorbent was virtually identical to the fresh sample, indicating that the structural integrity of the adsorbent was well preserved after diclofenac adsorption. Ganesan et al. (2020) demonstrated the desorption of sulfamethoxazole from Jatropha shell activated carbon by 0.1 M NaOH and regeneration up to 6 cycles, and the regeneration was achievable due to the existence of excess anions in alkaline media, resulting in significant electrostatic repulsion (Fan et al., 2012). Hoppen et al. (2019) evaluated the regeneration capability during three cycles of reuse, indicating an increase in adsorption and desorption efficiencies of 53.36% and 32.71%, respectively; even at the third cycle, the adsorption efficiency was approximately 93.13%. Khadir et al. (2020) demonstrated the reusability of *Luffa cylindrica* reinforced with polypyrrole in the removal of ibuprofen for up to five cycles with an ibuprofen removal percentage greater than 80%. According to Sekulic et al. (2019), acid catalyst-functionalized *Prunus domestica L.* can be reused after desorption in acidic solution for at least three cycles.

9.3 REMOVAL OF HEAVY METALS THROUGH BIOADSORPTION USING LIGNOCELLULOSIC MATERIALS

9.3.1 Occurrence and Fate of Heavy Metals in the Environment

According to most research, heavy metals are metals with an atomic number larger than 20 g/mol and specific gravity five times that of water, i.e., an elemental density greater than 5 g/cm^3 (Ali et al., 2019). Heavy metals, which are often found in the environment, pose a significant threat to ecosystems. In contrast to organic pollutants, heavy metals are highly resistant to environmental degradation like pharmaceuticals and tend to bioaccumulate, subsequently posing a great risk to all living creatures once they are converted from solid to ionic form or through biomethylation to organometallic fractions. The United States Environment Protection Agency (USEPA) designated trace metals such as arsenic (As), cadmium (Cd), chromium (Cr), copper (Cu), mercury (Hg), nickel (Ni), lead (Pb) and zinc (Zn) as priority control pollutants due to their toxicity, bioaccumulation and non-degradability. A number of heavy metals are necessary for plant growth up to a certain point, but beyond that point, they are extremely detrimental to living organisms. Similarly, Zamora-Ledezma et al. (2021) reported that trace amounts of Co (cobalt), Fe (iron), Mg (magnesium), Mn (manganese), Mo (molybdenum), Cu, Cr, Ni and Zn metals are required for metabolic activities in humans in order to avoid disease and syndromes. However, an excess of these heavy metals can cause serious human health issues such as muscular, physical and neurological degeneration (Kumar et al., 2019). The major sources of heavy metal pollution are summarized in Figure 9.9, along with some of the harmful effects of heavy metals on human health. Table 9.7 shows some common heavy metals, as well as their sources and health consequences, as well as detection limits.

Several indices have been proposed to quantify the level of heavy metal pollution level in the environment. Some extensively used indices are enrichment factor, contamination factor, pollution load index (PLI) and geoaccumulation index. To quantify heavy metal contamination, identify likely sources and assess potential ecological and health consequences, indices such as the ecological risk index (RI) and hazard index (HI) are utilized (Hu et al., 2017; Mehr et al., 2017; Singh and Kumar, 2017; Rehman et al., 2018). Hu et al. (2017) found that heavy metal concentrations in soil irrigated with wastewater in China ranged between 5.03 and 8.71 mg/kg for Fe, 101.34 and 259.32 mg/kg for Zn, 90.50 and 256.22 mg/kg for Cr, 23.93 and 52.75 mg/kg for Cu, 24.79 and 37.48 mg/kg for Pb, and 1.56 and 2.70 mg/kg for Cd. The soil samples in the research area were shown to pose a significant ecological risk, particularly to youngsters. In the Moroccan city of Marrakech, researchers discovered wastewater-irrigated soil contaminated with Cd, Cu, Pb and Zn (11.22 mg/kg, 17.70 mg/kg, 57.36 mg/kg and 112.71 mg/kg, respectively) (Chaoua et al., 2019). Singh and Kumar (2017) found that high concentrations of Cd, Pb and Fe in the Ajay River, India, resulted in heavy metal pollution index and pollution index values of <100 and >5, respectively. Sarkar et al. (2021) discovered microplastics as a possible heavy metal transporter in aquatic ecosystems. The wastewater and microplastics were found to carry approximately 1.03 to 345.6 μg/L and 2.03 to 191.01 μg/g, respectively.

9.3.2 Utilization of Lignocellulosic Biomass to Remove Heavy Metals

Lignocellulose can also be used to remove heavy metals by producing activated carbon, biochar, magnetic biochar, carbon nano- or microtubes, or by applying it directly. Sugarcane bagasse, plant sludge, rice husk, rice stalk, corn stalk, beet pulp, peanut hulls, coconut shell, nut shell, fruit bunch fibers, wood chippings, leaves, bark

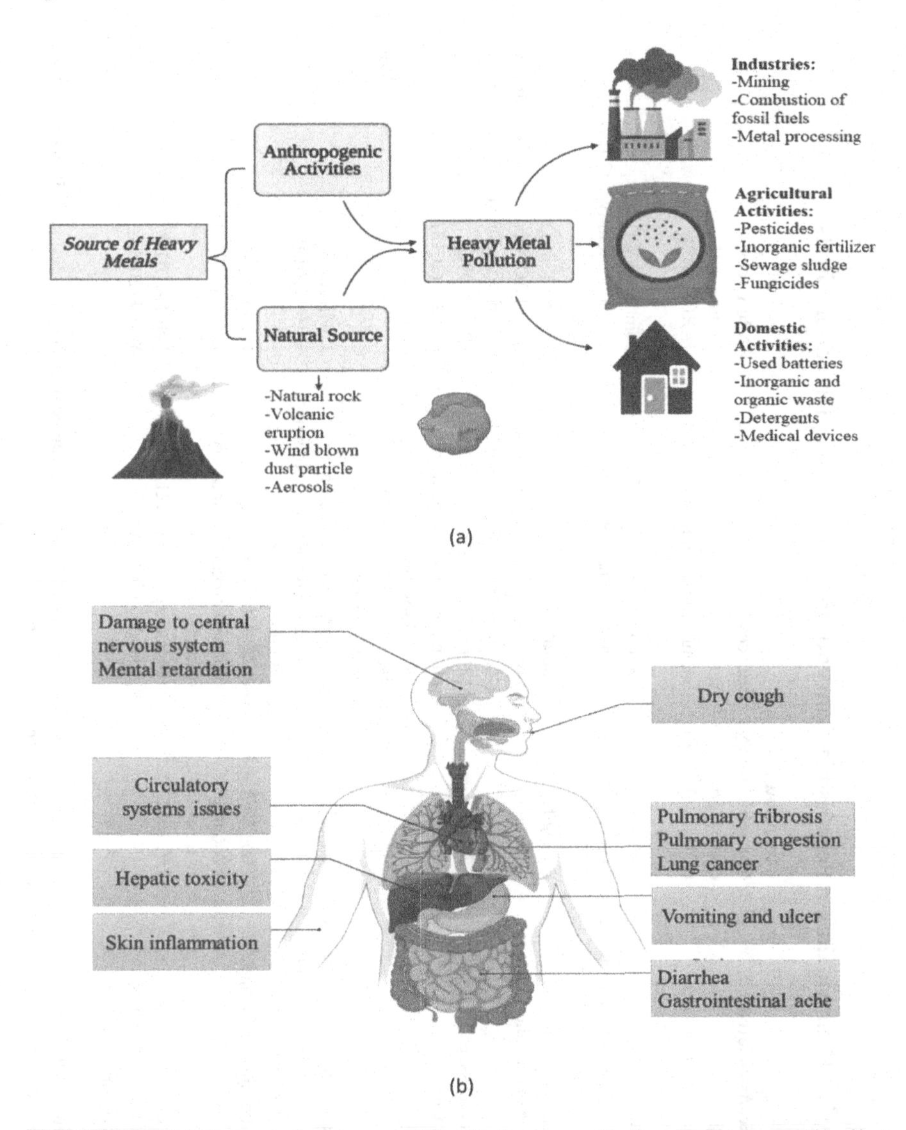

FIGURE 9.9 (a) Sources of environmental pollution caused by heavy metals and (b) effects of heavy metals in different vital organs of human health. (Reprinted with permission from Zamora-Ledezma et al., 2021.)

and seeds of many other plants have all been used successfully to remove metal ions (Gautam et al., 2014). Table 9.8 summarizes the performance evaluation of different low-cost bioadsorbents for metal removal. The kinetics of adsorption describes the rate of adsorbate uptake on adsorbent, controls the equilibrium time and is important for process control, as described in Section 9.2.4.1. In most studies, heavy metal kinetics was fitted to pseudo-second-order kinetics, followed by pseudo-first-order kinetics, as shown in Table 9.8.

TABLE 9.7
Health Effects of Heavy Metals with Their Maximum Permissible Concentration Sources and Detection Limits

Heavy Metals	Limit of Detection	Health Risks Associated	Max Allowable Concentration*	Sources	References
As	0.1 μg/L[a] 2 μg/L[b]	Carcinogenic May cause cancer in bladder, kidney, skin, lung, and liver	0.010 mg/L 0.010 mg/L	Naturally occurring, electronics manufacturing industry	Edition (2011) IARC (2012)
Cd	0.01 μg/L[a] 2 μg/L[b]	Cancer in lungs, prostate cancer, bone fractures, kidney dysfunction and hypertension can be fatal	0.005 mg/L 0.003 mg/L	Naturally occurring, steel and chemical industries	Joint FAO/WHO Expert Committee on Food Additives (2013) WHO (2019)
Cr	0.05–0.2 μg/L[b]	Allergic dermatitis, diarrhea, nausea, lung cancer and vomiting	0.1 mg/L 0.05 mg/L	Naturally occurring, tanning, pigments and paints, fungicide, ceramic, steel, and glass industries	USEPA (2011) Joint FAO/WHO Expert Committee on Food Additives (2013)
Cu	0.02–0.1 μg/L[a] 0.3 μg/L[c] 0.5 μg/L[b]	Gastrointestinal issues and liver or kidney damage	1.3 mg/L 2.0 mg/L	Naturally occurring, household plumbing system, insecticides, wood preservatives, electroplating, dye manufacturing industry and petroleum refining	IPCS (1998) USEPA (2018)
Pb	1 μg/L[b]	Plumbism, anemia, nephropathy, gastrointestinal colic and central nervous system symptoms	0.0 mg/L 0.01 mg/L	Battery industries, lead-based products and household plumbing systems	Levin (1997) WHO (2019)
Hg	5 μg/L[b] 0.05 μg/L[d]	Skin, nervous, lungs and kidney damage	0.002 mg/L 0.006 mg/L	Fossil fuel combustion, mining, electroplating, paper industry and batteries	WHO (2005) USEPA (2018)
Ni	0.1 μg/L[a] 0.5 μg/L[b]	Lung fibrosis, cancer of the respiratory tract and kidney and cardiovascular diseases	0.07 mg/L	Stainless steel and nickel alloy production industries	USEPA (2018) WHO (2019)

[a] Inductively coupled plasma mass spectrometry (ICP-MS) [b] Flame atomic absorption spectrometry (AAS) [c] Inductively coupled plasma (ICP) spectroscopy
[d] cold vapor AAS * Max allowable concentration in drinking water recommended by USEPA and WHO

TABLE 9.8
Adsorptive Behavior of Lignocellulosic Materials in Heavy Metal Removal

Adsorbent	Experimental Condition	Adsorption Isotherm, Kinetics	Adsorption Capacity or Removal %	Thermodynamic Elucidation	Heavy Metal	Reference
Coconut shell magnetic biochar	Dose: 2 g/L pH: 4.5 (for Pb(II)) and pH : 4.8 (for Cd(II)) Temp: 30 °C Reaction time: 20 min Initial concentration: 25–125 mg/L 0.5 g ($FeCl_3$: Biomass) Impregnation ratio	Pseudo-second order, Langmuir isotherm	4.96 mg/g 4.77 mg/g		Pb(II) Cd(II)	Yap et al. (2017)
Banana peel	Dose: 1 g/50mL pH: 2 Temp: 30°C Reaction time: 30 min Initial concentration: 5 µg/L	Pseudo-second order, Langmuir isotherm	41.66 mg/g 94%		Cr(VI)	Ahmed and Misganaw (2019)
Luffa cylindrica sponge	Dose: 2 g/L pH: 4–14 Temp: 30°C 200 rpm Reaction time: 3h Initial concentration: 50–400 mg/L	Pseudo-second order, Langmuir isotherm	75.853 mg/g	Exothermic Stable	Pb(II)	Adewuyi and Pereira (2017)
Corn straw	Dose: 1 g/L pH: 3 Temp: 25°C 200 rpm Reaction time: 16 h Initial concentration: 100–350 mg/L	Pseudo-second order, Langmuir isotherm	175.44 mg/g	Endothermic Spontaneous	Cr(VI)	Ma et al. (2019)

(Continued)

TABLE 9.8 *(Continued)*
Adsorptive Behavior of Lignocellulosic Materials in Heavy Metal Removal

Adsorbent	Experimental Condition	Adsorption Isotherm, Kinetics	Adsorption Capacity or Removal %	Thermodynamic Elucidation	Heavy Metal	Reference
Unmodified watermelon rind	Dose: 1 g/L pH: 5.2 Temp: 20°C 180 rpm Reaction time: 2 h Initial concentration: 50–400 mg/L	Pseudo-second order, Langmuir isotherm	85% 78%		As(III) As(V)	Shakoor et al. (2018)
Xanthated watermelon rind	Dose: 1 g/L pH: 8.2 (for As(III)) pH: 4.6 (for As(V)) Temp: 20°C 4000 rpm Reaction time: 2 h Initial concentration: 4 mg/L	Pseudo-second order, Langmuir isotherm	99% 98%	Endothermic Spontaneous	As(III) As(V)	Shakoor et al. (2018)
Citric acid modified watermelon rind	Dose: 1 g/L pH: 7.2 (for As(III)) and pH: 4.5 (for As(V)) Temp: 20°C 4000 rpm Reaction time: 2 h Initial concentration: 4 mg/L	Pseudo-second order, Langmuir isotherm	96% 97%	Endothermic Spontaneous	As(III) As(V)	Shakoor et al. (2018)
Coffee husk biomass waste	Dose: 0.5 g/25 mL Temp: room 300 rpm Reaction time: 1.5 h Initial concentration: 50–400 mg/L	Pseudo-second order, Freundlich isotherm	37.04 mg/g		Pb(II)	Alhogbi (2017)

(Continued)

TABLE 9.8 *(Continued)*

Adsorptive Behavior of Lignocellulosic Materials in Heavy Metal Removal

Adsorbent	Experimental Condition	Adsorption Isotherm, Kinetics	Adsorption Capacity or Removal %	Thermodynamic Elucidation	Heavy Metal	Reference
Corn stalk biochar	Dose: 2 g/L pH: 5.5 Temp: 28°C 150 rpm Reaction time: 12 h Initial concentration: 100–300 mg/L	Pseudo-second order, Langmuir isotherm	57.26 mg/g		Pb(II)	Liu et al. (2021)
Magnetic rice husk biochar modified with $KMnO_4$	Dose: 2.5 g/L pH: >2.5 (for Pb(II)) and >3.5 (for Cd(II)) Temp: 25°C 200 rpm Reaction time: 24 h Initial concentration: 10–500 mg/L	Pseudo-second order, Langmuir isotherm	148 mg/g 79 mg/g		Pb(II) Cd(II)	Sun et al. (2019)
Mg-Fe-layered double hydroxide – Rice husk ash	Dose: 0.1 g/L pH: 5 Temp: 25°C 200 rpm Reaction time: 24 h Initial concentration: 10–150 mg/L	Pseudo-second order, Sips isotherm	682.2 mg/g	Endothermic Spontaneous	Pb(II)	Yu et al. (2018)

(Continued)

TABLE 9.8 *(Continued)*
Adsorptive Behavior of Lignocellulosic Materials in Heavy Metal Removal

Adsorbent	Experimental Condition	Adsorption Isotherm, Kinetics	Adsorption Capacity or Removal %	Thermodynamic Elucidation	Heavy Metal	Reference
Rice husk biochar	Dose: 10 g/L pH: 7.37 Temp: 24°C 150 rpm Reaction time: 2 h Initial concentration: 1.82 mg/L (Cr(total)), 9.28 mg/L (Fe(II)) 1.59 mg/L (Pb(II))	Freundlich isotherm	90% 90% 65%		Pb(II) Cr(total) Fe(II)	Sanka et al. (2020)
Sugar cane bagasse	Dose: 5 g/L pH: 7 Temp: 25°C 400 rpm Reaction time: 96 h Initial concentration: 0.1 mM	Pseudo-second order, Langmuir isotherm	100% (96 h) 95.99% (15 min)		Cu(II)	Saleh et al. (2020)
Magnetic rice husk biochar	Dose: 0.4 g/L pH: 7 Temp: 25°C 150 rpm Reaction time: 8 h Initial concentration: 10–80 mg/L	Pseudo-second order, Langmuir isotherm	129 mg/g 118 mg/g	Endothermic Spontaneous	Pb(II) U(VI)	Wang et al. (2018)

(Continued)

TABLE 9.8 *(Continued)*
Adsorptive Behavior of Lignocellulosic Materials in Heavy Metal Removal

Adsorbent	Experimental Condition	Adsorption Isotherm, Kinetics	Adsorption Capacity or Removal %	Thermodynamic Elucidation	Heavy Metal	Reference
NaOH-modified peanut husk	Dose: 5 g/L pH: 6 Temp: 25 °C 250 rpm Contact time: 3 h Initial concentration: 20–100 mg/L	Langmuir isotherm	27.03 mg/g 11.36 mg/g 14.29 mg/g 6.10 mg/g 56.82 mg/g		Pb(II) Cd(II) Mn(II) Co(II) Ni(II)	Abdelfattah et al. (2016)
Natural coconut husk	Dose: 10 g/L pH: 6 Temp: 30°C 120 rpm Contact time: 40 min Initial concentration: 300 mg/L	Pseudo-second order, Langmuir, Freundlich, Temkin, Dubinin-Radushkevich and Flory-Huggins isotherms	443.0 mg/g 88.6% 338.0 mg/g 67.6% 404.5 mg/g 80.9% 362.2 mg/g 72.4%		Cu(II) Zn(II) Ni(II) Pb(II)	Malik et al. (2017)
Unmodified milled olive stones	Dose: 4 g/L pH: 6 Temp: 20°C Contact time: 1 h Initial concentration: 1–10 mg/L	Pseudo-second order, Langmuir isotherm	0.3 mg/g 0.557 mg/g 0.581 mg/g		Cd(II) Cu(II) Pb(II)	Amar et al. (2020)
Unmodified aloe vera waste	Dose: 1.5 g/L pH: 4 (for U(VI)) pH: 5 (for Cd(II)) Temp: 20°C Contact time: 24 h	Pseudo-second order, Freundlich isotherm	201.2 mg/g 70.4 mg/g	Endothermic Spontaneous	U(VI) Cd(II)	Noli et al. (2019)

(Continued)

TABLE 9.8 *(Continued)*
Adsorptive Behavior of Lignocellulosic Materials in Heavy Metal Removal

Adsorbent	Experimental Condition	Adsorption Isotherm, Kinetics	Adsorption Capacity or Removal %	Thermodynamic Elucidation	Heavy Metal	Reference
H_3PO_4-modified aloe vera waste	Dose: 1.5 g/L pH: 4 (for U(VI)) and pH: 5 (for Cd(II)) Temp: 20°C Contact time: 24 h	Pseudo-second order, Langmuir isotherm	208 mg/g 66.2 mg/g	Endothermic Spontaneous	U(VI) Cd(II)	Noli et al. (2019)
NaOH-modified aloe vera waste	Dose: 1.5 g/L pH: 4 (for U(VI)) and pH: 5 (for Cd(II)) Temp: 20°C Contact time: 24 h	Pseudo-second order, Langmuir isotherm	370.4 mg/g 104.2 mg/g	Endothermic Spontaneous	U(VI) Cd(II)	Noli et al. (2019)
Unmodified bark of *Platanus orientalis*	Dose: 2 g/L pH: 5 Temp: 25°C 200 rpm Reaction time: 5 h	Freundlich isotherm	13.42 mg/g 89.6 %		Cr(VI)	Akar et al. (2019)
Carboxylate parenchyma cellulose from bagasse pith	Dose: 10 g/L pH: 5 Temp: 25°C 2000 rpm Reaction time: 3 h Initial concentration: 100 mg/L	Pseudo-second order, Freundlich isotherm	263.16 mg/g	Exothermic Spontaneous	Cu(II)	Gao et al. (2018)

(Continued)

TABLE 9.8 *(Continued)*
Adsorptive Behavior of Lignocellulosic Materials in Heavy Metal Removal

Adsorbent	Experimental Condition	Adsorption Isotherm, Kinetics	Adsorption Capacity or Removal %	Thermodynamic Elucidation	Heavy Metal	Reference
Rice stalk	Dose: 2 g/L pH: 5.5 Temp: 25±2°C 150 rpm Reaction time: 24 h Initial concentration: 100 mg/L	Pseudo-second order, Langmuir isotherm	127.57 mg/g		Pb(II)	Liu et al. (2020)
Rice husk	Dose: 2 g/L pH: 5.5 Temp: 25±2°C 150 rpm Reaction time: 24 h Initial concentration: 100 mg/L	Pseudo-second order, Langmuir isotherm	30.40 mg/g		Pb(II)	Liu et al. (2020)
Saw dust	Dose: 2 g/L pH: 5.5 Temp: 25±2°C 150 rpm Reaction time: 24 h Initial concentration: 100 mg/L	Pseudo-second order, Freundlich isotherm	24.05 mg/g		Pb(II)	Liu et al. (2020)
Spent coffee grounds	Dose: 2 g/L pH: 6.5 Temp: 37°C 300 rpm Reaction time: 2 h Initial concentration: 30 mg/L	Pseudo-second order, Freundlich isotherm	22.75 mg/g	Endothermic Spontaneous	Cr(VI)	Krishna Mohan et al. (2019)

(Continued)

TABLE 9.8 *(Continued)*
Adsorptive Behavior of Lignocellulosic Materials in Heavy Metal Removal

Adsorbent	Experimental Condition	Adsorption Isotherm, Kinetics	Adsorption Capacity or Removal %	Thermodynamic Elucidation	Heavy Metal	Reference
Ultrasound-assisted xanthated alkali-treated camphor leaf	Dose: 2 g/L pH: 6 Temp: 30±3°C 150 rpm Equilibrium time: 0.5 h Initial concentration: 200 mg/L	Pseudo-second order, Langmuir isotherm	134.41 mg/g		Pb(II)	Wang et al. (2018)
Carboxymethylated pulp	pH: 5.0 and 6.0 Temp: 55°C 150 rpm Equilibrium time: 1 h Initial concentration: 50 mg/L	Pseudo-second order, Langmuir isotherm	89.89 mg/g 94.88 mg/g	Endothermic Spontaneous	Cu(II) Pb(II)	Wang et al. (2019)
Walnut shell carbon microsphere	Dose: 0.25 g/L pH: 5 Temp: 40°C 150 rpm Contact time: 6 h Initial concentration: 50 mg/L	Pseudo-second order, Langmuir isotherm	345 mg/g 94.2% 638 mg/g 97% 792 mg/g 99.8% 574 mg/g 98.8%		Cu(II) Pb(II) Cr(III) Cd(II)	Zbair et al. (2019)

(Continued)

TABLE 9.8 *(Continued)*
Adsorptive Behavior of Lignocellulosic Materials in Heavy Metal Removal

Adsorbent	Experimental Condition	Adsorption Isotherm, Kinetics	Adsorption Capacity or Removal %	Thermodynamic Elucidation	Heavy Metal	Reference
Comamonas testosteroni FJ17-modified rice straw	Dose: 12.6 g/L pH: 5 Temp: 40°C 150 rpm Contact time: 4 h Initial concentration: 100 mg/L	Pseudo-second order, Freundlich isotherm	28.4 mg/g	Endothermic Spontaneous	Cu(II)	Xue et al. (2020)
Citrus limetta peel	Dose: 12.6 g/L pH: 3–10 Temp: 25°C 90 rpm Contact time: 6 h Initial concentration: 0.05-2 mg/L	Pseudo-second order, Langmuir isotherm	714.28 μg/g 2000 μg/g	Exothermic Endothermic, Spontaneous	As(III) As(V)	Verma et al. (2019)
Citrus limetta pulp	Dose: 12.6 g/L pH: 3-10 Temp: 25°C 90 rpm Contact time: 6 h Initial concentration: 0.05–2 mg/L	Pseudo-second order, Freundlich isotherm	526.31 μg/g 2000 μg/g	Endothermic Spontaneous	As(III) As(V)	Verma et al. (2019)

(Continued)

TABLE 9.8 *(Continued)*
Adsorptive Behavior of Lignocellulosic Materials in Heavy Metal Removal

Adsorbent	Experimental Condition	Adsorption Isotherm, Kinetics	Adsorption Capacity or Removal %	Thermodynamic Elucidation	Heavy Metal	Reference
Unmodified sugarcane bagasse	Dose: 5 g/L pH: 5 Temp: 25°C 150–160 rpm Contact time: 1 h Initial concentration: 10 mg/L	Langmuir isotherm	4.89 mg/g 88.9%	Endothermic	Cu(II)	Gupta et al. (2018)
Citric acid-modified sugarcane bagasse	Dose: 5 g/L pH: 5 Temp: 25°C 150–160 rpm Contact time: 1 h Initial concentration: 10 mg/L	Langmuir isotherm	5.35 mg/g 96.9%	Endothermic	Cu(II)	Gupta et al. (2018)
NaOH-modified sugarcane bagasse	Dose: 5 g/L pH: 5 Temp: 25°C 150–160 rpm Contact time: 1 h Initial concentration: 10 mg/L	Langmuir isotherm	2.06 mg/g 94.8%	Endothermic	Cu(II)	Gupta et al. (2018)

(Continued)

TABLE 9.8 *(Continued)*
Adsorptive Behavior of Lignocellulosic Materials in Heavy Metal Removal

Adsorbent	Experimental Condition	Adsorption Isotherm, Kinetics	Adsorption Capacity or Removal %	Thermodynamic Elucidation	Heavy Metal	Reference
Pea pod peels	Dose: 10 g/L pH: 2.5–3.5 Temp: 25°C 80–100 rpm Contact time: 0.75 h		4.33 mg/g		Cr(VI)	Sharma and Ayub (2019)
Tea and ginger mix	Dose: 10 g/L pH: 2.5–3.5 Temp: 25°C 80–100 rpm Contact time: 0.75 h		7.29 mg/g		Cr(VI)	Sharma and Ayub (2019)
Banana peels	Dose: 10 g/L pH: 2.5–3.5 Temp: 25°C 80-100 rpm Contact time: 0.75 h		10 mg/g		Cr(VI)	Sharma and Ayub (2019)

Adsorption isotherms, on the other hand, characterize the interaction of an adsorbate and an adsorbent in any system. The isotherm models are based on the adsorption mechanism of metal ions and are capable of not only representing but also explaining and predicting experimental behavior. It describes the equilibrium between the concentration of a solute absorbed in the solid phase and the concentration in solution. The Langmuir, Freundlich, Dubinin-Radushkevich, Sips, Temkin and Redlich-Peterson isotherms are some of the more often utilized adsorption isotherms (Table 9.8). These models can be used to determine the metal adsorption capacity of various species and the differences in metal uptake between them.

9.3.3 Effects of Process Variables on Heavy Metal Removal

The main factors of adsorption for heavy metals are solution pH, temperature, contact time, adsorbent dose, initial concentration and coexisting ions, all of which affect the performance of heavy metal removal (Shang et al., 2017; Shakoor et al., 2018; Sharma and Ayub, 2019). Aside from these factors, mixing speed, flow rate and bed height in column studies all have an impact on the adsorption process (Yahya et al., 2020; Yu et al., 2020).

9.3.3.1 Effect of pH

The pH value reflects the concentration of H^+ in water and is an important parameter in the sorption uptake of metals by the majority of biomasses, as ion exchange and electrostatic interactions are greatly influenced by H^+ concentration. The total charge of the biosorbent, which can retain or release protons to the medium depending on its acidity constant, and the solubility and distribution of metal ion species in solution are all greatly affected by solution pH. The pH_{ZPC} values of various lignocellulosic materials vary. It is known that the surface of any given particle and its amphoteric character are neutral at $pH = pH_{ZPC}$, negatively charged at $pH > pH_{ZPC}$ and positively charged at $pH < pH_{ZPC}$. $pH < pH_{ZPC}$ refers to increased H^+ exchange with heavy metal cations and protonation of surface functional groups (e.g., $-NH_2$ and $-OH$) of carbon-based materials, resulting in an increase in the number of surface positive charges (Ma et al., 2019; Yang et al., 2019). This results in decreased adsorption of heavy metal cations in water (Duan et al., 2020), but increased anionic adsorption. When pH exceeds pH_{ZPC}, deprotonation restores the negative charge to functional groups, while cations and anions experience the opposite effect (Ma et al., 2019). As a result, in cation adsorption, the pH_{ZPC} of lignocellulosic adsorbents should be more acidic for a wider operational pH range; conversely, for anionic heavy metals, the pH_{ZPC} should be more alkaline (Sharma and Ayub, 2019). The different ionic forms of heavy metal ions have different affinity to functional groups at different pH levels (Shang et al., 2017), affecting the adsorption capacity. According to Shakoor et al. (2018), increasing the pH from 3 to 8.2 leads to an increase in the percentage removal of As(III), and increasing the pH further leads to a decrease in the removal efficiency by xanthated water melon rind (X-WMR) at a higher rate. However, above pH 6.3, As(V) sorption decreased significantly for the sorbent X-WMR by up to 78%. Liu et al. (2021) discovered

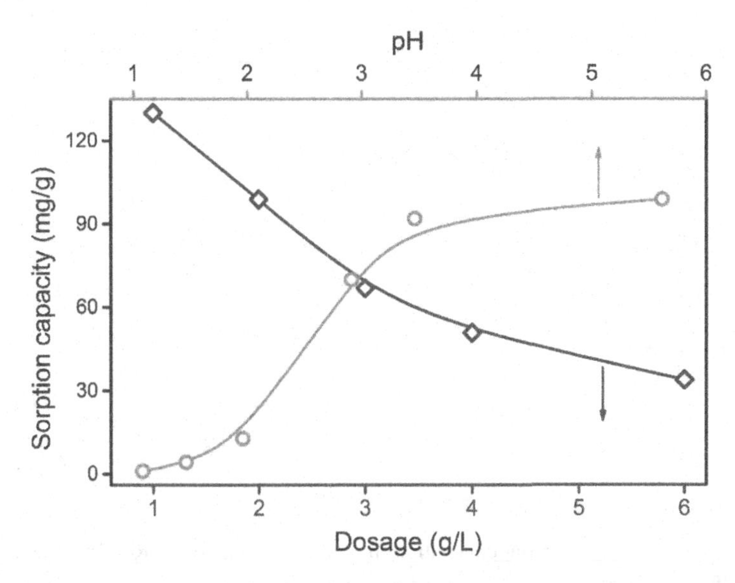

FIGURE 9.10 Effect of solution pH on the adsorption capacity for the removal of Pb(II) ions. (Reprinted with permission from Wang et al., 2018.)

that increasing the pH of corn stalk biochar and rice husk biochar increased their adsorption capacities up to a pH of 4, but then decreased. Sharma and Ayub (2019) discovered that a pH range of 2 to 3 was optimal for Cr(VI) removal by agro-horticultural waste. Similarly, Wang et al. (2018) discovered that Pb(II) adsorption capacity was pH dependent and decreased with decreasing pH, particularly below 3, as shown in Figure 9.10.

9.3.3.2 Effect of Temperature

The influence of temperature on adsorption is related to the thermodynamic nature of the adsorption process and various parameters, such as Gibbs free energy change (ΔG, kJ/mol), enthalpy change (ΔH, kJ/mol) and entropy (ΔS, J/mol K), which can be calculated using thermodynamic equations (Equations. 9.13 and 9.14) as stated in Section 9.2.4.3. Shakoor et al. (2018) discovered $\Delta G < 0$, $\Delta H > 0$ and $\Delta S > 0$ for As(III) and As(V) adsorption by the most potential adsorbents, X-WMR and CA-WMR, indicating an endothermic, spontaneous adsorption process and an increase in randomness at the solid-liquid interface, as well as structural changes in As(III)/As(V).

9.3.3.3 Effect of Initial Concentration

The initial concentration is an essential driving force in the process of improving heavy metal adsorption. Several studies have discovered that, within a certain range, the adsorption capacity for heavy metals increases with increasing initial concentration, but then begins to decrease (Alhogbi, 2017). As a result, higher initial concentrations contribute to increased power transfer from solution to adsorbent surface (Hayati et al., 2017); however, as the initial concentration increases further, all adsorption sites are occupied by the firstly adsorbed ions, resulting in a hindered

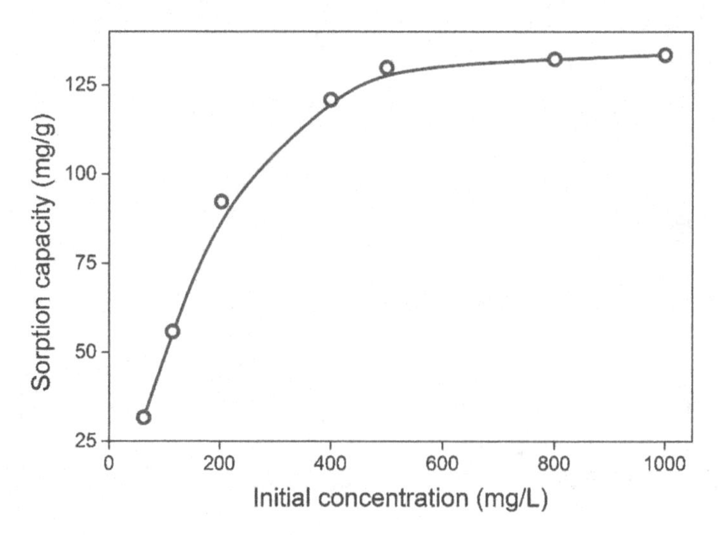

FIGURE 9.11 Effect of initial concentration on Pb(II) adsorption. (Reprinted with permission from Wang et al., 2018.)

interaction between those unadsorbed ions and the adsorbent. Many studies have shown that increasing the initial concentration of heavy metal reduces adsorption efficiency (Shakoor et al., 2018). Aside from that, it was discovered that changing the target metal concentration affected the adsorption isotherm model by causing the adsorption transition from single molecular layer to multi-molecular layer as the initial concentration of adsorbates increased.

Wang et al. (2018) investigated the adsorption capability of ultrasound-assisted xanthated alkali-treated camphor leaf in a batch adsorption system for the removal of Pb(II). The adsorption capacity of Pb(II) ions was analyzed as a function of initial Pb(II) concentrations ranging from 50 to 1000 mg/L. Figure 9.11 shows that more metal was taken up at higher concentrations, which could be because there was more mass transfer driving force.

9.3.3.4 Effect of Contact Time

Contact time is an important variable to consider when designing a cost-effective bioadsorbent for a heavy metal effluent treatment system. The data from the contact time studies can be used to calculate the rate at which heavy metal ions are removed from the effluent. As more time is required to fully adsorb the heavy metal ions onto the adsorbent, the adsorption capacity increases rapidly and reaches a stable position. Figures 9.12 A, B depict the relationship between adsorption capacity and contact time for the removal of Pb(II) ions, which was studied by two distinct groups of researchers. The maximum adsorption capacity was achieved in a short period of time, which could be attributed to charge neutralization. As contact time was increased, a gradual improvement in adsorption capacity was observed, which could be attributed to physical adsorption, and finally, equilibrium was achieved between the heavy metal ions and the bioadsorbent.

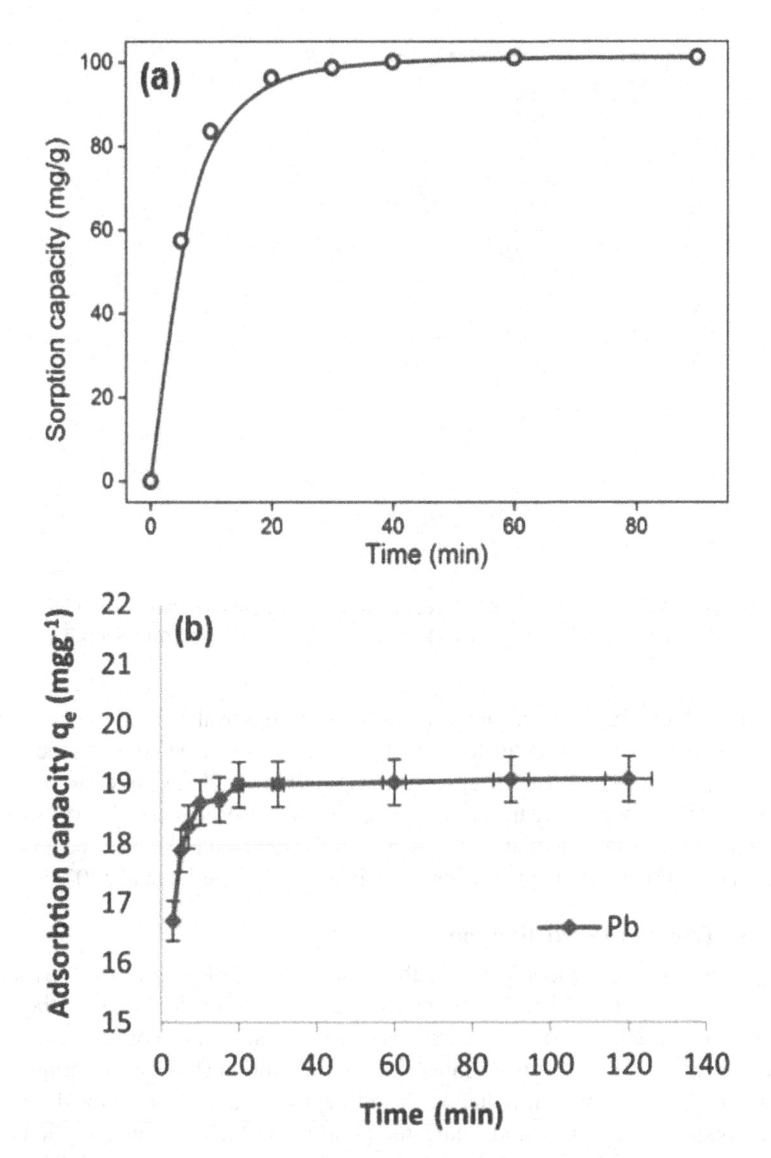

FIGURE 9.12 Effects of contact time on the adsorption of (a) Pb(II) onto ultrasound-assisted xanthated alkali-treated camphor leaf (reprinted with permission from Wang et al., 2018), and (b) Pb(II) onto coffee husk biomass waste. (Reprinted with permission from Alhogbi, 2017.)

9.3.3.5 Effect of Adsorbent Dose

Adsorbent dose is directly related to removal efficiency and adsorption capacity up to a certain limit by increasing the availability of active sites; however, higher doses can cause particle agglomeration and a trimming of active surface area, affecting the availability of binding sites and reducing removal efficiency. Some studies also found that

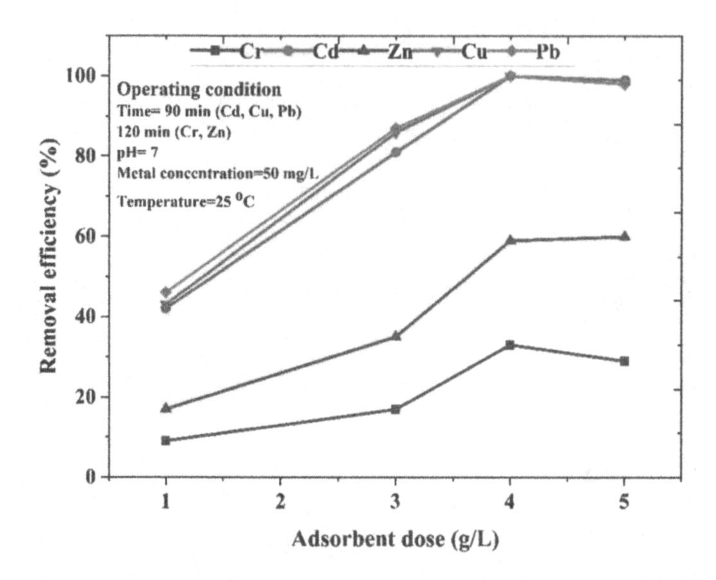

FIGURE 9.13 Effect of adsorbent dose on the removal efficiencies of Cr(VI), Cd(II), Zn(II), Cu(II) and Pb(II) by *Allium cepa* seeds. (Reprinted with permission from Sheikh et al., 2021.)

increasing adsorbent doses resulted in a decrease in removal efficiency or adsorption capacity, which can be attributed to unsaturation adsorption due to the presence of multiple adsorption sites and particle aggregation (Kadir et al., 2012). As shown in Figures 9.10 and 9.13, the sorption capacity decreased as the adsorbent dosage was increased for Pb(II) removal (Wang et al., 2018), whereas the removal efficiency increased with increasing adsorbent dosage for various metal removals (Sheikh et al., 2021).

9.3.3.6 Effect of Coexisting Ions

In solution, ionic strength represents the concentration of ions in the surrounding environment, which influences the interface potential and thickness of the double layer (Liu et al., 2020). Coexisting cations or anions play an important role in heavy metal ion removal by creating competition for adsorption sites and forming coordination ions or precipitation, reducing the adsorption capacity or removal efficiency of adsorbent. When two or more heavy metals are present, high concentrations of K^+, Na^+, Ca^{2+} and Mg^{2+} perform a relative decrease based on ionic potential, ionic radius and ion species of those cations when compared to a single system. The presence of anions would result in interactions with heavy metal cations to form coordination ions or precipitate, as well as competition for adsorption sites with heavy metal anions. Liu et al. (2020) discovered that increasing ionic strength due to Na^+ had an inhibitory effect on Pb(II) adsorption. Additionally, Krishna Mohan et al. (2019) discovered that anions interfere in the following order: $SO_4^{2-} > NO_3^- > Cl^- = HCO_3^- > PO_4^{3-}$, while cations interfere in the following order: $Zn^{2+} > Cu^{2+} > Mg^{2+} > Ca^{2+} > Fe^{3+}$. As shown in Figure 9.14, Wang et al. (2018) also demonstrated the effect of coexisting ions (K^+, Mg^{2+} and Cu^{2+}) on Pb(II) sorption. The authors discovered that the K^+ ion has little effect on the sorption capacity of Pb(II), whereas divalent Mg^{2+}

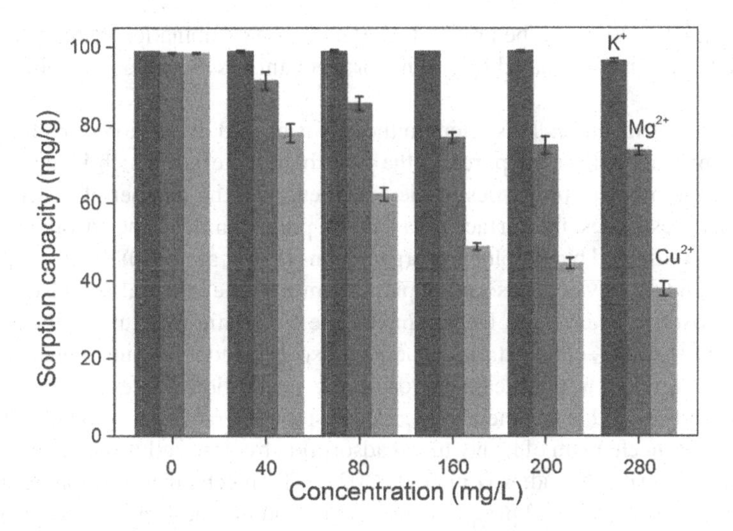

FIGURE 9.14 Effect of coexisting ions on the adsorption of Pb(II). (Reprinted with permission from Wang et al., 2018.)

and Cu^{2+} ions gradually decrease the sorption capacity, particularly Cu^{2+}, indicating competitive sorption between divalent coexisting ions and Pb(II).

Sanka et al. (2020) used real wastewater in their study, with initial pollutant concentrations of 1.82 mg/L for Cr(total), 9.28 mg/L for Fe(II) and 1.59 mg/L for Pb(II), all of which were far above the WHO thresholds, and other wastewater characteristics included the following: temperature 24°C, EC 2674 μS/cm, TDS value 1472 mg/L, initial wastewater pH 7.37, wastewater turbidity 8480 NTU and dissolved oxygen 0.6 mg/L. Using rice husk and corn husk biochar pyrolyzed at different temperatures, removal efficiencies of 50–65% and 20–40% for Cr(total), respectively, 70–90% for Fe(II) and approximately 90% for Pb(II) were found in all cases. Similarly, Wang et al. (2020) investigated the influence of coexisting anions like NO_3^- and PO_4^{3-} in the range of 0.001–0.1 M on the adsorption of Pb(II) and U(VI) using rice husk biochar in batch adsorption studies.

9.3.4 Adsorption Mechanisms Associated with the Adsorption of Metal Ions

Depending on the nature of the adsorbent, its functional groups and the solution pH, a variety of mechanisms, such as acid-base, electrostatic, van der Waals, hydrophobic, ion exchange, surface complexation, precipitation and H-bonds, as well as n-π and π-π EDA interactions, can be used to describe the adsorption of heavy metal contaminants from aqueous solutions onto the surface of carbon-based adsorbents. Depending on the type of bioadsorbent used, various functional groups such as amide, amine, carbonyl, hydroxyl, imine, sulfonate, carboxyl, phosphate thioether, phenolic and phosphodiester groups can influence physical or chemical adsorption efficiency by attracting metal ions to the surfaces. Important analyses such as pH_{ZPC}, FTIR analysis, kinetic and isotherm studies help in understanding the mechanisms involved. Adsorption can be physical, chemical or a combination of the two. Heavy metal adsorption is mediated by five major mechanisms, which are as follows:

Physical adsorption: This refers to the diffusion and deposition of metal contaminants within the pores of the adsorbent material, which is dependent on the porosity properties of the adsorbent, i.e., the number of micropores and mesopores, the surface area and the pore volume of the adsorbent, and is not affected by dissoluble inorganic ions (Liu et al., 2020). In the majority of studies, physical adsorption played a minor role in metal ion adsorption.

Electrostatic interaction: Depending on the electronic structure of the adsorbent material, electrostatic attraction, also referred to as outer sphere complexation, is a significant factor in the adsorption process between the adsorbent and heavy metal ions. Electrostatic interaction was identified as a major mechanism of heavy metal adsorption by lignocellulosic materials in the majority of studies (Liu et al., 2021). The mechanism is strongly influenced by the pH and pH_{ZPC} of adsorbents, and interactions between protonated or deprotonated functional groups present in adsorbents and heavy metal ions can result in electrostatic attraction.

Ion exchange: Ion exchange occurs most frequently between metal cations and the proton of oxygen-containing functional groups, such as –COOH and –OH), and/or metal mineral ions, such as K^+, Na^+, Ca^{2+} and Mg^{2+}. Liu et al. (2020) used an acid-wash process with a 1 M HCl solution to demineralize biochar samples and remove the Pb-containing minerals formed during ion exchange.

Surface complexation: Surface complexation refers to the binding of heavy metals and functional groups (e.g., –OH, –COOH, –O–, and –CO-NH–) on the adsorbent surface. In reaction processes, surface complexation results in the formation of multi-atom structures. Certain reports indicated that Cd^{2+}, Cu^{2+} and Ni^{2+} complexed with –OH and –COOH. The XPS peak at 139.6 eV and the XRD analysis indicated that the adsorption mechanism was surface complexation (Wang et al., 2018).

Precipitation/coprecipitation: Heavy metal ions have the potential to form solid precipitates or coprecipitations with anions (e.g., CO_3^{2-}, PO_4^{3-} and OH^-) and function groups present on the adsorbent surface. XPS measurements revealed the presence of a main peak at –139 eV (Pb-O), indicating that precipitation, in addition to cation release, is the dominant adsorption mechanism (Wang et al., 2018).

9.3.5 Regeneration and Reusability Possibilities of Exhausted Bioadsorbents Loaded with Metal Ions

The ability of an adsorbent to regenerate means that it can desorb adsorbates from its surface and reuse it. Depending on the quality of the water or effluent, the type of adsorbent, further treatment goals and regeneration costs, various regeneration techniques such as ball milling, steam desorption, solvent extraction, electrochemical treatment, pH adjustment, ultrasonic irradiation and microbiological methods may

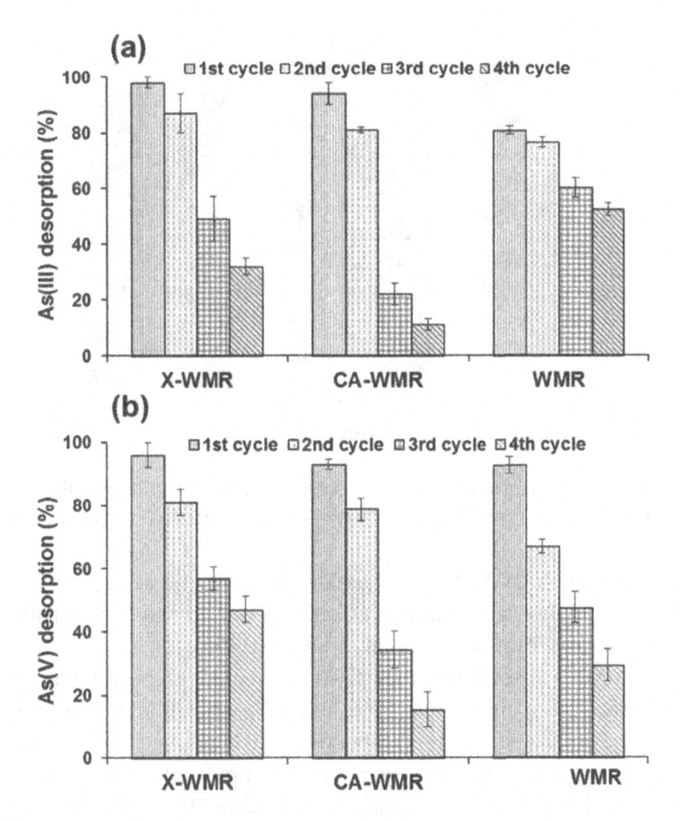

FIGURE 9.15 Desorption efficiencies of (a) As(III) and (b) As(V) from the WMR, X-WMR and CA-WMR sorbents. (Reprinted with permission from Shakoor et al., 2018.)

be used. Shakoor et al. (2018) conducted desorption studies with 0.1 M NaOH and discovered that the desorption efficiencies for As(III) were 98% for xanthated water melon rind (X-WMR), 94% for citric acid-modified water melon rind (CA-WMR) and 97% percent for unmodified water melon rind (WMR) after the first adsorption-desorption cycle. Desorption efficiencies for As(V) were 96%, 93% and 98%, respectively (Figure 9.15). Additionally, the bioadsorbent was regenerated using sodium hydroxide (NaOH) and the disodium salt of ethylenediaminetetraacetic acid (EDTA) as desorbing agents (Gao et al., 2018).

According to Noli et al. (2019), EDTA as a desorption agent recovered 95–98% of Cd(II) and U(VI) from aloe vera and 10–14% from modified aloe vera, whereas H_2SO_4 recovered 45–49% and 97–99%, respectively. The regeneration study by Zbair et al. (2019) with 100 mL 0.5 M HCl resulted in a slight decrease in removal efficiency of 0.97%, 1.22%, 0.32% and 0.1% even after four cycles of adsorption, and the resulting adsorption efficiencies were around 92.4%, 96.4%, 92.8% and 99.1% for Cu(II), Cd(II), Pb(II) and Cr(III), respectively. Krishna Mohan et al. (2019) discovered that 0.1 M NaOH was effective in the regeneration of the spent coffee ground bioadsorbent for Cr(VI) removal. Even after five generations, the percentage removal was found to be more than 81.0%. Similarly, Wang et al. (2018) discovered

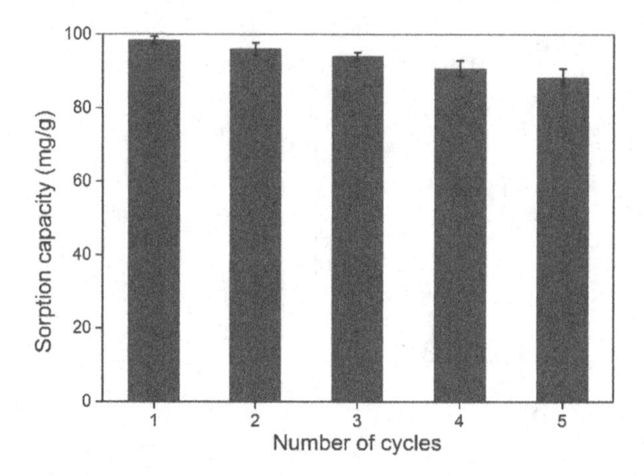

FIGURE 9.16 Cycle performance of an ultrasound-assisted xanthated alkali-treated camphor leaf for Pb(II) adsorption. (Reprinted with permission from Wang et al., 2018.)

that when treated with 0.05 M ethylenediaminetetraacetic acid disodium salt solution, xanthated camphor leaf could be effectively reused and high sorption capacity was retained after five cycles, as shown in Figure 9.16.

Sun et al. (2019) discovered that $KMnO_4$-treated magnetic biochar retained adsorption capacities of more than 50% and 87% after four cycles of adsorption-desorption for Pb(II) and Cd(II), respectively, where regeneration was done by CH_3COONa. The adsorption capacity of Pb(II) on magnetic rice husk biochar with HCl as a reagent showed a slight decrease from 47.3 to 40.8 mg/g even after five consecutive adsorption-desorption cycles, but not for U(VI), which showed a significant decrease from 28.1 to 13.7 mg/g (Wang et al., 2018).

9.4 COST ANALYSIS

Lignocellulosic materials are significantly less expensive than commercially available activated carbon adsorbents. Many researchers have conducted economic analyses to demonstrate the cost-effectiveness. The following items are included in the cost analysis:

Cost of raw material: This is the cost of obtaining raw materials for the preparation of adsorbents. Most studies have only included the purchase price because they are primarily based on lab studies. The cost of transportation should also be considered here.

Cost of reagents: This includes the cost of various acids, bases and salt reagents used for preparation, pH or other process optimization, and adsorbent regeneration.

Other costs include drying costs, grinding costs, heating costs and other overhead costs (10–15% is taken generally). Tables 9.9 and 9.10 show a rough cost estimate for the preparation of lignocellulosic bioadsorbents and a comparison with commercial activated carbon.

TABLE 9.9
Cost Estimation of Some Lignocellulosic Bioadsorbents

	PPP		T&G		BP		SG		ASG		BSG	
Item	Amount	Cost (INR)	Amount	Cost (INR)	Amount	Cost (INR)	Amount	Cost (INR)	Amount	Cost (INR)	Amount	Cost (INR)
Raw material	1 kg	–	1 kg	–	1 kg	–	1 kg	1.8	1 kg	1.8	1 kg	1.8
H_2SO_4	1.5 L	9.00	1.4 L	8.40	1.2 L	7.20	–	–	–	–	–	–
NaOH	5 L	8.00	4 L	6.40	3 L	4.80	–	–	–	–	0.1 kg	49.2
Citric acid	–	–	–	–	–	–	–	–	0.526 kg	236.7	–	–
Drying cost	1 kWh (10 h)	5.80	1 kWh (10 h)	5.80	1 kWh (10 h)	5.80	1 kg	30	1 kg	30	1 kg	30
Heating cost	1.2 kWh (10 h)	6.96	1.2 kWh (10 h)	6.96	1.2 kWh (10 h)	6.96	–	–	–	–	–	–
Grinding cost	0.8 kWh (1 h)	4.64	0.4 kWh (1 h)	2.32	0.4 kWh (1 h)	2.32	1 kg	13	1 kg	13	1 kg	13
Net cost (INR)	–	34.40	1 kg	29.88	1 kg	27.08	1 kg	44.80	1 kg	281.5	1 kg	94
Other overhead cost (INR)	15% of net cost	5.16	15% of net cost	4.48	15% of net cost	4.06	10% of net cost	4.48	10% of net cost	28.15	1 kg	9.4
Total cost (INR)	–	39.56	–	34.36	–	31.14	–	–	–	309.65	–	103.40
Reference	Sharma and Ayub (2019)						Gupta et al. (2018)					

ASG: Acid-modified sugarcane bagasse; BP: Banana peel waste; BSG: Base-modified sugarcane bagasse; PPP: Pea pod peel waste; SG: Sugarcane bagasse; T&G: Used tea and ginger waste.

TABLE 9.10
Cost of Bioadsorbent for the Removal of 1 g of Heavy Metal Ions

Bioadsorbent	Sorption Capacity mg/g	Cost of Bioadsorbent (INR/kg)	Cost of Bioadsorbent for Removing 1 g of Heavy Metal (INR)	Reference
PPP	4.33	39.56	9.14 (Cr(VI))	Sharma and Ayub (2019)
T&G	7.30	34.36	4.71 (Cr(VI))	Sharma and Ayub (2019)
BP	10.00	31.14	3.11 (Cr(VI))	Sharma and Ayub (2019)
SG	4.84	49.28	10.18 (Cu(II))	Gupta et al. (2018)
ASG	5.35	309.65	57.88 (Cu(II))	Gupta et al. (2018)
BSG	2.06	103.40	50.19 (Cu(II))	Gupta et al. (2018)
AC	5.62	3650	649.47 (Cu(II))	Gupta et al. (2018)

AC: Activated carbon; ASG: Acid-modified sugarcane bagasse; BP: Banana peel waste; BSG: Base-modified sugarcane bagasse; PPP: pea pod peel waste; SG: Sugarcane bagasse; T&G: Used tea and ginger waste.

Sharma and Ayub (2019) prepared the adsorbents discussed in Table 9.9 by sun drying them for 15 d, then drying them in an oven for 5 h daily at 95°C for 3 d and finally drying them in the sunlight for a week after a repeated distilled water wash. After sun drying, the adsorbents were powdered and washed twice with 0.1 M NaOH and 0.1 M H_2SO_4 to remove lignin and alkalinity. The powdered adsorbents were dried and stored in desiccators after being washed with distilled water. These ready sorbents were sieved through an Indian standard sieve to separate the grains before use. The authors then conducted batch adsorption experiments of chromium using the standard method at 25–28°C to study the effects of various parameters such as adsorbent dose, pH, grain size, contact time, temperature and the mixing speed using chromium standard solution for different chromium concentrations, and thus optimum values of all parameters were determined to be used for adsorption isotherms. The samples were filtered using Whatman No.1 filter paper for all optimized affecting parameters, and the filtrates were analyzed for residual chromium using atomic absorption spectroscopy (AAS). The adsorption isotherm was performed under the fixed environmental conditions demonstrated in the preceding procedures. The Langmuir and Freundlich models were used to determine which isotherm was best for chromium adsorption. The maximum monolayer capacity was determined by correlating the Langmuir isotherms for PPP, T&G and BW, which were 4.33 mg/g, 7.30 mg/g and 10.00 mg/g, respectively.

Gupta et al. (2018) collected the sugarcane bagasse from the local market, cut it into pieces and washed it several times with tap water before washing it with distilled water to remove dirt. The wet bagasse was sun dried to remove the water before being oven dried for 48 h at 60°C. The dried bagasse was ground and sieved to achieve particle sizes ranging from 150 to 300 μm. The adsorbent created in this manner was known as SG. A portion of SG was modified with citric acid by soaking it in a 0.1 M citric acid solution for 24 h and then in distilled water for 4 h to remove the free citric acid and finally drying it in an oven at 60°C for 48 h. The adsorbent created in this manner was designated as ASG. BSG was sodium hydroxide modified SG.

It was prepared by soaking SG in a 0.1 M NaOH solution for 24 h, followed by soaking in distilled water and repeated washing until the wash water became almost colorless. BSG was then dried in an oven at 60°C for 48 h before being used in experiments. Following that, the authors placed a known amount of adsorbent and 50 mL of a known concentration of Cu(II) solution in 250-mL flasks. A water bath shaker was used to agitate the contents of the flasks at 150–160 rpm. The pH of the solution was adjusted with either 0.1 M NaOH or 0.1 M HCl. The samples were drawn as directed by the experiment and filtered first with coarse filter paper and then with Whatman No. 42 filter paper. The concentration of Cu(II) in the filtrate was determined using a UV-Vis spectrophotometer following cuprethol method. The adsorption studies with industrial wastewater were carried out under the same conditions as the batch adsorption studies with synthetic effluent, namely at pH 5, contact time 60 min and adsorbent dose 5 g/L. The Cu(II) uptake capacity of AC, SG, ASG and BSG was 5.62 mg/g, 4.84 mg/g, 5.35 mg/g and 2.06 mg/g, respectively.

The use of these lignocellulosic wastes to adsorb metal ions from aqueous solutions aids in solid biowaste management. These wastes not only adsorb metal ions, but they are also recycled and revalued, providing an indirect benefit to society and the environment; otherwise, they are discarded due to the high cost of transportation, which is also a serious environmental concern. It is extracted at a low cost, which provides a direct financial benefit to the metal-handling industry. Because treated water is of such high quality, it can be reused for a variety of industrial purposes. This advantage also reduces the overall cost of their processes and products. The benefit to the environment is that metals are eliminated from the industry without creating any disturbance to the ecosystem. As a result, in addition to the environmental and monetary benefits, the metal-handling industries play an important role in providing employment and environmental protection at a low cost. This research lays the groundwork for the long-term development of industries, communities and nations. The most important thing about this study is that it meets all of the requirements for sustainable development, which include social, environmental and economic factors as well as other factors.

9.5 CONCLUSION AND FUTURE PROSPECTS

Pharmaceutical pollutants have emerged as a serious concern all over the world, and lignocellulosic waste materials are not only a viable bioadsorbent, but their use will also solve the disposal problem that would otherwise exist. Despite the growing number of studies on lignocellulosic bioadsorbents for the removal of pharmaceutical pollutants, there is a dearth of understanding regarding their practical applications. More research is needed on the toxicity detection and persistence of lesser-known pharmaceutical compounds, particularly to provide a solid foundation and aid decision-making, as well as to assess the environmental impact of various pharmaceutical pollutants and the best treatment technology to ensure better removal. Several studies focusing on practical application in treating real wastewater and regeneration of these adsorbents are still required. Studies on the actual cost of producing bioadsorbent should also be conducted. Heavy metals are dangerous pollutants that can persist in the environment and are resistant to conventional wastewater and drinking water treatment methods. Continuous discharge of these hazardous contaminants poses a significant threat to all living creatures on the planet. Among all available treatment methods, adsorption stands out as a promising, long-term, low-cost, energy-efficient and simple treatment method. The use of lignocellulosic material as a bioadsorbent for heavy metal removal has numerous advantages, including easy availability, improved process control and nutrient independence. In most studies, heavy metal adsorption by lignocellulosic materials followed pseudo-second-order kinetics, with some studies using pseudo-first-order kinetics. Majority of the studies fitted Langmuir and Freundlich isotherm models. Surface complexation, physical adsorption, ion exchange and electrostatic interaction were discovered to be the mechanisms of heavy metal adsorption by lignocellulosic materials. Adsorbent regeneration with various acids and alkalis has been shown to be beneficial even after four to five cycles of reuse. A cost analysis conducted by a few researchers confirmed the cost-effectiveness of bioadsorbents when compared to others. Despite numerous studies utilizing various agro-industrial lignocellulosic waste for heavy metal removal, there is still a research gap. Majority of studies are conducted on synthetic solutions. Despite the fact that some researchers were able to test real industrial effluents, there is still a lack of practical application rather than lab-scale studies. Unmodified bioadsorbents performed less efficiently than modified bioadsorbents, but the latter makes the process more complicated and costly, and it may cause toxicity in the system. There is still a lack of research for the subsequent study, i.e., what the fate of these contaminants bearing adsorbents would be. Toxicity analysis of lignocellulosic material applications has not yet been addressed. Regeneration by acids or alkalis has the potential to cause mass loss, strength loss, functional group deactivation and adsorption capacity loss. As a result, better regeneration methods have yet to be discovered. There were very few articles related to cost analysis that were discovered. A detailed cost analysis for industrial applications is still lacking. Secondary pollution caused by the use of lignocellulosic materials has yet to be investigated.

REFERENCES

Abdelfattah, I., A. A. Ismail, F. Al Sayed, A. Almedolab and K. M. Aboelghait. 2016. Biosorption of heavy metals ions in real industrial wastewater using peanut husk as efficient and cost effective adsorbent. *Environmental Nanotechnology, Monitoring & Management* 6: 176–183. https://doi.org/10.1016/j.enmm.2016.10.007

Abdolali, A., W. S. Guo, H. H. Ngo, S. S. Chen, N. C. Nguyen and K. L. Tung. 2014. Typical lignocellulosic wastes and by-products for biosorption process in water and wastewater treatment: A critical review. *Bioresource Technology* 160: 57–66. https://doi.org/10.1016/j.biortech.2013.12.037

Abo El Naga, A. O., M. El Saied, S. A. Shaban and F. Y. El Kady. 2019. Fast removal of diclofenac sodium from aqueous solution using sugar cane bagasse-derived activated carbon. *Journal of Molecular Liquids* 285: 9–19. https://doi.org/10.1016/j.molliq.2019.04.062

Adewuyi, A. and F. V. Pereira. 2017. Underutilized *Luffa cylindrica* sponge: A local bio-adsorbent for the removal of Pb(II) pollutant from water system. *Beni-Suef University Journal of Basic and Applied Sciences* 6: 118–126. https://doi.org/10.1016/j.bjbas.2017.02.001

Afsa, S., K. Hamden, P. A. L. Martin and H. B. Mansour. 2020. Occurrence of 40 pharmaceutically active compounds in hospital and urban wastewaters and their contribution to Mahdia coastal seawater contamination. *Environmental Science and Pollution Research* 27: 1941–1955. https://doi.org/10.1007/s11356-019-06866-5

Ahmed, M. A. and W. Misganaw. 2019. Chemically modified banana peel as a potential bio-adsorbent to remove chromium (VI) from effluent. *International Journal for Research in Engineering Application & Management* 5(6): 32–38.

Ahmed, M. J. and S. K. Theydan. 2014. Fluoroquinolones antibiotics adsorption onto micro-porous activated carbon from lignocellulosic biomass by microwave pyrolysis. *Journal of the Taiwan Institute of Chemical Engineers* 45(1): 219–226. https://doi.org/10.1016/j.jtice.2013.05.014

Aitken, M., M. Kleinrock, A. Simorellis and D. Nass. 2019. The Global Use of Medicine in 2019 and Outlook to 2023: Forecasts and Areas to Watch. IQVIA Institute for Human Data Science, Parsippany, NJ.

Akar, S., B. Lorestani, S. Sobhanardakani, M. Cheraghi and O. Moradi. 2019. Surveying the efficiency of *Platanus orientalis* bark as biosorbent for Ni and Cr(VI) removal from plating wastewater as a real sample. *Environmental Monitoring and Assessment* 191: 373. https://doi.org/10.1007/s10661-019-7479-z

Alhogbi, B. G. 2017. Potential of coffee husk biomass waste for the adsorption of Pb(II) ion from aqueous solutions. *Sustainable Chemistry and Pharmacy* 6: 21–25. https://doi.org/10.1016/j.scp.2017.06.004

Ali, H., E. Khan and I. Ilahi. 2019. Environmental chemistry and ecotoxicology of hazardous heavy metals: Environmental persistence, toxicity, and bioaccumulation. *Journal of Chemistry* 2019: 6730305. https://doi.org/10.1155/2019/6730305

Al-Qarni, H., P. Collier, J. O'Keeff and J. Akunna. 2016. Investigating the removal of some pharmaceutical compounds in hospital wastewater treatment plants operating in Saudi Arabia. *Environmental Science and Pollution Research* 23: 13003–13014. https://doi.org/10.1007/s11356-016-6389-7

Álvarez-Torrellas, S., A. Rodríguez, G. Ovejero and J. García. 2016. Comparative adsorption performance of ibuprofen and tetracycline from aqueous solution by carbonaceous materials. *Chemical Engineering Journal* 283: 936–947. http://dx.doi.org/10.1016/j.cej.2015.08.023

Amar, M. B., K. Walha and V. Salvadó. 2020. Evaluation of olive stones for Cd(II), Cu(II), Pb(II) and Cr(VI) biosorption from aqueous solution: Equilibrium and kinetics. *International Journal of Environmental Research* 14: 193–204. https://doi.org/10.1007/s41742-020-00246-5

Archana, G., R. Dhodapakar and A. Kumar. 2016. Offline solid-phase extraction for preconcentration of pharmaceuticals and personal care products in environmental water and their simultaneous determination using the reversed phase high-performance liquid chromatography method. *Environmental Monitoring and Assessment* 188: 512. https://doi.org/10.1007/s10661-016-5510-1

Aus der Beek, T., F. A. Weber, A. Bergmann, S. Hickmann, I. Ebert, A. Hein and A. Küster. 2016. Pharmaceuticals in the environment-global occurrences and perspectives. *Environmental Toxicology and Chemistry* 35(4): 823–835. https://doi.org/10.1002/etc.3339

Avrami, M. 1939. Kinetics of phase change. I General theory. *The Journal of Chemical Physics* 7(12): 1103–1112. https://doi.org/10.1063/1.1750380

Baccar, R., M. Sarrà, J. Bouzid, M. Feki and P. Blánquez. 2012. Removal of pharmaceutical compounds by activated carbon prepared from agricultural by-product. *Chemical Engineering Journal* 211–212: 310–317. https://doi.org/10.1016/j.cej.2012.09.099

Balakrishna, K., A. Rath, Y. Praveenkumarreddy, K. S. Guruge and B. Subedi. 2017. A review of the occurrence of pharmaceuticals and personal care products in Indian water bodies. *Ecotoxicology and Environmental Safety* 137: 113–120. https://doi.org/10.1016/j.ecoenv.2016.11.014

Bamdad, S., K. Hawboldt and S. MacQuarrie. 2018. A review on common adsorbents for acid gases removal: Focus on biochar. *Renewable and Sustainable Energy Reviews* 81: 1705–1720. https://doi.org/10.1016/j.rser.2017.05.261

Bashir, A., L. A. Malik, S. Ahad, T. Manzoor, M. A. Bhat, G. N. Dar and A. H. Pandith. 2019. Removal of heavy metal ions from aqueous system by ion-exchange and biosorption methods. *Environmental Chemistry Letters* 17: 729–754. https://doi.org/10.1007/s10311-018-00828-y

Behera, S. K., H. W. Kim, J. E. Oh and H. S. Park. 2011. Occurrence and removal of antibiotics, hormones and several other pharmaceuticals in wastewater treatment plants of the largest industrial city of Korea. *Science of the Total Environment* 409(20): 4351–4360. https://doi.org/10.1016/j.scitotenv.2011.07.015

Boudrahem, F., I. Yahiaoui, S. Saidi, K. Yahiaoui, L. Kaabache, M. Zennache and F. Aissani-Benissad. 2019. Adsorption of pharmaceutical residues on adsorbents prepared from olive stones using mixture design of experiments model. *Water Science and Technology* 80(5): 998–1009. https://doi.org/10.2166/wst.2019.346

Brodin, T., S. Piovano, J. Fick, J. Klaminder, M. Heynen and M. Jonsson. 2014. Ecological effects of pharmaceuticals in aquatic systems-impacts through behavioural alterations. *Philosophical Transactions of the Royal Society B: Biological Sciences* 369(1656): 20130580. https://doi.org/10.1098/rstb.2013.0580

Cabrita, I., B. Ruiz, A. S. Mestre, I. M. Fonseca, A. P. Carvalho and C. O. Ania. 2010. Removal of an analgesic using activated carbons prepared from urban and industrial residues. *Chemical Engineering Journal* 163(3): 249–255. https://doi.org/10.1016/j.cej.2010.07.058

Camacho-Muñoz, D., J. Martín, J. L. Santos, I. Aparicio and E. Alonso. 2014. Concentration evolution of pharmaceutically active compounds in raw urban and industrial wastewater. *Chemosphere* 111: 70–79. https://doi.org/10.1016/j.chemosphere.2014.03.043

Castro, L., M. L. Blázquez, F. González, J. A. Muñoz and A. Ballester. 2017. Biosorption of Zn(II) from industrial effluents using sugar beet pulp and *F. vesiculosus*: From laboratory tests to a pilot approach. *Science of the Total Environment* 598: 856–866. https://doi.org/10.1016/j.scitotenv.2017.04.138

Chaoua, S., S. Boussaa, A. E. Gharmali and A. Boumezzough. 2019. Impact of irrigation with wastewater on accumulation of heavy metals in soil and crops in the region of Marrakech in Morocco. *Journal of the Saudi Society of Agricultural Sciences* 18(4): 429–436. https://doi.org/10.1016/j.jssas.2018.02.003

Cherik, D. and K. Louhab. 2018. A kinetics, isotherms, and thermodynamic study of diclofenac adsorption using activated carbon prepared from olive stones. *Journal of Dispersion Science and Technology* 39(6): 814–825. https://doi.org/10.1080/01932691.2017.1395346

Chitongo, R., B. O. Opeolu and O. S. Olatunji. 2018. Abatement of amoxicillin, ampicillin, and chloramphenicol from aqueous solutions using activated carbon prepared from grape slurry. *CLEAN – Soil, Air, Water* 47(2): 1800077. https://doi.org/10.1002/clen.201800077

Christou, A., A. Agüera, J. M. Bayona, E. Cytryn, V. Fotopoulos, D. Lambropoulou, C. M. Manaia, C. Michael, M. Revitt, P. Schröder and D. Fatta-Kassinos. 2017. The potential implications of reclaimed wastewater reuse for irrigation on the agricultural environment: The knowns and unknowns of the fate of antibiotics and antibiotic resistant bacteria and resistance genes - A review. *Water Research* 123: 448–467. https://doi.org/10.1016/j.watres.2017.07.004

Czech, B., K. Shirvanimoghaddam, E. Trojanowska and M. Naebe. 2020. Sorption of pharmaceuticals and personal care products (PPCPs) onto a sustainable cotton based adsorbent. *Sustainable Chemistry and Pharmacy* 18: 100324. https://doi.org/10.1016/j.scp.2020.100324

Daneshfozoun, S., M. A. Abdullah and B. Abdullah. 2017. Preparation and characterization of magnetic biosorbent based on oil palm empty fruit bunch fibers, cellulose and *Ceiba pentandra* for heavy metal ions removal. *Industrial Crops and Products* 105: 93–103. https://doi.org/10.1016/j.indcrop.2017.05.011

Daughton, C. G. 2013. Pharmaceuticals in the environment: Sources and their management. *Comprehensive Analytical Chemistry* 62: 37–69. https://doi.org/10.1016/B978-0-444-62657-8.00002-1

de Jesus Gaffney, V., C. M. M. Almeida, A. Rodrigues, E. Ferreira, M. J. Benoliel and V. V. Cardoso. 2015. Occurrence of pharmaceuticals in a water supply system and related human health risk assessment. *Water Research* 72: 199–208. https://doi.org/10.1016/j.watres.2014.10.027

Duan, C., T. Ma, J. Wang and Y. Zhou. 2020. Removal of heavy metals from aqueous solution using carbon-based adsorbents: A review. *Journal of Water Process Engineering* 37: 101339. https://doi.org/10.1016/j.jwpe.2020.101339

Dubinin, M. M. 1960. The potential theory of adsorption of gases and vapors for adsorbents with energetically non-uniform surface. *Chemical Reviews* 60(2): 235–241. https://doi.org/10.1021/cr60204a006

Dubinin, M. M. and L. V. Radushkevich. 1947. Equation of the characteristic curve of activated charcoal. *Proceedings of the Academy of Sciences, Physical Chemistry Section USSR* 55: 331–333.

Edition, F. 2011. Guidelines for drinking-water quality. *WHO Chronicle* 38: 104–108.

Elizalde-González, M. P. and V. Hernández-Montoya. 2009. Guava seed as an adsorbent and as a precursor of carbon for the adsorption of acid dyes. *Bioresource Technology* 100(7): 2111–2117. https://doi.org/10.1016/j.biortech.2008.10.056

Ezemonye, L. I., P. O. Adebayo, A. A. Enuneku, I. Tongo and E. Ogbomida. 2019. Potential health risk consequences of heavy metal concentrations in surface water, shrimp (*Macrobrachium macrobrachion*) and fish (*Brycinus longipinnis*) from Benin River, Nigeria. *Toxicology Reports* 6: 1–9. https://doi.org/10.1016/j.toxrep.2018.11.010

Fang, L., L. Li, Z. Qu, H. Xu, J. Xu and N. Yan. 2018. A novel method for the sequential removal and separation of multiple heavy metals from wastewater. *Journal of Hazardous Materials* 342: 617–624. https://doi.org/10.1016/j.jhazmat.2017.08.072

Fan, L., Y. Zhang, X. Li, C. Luo, F. Lu and H. Qiu. 2012. Removal of alizarin red from water environment using magnetic chitosan with alizarin red as imprinted molecules. *Colloids and Surfaces B: Biointerfaces* 91: 250–257. https://doi.org/10.1016/j.colsurfb.2011.11.014

Fernandes, M. J., M. M. Moreira, P. Paíga, D. Dias, M. Bernardo, M. Carvalho, N. Lapa, I. Fonseca, S. Morais, S. Figueiredo and C. Delerue-Matos. 2019. Evaluation of the adsorption potential of biochars prepared from forest and agri-food wastes for the removal of fluoxetine. *Bioresource Technology* 292: 121973. https://doi.org/10.1016/j.biortech.2019.121973

Fernández-López, C., J. M. Guillén-Navarro, J. J. Padilla and J. R. Parsons. 2016. Comparison of the removal efficiencies of selected pharmaceuticals in wastewater treatment plants in the region of Murcia, Spain. *Ecological Engineering* 95: 811–816. https://doi.org/10.1016/j.ecoleng.2016.06.093

Freundlich, H. M. F. 1906. Over the adsorption in solution. *Journal of Physical Chemistry* 57: 385–470.

Ganesan, S., K. Karthick, C. Namasivayam, L. A. Pragasan, V. S. Kirankumar, S. Devaraj and V. K. Ponnusamy. 2020. Discarded biodiesel waste-derived lignocellulosic biomass as effective biosorbent for removal of sulfamethoxazole drug. *Environmental Science and Pollution Research* 27: 17619–17630. https://doi.org/10.1007/s11356-019-07022-9

Gao, X., H. Zhang, K. Chen, J. Zhou and Q. Liu. 2018. Removal of heavy metal and sulfate ions by cellulose derivative-based biosorbents. *Cellulose* 25: 2531–2545. https://doi.org/10.1007/s10570-018-1690-x

Garcia-Ivars, J., J. Durá-María, C. Moscardó-Carreño, C. Carbonell-Alcaina, M. I. Alcaina-Miranda and M. I. Iborra-Clar. 2017a. Rejection of trace pharmaceutically active compounds present in municipal wastewaters using ceramic fine ultrafiltration membranes: Effect of feed solution pH and fouling phenomena. *Separation and Purification Technology* 175: 58–71. https://doi.org/10.1016/j.seppur.2016.11.027

Garcia-Ivars, J., L. Martella, M. Massella, C. Carbonell-Alcaina, M. I. Alcaina-Miranda and M. I. Iborra-Clar. 2017b. Nanofiltration as tertiary treatment method for removing trace pharmaceutically active compounds in wastewater from wastewater treatment plants. *Water Research* 125: 360–373. https://doi.org/10.1016/j.watres.2017.08.070

Gautam, R. K., A. Mudhoo, G. Lofrano and M. C. Chattopadhyaya. 2014. Biomass-derived biosorbents for metal ions sequestration: Adsorbent modification and activation methods and adsorbent regeneration. *Journal of Environmental Chemical Engineering* 2(1): 239–259. https://doi.org/10.1016/j.jece.2013.12.019

Giles, C. H., T. H. MacEwan, S. N. Nakhwa and D. Smith. 1960. Studies in adsorption. Part XI. A system of classification of solution adsorption isotherms, and its use in diagnosis of adsorption mechanisms and in measurement of specific surface areas of solids. *Journal of the Chemical Society* 3: 3973–3993.

Gong, X. Y., Z. H. Huang, H. Zhang, W. L. Liu, X. H. Ma, Z. L. Xu and C. Y. Tang. 2020. Novel high-flux positively charged composite membrane incorporating titanium-based MOFs for heavy metal removal. *Chemical Engineering Journal* 398: 125706. https://doi.org/10.1016/j.cej.2020.125706

Goodman, B. A. 2020. Utilization of waste straw and husks from rice production: A review. *Journal of Bioresources and Bioproducts* 5(3): 143–162. https://doi.org/10.1016/j.jobab.2020.07.001

Goyer, R. A. 1997. Toxic and essential metal interactions. *Annual Review of Nutrition* 17(1): 37–50. https://doi.org/10.1146/annurev.nutr.17.1.37

Guo, S., X. Dong, T. Wu and C. Zhu. 2016. Influence of reaction conditions and feedstock on hydrochar properties. *Energy Conversion and Management* 123: 95–103. https://doi.org/10.1016/j.enconman.2016.06.029

Gupta, A., A. Dutta, M. K. Panigrahi and P. Sar. 2020. Geomicrobiology of mine tailings from Malanjkhand copper project, India. *Geomicrobiology Journal* 38(2): 97–114. https://doi.org/10.1080/01490451.2020.1817197

Gupta, M., H. Gupta and D. S. Kharat. 2018. Adsorption of Cu(II) by low cost adsorbents and the cost analysis. *Environmental Technology & Innovation* 10: 91–101. https://doi.org/10.1016/j.eti.2018.02.003

Güzel, F. and H. Sayğılı. 2016. Adsorptive efficacy analysis of novel carbonaceous sorbent derived from grape industrial processing wastes towards tetracycline in aqueous solution. *Journal of the Taiwan Institute of Chemical Engineers* 60: 236–240. https://doi.org/10.1016/j.jtice.2015.10.003

Halysh, V., O. Sevastyanova, S. Pikus, G. Dobele, B. Pasalskiy, V. M. Gun'ko and M. Kartel. 2020. Sugarcane bagasse and straw as low-cost lignocellulosic sorbents for the removal of dyes and metal ions from water. *Cellulose* 27: 8181–8197. https://doi.org/10.1007/s10570-020-03339-8

Hayati, B., A. Maleki, F. Najafi, H. Daraei, F. Gharibi and G. McKay. 2017. Super high removal capacities of heavy metals (Pb^{2+} and Cu^{2+}) using CNT dendrimer. *Journal of Hazardous Materials* 336: 146–157. https://doi.org/10.1016/j.jhazmat.2017.02.059

Herojeet, R., M. S. Rishi and N. Kishore. 2015. Integrated approach of heavy metal pollution indices and complexity quantification using chemometric models in the Sirsa Basin, Nalagarh valley, Himachal Pradesh, India. *Chinese Journal of Geochemistry* 34(4): 620–633. https://doi.org/10.1007/s11631-015-0075-1

Hoppen, M. I., K. Q. Carvalho, R. C. Ferreira, F. H. Passig, I. C. Pereira, R. C. P. Rizzo-Domingues, M. K. Lenzi and R. C. R. Bottini. 2019. Adsorption and desorption of acetylsalicylic acid onto activated carbon of babassu coconut mesocarp. *Journal of Environmental Chemical Engineering* 7(1): 102862. https://doi.org/10.1016/j.jece.2018.102862

Hu, B., J. Wang, B. Jin, Y. Li and Z. Shi. 2017. Assessment of the potential health risks of heavy metals in soils in a coastal industrial region of the Yangtze River Delta. *Environmental Science and Pollution Research* 24: 19816–19826. https://doi.org/10.1007/s11356-017-9516-1

Hu, B., J. Zhou, L. Liu, W. Meng and Z. Wang. 2017. Assessment of heavy metal pollution and potential ecological risk in soils of Tianjin sewage irrigation region, North China. *Journal of Environmental & Analytical Toxicology* 7(1): 1000425. https://doi.org/10.4172/2161-0525.1000425

IARC. 2012. Arsenic and arsenic compounds. *Monographs on the Evaluation of Carcinogenic Risks to Humans* 100C: 41–93.

Imbimbo, B. P., V. Solfrizzi and F. Panza. 2010. Are NSAIDs useful to treat Alzheimer's disease or mild cognitive impairment? *Frontiers in Aging Neuroscience* 2: 19. http://doi:10.3389/fnagi.2010.00019

IPCS (International Programme on Chemical Safety). 1998. Copper: Environmental Health Criteria 200. World Health Organisation, Geneva, Switzerland.

Israel, A. U., R. E. Ogali, O. Akaranta and I. B. Obot. 2011. Extraction and characterization of coconut (*Cocos nucifera L.*) coir dust. *Songklanakarin Journal of Science and Technology* 33(6): 717–724.

Jan, A. T., M. Azam, K. Siddiqui, A. Ali, I. Choi and Q. M. R. Haq. 2015. Heavy metals and human health: Mechanistic insight into toxicity and counter defense system of antioxidants. *International Journal of Molecular Sciences* 16(12): 29592–29630. https://doi.org/10.3390/ijms161226183

Jaria, G., V. Calisto, M. V. Gil, M. Otero and V. I. Esteves. 2015. Removal of fluoxetine from water by adsorbent materials produced from paper mill sludge. *Journal of Colloid and Interface Science* 448: 32–40. https://doi.org/10.1016/j.jcis.2015.02.002

Jaria, G., C. P. Silva, J. A. B. P. Oliveira, S. M. Santos, M. V. Gil, M. Otero, V. Calisto and V. I. Esteves. 2019. Production of highly efficient activated carbons from industrial wastes for the removal of pharmaceuticals from water-A full factorial design. *Journal of Hazardous Materials* 370: 212–218. https://doi.org/10.1016/j.jhazmat.2018.02.053

Jiang, M., W. Yang, Z. Zhang, Z. Yang and Y. Wang. 2015. Adsorption of three pharmaceuticals on two magnetic ion-exchange resins. *Journal of Environmental Sciences* 31(1): 226–234. https://doi.org/10.1016/j.jes.2014.09.035

Joint FAO/WHO Expert Committee on Food Additives. 2013. Evaluation of certain food additives and contaminants: Seventy-seventh report of the joint FAO/WHO expert committee on food additives. World Health Organization, Geneva, Switzerland.

Kadir, A. A., N. Othman and N. A. M. Azmi. 2012. Potential of using rosa centifolia to remove iron and manganese in groundwater treatment. *International Journal of Sustainable Construction Engineering and Technology* 3(2): 70–82.

Katakojwala, R. and S. V. Mohan. 2020. Microcrystalline cellulose production from sugarcane bagasse: Sustainable process development and life cycle assessment. *Journal of Cleaner Production* 249: 119342. https://doi.org/10.1016/j.jclepro.2019.119342

Khadir, A., M. Negarestani and A. Mollahosseini. 2020. Sequestration of a non-steroidal anti-inflammatory drug from aquatic media by lignocellulosic material (*Luffa cylindrica*) reinforced with polypyrrole: Study of parameters, kinetics, and equilibrium. *Journal of Environmental Chemical Engineering* 8(3): 103734. https://doi.org/10.1016/j.jece.2020.103734

Kibuye, F. A, H. E. Gall, K. R. Elkin, B. Ayers, T. L. Veith, M. Miller, S. Jacob, K. R. Hayden, J. E. Watson and H. A. Elliott. 2019. Fate of pharmaceuticals in a spray-irrigation system: From wastewater to groundwater. *Science of the Total Environment* 654: 197–208. https://doi.org/10.1016/j.scitotenv.2018.10.442

Kleywegt, S., V. Pileggi, Y. M. Lam, A. Elises, A. Puddicomb, G. Purba, J. D. Caro and T. Fletcher. 2016. The contribution of pharmaceutically active compounds from healthcare facilities to a receiving sewage treatment plant in Canada. *Environmental Toxicology and Chemistry* 35(4): 850–862. https://doi.org/10.1002/etc.3124

Krishna Mohan, G. V., A. Naga Babu, K. Kalpana and K. Ravindhranath. 2019. Removal of chromium(VI) from water using adsorbent derived from spent coffee grounds. *International Journal of Environmental Science and Technology* 16: 101–112. https://doi.org/10.1007/s13762-017-1593-7

Kumari, M., S. S. Khan, S. Pakrashi, A. Mukherjee and N. Chandrasekaran. 2011. Cytogenetic and genotoxic effects of zinc oxide nanoparticles on root cells of *Allium cepa*. *Journal of Hazardous Materials* 190(1–3): 613–621. https://doi.org/10.1016/j.jhazmat.2011.03.095

Kumar, S., S. Prasad, K. K. Yadav, M. Shrivastava, N. Gupta, S. Nagar, Q. V. Bach, H. Kamyab, S. A. Khan, S. Yadav and L. C. Malav. 2019. Hazardous heavy metals contamination of vegetables and food chain: Role of sustainable remediation approaches – A review. *Environmental Research* 179: 108792. https://doi.org/10.1016/j.envres.2019.108792

Langmuir, I. 1916. The constitution and fundamental properties of solids and liquids. Part I. Solids. *Journal of the American Chemical Society* 38(11): 2221–2295. https://doi.org/10.1021/ja02268a002

Largitte, L. and R. Pasquier. 2016. A review of the kinetics adsorption models and their application to the adsorption of lead by an activated carbon. *Chemical engineering research and design* 109: 495–504. https://doi.org/10.1016/j.cherd.2016.02.006

Larous, S. and A. H. Meniai. 2016. Adsorption of diclofenac from aqueous solution using activated carbon prepared from olive stones. *International Journal of Hydrogen Energy* 41(24): 10380–10390. https://doi.org/10.1016/j.ijhydene.2016.01.096

Levin, R. 1997. Lead in drinking water. In R. D. Morgenstern (ed.), Economic Analyses at EPA: Assessing Regulatory Impact Resources for the Future (pp. 205–232). Washington, DC.

Lima, D. R., A. Hosseini-Bandegharaei, P. S. Thue, E. C. Lima, Y. R. T. de Albuquerque, G. S. dos Reis, C. S. Umpierres, S. L. P. Dias and H. N. Tran. 2019. Efficient acetaminophen removal from water and hospital effluents treatment by activated carbons derived from Brazil nutshells. *Colloids and Surfaces A: Physicochemical and Engineering Aspects* 583: 123966. https://doi.org/10.1016/j.colsurfa.2019.123966

Lima, D. R., E. C. Lima, C. S. Umpierres, P. S. Thue, G. A. El-Chaghaby, R. S. da Silva, F. A. Pavan, S. L. P. Dias and C. Biron. 2019. Removal of amoxicillin from simulated hospital effluents by adsorption using activated carbons prepared from capsules of cashew of Para. *Environmental Science and Pollution Research* 26: 16396–16408. https://doi.org/10.1007/s11356-019-04994-6

Li, Y., M. A. Taggart, C. McKenzie, Z. Zhang, Y. Lu, S. Pap and S. Gibb. 2019. Utilizing low-cost natural waste for the removal of pharmaceuticals from water: Mechanisms, isotherms and kinetics at low concentrations. *Journal of Cleaner Production* 227: 88–97. https://doi.org/10.1016/j.jclepro.2019.04.081

Liu, L., Y. Huang, J. Cao, H. Hu, L. Dong, J. Zha, Y. Su, R. Ruan and S. Tao. 2021. Qualitative and relative distribution of Pb^{2+} adsorption mechanisms by biochars produced from a fluidized bed pyrolysis system under mild air oxidization conditions. *Journal of Molecular Liquids* 323: 114600. https://doi.org/10.1016/j.molliq.2020.114600

Liu, L., Y. Huang, Y. Meng, J. Cao, H. Hu, Y. Su, L. Dong, S. Tao and R. Ruan. 2020. Investigating the adsorption behavior and quantitative contribution of Pb^{2+} adsorption mechanisms on biochars by different feedstocks from a fluidized bed pyrolysis system. *Environmental Research* 187: 109609. https://doi.org/10.1016/j.envres.2020.109609

Liu, H., J. Zhang, N. Bao, C. Cheng, L. Ren and C. Zhang. 2012. Textural properties and surface chemistry of lotus stalk-derived activated carbons prepared using different phosphorus oxyacids: Adsorption of trimethoprim. *Journal of Hazardous Materials* 235–236: 367–375. https://doi.org/10.1016/j.jhazmat.2012.08.015

Liu, W., J. Zhang, C. Zhang and L. Ren. 2011. Sorption of norfloxacin by lotus stalk-based activated carbon and iron-doped activated alumina: Mechanisms, isotherms and kinetics. *Chemical Engineering Journal* 171(2): 431–438. https://doi.org/10.1016/j.cej.2011.03.099

Luo, J., X. Luo, J. Crittenden, J. Qu, Y. Bai, Y. Peng and J. Li. 2015. Removal of antimonite (Sb(III)) and antimonate (Sb(V)) from aqueous solution using carbon nanofibers that are decorated with zirconium oxide (ZrO_2). *Environmental Science & Technology* 49(18): 11115–11124. https://doi.org/10.1021/acs.est.5b02903

Magioglou, E., Z. Frontistis, J. Vakros, I. D. Manariotis and D. Mantzavinos. 2019. Activation of persulfate by biochars from valorized olive stones for the degradation of sulfamethoxazole. *Catalysts* 9(5): 419. https://doi.org/10.3390/catal9050419

Maigari, A. U., E. O. Ekanem, I. H. Garba, A. Harami and J. C. Akan. 2016. Health risk assessment for exposure to some selected heavy metals via drinking water from Dadinkowa dam and river Gombe Abba in Gombe state, Northeast Nigeria. *World Journal of Analytical Chemistry* 4(1): 1–5.

Majumdar, A. and Pal, A., 2020. Recent advancements in visible-light-assisted photocatalytic removal of aqueous pharmaceutical pollutants. *Clean Technologies and Environmental Policy* 22: 11–42. https://doi.org/10.1007/s10098-019-01766-1

Malik, R., S. Dahiya and S. Lata. 2017. An experimental and quantum chemical study of removal of utmostly quantified heavy metals in wastewater using coconut husk: A novel approach to mechanism. *International Journal of Biological Macromolecules* 98: 139–149. https://doi.org/10.1016/j.ijbiomac.2017.01.100

Martins, A. C., O. Pezoti, A. L. Cazetta, K. C. Bedin, D. A. S. Yamazaki, G. F. G. Bandoch, T. Asefa, J. V. Visentainer and V. C. Almeida. 2015. Removal of tetracycline by NaOH-activated carbon produced from macadamia nut shells: Kinetic and equilibrium studies. *Chemical Engineering Journal* 260: 291–299. https://doi.org/10.1016/j.cej.2014.09.017

Marzbali, M. H., M. Esmaieli, H. Abolghasemi and M. H. Marzbali. 2016. Tetracycline adsorption by H_3PO_4-activated carbon produced from apricot nut shells: A batch study. *Process Safety and Environmental Protection* 102: 700–709. https://doi.org/10.1016/j.psep.2016.05.025

Mashayekh-Salehi, A. and G. Moussavi. 2016. Removal of acetaminophen from the contaminated water using adsorption onto carbon activated with NH_4Cl. *Desalination and Water Treatment* 57(27): 12861–12873. https://doi.org/10.1080/19443994.2015.1051588

Maszkowska, J., S. Stolte, J. Kumirska, P. Łukaszewicz, K. Mioduszewska, A. Puckowski, M. Caban, M. Wagil, P. Stepnowski and A. Białk-Bielińska. 2014a. Beta-blockers in the environment: Part I. Mobility and hydrolysis study. *Science of the Total Environment* 493: 1112–1121. https://doi.org/10.1016/j.scitotenv.2014.06.023

Maszkowska, J., S. Stolte, J. Kumirska, P. Łukaszewicz, K. Mioduszewska, A. Puckowski, M. Caban, M. Wagil, P. Stepnowski and A. Białk-Bielińska. 2014b. Beta-blockers in the environment: Part II. Ecotoxicity study. *Science of the Total Environment* 493: 1122–1126. https://doi.org/10.1016/j.scitotenv.2014.06.039

Ma, H., J. Yang, X. Gao, Z. Liu, X. Liu and Z. Xu. 2019. Removal of chromium (VI) from water by porous carbon derived from corn straw: Influencing factors, regeneration and mechanism. *Journal of Hazardous Materials* 369: 550–560. https://doi.org/10.1016/j.jhazmat.2019.02.063

Ma, X., H. Zuo, M. Tian, L. Zhang, J. Meng, X. Zhou, N. Min, X. Chang and Y. Liu. 2016. Assessment of heavy metals contamination in sediments from three adjacent regions of the yellow river using metal chemical fractions and multivariate analysis techniques. *Chemosphere* 144: 264–272. https://doi.org/10.1016/j.chemosphere.2015.08.026

Mazzeo, D. E. C., T. C. C. Fernandes, C. E. Levy, C. S. Fontanetti and M. A. Marin-Morales. 2015. Monitoring the natural attenuation of a sewage sludge toxicity using the *Allium cepa* test. *Ecological Indicators* 56: 60–69. https://doi.org/10.1016/j.ecolind.2015.03.026

Mehr, M. R., B. Keshavarzi, F. Moore, R. Sharifi, A. Lahijanzadeh and M. Kermani. 2017. Distribution, source identification and health risk assessment of soil heavy metals in urban areas of Isfahan province, Iran. *Journal of African Earth Sciences* 132: 16–26. https://doi.org/10.1016/j.jafrearsci.2017.04.026

Mestre, A. S., R. A. Pires, I. Aroso, E. M. Fernandes, M. L. Pinto, R. L. Reis, M. A. Andrade, J. Pires, S. P. Silva and A. P. Carvalho. 2014. Activated carbons prepared from industrial pre-treated cork: Sustainable adsorbents for pharmaceutical compounds removal. *Chemical Engineering Journal* 253: 408–417. https://doi.org/10.1016/j.cej.2014.05.051

Mestre, A. S., E. Tyszko, M. A. Andrade, M. Galhetas, C. Freire and A. P. Carvalho. 2015. Sustainable activated carbons prepared from a sucrose-derived hydrochar: Remarkable adsorbents for pharmaceutical compounds. *RSC Advances* 5: 19696–19707. https://doi.org/10.1039/c4ra14495c

Mezzelani, M., S. Gorbi and F. Regoli. 2018. Pharmaceuticals in the aquatic environments: Evidence of emerged threat and future challenges for marine organisms. *Marine Environmental Research* 140: 41–60. https://doi.org/10.1016/j.marenvres.2018.05.001

Mohapatra, S., C. H. Huang, S. Mukherji and L. P. Padhye. 2016. Occurrence and fate of pharmaceuticals in WWTPs in India and comparison with a similar study in the United States. *Chemosphere* 159: 526–535. https://doi.org/10.1016/j.chemosphere.2016.06.047

Moro, T. R., F. R. Henrique, L. C. Malucelli, C. M. R. de Oliveira, M. A. da Silva Carvalho Filho and E. C. de Vasconcelos. 2017. Adsorption of pharmaceuticals in water through lignocellulosic fibers synergism. *Chemosphere* 171: 57–65. https://doi.org/10.1016/j.chemosphere.2016.12.040

Moslah, B., E. Hapeshi, A. Jrad, D. Fatta-Kassinos and A. Hedhili. 2018. Pharmaceuticals and illicit drugs in wastewater samples in north-eastern Tunisia. *Environmental Science and Pollution Research* 25: 18226–18241. https://doi.org/10.1007/s11356-017-8902-z

Mouchtari, E. M. E., C. Daou, S. Rafqah, F. Najjar, H. Anane, A. Piram, A. Hamade, S. Briche and P. Wong-Wah-Chung. 2020. TiO_2 and activated carbon of *Argania Spinosa* tree nutshells composites for the adsorption photocatalysis removal of pharmaceuticals from aqueous solution. *Journal of Photochemistry and Photobiology A: Chemistry* 388: 112183. https://doi.org/10.1016/j.jphotochem.2019.112183

Moussavi, G., A. Alahabadi, K. Yaghmaeian and M. Eskandari. 2013. Preparation, characterization and adsorption potential of the NH_4Cl-induced activated carbon for the removal of amoxicillin antibiotic from water. *Chemical Engineering Journal* 217: 119–128. https://doi.org/10.1016/j.cej.2012.11.069

N'diaye, A. D. and M. S. Kankou. 2019. Valorization of *Balanites aegyptiaca* seeds from Mauritania: Modeling of adsorption isotherms of caffeine from aqueous solution. *Journal of Environmental Treatment Techniques* 7(3): 450–455.

Nayak, A. K. and A. Pal. 2019. Development and validation of an adsorption kinetic model at solid-liquid interface using normalized Gudermannian function. *Journal of Molecular Liquids* 276: 67–77. https://doi.org/10.1016/j.molliq.2018.11.089

Nazari, G., H. Abolghasemi and M. Esmaieli. 2016. Batch adsorption of cephalexin antibiotic from aqueous solution by walnut shell-based activated carbon. *Journal of the Taiwan Institute of Chemical Engineers* 58: 357–365. https://doi.org/10.1016/j.jtice.2015.06.006

Nche, N. G., A. Bopda, D. R. T. Tchuifon, C. S. Ngakou, I. T. Kuete and A. S. Gabche. 2017. Removal of paracetamol from aqueous solution by adsorption onto activated carbon prepared from rice husk. *Journal of Chemical and Pharmaceutical Research* 9(3): 56–68.

Noli, F., E. Kapashi and M. Kapnisti. 2019. Biosorption of uranium and cadmium using sorbents based on aloe vera wastes. *Journal of Environmental Chemical Engineering* 7(2): 102985. https://doi.org/10.1016/j.jece.2019.102985

Okon, O., U. Eduok and A. Israel. 2012. Characterization and phytochemical screening of coconut (*Cocos nucifera L.*) coir dust as a low cost adsorbent for waste water treatment. *Elixir Applied Chemistry* 47: 8961–8968.

Oliveira, G., V. Calisto, S. M. Santos, M. Otero and V. I. Esteves. 2018. Paper pulp-based adsorbents for the removal of pharmaceuticals from wastewater: A novel approach towards diversification. *Science of the Total Environment* 631–632: 1018–1028. https://doi.org/10.1016/j.scitotenv.2018.03.072

Palli, L., F. Spina, G. C. Varese, M. Vincenzi, M. Aragno, G. Arcangeli, N. Mucci, D. Santianni, S. Caffaz and R. Gori. 2019. Occurrence of selected pharmaceuticals in wastewater treatment plants of Tuscany: An effect-based approach to evaluate the potential environmental impact. *International Journal of Hygiene and Environmental Health* 222(4): 717–725. https://doi.org/10.1016/j.ijheh.2019.05.006

Paredes-Laverde, M., J. Silva-Agredo and R. A. Torres-Palma. 2018. Removal of norfloxacin in deionized, municipal water and urine using rice (*Oryza sativa*) and coffee (*Coffea arabica*) husk wastes as natural adsorbents. *Journal of Environmental Management* 213: 98–108. https://doi.org/10.1016/j.jenvman.2018.02.047

Paunovic, O., S. Pap, S. Maletic, M. A. Taggart, N. Boskovic and M. T. Sekulic. 2019. Ionisable emerging pharmaceutical adsorption onto microwave functionalised biochar derived from novel lignocellulosic waste biomass. *Journal of Colloid and Interface Science* 547: 350–360. https://doi.org/10.1016/j.jcis.2019.04.011

Pezoti, O., A. L. Cazetta, K. C. Bedin, L. S. Souza, A. C. Martins, T. L. Silva, O. O. Santos Júnior, J. V. Visentainer and V. C. Almeida. 2016. NaOH-activated carbon of high surface area produced from guava seeds as a high-efficiency adsorbent for amoxicillin removal: Kinetic, isotherm and thermodynamic studies. *Chemical Engineering Journal* 288: 778–788. https://doi.org/10.1016/j.cej.2015.12.042

Phan, N. H., S. Rio, C. Faur, L. L. Coq, P. L. Cloirec and T. H. Nguyen. 2006. Production of fibrous activated carbons from natural cellulose (jute, coconut) fibers for water treatment applications. *Carbon* 44(12): 2569–2577. https://doi.org/10.1016/j.carbon.2006.05.048

Quesada, H. B., L. F. Cusioli, C. de Oliveira Bezerra, A. T. A. Baptista, L. Nishi, R. G. Gomes and R. Bergamasco. 2019. Acetaminophen adsorption using a low-cost adsorbent prepared from modified residues of *Moringa oleifera* Lam. seed husks. *Journal of Chemical Technology and Biotechnology* 94(10): 3147–3157. https://doi.org/10.1002/jctb.6121

Qu, Z., L. Fang, D. Chen, H. Xu and N. Yan. 2017. Effective and regenerable Ag/graphene adsorbent for Hg(II) removal from aqueous solution. *Fuel* 203: 128–134. https://doi.org/10.1016/j.fuel.2017.04.105

Redlich, O. and D. L. Peterson. 1959. A useful adsorption isotherm. *The Journal of Physical Chemistry* 63(6): 1024. https://doi.org/10.1021/j150576a611

Rehman, I. U, M. Ishaq, L. Ali, S. Khan, I. Ahmad, I. U. Din and H. Ullah. 2018. Enrichment, spatial distribution of potential ecological and human health risk assessment via toxic metals in soil and surface water ingestion in the vicinity of Sewakht mines, district Chitral, Northern Pakistan. *Ecotoxicology and Environmental Safety* 154: 127–136. https://doi.org/10.1016/j.ecoenv.2018.02.033

Richards, S., J. Dawson and M. Stutter. 2019. The potential use of natural vs commercial biosorbent material to remediate stream waters by removing heavy metal contaminants. *Journal of Environmental Management* 231: 275–281. https://doi.org/10.1016/j.jenvman.2018.10.019

Rivera-Jaimes, J. A., C. Postigo, R. M. Melgoza-Alemán, J. Aceña, D. Barceló and M. L. de Alda. 2018. Study of pharmaceuticals in surface and wastewater from Cuernavaca, Morelos, Mexico: Occurrence and environmental risk assessment. *Science of the Total Environment* 613–614, 1263–1274. https://doi.org/10.1016/j.scitotenv.2017.09.134

Rodríguez-Navas, C., E. Björklund, S. A. Bak, M. Hansen, K. A. Krogh, F. Maya, R. Forteza and V. Cerdà. 2013. Pollution pathways of pharmaceutical residues in the aquatic environment on the island of Mallorca, Spain. *Archives of Environmental Contamination and Toxicology* 65: 56–66. https://doi.org/10.1007/s00244-013-9880-x

Roginsky, S. and Y. B. Zeldovich. 1934. The catalytic oxidation of carbon monoxide on manganese dioxide. *Acta Physicochimica USSR* 1: 554.

Rosa, M. A., J. A. Egido and M. C. Márquez. 2017. Enhanced electrochemical removal of arsenic and heavy metals from mine tailings. *Journal of the Taiwan Institute of Chemical Engineers* 78: 409–415. https://doi.org/10.1016/j.jtice.2017.06.046

Saldarriaga, J. F., N. A. Montoya, I. Estiati, A. T. Aguayo, R. Aguado and M. Olazar. 2020. Unburned material from biomass combustion as low-cost adsorbent for amoxicillin removal from wastewater. *Journal of Cleaner Production* 284: 124732. https://doi.org/10.1016/j.jclepro.2020.124732

Saleem, J., U. B. Shahid, M. Hijab, H. Mackey and G. McKay. 2019. Production and applications of activated carbons as adsorbents from olive stones. *Biomass Conversion and Biorefinery* 9: 775–802. https://doi.org/10.1007/s13399-019-00473-7

Saleh, M. E., Y. A. El-Damarawy, F. F. Assad, A. A. Abdesalam and R. A. Yousef. 2020. Removal of copper metal ions by sugarcane bagasse and rice husk biochars from contaminated aqueous solutions. *Mediterranean Journal of Soil Science* 1(1): 1–17.

Sanka, P. M., M. J. Rwiza and K. M. Mtei. 2020. Removal of selected heavy metal ions from industrial wastewater using rice and corn husk biochar. *Water, Air, & Soil Pollution* 231: 244. https://doi.org/10.1007/s11270-020-04624-9

Sarkar, D. J., S. D. Sarkar, B. K. Das, B. K. Sahoo, A. Das, S. K. Nag, R. K. Manna, B. K. Behera and S. Samanta. 2021. Occurrence, fate and removal of microplastics as heavy metal vector in natural wastewater treatment wetland system. *Water Research* 192: 116853. https://doi.org/10.1016/j.watres.2021.116853

Say ğılı, H. and F. Güzel. 2016. Effective removal of tetracycline from aqueous solution using activated carbon prepared from tomato (*Lycopersicon esculentum* Mill.) industrial processing waste. *Ecotoxicology and Environmental Safety* 131: 22–29. https://doi.org/10.1016/j.ecoenv.2016.05.001

Sekulic, M. T., N. Boskovic, A. Slavkovic, J. Garunovic, S. Kolakovic and S. Pap. 2019. Surface functionalised adsorbent for emerging pharmaceutical removal: Adsorption performance and mechanisms. *Process Safety and Environmental Protection* 125: 50–63. https://doi.org/10.1016/j.psep.2019.03.007

Shahrokhi-Shahraki, R., C. Benally, M. G. El-Din and J. Park. 2021. High efficiency removal of heavy metals using tire-derived activated carbon vs commercial activated carbon: Insights into the adsorption mechanisms. *Chemosphere* 264: 128455. https://doi.org/10.1016/j.chemosphere.2020.128455

Shakoor, M. B., N. K. Niazi, I. Bibi, M. Shahid, F. Sharif, S. Bashir, S. M. Shaheen, H. Wang, D. C. W. Tsang, Y. S. Ok and J. Rinklebe. 2018. Arsenic removal by natural and chemically modified water melon rind in aqueous solutions and groundwater. *Science of the Total Environment* 645: 1444–1455. https://doi.org/10.1016/j.scitotenv.2018.07.218

Shan, D., S. Deng, T. Zhao, B. Wang, Y. Wang, J. Huang, G. Yu, J. Winglee and M. R. Wiesner. 2016. Preparation of ultrafine magnetic biochar and activated carbon for pharmaceutical adsorption and subsequent degradation by ball milling. *Journal of Hazardous Materials* 305: 156–163. https://doi.org/10.1016/j.jhazmat.2015.11.047

Shang, J., M. Zong, Y. Yu, X. Kong, Q. Du and Q. Liao. 2017. Removal of chromium (VI) from water using nanoscale zerovalent iron particles supported on herb-residue biochar. *Journal of Environmental Management* 197: 331–337. https://doi.org/10.1016/j.jenvman.2017.03.085

Sharma, P. K. and S. Ayub. 2019. The cost analysis and economic feasibility of agro wastes to adsorb chromium (VI) from wastewater. *International Journal of Civil Engineering and Technology* 10(2): 2387–2402.

Sheikh, Z., M. Amin, N. Khan, M. N. Khan, S. K. Sami, S. B. Khan, I. Hafeez, S. A. Khan, E. M. Bakhsh and C. K. Cheng. 2021. Potential application of *Allium cepa* seeds as a novel biosorbent for efficient biosorption of heavy metals ions from aqueous solution. *Chemosphere* 279: 130545. https://doi.org/10.1016/j.chemosphere.2021.130545

Silva, C. P., G. Jaria, M. Otero, V. I. Esteves and V. Calisto. 2019. Adsorption of pharmaceuticals from biologically treated municipal wastewater using paper mill sludge-based activated carbon. *Environmental Science and Pollution Research* 26: 13173–13184. https://doi.org/10.1007/s11356-019-04823-w

Singh, R. R., L. F. Angeles, D. M. Butryn, J. W. Metch, E. Garner, P. J. Vikesland and D. S. Aga. 2019. Towards a harmonized method for the global reconnaissance of multi-class antimicrobials and other pharmaceuticals in wastewater and receiving surface waters. *Environment International* 124: 361–369. https://doi.org/10.1016/j.envint.2019.01.025

Singh, U. K. and B. Kumar. 2017. Pathways of heavy metals contamination and associated human health risk in Ajay River basin, India. *Chemosphere* 174: 183–199. https://doi.org/10.1016/j.chemosphere.2017.01.103

Sips, R. 1948. On the structure of a catalyst surface. *The Journal of Chemical Physics* 16(5): 490–495. https://doi.org/10.1063/1.1746922

Smith, D. J. 2007. Characterization of nanomaterials using transmission electron microscopy. In J. Hutchison & A. Kirkland (Eds.), Nanocharacterisation (pp. 1–27). RSC Publishing.

Šoštarić, T. D., M. S. Petrović, F. T. Pastor, D. R. Lonćarević, J. T. Petrović, J. V. Milojković and M. D. Stojanović. 2018. Study of heavy metals biosorption on native and alkali-treated apricot shells and its application in wastewater treatment. *Journal of Molecular Liquids* 259: 340–349. https://doi.org/10.1016/j.molliq.2018.03.055

Subedi, B., K. Balakrishna, D. I. Joshua and K. Kannan. 2017. Mass loading and removal of pharmaceuticals and personal care products including psychoactives, antihypertensives, and antibiotics in two sewage treatment plants in southern India. *Chemosphere* 167: 429–437. https://doi.org/10.1016/j.chemosphere.2016.10.026

Suhas, P. J. M. Carrott and M. M. L. Ribeiro Carrott. 2007. Lignin-from natural adsorbent to activated carbon: A review. *Bioresource Technology* 98(12): 2301–2312. https://doi.org/10.1016/j.biortech.2006.08.008

Sun, C., T. Chen, Q. Huang, J. Wang, S. Lu and J. Yan. 2019. Enhanced adsorption for Pb(II) and Cd(II) of magnetic rice husk biochar by $KMnO_4$ modification. *Environmental Science and Pollution Research* 26: 8902–8913. https://doi.org/10.1007/s11356-019-04321-z

Tabak, A., K. Sevimli, M. Kaya and B. Çağlar. 2019. Preparation and characterization of a novel activated carbon component via chemical activation of tea woody stem. *Journal of Thermal Analysis and Calorimetry* 138: 3885–3895. https://doi.org/10.1007/s10973-019-08387-2

Tahir, M. B., H. Kiran and T. Iqbal. 2019. The detoxification of heavy metals from aqueous environment using nano-photocatalysis approach: A review. *Environmental Science and Pollution Research* 26: 10515–10528. https://doi.org/10.1007/s11356-019-04547-x

Tahmasebi, E., M. Y. Masoomi, Y. Yamini and A. Morsali. 2015. Application of mechanosynthesized azine-decorated zinc(II) metal-organic frameworks for highly efficient removal and extraction of some heavy-metal ions from aqueous samples: A comparative study. *Inorganic Chemistry* 54(2): 425–433. https://doi.org/10.1021/ic5015384

Teixeira, S., C. Delerue-Matos and L. Santos. 2019. Application of experimental design methodology to optimize antibiotics removal by walnut shell based activated carbon. *Science of the Total Environment* 646: 168–176. https://doi.org/10.1016/j.scitotenv.2018.07.204

Temkin, M. J. and V. Pyzhev. 1940. Kinetics of ammonia synthesis on promoted iron catalyst. *Acta Physicochimica USSR* 12: 327–356.

Tran, H. N., S. J. You, T. V. Nguyen and H. P. Chao. 2017. Insight into the adsorption mechanism of cationic dye onto biosorbents derived from agricultural wastes. *Chemical Engineering Communications* 204(9): 1020–1036. https://doi.org/10.1080/00986445.2017.1336090

Tunç, Ö., H. Tanacı and Z. Aksu. 2009. Potential use of cotton plant wastes for the removal of Remazol Black B reactive dye. *Journal of Hazardous Materials* 163(1): 187–198. https://doi.org/10.1016/j.jhazmat.2008.06.078

Uçar, S., M. Erdem, T. Tay and S. Karagöz. 2009. Preparation and characterization of activated carbon produced from pomegranate seeds by $ZnCl_2$ activation. *Applied Surface Science* 255(21): 8890–8896. https://doi.org/10.1016/j.apsusc.2009.06.080

US Food and Drug Administration. 2018. Novel drug approvals for 2018. https://www.fda.gov/drugs/new-drugs-fda-cders-new-molecular-entities-and-new-therapeutic-biological-products/novel-drug-approvals-2018

US Food and Drug Administration. 2019. Novel drug approvals for 2019. https://www.fda.gov/drugs/new-drugs-fda-cders-new-molecular-entities-and-new-therapeutic-biological-products/novel-drug-approvals-2019

USEPA. 2011. Water: Chromium in Drinking Water. United States Environmental Protection Agency, Washington, DC.

USEPA. 2018. Edition of the Drinking Water Standards and Health Advisories Tables, EPA 822-F-18-001. United States Environmental Protection Agency, Washington, DC.

Valdés, J., M. Guiñez, A. Castillo and S. E. Vega. 2014. Cu, Pb, and Zn content in sediments and benthic organisms from San Jorge Bay (northern Chile): Accumulation and biotransference in subtidal coastal systems. *Ciencias Marinas* 40(1): 45–58. https://doi.org/10.7773/cm.v40i1.2318

Verma, L., M. A. Siddique, J. Singh and R. N. Bharagava. 2019. As(III) and As(V) removal by using iron impregnated biosorbents derived from waste biomass of *Citrus limmeta* (peel and pulp) from the aqueous solution and ground water. *Journal of Environmental Management* 250: 109452. https://doi.org/10.1016/j.jenvman.2019.109452

Villaescusa, I., N. Fiol, J. Poch, A. Bianchi and C. Bazzicalupi. 2011. Mechanism of paracetamol removal by vegetable wastes: The contribution of π-π interactions, hydrogen bonding and hydrophobic effect. *Desalination* 270(1–3): 135–142. https://doi.org/10.1016/j.desal.2010.11.037

Wang, L., N. S. Bolan, D. C. W. Tsang and D. Hou. 2020. Green immobilization of toxic metals using alkaline enhanced rice husk biochar: Effects of pyrolysis temperature and KOH concentration. *Science of the Total Environment* 720: 137584. https://doi.org/10.1016/j.scitotenv.2020.137584

Wang, S., W. Guo, F. Gao, Y. Wang and Y. Gao. 2018. Lead and uranium sorptive removal from aqueous solution using magnetic and nonmagnetic fast pyrolysis rice husk biochars. *RSC Advances* 8(24): 13205–13217. https://doi.org/10.1039/C7RA13540H

Wang, J., N. Hu, M. Liu, J. Sun and Y. Xu. 2019. A novel core–shell structured biosorbent derived from chemi-mechanical pulp for heavy metal ion removal. *Cellulose* 26: 8789–8799. https://doi.org/10.1007/s10570-019-02693-6

Wang, C., H. Wang and G. Gu. 2018. Ultrasound-assisted xanthation of cellulose from lignocellulosic biomass optimized by response surface methodology for Pb(II) sorption. *Carbohydrate Polymers* 182: 21–28. https://doi.org/10.1016/j.carbpol.2017.11.004

Weber Jr., W. J. and J. C. Morris. 1963. Kinetics of adsorption on carbon from solution. *Journal of the Sanitary Engineering Division* 89(2): 31–59. https://doi.org/10.1061/JSEDAI.0000430

Wee, S. Y. and A. Z. Aris. 2017. Endocrine disrupting compounds in drinking water supply system and human health risk implication. *Environment International* 106: 207–233. https://doi.org/10.1016/j.envint.2017.05.004

WHO. 2005. Mercury in Drinking-Water. Background document for development of WHO Guidelines for drinking-water quality. World Health Organization, Geneva, Switzerland.

WHO. 2019. Safer Water, Better Health. World Health Organization. https://apps.who.int/iris/handle/10665/329905. License: CC BY-NC-SA 3.0 IGO

Xue, C., Q. Zhang, G. Owens and Z. Chen. 2020. A cellulose degrading bacterial strain used to modify rice straw can enhance Cu(II) removal from aqueous solution. *Chemosphere* 256: 127142. https://doi.org/10.1016/j.chemosphere.2020.127142

Yahya, M. D., A. S. Aliyu, K. S. Obayomi, A. G. Olugbenga and U. B. Abdullahi. 2020. Column adsorption study for the removal of chromium and manganese ions from electroplating wastewater using cashew nutshell adsorbent. *Cogent Engineering* 7(1): 1748470. https://doi.org/10.1080/23311916.2020.1748470

Yang, C. 1993. Statistical mechanical aspects of adsorption systems obeying the Temkin isotherm. *The Journal of Physical Chemistry* 97(27): 7097–7101. https://doi.org/10.1021/j100129a029

Yang, X., Y. Wan, Y. Zheng, F. He, Z. Yu, J. Huang, H. Wang, Y. S. Ok, Y. Jiang and B. Gao. 2019. Surface functional groups of carbon-based adsorbents and their roles in the removal of heavy metals from aqueous solutions: A critical review. *Chemical Engineering Journal* 366: 608–621. https://doi.org/10.1016/j.cej.2019.02.119

Yan, L., Y. Liu, Y. Zhang, S. Liu, C. Wang, W. Chen, C. Liu, Z. Chen and Y. Zhang. 2020. $ZnCl_2$ modified biochar derived from aerobic granular sludge for developed microporosity and enhanced adsorption to tetracycline. *Bioresource Technology* 297: 122381. https://doi.org/10.1016/j.biortech.2019.122381

Yap, M. W., N. M. Mubarak, J. N. Sahu and E. C. Abdullah. 2017. Microwave induced synthesis of magnetic biochar from agricultural biomass for removal of lead and cadmium from wastewater. *Journal of Industrial and Engineering Chemistry* 45: 287–295. https://doi.org/10.1016/j.jiec.2016.09.036

Yu, P., X. Wang, K. Zhang, M. Wu, Q. Wu, J. Liu, J. Yang and J. Zhang. 2020. Continuous purification of simulated wastewater based on rice straw composites for oil/water separation and removal of heavy metal ions. *Cellulose* 27(9): 5223–5239. https://doi.org/10.1007/s10570-020-03135-4

Yu, J., Z. Zhu, H. Zhang, Y. Qiu and D. Yin. 2018. Mg-Fe layered double hydroxide assembled on biochar derived from rice husk ash: Facile synthesis and application in efficient removal of heavy metals. *Environmental Science and Pollution Research* 25: 24293–24304. https://doi.org/10.1007/s11356-018-2500-6

Zamora-Ledezma, C., D. Negrete-Bolagay, F. Figueroa, E. Zamora-Ledezma, M. Ni, F. Alexis and V. H. Guerrero. 2021. Heavy metal water pollution: A fresh look about hazards, novel and conventional remediation methods. *Environmental Technology & Innovation* 22: 101504. https://doi.org/10.1016/j.eti.2021.101504

Zbair, M., H. A. Ahsaine, Z. Anfar and A. Slassi. 2019. Carbon microspheres derived from walnut shell: Rapid and remarkable uptake of heavy metal ions, molecular computational study and surface modeling. *Chemosphere* 231: 140–150. https://doi.org/10.1016/j.chemosphere.2019.05.120

Zhang, H., B. Huang, L. Dong, W. Hu, M. S. Akhtar and M. Qu. 2017. Accumulation, sources and health risks of trace metals in elevated geochemical background soils used for greenhouse vegetable production in southwestern China. *Ecotoxicology and Environmental Safety* 137: 233–239. https://doi.org/10.1016/j.ecoenv.2016.12.010

10 Valorization of Industrial Solid Waste to Green Sustainable Products

Nermine El Sayed Maysour Refaat,
Shyma Mohammed El Saeed,
and El Sayed Gamal Zaki

10.1 TYPES OF WASTES

This chapter focuses on types of wastes and how disposing these wastes is effective and crucial for having a healthy environment. Quantities and characteristics of wastes are the key to an effective solid waste (SW) management strategy. Defining those components will help in a great way in deciding the techniques of getting out the highest advantage of managing wastes. These will differ in regions and countries [1].

There are different types of wastes; some are as follows:

- Residential waste from food, paper, plastics, glass, and other single dwellings.
- Medical waste which are generated from hospitals and any medical institute.
- Hazardous waste like motor oil, batteries, paints, and chemicals used.
- Industrial waste may include all the above types.

Referring to the administrative regulations, the term of solid waste and the solid waste management are not the same in many countries.

10.2 SOLID WASTE (SW)

According to the definition of SWs in the Council of European Union's Waste Directive (75/442/EEC) released in 1975 [1], "waste" refers to "materials or substances that have been discarded, are about to be thrown away, or must be disposed of by owners." These definitions demonstrate that the term "solid waste" encompasses two categories of waste. The first is "waste," which refers to materials that have outlived their original utility, such as plastics, scrap cars, and the vast majority of municipal SW. They also include the majority of wastes that come from the industry, such as water treatment fly ash and sludge, as well as substances with no obvious production targets or functions during manufacturing processes, as well as by-products of specific goods in the manufacturing process.

DOI: 10.1201/9781003354475-10

The second concept is "abandoned," which means that the owners have abandoned these substances. By different meaning, it is not possible to employ these materials for their intended purpose.

The SWs are considered garbage substances by any person and may be considered a benefit by another. Therefore, a substance that is considered "waste" may have great utility in another. Today's waste could be tomorrow's resources. As a result, the waste has distinct geographical and temporal features. To put it another way, waste is misallocated resources. However, due to economic and technical constraints, not all SW can be utilized at this time. It concerns the "resource" characteristic of SW which is a natural one. The cost-effectiveness of SW should be examined so it could be used as a resource. Therefore, if any form of trash reuse results in increased energy and resource consumption or increased pollution, the value of this type of waste may be lost. It is difficult to identify SW since it has two characteristics. The distinction between "old" and "waste" should be made firstly. Many commodities may miss their utility to human kind, yet they continue to be useful to others. Some items may lose some of their functions, whereas others may gain new ones. However, they may be able to keep some or all of their functions after repair. Some people promote "reuse," which is defined as "repairing the old and utilizing the waste." It's the well-known 3Rs of SW management [2]; the three R principles for utilizing and reducing SW. Figure 10.1 shows Recycling, Reusability, and Reduction, which are still in use today.

The commodities listed are inaccurate, but rather "old commodities." The most important management concept for SWs is that waste and pollution reduction, as well as resource consumption and boosting societal wealth reduction, is dependent on the continued utilizing scrapped products for their outstanding function [1].

Based on the conditions, "waste" and "old commodities" must be identified as there is no clear distinction between them. On the other hand, "waste" and "raw materials" must be distinguished. When a material has totally lost its original use value, it could be reprocessed as another substance for other products. Hereby, examples of prevalent procedures, for steel manufacturing use recycled scrap steel, for oil refining use waste polymer, for paper manufacturing use waste paper as raw materials. Waste can be used as a raw material source for manufacturing in two ways. One

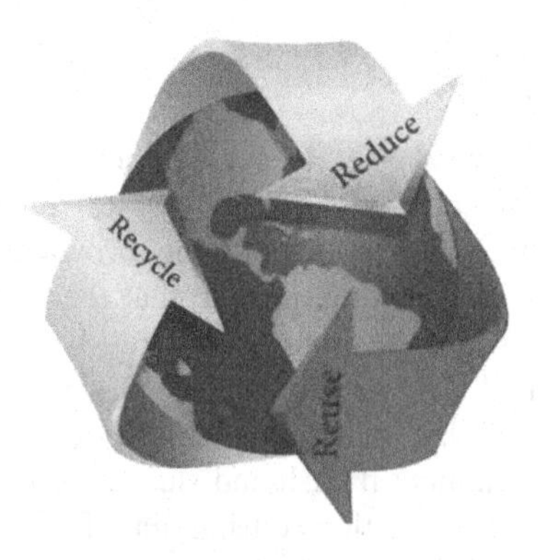

FIGURE 10.1 Diagram of the three R concepts for waste management.

option is to recycle waste and use it on the same manufacturing line as basic raw materials, like waste steel for smelting. Resources and energy investment, as well as pollution reduction, are the main advantages of using wastes as raw materials; compared to the traditional production where the primary raw materials are used.

10.3 INDUSTRIAL SOLID WASTE (ISW)

ISW is termed in the law on environmental pollution control as "solid waste generated in industrial production activities." Therefore, this encompasses a broad variety of ISW sources. Any undesired solid products, liquids, gases, or mixes released or discharged from any industrial operation are referred to as industrial waste. Industrial waste can come from a variety of sources, including chemicals, mining, coal, metallurgy, petroleum, lightening industry, manufacturing, communications, power, construction materials, electronics, glass and metal reforming, and medical sectors. Solid waste is produced in a variety of ways by various types of industrial activities. They primarily consist of waste intermediates, raw substances, equipment, nonconformities, offal, and offcut produced in manufacturing, as well as substances produced from facilities related to control pollution. Two categories of solid wastes related to the industry are classified according to the sources. The by-products from the manufacturing process are the first type. Invalid raw materials or products are another type. On the other hand, waste created in industrial enterprises comes from everyday life and office operations; one important point is that transportation-related waste is not considered ISW in general.

The industrial waste is considered as one of the most dangerous problems affecting the world because of the environmental toxicity depending on soil pollution [3]. The maximum pollution effects are determined from the textile dyeing factory wastes [4, 5]. The SW created from the manufacturing process separately is defined as ISW, which can pollute air, groundwater, and soil. ISW includes all kinds of the industrial wastes such as dangerous, nondangerous, recyclable, and nonrecyclable wastes [6, 7]. ISW production rises steeply, although the protections are regulated and controlled by its management. So, this problem has glowed commonly, and its management has blossomed into a thriving research field, attracting researchers from a variety of disciplines, including environmental, chemical, and civil ones.

10.4 SOURCES OF (ISW)

ISW is generated by a variety of industries or departments. Furthermore, even within the same industry, the quantity, contents, properties, and types of solid waste are so complicated because of the different kinds of products, production methods, and raw substances. Three main sources of ISW will be discussed below based on the industrial process's nature:

- *First type*: It deals with the products or the raw substance which had lost or had their original use value depleted but have not changed in form. Raw materials that have expired or become contaminated, as well as products that have been rejected or are nonconforming, fall into this category.

- *Second type*: The by-products of manufacturing processes. This category includes substances such as a variety of manufacturing processes' liquid waste, the dangerous waste residues, waste from raw material extraction, and a variety of chemical reactions' waste. The material balance principle should be followed by industrial production (or industries) in this process. This process generates waste gas, products, and wastewater, with the remainder being solid waste.
- *Third type*: The wide industrial facilities, equipment, and pollutants, such as waste dismantling, polluted goods, and heavily contaminated land.

10.5 CLASSIFICATION OF (ISW)

Because of the global industrialization and development, the ISW produced by industrial activities is regarded as a major real problem with a long-term impact. ISW includes many kinds and quantities which differ depending on the raw substances, industry type, manufacturing process techniques, and production capacity. ISW classification is considered as a critical step in defying the problem and also the environmental impact justification [8–10]. Based on their pollution, nature, industrial sectors, characteristics, and industrial processes, the ISW can be divided into four categories. Figure 10.2 shows the ISW classification diagram. Locally in Egypt, the construction materials industrial sector (marble cement, ceramic, granite bricks, and steel) generates the most solid waste.

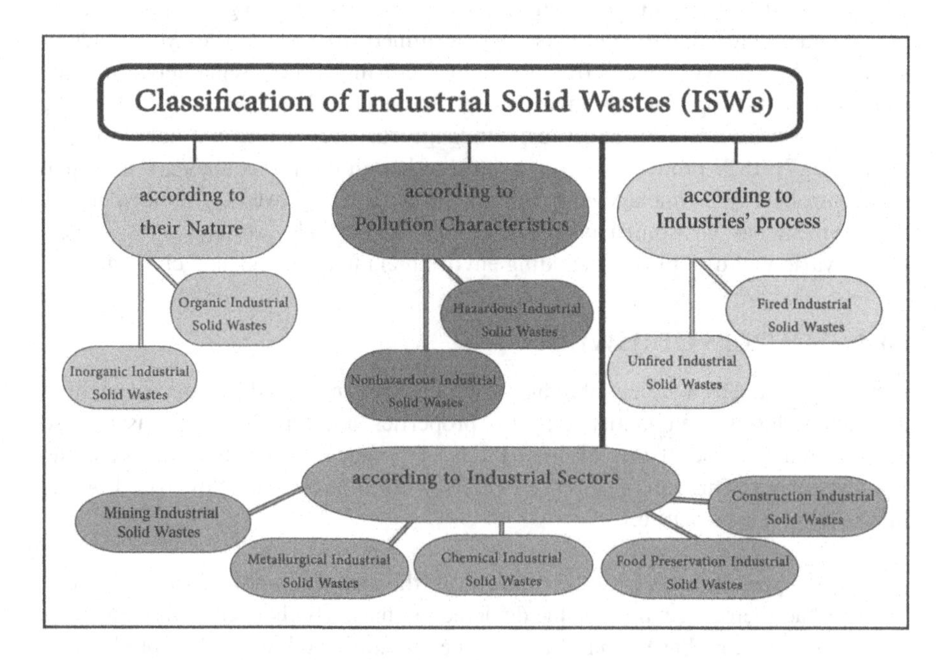

FIGURE 10.2 The classification of industrial solid waste.

10.5.1 The Classification of Industrial Solid Wastes Referring to Nature

The ISWs were classified into two major categories depending on their nature. The organic ISW is the first category which has an organic nature or its chemical composition has organic materials. The organic ISW is produced by a variety of industries, including food preservation, wood manufacturing, painting, oil extraction, tanning, water treatment, dyeing, and plastic manufacturing. The amount of organic ISW and its physical shape were determined by its source and the technologies used. Organic ISW sources can be arranged in the following order based on their capacities: wood mills, water treatment plants, oil extraction, food preservation, plastic, painting, tanning, and dyeing. In addition, the organic ISW was discovered in herbal environments such as food preservation, wood waste, and oil extraction factories [6–8, 11]. Inorganic ISW is the second category which has inorganic chemical materials in its chemical composition [9, 10]. The inorganic ISW was created by various industries such as ceramic, cement, and granite manufacturing. The most dangerous solid waste product is cement dust which produces in huge quantities. The inorganic ISWs have a greater environmental impact and hazard effect than the organic ISWs. The inorganic ISW considers composite materials that contain metal oxide and sulfate variations [9, 10, 12–14].

10.5.2 The Classification of ISW Referring to the Pollution Characteristics

The ISW was classified into hazardous and nonhazardous ISW based on pollution characteristics. Hazardous ISW is waste with hazardous properties, which requires special considerations in accordance with hazardous waste identification standards. However, hazardous waste is defined broadly as waste that, as a result of its transportation, disposal, storage, or treatment, causes severe illness, or raise in mortality, or poses a prospective hazard to both health and the environment because of its chemical, physical, concentration, and infection characteristics or quantity. Dangerous waste is the waste with toxicity, chemical activity, and erosion which can affect both the environment and human health [15]. Nonhazardous ISW is waste with no hazardous properties, which does not require special handling or removal. The Nonhazardous ISW includes waste that has been reused or recycled, such as textile industry waste, food industry waste, marble industry, and waste granite [8, 11].

10.5.3 The Classification of ISW Referring to the Industrial Sectors

The ISW is classified into many categories depending on the industry sectors. Mining industry solid waste is the first category that consists of the solid waste created through the mining processes like tailings and stones. Metallurgical industry solid waste is the second type of solid waste that consists of metallurgical slag and the metal and nonmetal processing [16]. Chemical industry solid waste is the third category of waste that consists of the outgrowth, unwanted products, unreacted raw materials, acidic and alkali slag, and impurities. Moreover, waste medicines, pharmaceutical waste, waste insecticide sludge, and waste pesticides are produced from the wastewater treatment [17, 18]. Food preservation industry solid waste is the fourth category; this waste includes roots, damaged seed, stock, dust, and soil. Solid waste produced from the construction industries is the last category, which produced an extensive series of wastes such as ceramic waste, cement dust, granite, steel waste, marble waste, and brick waste.

10.5.4 The Classification of ISW Referring to the Industrial Process

ISW is classified, according to the industrial process, into fired ISW that contains the waste from firing processes like fired brick waste, cement dust, fired ceramic waste, and fired steel waste [9, 10]. The second type is the unfired ISW that contains the nonfired waste produced from manufacturing processes like granite and marble waste, food preservation waste, paper industry waste, and unfired ceramic waste [8, 11].

10.6 INDUSTRIAL WASTE MANAGEMENT

To wrap up, industrial waste can be hazardous and nonhazardous. Both, however, can cause crucial damage to the environment if not properly managed. The industrial sector's waste should be effectively and properly managed because in this sector certainly, employees could be affected by hazardous waste.

Since a wide variety of wastes can technically be categorized as industrial waste, it's vital that the different types of raw materials businesses use be determined. By determination, it will be easy to know the ways to reuse and recycle those wastes into sustainable green energies [19]. Finding eco-friendly solution could save a lot of costs and will be then be reflected on economy and can shift the country into another area.

Every region has its own law of waste management as some substances are not considered hazardous in all countries or regions. Chemical waste is contemplated as the most harmful waste, and by harmful doesn't mean it is always hazardous unless it has toxicity characteristics. Making researches and finding new ways to valorize those waste is very important. Factories should be dedicated to use those ways by finding companies to prevent harms as also it will impact factory reputation. Putting standards and regulations will guarantee global safety and unpolluted community by reusing the waste into beneficial ways. That is why many industrial waste generators work with a reputable disposal company to help them manage this process and alleviate any issues that may arise from the transportation and disposal of their waste – especially once it leaves your facility.

10.6.1 Life Cycle Assessment

More information about sorting plants are shown in the following figures. This scenario (Figure 10.3) relies on a facility that can separate waste into several components, such as iron-bearing materials from nonferrous materials and the heavy fraction of waste. The organic fraction consists primarily of kitchen trash, while the inorganic fraction consists primarily of plastics, paper, cardboard, wood, textiles,

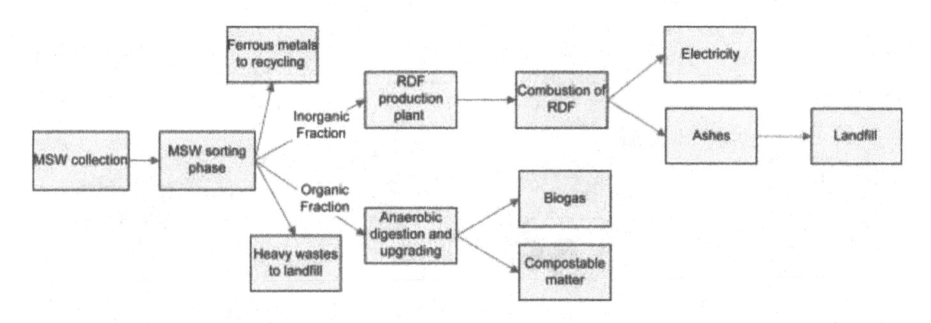

FIGURE 10.3 The main steps involved in Scenario 2 diagram [20].

and rubber. An exergy-based allocation is applied to each output following the sorting phase of municipal solid waste (MSW). As a result, the inorganic fraction bears 79% of the environmental costs, the organic sector 29%, and ferrous metals 2%. Heavy waste buried in landfills is not a product, and as a result, it cannot be tracked. There is no need for further allocation after this step, as all flows have been treated independently (biogas is assumed as the main product of the anaerobic digestion step, and no burdens are assigned to the compostable matter, as well as electricity is the main product of the combustion step and ashes are landfilled) [20]. In order to design a proper waste management system (WMS), an objective comparison of alternatives must be made between various processes, technologies, and methods. An LCA can be used to compare possible alternative scenarios and create an evaluation grid where environmental parameters are reported. It was the goal of this study to compare the environmental effects of four different technological and economic scenarios that had previously been examined. The base case scenario was based on a real waste management system implemented in Caserta Province, a southern Italian region with a population of 924,614 people in Campania.

As can be seen in Figures. 10.4 and 10.5, there are a total of five possible futures for the province of Tehran's MSW management practices, including those that are

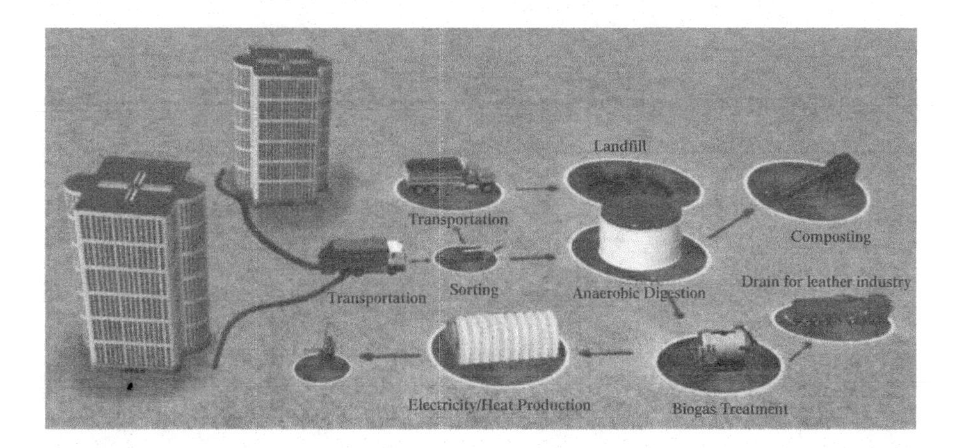

FIGURE 10.4 Diagram of scenario Sc-0 (currently in use) [21].

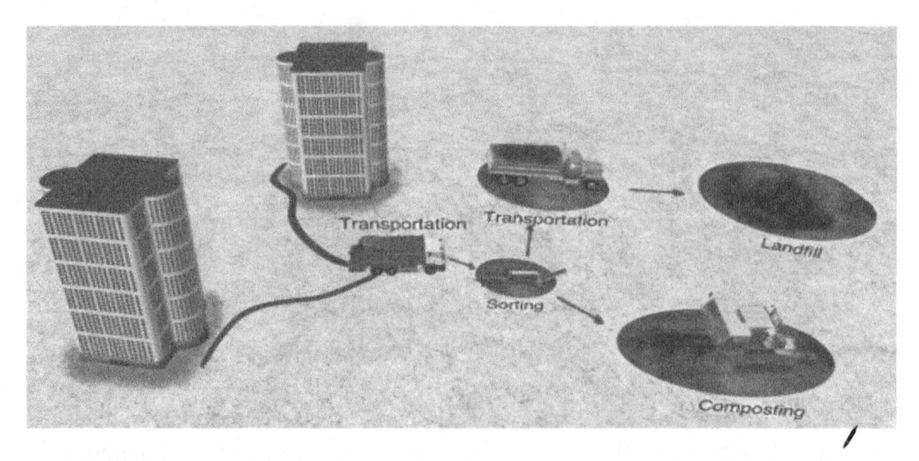

FIGURE 10.5 Diagram of the main steps encompassed in scenario Sc-1 (fading scenario) [21].

already in use and those that are yet to be implemented. Case Sc-0: currently in-use scenario. The AD process is used to manage MSW in this scenario. MSWs are first taken to the sorting plant at the center. Separation of recyclable materials such as aluminum and polyethylene terephthalate (PET), as well as glass, iron sheet, cardboard, and plastic bags, is carried out in this procedure (contribute 3.04 percent per tonne of waste). In addition, the AD plant receives organic materials that have been separated (contribute 48 percent per tonne of waste) [21].

10.6.2 Processing and Control

In Europe and around the world, district heating networks (DHNs) can play an important role in reducing reliance on fossil fuels and, as a result, helping to meet the 2050 climate goals. A dynamic model of the heating plant at Mo District Heating depicted in Figure 10.6 has been developed using the modeling language Modelic.

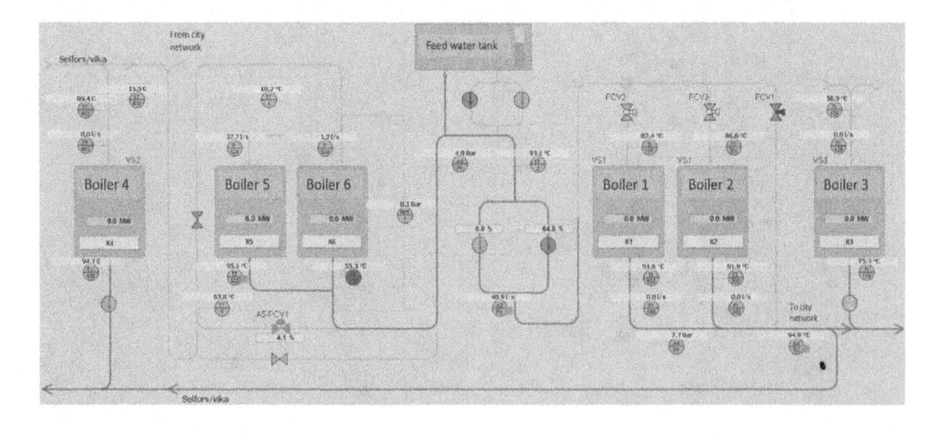

FIGURE 10.6 Human machine interface of the heating plant at Mo District Heating [22].

10.6.3 Treatment and Pretreatment

The increase in agriculture and food accompanied with the increase in demand for agricultural products has led Morocco to opt for increasing the cultivated area. However, the use of fertilizers remains moderate and currently does not exceed 33% of the nutrition requirements per hectare, according to the report "Use of fertilizers by crops in Morocco," published by the FAO in 2006.

It was found that the initial inoculums' lactic acid bacteria content had a significant impact on the saccharolytic efficiency, which was determined by tracking the enzymatic kinetics. After 30 hours of incubation with a YEA/LAB ratio of 20/80, the highest activity was measured at 43.900 IU/l. When YEA/LAB ratio is 20/80, pH drops to 4.14 for all inoculum, which corresponds to an increase in the rate of sucrose inversion (Figure 10.7) [23].

Biomass (Figure 10.8) was plotted on three-dimensional curves to show how each factor interacted with each other. Each graph shows the combined effects of two independent variables, one of which is set to zero. It is much easier to locate the best experimental conditions when using these three-dimensional curves to show the interaction between two variables. An infinite number of possible combinations of the two process variables resulted in each of these response surface curves, which each had the maximum response with the smallest ellipse. Using this ellipse as a reference point, we can determine the best values for a given designed variable. SQR-T037 was manufactured [24].

For the incorporation of nutrients into the soil, biochar is becoming more and more popular in recent years, recent Biochar "fine-tuning" studies for specific applications

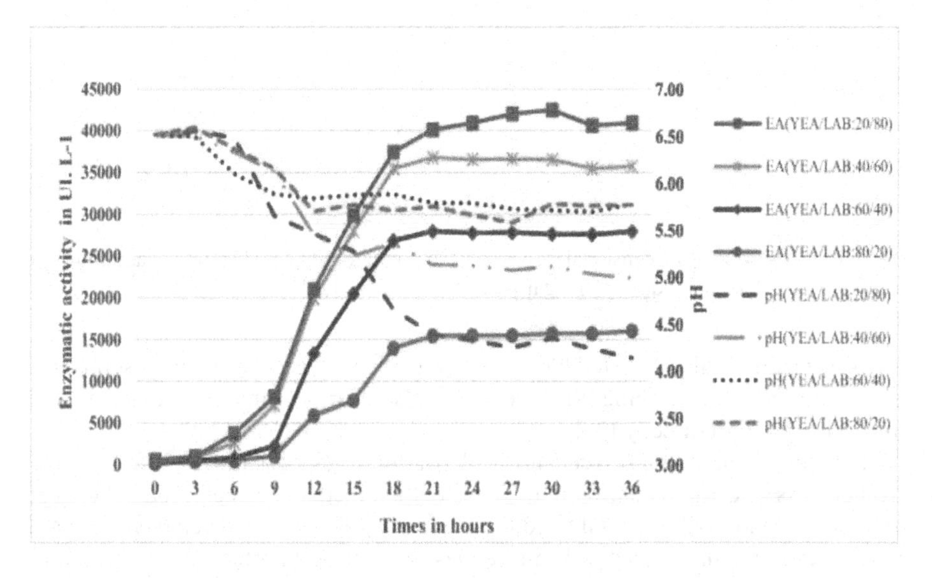

FIGURE 10.7 Changes in enzymatic activity of the biomass of different inocula [23].

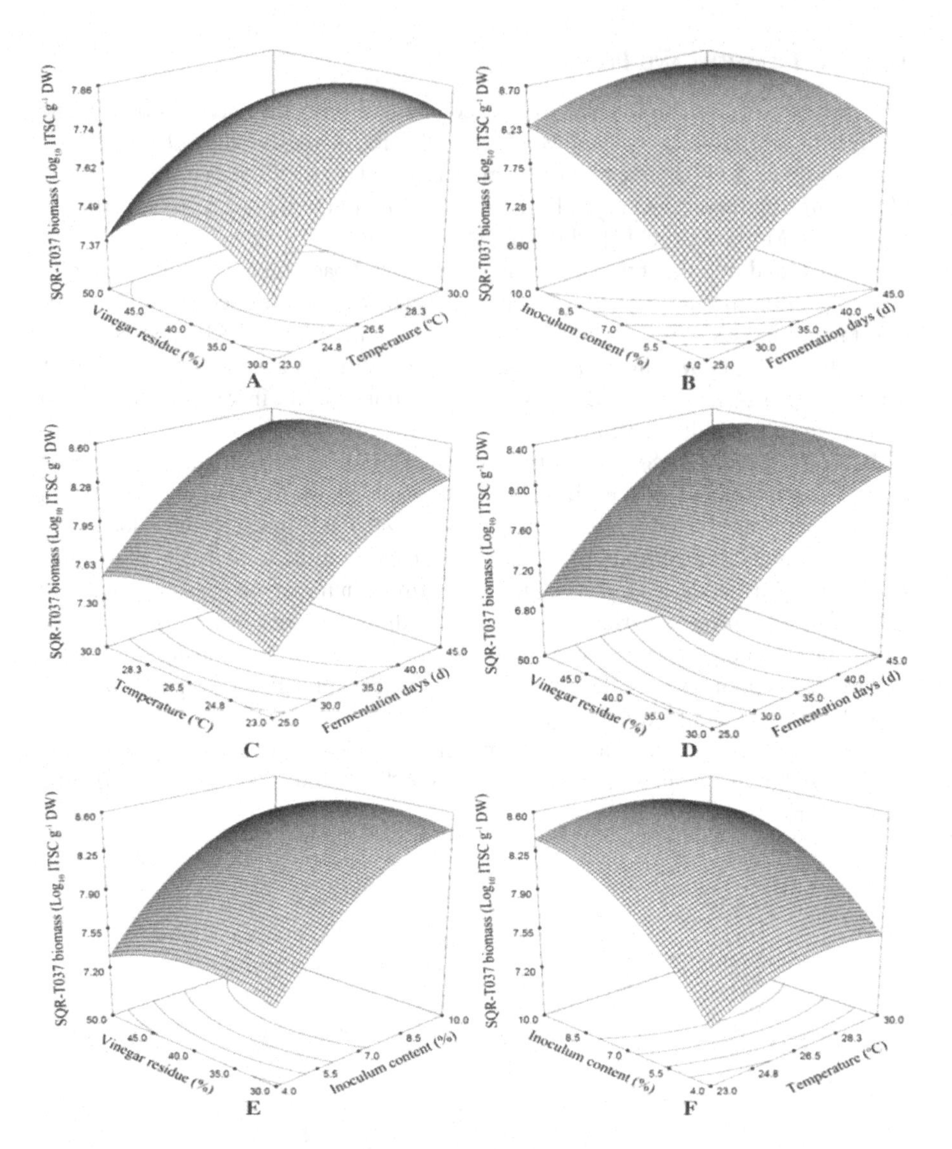

FIGURE 10.8 Three-dimensional response surface plot for SQR-T037 biomass as a function of variables. Vinegar residue = Vinegar production residue [24].

Nitrification is also affected by biochar (Figure 10.9). Biochar reduces soil NH_4+ concentrations by adsorbing NH_4+ ions and also affects ammonium oxidants and nitrifying bacteria directly [25].

Explain the underlying mechanism of bag-filter gas dust (BGD) effect on the AD process of cattle manure. First, biogas output, methane content, total chemical oxygen demand (TCOD), total solid (TS), and volatile solid (VS) removal rates are used to examine the impact of bag-filter gas BGD on AD performance. Second, to

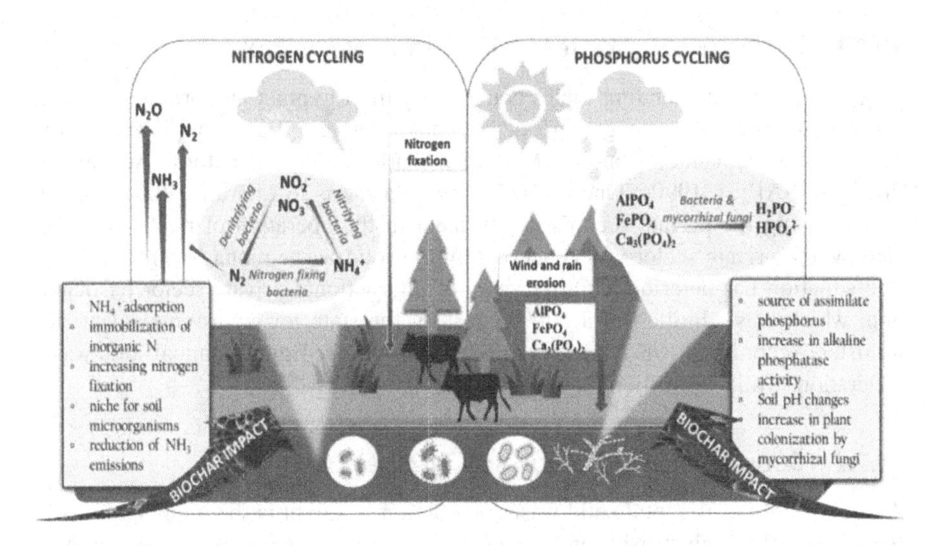

FIGURE 10.9 Biochar impact on nitrogen and phosphorus cycle [25].

examine the potential of residual digestate with BGD, thermal analysis and fertility tests are used [26–30].

Using crude glycerol as an alternative carbon fermentable substrate, evaluate a realistic technique for the cost-effective manufacture of yeast lipid and cell-bound lipase. In addition, to boost the output, fish waste hydrolysate and different surfactants such as gum Arabic, Tween20, and Tween80 were mixed in with the waste, with fuel characteristics calculated based on fatty acid composition in each combination [31–35].

The waste combinations were also evaluated to see whether there were any additional expenses associated with adding expensive other nutrients. In addition, a two-phase technique for cell development and lipid accumulation was used to try to boost the selected yeast's lipid output. Finally, the direct transesterification of yeast lipid into biodiesel was assessed and improved [36–39].

10.7 EGYPT GOES GREEN

As a developing country needs legal and institutional restructuring, Egypt, which is considered the 6th largest Arab world country with over 22 million tons of waste production annually, increasingly has the waste management challenge.

The sector's development has been hindered by a lack of effective national policies and programs, also among other things a lack of mandatory infrastructure for storage, collection, handling, processing, recycling, treatment, and using waste as an energy source. Recently, the world has recognized the need of directing all countries' efforts toward achieving the Sustainable Development Goals, which were accepted in 2015 by all United Nations members, as follows: "A blueprint to achieve a better and more sustainable future for all by 2030."

10.7.1 The Industrial Waste Management System in Egypt

Among the most demanding topics confronting the Egyptian authorities, the solid waste management challenges are of the major interest. Egypt implemented the International Monetary Fund's (IMF) and World Bank's Structural Adjustment Program (SAP) in 1990s. One of the reform goals was to liberalize the economy by increasing the private sector's involvement in the operation of public projects. Hereby, the private sector was concerned in the solid waste management in Egypt. The situation has deteriorated after years of international private sector participation, which was admitted in a 2009 report by the state for environmental affairs minister. According to the report, 75 million tons was the Egypt's annual solid waste generation, while 20 million tons was the municipal waste [26].

10.7.2 Solid Waste Production and Treatment in Egypt

A total of 25% of the total solid waste production in Egypt is the municipal waste. Because of the high density and a lack of management systems, the situation in Greater Cairo is critical [26]. Due to the lack of weighing amenities at removal spots and the lack of a tradition of waste selection and analysis, waste production and removal statistics in Egypt can be defective [27]. Furthermore, waste quantities and types effectively depend on the location and urban patterns, Figure 10.10.

10.7.3 The Egyptian Legal Framework for Solid Waste Management

Solid waste management legal framework is dispersed across multiple parts of law-making due to the Egyptian solid waste management law lacks. The two most important laws are Law No. 38/1967 for General Public Cleaning and Law No. 4/1994

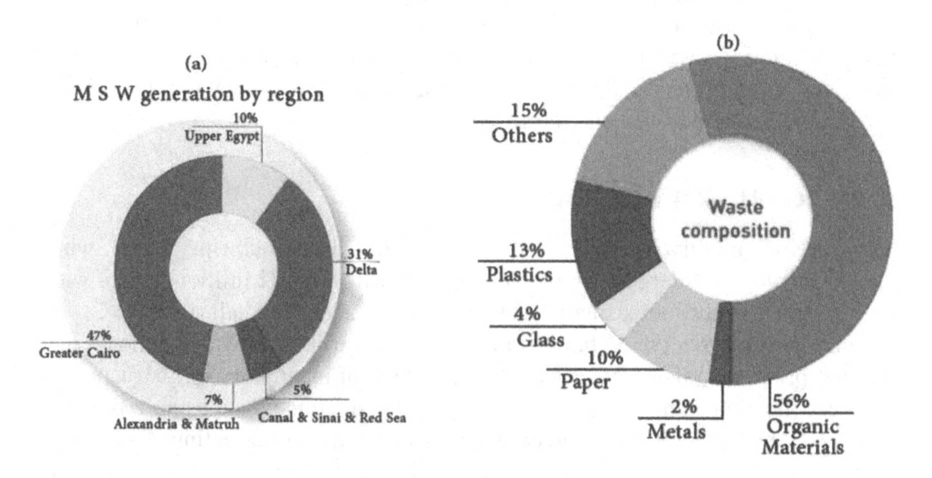

FIGURE 10.10 (a) Solid wastes in Egypt by region [28] (b) Solid waste composition in Egypt [28].

for Environmental Protection. Additional legislative updates from 2005 to 2010 are listed below [28]:

- Law No. 10/2005, establishing a solid waste collection fee system.
- Prime Minister Decree No. 1741/2005, amending the Executive Regulations of Law 4/1994, and covering regulations for the selection of sites for recycling, landfilling, and equipment requirements for waste collection and transfer.
- Law No. 9/2009, amending Law 4/1994 and regulating disposal of hazardous waste.
- Presidential Decree No. 86/2010, regulating the closure of existing dumping sites and landfills at Greater Cairo and allocating five new sites outside the belt of Greater Cairo.

10.7.4 Concerns About the Situation of Solid Waste Management in Egypt

- Egypt's total annual municipal solid waste generation has increased by more than 36% since 2000 [28].
- Less than 60% of waste generated is managed by the public and private sectors. The remainder accumulates on streets and illegal dumping grounds, indicating that the management system is largely ineffective [28].
- More than 80% of municipal solid waste generated in Egypt is simply dumped, as the overall recovery rate is less than 11.5 percent [28].
- In Egypt, solid waste management is dispersed across multiple ministries that lack a planning and cooperation concept. So, separately, every ministry achieves its management [26].
- Although the real efforts returned to the Egyptian country and citizens have so tiny progress, the Egyptian government applied many initiatives to improve the waste management sector by the turn of the millennium [28].
- The issue has been exacerbated by the absence of effective regulations and a clear and straight legal framework, besides the limited money resources and municipal authorities' powerlessness to offer cost-effective dependable facilities.
- The authorities never considered or directly addressed the problem of public consciousness and citizen attitude toward the management of solid waste [26].
- The situation is having a negative impact on the environment. Indeed, improper solid waste disposal in waterways and drains has contaminated water supplies, putting Egypt's natural resources and public health at risk.
- As a result of Egypt's poor solid waste management performance over the last decade, the level of street cleanliness has deteriorated significantly, and pollution from garbage incineration has increased significantly [26].

10.7.5 Sustainable Strategy for Solid Waste Management in Egypt

Based on the previously researched leading models in various countries, as well as the current situation in Egypt, an integrated and sustainable solid waste management strategy in Egypt can be developed as shown in Table 10.1.

TABLE 10.1
Sustainable Strategy for Solid Waste Management in Egypt

Main Strategic Axes	Description
National planning	• Developing and implementing a well defined national policy with the primary goal of reducing waste generation. • Establishing a national source separation programme and a reliable collection system.
Legal framework	• Enacting a direct and unified national solid waste management law.
Private sector incorporation	• Adopting appropriate mechanisms to increase private sector participation in the solid waste management system, which must be a major policy within the national strategy. • Finding appropriate mechanism to integrate the informal sector (garbage dealers) in the privatization process.
Finance and cost recovery	• Using the "polluter pays" principle as another important policy of the strategy, in addition to the expanded "producer's responsibility" principle, as a means to improve waste recycling. • The government allocating an annual budget for the solid waste management sector until a sufficient and more advanced cost recovery mechanism is achieved. • Allocate a sufficient budget for investments in waste removal, collection system improvement, and the establishment of transfer stations, recycling centres, and sanitary landfills. • Subsidizing solid waste management projects that reduce the generation of pollutants and greenhouse gas emissions. • Investigating and implementing the potential revenue generated by existing government-owned composting facilities following rehabilitation.
Management and monitoring	• Establishing a solid waste management monitoring and evaluation system in each governorate, with specific roles and responsibilities. • Improving the managerial and marketing skills for waste management staff. • Developing a dependable database by applying precise techniques for measuring the annual generated quantities of various waste types.
Training and capacity building	• Developing and implementing capacity-building and training programmes for municipal staff across the country in the areas of planning, contracting, implementing, and monitoring solid waste management services. • Creating a national capacity building programme to integrate the technical, conceptual, and social skills needed to facilitate multi-stakeholder participation. • Putting in place human resource development programmes for those in charge of operating and maintaining solid waste management equipment.

An action plan must be created in cooperation with the abovementioned strategies.

Firstly, an immediate framework that addresses the key influential issues should be shown in Table 10.2.

TABLE 10.2
Primary Framework for Sustainable Solid Waste Management in Egypt

Main Target	Proposed Actions
Enhancement of solid waste management systems	• Intensifying efforts to remove solid waste accumulation in all urban areas. • As a priority, extend solid waste management services to rural areas, involving nongovernmental organizations and local contractors. • Reducing the performance gap between current performance and strategy targets by increasing collection coverage, closing existing improper dumpsites and establishing sanitary landfills for disposal, improving waste recovery procedures, and implementing source separation and source reduction mechanisms.
Waste valorization	• Developing a formal recycling sector by establishing national central recycling centresl, and setting up standards for recycling industries and recycled products. • Implementing new treatment technologies such as biogas and waste-to-energy projects, as well as establishing recycling centres for specific waste types such as electronic waste.
Public awareness	• Adopting a national communication plan in collaboration with the public media to increase participation and raise awareness, particularly about hazardous waste.
Networking and partnership	• Efficiently restructuring the solid waste management sector (steering committees, terms of reference, members and roles, regular reporting and assessment systems, etc). • Preparing and updating a database of contacts, areas of expertise, and any relevant information about solid waste management institutions and practitioners. • Sharing technical expertise, information, educational resources, and opportunities between the network members and organizations. • Establishing a data hub for Egypt's waste management. • Developing a set of indicators to track network progress. • Establishing an electronic environment for knowledge exchange.
Technical support	• Extending logistical support for the national network as far as possible. • Designing and updating manuals and guidelines on solid waste management best practices. • Assisting in the organization of regular workshops and seminars (to be presented by the network experts). • Providing technical support to fill gaps and assist in the development of a national database that covers all activities in the solid waste management sector. • Assisting with any other programmes aimed at increasing awareness and capacity building.

10.7.6 Major Government Projects

Egypt has begun to direct its efforts toward developing the waste system's infrastructure through the establishment of waste stations, having established both a legal and institutional framework.

On 15 March 2021, the Egyptian government inaugurated Al-Baragil Intermediate Waste Station at a cost of around 23 million Egyptian pounds. The station was designed to hold 2,000 tons of waste per day serving about 500,000 residents.

The Baragil Waste Station is just one of the projects that will help Egypt achieve its goal of establishing an effective waste management system. Similarly, the Egyptian government planned to build nine more fixed intermediate stations in six governorates the previous year, as well as finish the remaining 20% of the infrastructure for Cairo governorate's waste system.

On the other hand, the Egyptian government intends to use the New Administrative Capital as a model for sustainable urban planning by utilizing solar energy, recycling waste, and increasing green areas, thereby avoiding the causes of environmental hazards that plague Greater Cairo and many other governorates.

The megaproject, announced in March 2015, is part of the government's plan to expand urban areas in order to deal with the country's rapid population growth and to improve the country's infrastructure.

To that aim, the Egyptian government contracted with a sustainability and environmental services company, which prepared a work plan for waste diversion using comprehensive and integrated waste management solutions rather than the traditional methods of resorting to landfills.

As a result, with the doors wide open for investment opportunities in the waste management sector, the question remains whether these efforts will be followed by additional steps to achieve the necessary reform and create a better, more sustainable future for the nation.

10.8 CONCLUSION

Industrial waste should still be classified into two classes based on its economic value, despite the fact that this classification has already been made. Economically, the first type has already proven its worth, while the second type of agro-industrial residues has the potential for great scientific research in the years to come.

It can be concluded that biochar-based slow release fertilizer is possible. An abundance of functional groups can be incorporated into biochar through its porous structure and numerous functional groups on its surface. It is possible to use biochar as a fertilizer in and of itself, but the amount of nutrients it contains depends on the type of feedstock and the conditions of the pyrolysis.

REFERENCES

1. Cossu R, Stegmann R, 13.2 - Landfill Planning and Design, Solid Waste Landfilling, Concepts, Processes, Technologies, 2018, Pages 755–772.
2. Memon MA. Integrated solid Waste manage based on the 3R approach. J Mater Cycles Waste 2010,12:30–40.

3. Iqbal M, Abbas M, Nisar J, Nazir A, Qamar A. Bioassays based on higher plants as excellent dosimeters for ecotoxicity monitoring: a review. Chem Int 2019,5:1–80.
4. Iqbal M. Vicia faba bioassay for environmental toxicity monitoring: a review. Chemosphere 2016,144:785–802.
5. Abbas M, Adil M, Ehtisham-ul-Haque S, Munir B, Yameen M, Ghaffar A, et al. Vibrio fischeri bioluminescence inhibition assay for ecotoxicity assessment: a review. Sci Total Environ 2018,626:1295–309.
6. Maczulak AE. Pollution: treating environmental toxins. Infobase Publishing; 2010.
7. U.S. Environmental Protection Agency W., DC, Document no. EPA-833-B-11-001. pp. 1-1, 1-2 Introduction to the national pretreatment program; 2011.
8. Gaafar AA, Ibrahim EA, Asker MS, Moustafa AF, Salama ZA. Characterization of polyphenols, polysaccharides by HPLC and their antioxidant, antimicrobial and anti-inflammatory activities of defatted Moringa (Moringa oleifera L.). Meal Extractnt J Pharmaceut Clin Res 2016,8:565–73.
9. Abo-Almaged H, Moustafa A, Ismail A, Amin S, Abadir M. Hydrothermal treatment management of high alumina waste for synthesis of nanomaterials with new morphologies. Interceram-International Ceramic Review 2017,66:172–9.
10. Amin SK, Roushdy M, Abdallah H, Moustafa A, Abadir M. Preparation and characterization of ceramic nano–filtration membrane prepared from hazardous industrial waste. Int J Appl Ceram Technol 2020,17:162–74.
11. Khamis Soliman N, Moustafa AF, Aboud AA, Halim KSA. Effective utilization of Moringa seeds waste as a new green environmental adsorbent for removal of industrial toxic dyes. J Mater Res Technol 2019,8(2):1798–808.
12. Ahmed HY, Othman AM, Mahmoud AA. Effect of using waste cement dust as a mineral filler on the mechanical properties of hot mix asphalt. Ass Univ Bull Environ Res 2006,9:51–60.
13. Taha R, Al-Rawas A, Al-Harthy A, Al-Siyabi H. Use of cement by-pass dust in soil stabilization. Eng Univ Qatar 2001,14:61–76.
14. Mostafa HM, Rashed EM, Mostafa A. Utilizations of by-pass kiln dust for treatment of tanneries effluent wastewater. 9th International Water Technology Conference, WTC9; 2005, p. 133–41.
15. Duan H, Huang Q, Wang Q, Zhou B, Li J. Hazardous waste generation and management in China: a review. J Hazard Mater 2008,158:221–7.
16. Habib A, Bhatti HN, Iqbal M. Metallurgical processing strategies for metals recovery from industrial slags. Z Phys Chem (N F) 2020,234:201–31.
17. Ukpaka C, Izonowei T. Model prediction on the reliability of fixed bed reactor for ammonia production. Chem Int 2017,3:46–57.
18. Ukpaka CP. Reactors functional parameters for sodium benzoate production: a comparative study. Chem Int 2019,5(2):143–57.
19. Karam MA, Sadan M, Farzadkia M, Mirzaei N, Asadi A. System analysis of industrial waste management: a case study of industrial plants located between Tehran and Karaj. Int J Environ Health Eng 2015,4(1):13.
20. Cherubinia F, Bargiglib S, Ulgiatic S. Life cycle assessment (LCA) of waste management strategies: Landfilling, sorting plant and incineration. Energy 2009,34(12): 2116–2123.
21. Rajaeifar MA, Tabatabaei M, Ghanavati H, Khoshnevisan B, Rafiee S. Comparative life cycle assessment of different municipal solid waste management scenarios in Iran. Renewable Sustainable Energy Rev, 2015,51:886–898.
22. Knudsen BR, Rohde D, Kauko H. Thermal energy storage sizing for industrial waste-heat utilization in district heating: a model predictive control approach. Energy 2021, 234:121200.

23. Atfaoui K, Ettouil A, Fadil M, Asmaa O, Inekach S, Ouhssine M, Zarrouk A. Controlled fermentation of food industrial wastes to develop a bioorganic fertilizer by using experimental design methodology. Journal of the Saudi Society of Agricultural Sciences, 2021,20, (8):544–552.
24. Chen L, Yang X, Raza W, Luo J, Zhang F, Shen Q. Solid-state fermentation of agro-industrial wastes to produce bioorganic fertilizer for the biocontrol of Fusarium wilt of cucumber in continuously cropped soil. Bioresour Technol 2011,102 (4):3900–3910.
25. Marcinczyk M, Oleszczuk P. Biochar and engineered biochar as slow- and controlled-release fertilizers. J Clean Prod 2022,339:130685.
26. Wang K, Yun S, Ke T, An J, Abbas Y, Liu X, Zou M, Liu L, Liu J. Use of bag-filter gas dust in anaerobic digestion of cattle manure for boosting the methane yield and digestate utilization. Bioresour Technol 2022,348:126729.
27. Duan X, Chen Y, Yan Y, Feng L, Chen Y, Zhou Q. New method for algae comprehensive utilization: Algae-derived biochar enhances algae anaerobic fermentation for short-chain fatty acids production. Bioresour Technol 2019,289:121637.
28. Guo X, Sun C, Lin R, Xia A, Huang Y, Zhu X, Show P-L, Murphy JD. Effects of foam nickel supplementation on anaerobic digestion: direct interspecies electron transfer. J Hazard Mater 2020,399:122830.
29. Jia B, Yun S, Shi J, Han F, Wang Z, Chen J, Abbas Y, Xu H, Wang K, Xing T. Enhanced anaerobic mono- and co-digestion under mesophilic condition: Focusing on the magnetic field and Ti-sphere core–shell structured additives. Bioresour Technol 2020,310:123450.
30. Jiang S, Park S, Yoon Y, Lee J-H, Wu W-M, PhuocDan N, Sadowsky MJ, Hu H-G. Methanogenesis facilitated by geobiochemical iron cycle in a novel syntrophic methanogenic microbial community. Environ Sci Technol 2013,47 (17):10078–10084.
31. Louhasakula Y, Cheirsilp B. Potential use of industrial by-products as promising feedstock for microbial lipid and lipase production and direct transesterification of wet yeast into biodiesel by lipase and acid catalysts. Bioresour Technol 2022, 348:126742.
32. Carota E, Petruccioli M, D'Annibale A, Gallo AM, Crognale S. Orange peel waste–based liquid medium for biodiesel production by oleaginous yeasts. Appl Microbiol Biotechnol 2020,104(10):4617–4628.
33. Chamola R, Khan MF, Raj A, Verma M, Jain S. Response surface methodology based optimization of in situ transesterification of dry algae with methanol, H_2SO4 and NaOH. Fuel 2019,239:511–520.
34. Cheirsilp B, Louhasakul Y. Industrial wastes as a promising renewable source for production of microbial lipid and direct transesterification of the lipid into biodiesel. Bioresour. Technol. 2013,142:329–337.
35. Di Fidio N, Dragoni F, Antonetti C, De Bari I, Raspol Galletti AM, Ragaglini G. From paper mill waste to single cell oil: Enzymatic hydrolysis to sugars and their fermentation into microbial oil by the yeast Lipomyces starkeyi. Bioresour Technol 2020,315:123790.
36. An JY, Sim SJ, Lee JS, Kim BW. Hydrocarbon production from secondarily treated piggery wastewater by the green alga Botryococcus braunii. J Appl Phycol 2003,15:185–191.
37. Cheirsilp B, H-Kittikun A, Limkatanyu S. Impact of transesterification mechanisms on the kinetic modeling of biodiesel production by immobilized lipase. Biochem Eng J 2008,42:261–26.

38. Li Y, Zhao Z, Bai F. High-density cultivation of oleaginous yeast Rhodosporidium toruloides Y4 in fed-batch culture. Enzyme Microb Technol 2007,41:312–317.
39. Li M, Liu GL, Chi Z, Chi ZM. Single cell oil production from hydrolysate of cassava starch by marine-derived yeast Rhodotorula mucilaginosa TJY15a. Biom Bioenergy 2010,34:101–107.

11 Reclaimed Irrigation Water Affect Soil Properties and Lettuce (*Lactuca Sativa* L.) Growth, Yield and Quality

Pierre G. Tovihoudji, G. Esaie Kpadonou, Sissou Zakari, and P.B. Irénikatché Akponikpè

11.1 INTRODUCTION

Water is an important factor for crop production and yield improvement, because of its roles in several plant physiological processes. It is, however, a limited resource diversely distributed around the world, of which about 70% is used for agricultural purposes (Sato et al., 2013). The gain in crop productivity is usually attributed to (i) improved crops' genetics, (ii) application of modern irrigation systems and (iii) better agronomic and irrigation management practices (Bennett and Harms, 2011). Hence, the inability to satisfy crop water requirement either by direct rainfall or irrigation constitutes a major cause of crops' yield reduction. The FAO projected an increase of about 50% of the global water requirements for agricultural production by 2050 (Gerland et al., 2014), which is mainly due to the population growth and the negative impacts of climate change (FAO 2017). Such situation has led farmers, especially vegetable crop growers, to look for other water sources, including reclaim water.

Reclaimed water is a wastewater that has been fully or partially treated and ready to be reused for other purposes. It is a common alternative for irrigation in urban agriculture, particularly for vegetable crop production during dry seasons in developing countries (Ungureanu et al., 2020). To date, about 680–960 million cubic meter wastewater is produced daily in the world, and only 5% is reclaimed (treated or partially treated), while the rest is wasted (Lautze et al., 2014). Less than 30% of the wastewater is currently treated in the megacities of the Sub-Saharan Africa (Nyenje et al., 2010), and even less for smaller cities. Wastewater treatment can be regarded as a good alternative for local government in cities with scant water resources, which make the vital resource available for their citizens using constructed wetlands as affordable treatment systems. Reclaim water contains a relative amount of nutrients that help to improve soil fertility and enhance crop yield (Akponikpè et al., 2011;

DOI: 10.1201/9781003354475-11

Castro et al., 2013; Matheyarasu et al., 2016; Börjesson and Kätterer, 2018). For instance, Xu et al. (2010) reported that the yield of leafy vegetables increases by 23% under reclaim water than freshwater irrigation. For fruity vegetables, reclaim water irrigation enhances the yield of tomato, cucumber, and eggplant by 15, 24 and 61%, respectively, compared to freshwater irrigation (Wu et al., 2010). Hence, urban gardeners usually take advantage of the nutrients in reclaim water to boost their vegetable production (da Fonseca et al., 2007; Kalavrouziotis et al., 2008; Matheyarasu et al., 2016). However, the use of untreated wastewater in vegetable crop production may have adverse effects on soils quality, and on the quality of harvested fresh products (leaves, fruits, roots, etc.) that can threaten human health (Kalavrouziotis et al., 2008; Su et al., 2012; Chau et al., 2014). The harmful effect increases with the raising chemical contents and abundance of pathogenic microorganisms in plants' edible parts. Therefore, the use of untreated water may have negative impacts on the quality of crop yields and groundwater.

Wastewater irrigation with salt content ranging between 600 and 1700 $\mu S\ cm^{-1}$ can cause soil salinity and alkalinity issues (Feigin et al., 1991; Jalali et al., 2008; Castro et al., 2013). Moreover, high concentrations of Cd, Ni, Cu and Pb occur in topsoils than reference soils after 30 years of raw wastewater irrigation (Zhang et al., 2008). Reclaim water is also a threat to human health because it usually contains helminth eggs and pathogenic microorganisms (viruses, bacteria and parasites) that are transferred to fruits, roots, leaves and other edible parts of vegetable crops (Hamilton et al., 2006; Wang et al., 2017). In fact, the annual infection risk can range from 10^{-3} to 10^{-1} when irrigation with reclaim waters ceases one day before harvesting the fresh vegetables for consumption and from 10^{-9} to 10^{-3} when the irrigation ceases two weeks before harvesting (Hamilton et al., 2006). Parasites such as *Ascaris lumbricoides* and *Trichuris trichiura* infest leafy and fruity vegetables, with higher infestation level in leafy vegetables than fruity vegetables (Nasiru et al., 2015). *Strongyloides stercoralis* and *Entamoeba histolytica/E.dipaar* are identified as common parasites that affect ten species of vegetables (Abebe et al., 2020), and *Ascaris lumbricoides* and *Giardia lamblia* are more prevalent parasites of fresh vegetables (Amissah-Reynolds et al., 2020).

With regard to its beneficial effects, the use of reclaim water for vegetable crop production is a potential climate smart agricultural (CSA) practice in the West Africa (Kpadonou et al., 2019a). This water source has been used as an alternative for vegetable crop production in Parakou, Republic of Benin (West Africa). The reclaim water in Parakou city consists of untreated wastewater, raw effluent water generated from the municipal market activities or treated and/or partially treated wastewater from the wetland, which have harmful contaminants' contents (Djaouga et al., 2016), especially fecal coliforms including *E. coli* and *S. enterica* (Agossou et al., 2014; and Sare et al., 2021). It is therefore necessary to know the quality of these water sources and understand their influence on the grown vegetable crops, because most of the vegetable growers in Parakou city rely on these sources of water.

The present chapter identifies and quantifies the infestation levels and the water sources' properties and then analyzes their effects on soil quality and vegetable crops. It first evaluates the quality of irrigation waters from various sources for vegetable crop production in Parakou and then analyzes the influence of these water sources on the growth, leaf yield and quality of lettuce (*Lactuca sativa* L.), and on the soil properties. The findings will help to better inform decision-makers and to reduce farmers' and consumers' contamination by untreated water and infested vegetables, respectively.

11.2 MATERIALS AND METHODS

11.2.1 Experimental Site

The study was conducted in the vegetable crop production area, nearby the market wastewater treatment station of Parakou's municipality in northern Benin (09°20'09.5" N and 02°37'38.8" E). The experimental site was a two year fallow before the experiment settled. The climate is characterized by a single rainy season from May to October. The average annual rainfall ranges between 900 and 1200 mm, and the average daily temperature is about 27.5°C. The soil is classified as ferruginous tropical in the French soil classification system which corresponds to Acrisols or Lixisols according to the World Reference Base (Youssouf and Lawani, 2002). Table 11.1 presents the initial properties of the soil. The soil texture is sandy loam (70% of sand, 17% of silt and 13% of clay), and it has a pH of 6.7. The total organic carbon and total nitrogen are 10.

11.2.2 Experimental Design and Plot Management

The experimental design was a randomized complete block with three replications. Two factors were considered: (*i*) sources of irrigation water, with three levels (groundwater, GW; untreated municipal wastewater, MW; and treated wastewater, TW) and (*ii*) fertilization with two levels (inorganic fertilizer at a recommended dose, +F, and the control, without fertilizer, –F). Treated wastewater was collected at

TABLE 11.1
Initial Characteristics of the Topsoil from the Experimental Field

Characteristic	0–20 cm
Sand	70.0
Silt	17.0
Clay	13.0
pH-H_2O	6.7
Electrical conductivity (mS m^{-1})	0.44
CEC ($cmol^+$ kg^{-1})	6.4
Available P-Bray1 (mg/kg)	9.1
Total carbon (g/kg)	5.9
Total nitrogen (g/kg)	0.8
C/N	7.4

the outlet of the microphytes pond of the Parakou's constructed wetland. Wastewater coming from the market of Parakou municipality is channeled to the water treating system to be used as input. The wastewater was first submitted to a primary decantation treatment and then to two levels of secondary microphyte treatment before its use for irrigation. Besides, influent (untreated) municipal wastewater was collected from the main sewer of Parakou. Groundwater was used as a control treatment in this study and was collected from a 1-m-depth (up to the surface of water) well located at the experimental site.

The experiment was settled in the dry season, using lettuce (*Lactuca sativa* L.) as a test crop because it is one of the most consumed exotic fresh leafy vegetable crops in Parakou. Plants at the four to six true-leaf stage were transplanted from a nursery plot (outside the experiment site) to the experimental plots (4 m × 1.5 m each) with 0.30 m × 0.30 m spacing between plants.

The irrigation water amount and frequency were similar for all treatments, following farmer's normal irrigation scheme: four watering cans with a capacity of 10 liters were sprayed twice a day per plot by direct watering, making a total of 13.33 mm of water per day. For treatments with fertilization (+F), 80 N kg ha^{-1} was applied using urea fertilizer (46% N).

11.2.3 Data Collection and Chemical Analysis

During the experiment, the quality of the three irrigation waters was monitored almost every week for five weeks. The chemical properties (biochemical oxygen demand, BOD5; chemical oxygen demand, COD; electrical conductivity, EC; pH; nitrate; ammonium and other cations) and biological characteristics (fecal coliforms, fecal streptococci and helminth eggs) were assessed using standard procedures according to the recommendations of the American Public Health Association APHA (1998) and the French Standard Association AFNOR (1990). Nutrients supplied were estimated as the average nutrient concentration in the irrigation water per ions (NH^{4+}, NO^{3-}, PO_4^{3-} and K^+) and per elements during the growing cycle in terms of total elements (N, P and K). The present study did not account for organic N or P supplied. At the beginning of the irrigation period, nine soil samples were taken in random from the entire experimental field, up to 20 cm depth to form a composite sample for soil characterization (Table 11.1). At the end of the experiment, soil samples were randomly taken from each experimental plot up to 20 cm depth and bulked (composite sample) for physico-chemical and microbiological analysis. For physico-chemical analysis, the soil samples were air-dried and sieved through a 2-mm sieve. The particle size distribution was determined using the pipette method (Gee and Or, 2002). pH (H_2O) was measured potentiometrically in a 1:2.5 soil:distilled water suspension (van Reeuwijk, 1993). Organic carbon was determined using the method described by Walkley and Black (1934). Total N was determined using the Kjeldahl method (Houba et al., 1995) and available phosphorus using the Bray-1 method (van Reeuwijk, 1993). Exchangeable cations (calcium, magnesium, potassium, and sodium) and cation exchange capacity (CEC) were determined after extraction by using ammonium acetate (NH4OAc) at pH 7. The extraction method used was the one described by van Reeuwijk (1993).

The sodium adsorption ratio (SAR) was determined as the ratio of exchangeable sodium divided by the square root of calcium plus magnesium (Ayers and Westcot, 1985). The concentration of heavy metals was determined using the diethylenetriaminepentaacetic acid (DTPA) method. The preparation and extracting procedures were carried out according to Page et al. (1982). Biological characteristics of the soil samples were assessed using the procedure of the French Standard Association AFNOR (1990).

Five weeks after transplantation, plants were harvested by cutting the stems at the soil level. Five plants were taken randomly from each plot, and parameters such as fresh weight, height and number of leaves were measured. Later, three whole plants were oven-dried at 60°C until constant weight for 72 h to determine the dry weight biomass and for further laboratory analysis. Sub-samples of the dried plant material were milled for total N, P, K, Ca, and Mg analysis. Samples were digested with sulfuric acid (H_2SO_4) + salicylic acid + hydrogen peroxide (H_2O_2) + selenium. The total N was determined using an auto-analyzer (Pulse Instrumentation Ltd., Saskatoon, Saskatchewan, Canada) with a colorimetric method based on the Berthelot reaction (Houba et al., 1995). Total P was determined using the colorimetric method based on the phosphomolybdate complex, reduced with ascorbic acid (Houba et al., 1995), and total K was determined by flame emission spectrophotometry.

11.2.4 Statistical Data Analysis

Prior to the statistical data analysis, data were checked for normal distribution using the Anderson-Darling test, and homogeneity of variance was assessed using Levene's test. Thereafter, data were subjected to analysis of variance (ANOVA) using SPSS 21 statistical software. Separation of means was performed using the honestly significant difference (HSD)/Tukey's test at an error probability $P < 0.05$.

11.3 RESULTS

11.3.1 Physico-Chemical Quality of Irrigation Water

The physico-chemical properties of the irrigation water are presented in Table 11.2. Most of the water properties were higher in the TW and MW compared to the GW. The physical properties (EC and pH) of the three water sources were within the recommended ranges of FAO (1–3 mS/cm and 6.5–8.5). The chemical properties were higher in TW and MW compared to GW, whose quality was within the acceptable ranges while TW and MW were not. The nitrate (NO_3^-) and exchangeable cation contents of the irrigation waters exceeded the FAO limits, whereas the ammonium (NH_4^+) content of the three irrigation waters was within the recommended range. Sodium adsorption ratio (SAR) was low in the three waters, and the heavy metals' concentration, i.e., iron, lead, cadmium, and copper, was within the acceptable ranges of FAO guidelines for crop production.

The microbial quality of wastewater is usually measured using the concentration of the two primary sources of water-borne fecal: coliforms and helminth eggs.

TABLE 11.2
Physico-Chemical Properties of the Irrigation Water (± Standard Error, n = 5)

Properties	Units	TW	MW	GW	Normal Range
EC*	mS/cm	0.74 ± 0.22	0.93 ± 0.04	0.64 ± 0.09	1–3[a]
pH*	–	8.04 ± 0.42	7.95 ± 0.86	7.35 ± 0.72	6.5–8.5[a]
BOD5	mg/l	33.9 ± 18.6	17.5 ± 7.5	10.9 ± 3.7	<25[a]
COD	mg/l	82.7 ± 46.9	30.3 ± 16.9	25.3 ± 10.6	<30–160[b]
NO_3^-	mg/l	166.3 ± 82.1	96.1 ± 32.1	16.4 ± 1.2	<10[b]
NO_3-N	mg/l	36.6 ± 18.1	21.1 ± 7.1	3.6 ± 0.3	
NH_4^+	mg/l	6.5 ± 5.3	0.6 ± 0.2	0.5 ± 0.3	1–40[b]
NH_4-N	mg/l	5.0 ± 4.1	0.5 ± 0.2	0.4 ± 0.2	
Mineral N	mg/l	41.6 ± 0.0	21.6	4.0	
Total P	mg/l	16.8 ± 2.6	6.9 ± 0.1	3.2 ± 0.2	
PO_4^{3-}	mg/l	51.6 ± 8.9	21.1 ± 0.2	9.9 ± 0.5	–
K+	mg/l	11.8 ± 2.1	28.5 ± 3.6	17.07 ± 1.2	10–40[b]
Na+	mg/l	48.4 ± 10.3	53.4 ± 0.7	26.9 ± 9.5	50–250[b]
Ca^{2+}	mg/l	34.3 ± 7.2	78.5 ± 8.5	54.1 ± 6.8	20–120[b]
Mg^{2+}	mg/l	21.8 ± 6.8	32.3 ± 3.5	16.3 ± 7.8	10–50[b]
Fe	mg/l	468.1 ± 114.4	431.7 ± 208.2	469.9 ± 220.6	–
Pb	mg/l	78.0 ± 20.0	62.8 ± 30.6	82 ± 14.7	0.05[b]
Cd	mg/l	3.3 ± 0.9	16.0 ± 11.5	6.3 ± 2.3	0.01[b]
Cu	mg/l	< 3	<3	<3	0.05[b]
SAR	mq/l	2.3 ± 0.0	1.2	1.8	4.5–7.9[b]

BOD: Biochemical oxygen demand; COD: Chemical oxygen demand; EC: Electric conductivity; GW: Groundwater; MW: Municipal wastewater; SAR: Sodium adsorption ratio = $Na^+/[0.5(Ca^{2+} + Mg^{2+})]^{0.5}$; TW: Treated wastewater.

* Physical properties

[a] Pescod, 1992 (FAO)

[b] Feigin et al. (1991)

Figure 11.1 illustrates the average rates of fecal coliforms and fecal streptococci in groundwater, municipal water and treated wastewater. The average load in fecal coliforms and streptococci of GW was conformed to WHO recommendation for irrigation water for fresh consumed crops like lettuce ($\leq 10^3$ fecal coliforms in 100 ml, i.e., 3 decimal logarithmic units per 100 ml or log (FCU/100 ml)) (Blumenthal et al., 2000; WHO, 2006). The average load of both microorganisms in TW and MW was higher than FAO standards (3.2 ± 0.3 and 3.9 ± 0.5 log FCU/100 ml for fecal coliforms and 3.5 ± 0.4 and 3.8 ± 0.5 for fecal streptococci in TW and MW, respectively). These values were significantly higher for TW and MW than GW ($p \leq 0.05$).

Table 11.3 shows the abundance of helminth eggs in the irrigation waters, with 10.6 ± 3.2 and 8.2 ± 2.1 eggs/l of helminth eggs for MW and TW, respectively.

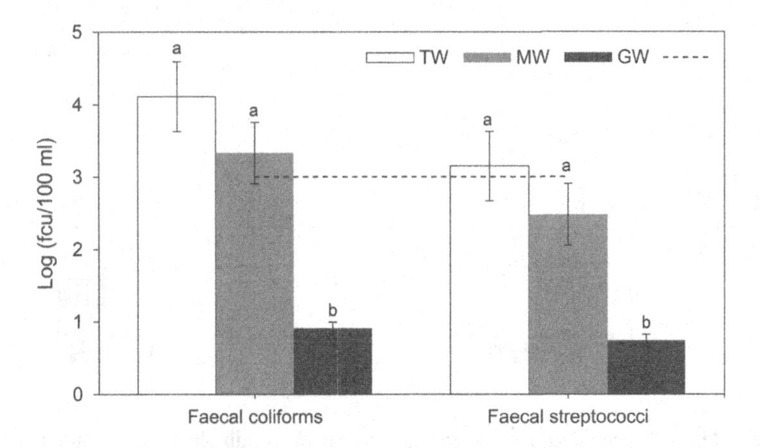

FIGURE 11.1 Average rates of fecal coliforms and fecal streptococci in groundwater (GW), municipal water (MW) and treated wastewater (TW). Error bars denote standard error. For each parameter, different letters mean there is significant difference between treatments ($P < 0.05$) according to HSD test. Horizontal bare denotes WHO standards. (Blumenthal et al., 2000).

TABLE 11.3
Average Helminth Eggs and Presence of *Salmonella* (± Standard Error, n = 5) in Groundwater (GW), Municipal Wastewater (MW) and Treated Wastewater (TW)

Irrigation Water	Helminth Eggs/100 ml (s.e.d.)	*Salmonella* Frequency
TW	8.2 ± 2.1a	5/5
MW	10.6 ± 3.2a	5/5
GW	5.1 ± 2.2b	2/5
WHO revised standard	<0.1	–

Note: Mean followed by similar letters in each column is not significantly different according to HSD test.

These abundances are higher than the accepted limits of WHO (less than 0.1 egg/l) for fresh consumed crops such as lettuce. The *Salmonella* occurrence in the samples was higher (5/5 times) in MW and TW than the GW (2/5 times).

11.3.2 Nutrients Supplied by the Irrigation Waters

Figure 11.2 shows nutrient supply of the three water sources. Overall, N and P contents from GW were lower compared to MW and TW and followed the order TW>MW>GW. However, the K content was higher in MW and followed the order MW>GW>TW. The N, P and K contents in TW were 243%, 65% and 46%, respectively, those required by lettuce crops, against 126%, 27%, and 111% in MW.

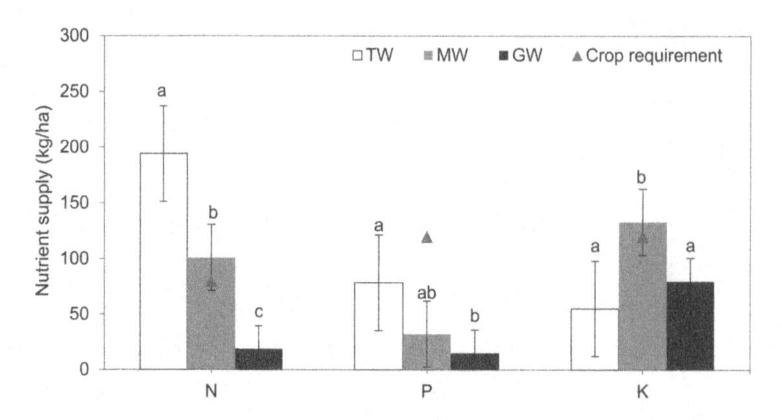

FIGURE 11.2 Nutrient supply by groundwater (GW), municipal wastewater (MW) and treated wastewater (TW), and lettuce nutrient requirement. Error bars denote standard error. For each parameter, different letters mean significant differences between treatments ($P < 0.05$) according to HSD test.

Only N from TW and MW and the K from MW satisfied the crop requirements of these nutrients. In short, the reclaimed waters (MW and TW) contained more nutrients than GW.

11.3.3 Effects of Irrigation Waters on Growth, Yield and Quality of Lettuce Crop

11.3.3.1 Growth and Yields

The growth and fresh yield (marketable leaves) of lettuce significantly increased under wastewater irrigation and inorganic fertilizer application ($P < 0.05$), as shown in Table 12.4. The use of MW and TW significantly enhanced the number of leaves and plant height compared to GW, with the highest leaves and plant height occurring under TW (16 and 24.2 cm, respectively). Moreover, the number of leaves was 58% higher under mineral fertilizer application compared to no application, and the plant height was only 27% higher. Only the plant height was significantly affected by the interaction irrigation water vs. fertilizer ($P \leq 0.01$), with the highest plant height obtained for TW + F and MW + F interaction treatments.

Table 11.4 also presents the effect of the different irrigation waters on fresh biomass of lettuce crop. The fresh biomass of lettuce was 255% higher under TW compared to GW. The addition of mineral fertilizer further significantly improved lettuce fresh biomass ($P \leq 0.05$) by 92% compared to the treatment without fertilizer. However, the fresh biomass was not significantly affected by the interaction irrigation water vs. fertilizer.

11.3.3.2 Quality of the Harvested Leaves: Nutrients (N, P, K, Ca and Mg) Content in the Leaves

Figure 11.3 illustrates the effect of irrigation water on the major nutrient content of lettuce leaves. The N content in the leaves increased by 204% and 118% under TW

TABLE 11.4
Lettuce Growth and Leaf Yield (Average ± Standard Error, n = 3) as Affected by Irrigation Water Type and Chemical Fertilizer

Factor	Number of Leaves	Plant Height (cm)	Fresh Yield (t/ha)
Irrigation water			
TW	16 ± 1b	24.2 ± 3.1b	6.4 ± 0.6b
MW	12 ± 2b	18.7 ± 3.9b	2.6 ± 0.7a
GW	8 ± 2a	11.5 ± 4.1a	1.8 ± 0.6a
Fertilizer application			
With (+)	19 ± 0.4a	25.5 ± 0.60a	7.3 ± 2.0a
Without (–)	12 ± 0.5b	20.1 ± 0.64b	3.8 ± 2.7b
ANOVA (P-values)			
Irrigation water (W)	**	*	***
Fertilizer application (F)	***	***	***
Interaction (W × F)	*ns*	**	*ns*

Note: For each factor and parameter, different letters mean significant differences between treatments: GW: Groundwater; ns: Non-significant; MW: Municipal wastewater; TW: Treated wastewater.
* Significant ($P \leq 0.05$)
** High significant ($P \leq 0.01$)
*** Very high significant ($P \leq 0.001$) according to HSD test.

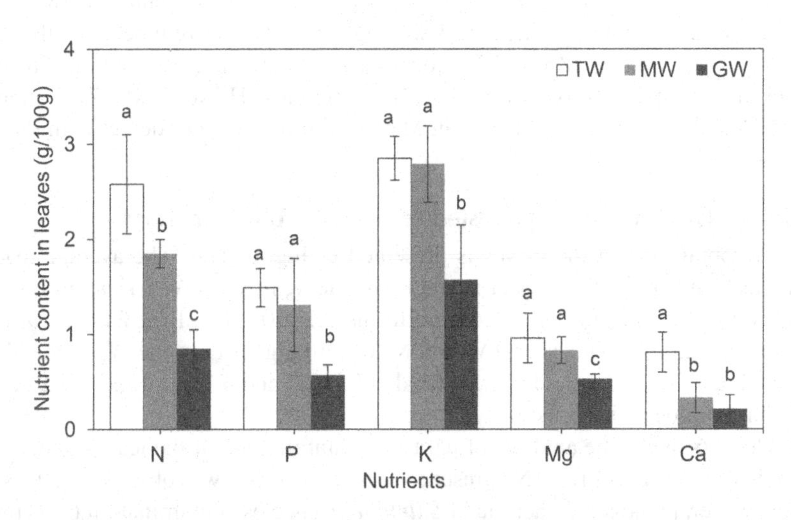

FIGURE 11.3 Macro- and microelement contents (dry matter basis) of lettuce leaves under treated wastewater (TW), municipal wastewater (MW) and groundwater (GW) irrigations. Error bars denote standard error, n = 3. For each parameter, different letters mean significant differences between treatments ($P < 0.05$) according to HSD test.

TABLE 11.5
Concentration of Heavy Metals (Mean ± Standard Error, n = 3) in Lettuce Irrigated with Treated Wastewater (TW), Municipal Wastewater (MW) and Groundwater (GW)

Irrigation Water	Cu (mg/kg)	Fer (mg/kg)	Zn (mg/kg)	Pb (mg/kg)	Cd (mg/kg)
TW	3.5 ± 0.4b	425.2 ± 14.9	28.9 ± 3.0b	3.6 ± 0.5b	ND
MW	3.2 ± 0.1b	477.9 ± 52.4	26.3 ± 2.6b	4.5 ± 0.8b	ND
GW	1.1 ± 0.6a	332.4 ± 35.1	18.3 ± 3.1a	1.3 ± 0.2a	ND
P-values	**	ns	*	**	–

ND: Not detected
Note: For each parameter, different letters mean significant differences between treatments:
ns: not significant ($P > 0.05$)
* Significant ($P \leq 0.05$)
** High significant ($P \leq 0.01$)
*** Very high significant ($P \leq 0.001$) according to HSD test

and MW compared to GW, and the P, K, Ca, and Mg raised by 161% and 130%, 82% and 78%, 81% and 57%, and 286% and 57%, respectively.

11.3.3.3 Quality of the Harvested Leaves: Heavy Metal (Cu, Fer, Zn, Pb and Cd) Content in the Leaves

The heavy metal content in lettuce leaves per treatment is presented in Table 11.5. The heavy metal content in the leaves was significantly different between the three water sources. Pb, Cu, Zn and Fe contents in the dry leaves were significantly higher in MW and TW irrigations than GW irrigation. However, the Zn content in the dry leaves was higher in TW than MW. Cadmium was not detected in the leaf samples.

11.3.3.4 Quality of the Harvested Leaves: Microbial Analyses

The microbial load in the leaves is presented in Figure 11.4. The average load of fecal coliforms and streptococci in the lettuce leaves from GW irrigation fits with the WHO standard limits ($\leq 10^3$ fecal coliforms per 100 g, i.e., 1 log fcu/g). However, the average microbial loads in TW and MW were higher than the WHO standards (4.1 ± 0.2 and 3.3 ± 0.3 log fcu/g for fecal coliforms and 3.1 ± 0.1 and 2.5 ± 0.1 log fcu/g for streptococci, respectively).

Table 11.6 shows the analysis of parasite contamination of the leaves under various irrigation water sources. No presence of helminth eggs was observed in the samples; however, pathogenic bacteria of *Salmonella* species contamination occurred in all the leaf samples under MW and TW irrigations, and *Salmonella* was found only twice over five in leaves under GW irrigation.

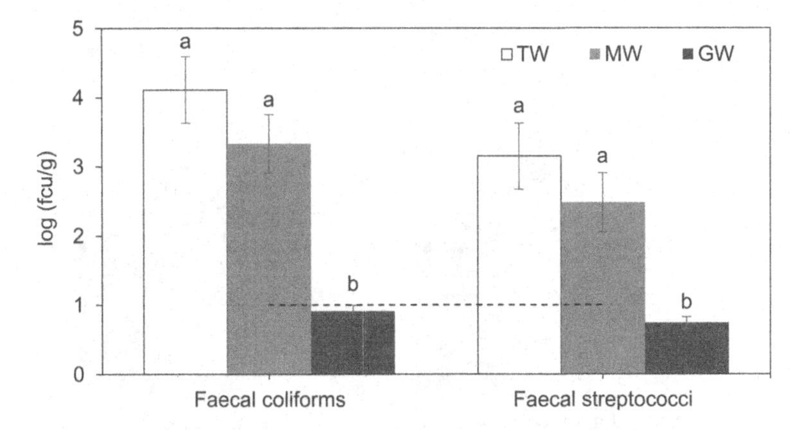

FIGURE 11.4 Average rates of fecal coliforms and fecal streptococci in lettuce leaves irrigated with groundwater (GW), municipal wastewater (MW) and treated wastewater (TW). Error bars denote standard error. For each parameter, different letters mean significant differences between treatments ($P < 0.05$) according to HSD test. Horizontal bar denotes WHO standards. (Blumenthal et al., 2000.)

TABLE 11.6
Presence of Helminth Eggs and *Salmonella* Contamination of Lettuce under Treated Wastewater (TW), Municipal Wastewater (MW) and Groundwater (GW) Irrigations (n = 5)

Lettuce Leaves	Helminth Eggs/g	*Salmonella* (n = 5)
TW	Not found	5/5
MW	Not found	5/5
GW	Not found	2/5

11.3.4 Effect of Irrigation Waters on Soil Nutrients and Microbial Contamination

11.3.4.1 Macronutrients (N, P and K) and Heavy Metal Content in the Soil

The N, P and K contents in soil were significantly higher under TW and MW irrigations than GW irrigation, as shown in Figure 11.3. The concentration of total soil Pb and Cd followed a similar trend, with higher values under TW and MW compared to GW irrigation. However, the Cu content was not significantly different among the soils of the three irrigation water sources.

11.3.4.2 Microbial Contamination of the Irrigated Soils

Figure 11.5 shows the rates of fecal coliforms and streptococci, and Table 11.8 presents the occurrence rates of helminth eggs and salmonella in the soil under various

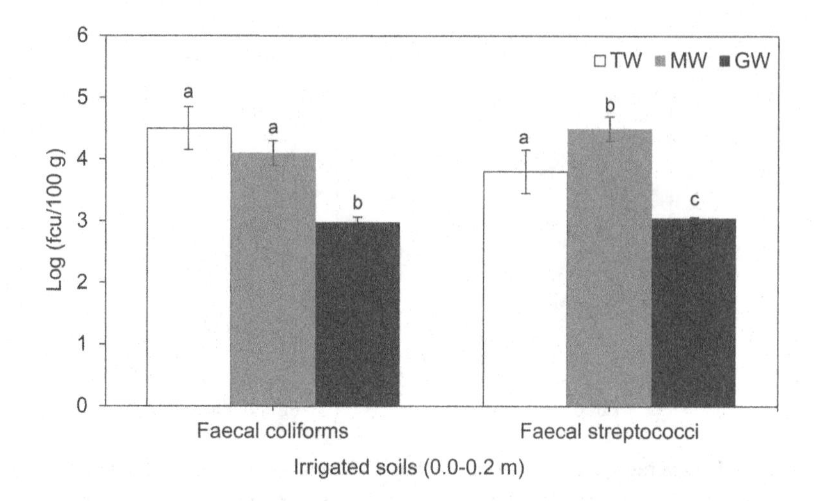

FIGURE 11.5 Average rates of fecal coliforms and fecal streptococci in soils irrigated with treated wastewater (TW), municipal wastewater (MW) and groundwater (GW). Error bars denote standard error. For each parameter, different letters mean significant difference between treatments (P < 0.05) according to HSD test.

irrigation water sources. The content of fecal coliforms and streptococci were higher in MW and TW compared to GW. Moreover, the helminth eggs were lower in soil under GW irrigation (31 ± 21 egg/100 g) than MW and TW irrigations (56 ± 40 and 46 ± 33 egg/100 g, respectively) after the experiment. Salmonella occurred in soil from the three irrigation water sources during the experiment.

TABLE 11.7
Chemical Properties of the 0–20-cm Soil Layer as Function of Irrigation Water Type: TW, MW, GW (n = 3)

Parameters	TW	MW	GW	*P-value*
N (%)	0.11 ± 0.015b	0.09 ± 0.005b	0.05 ± 0.004a	0.001**
P (mg/kg)	89.9 ± 27.0b	91.9 ± 15.6b	31.0 ± 8.2a	0.011*
K (mg/kg)	810.5 ± 22.6b	668.8 ± 13.6b	256.6 ± 46.8a	0.006*
Cd (mg/kg)	4.3 ± 0.7b	3.4 ± 0.7ab	1.7 ± 0.5a	*
Pb (mg/kg)	2.8 ± 1.1b	2.3 ± 1.0b	0.9 ± 1.2a	*
Cu (mg/kg)	4.4 ± 0.6	5.5 ± 1.2	4.5 ± 0.5	ns

TW: Treated Wastewater; MW: Municipal Wastewater; GW: Groundwater.
Note: For each parameter, different letters mean significant differences between treatments:
ns: non-significant (P > 0.05)
* Significant (P ≤ 0.05)
** High significant (P ≤ 0.01)
*** Very high significant (P ≤ 0.001) according to HSD test.

TABLE 11.8
Presence of Helminth Eggs and *Salmonella* in Soils under Treated Wastewater (TW), Municipal Wastewater (MW) and Groundwater (GW) Irrigations (Mean ± Standard Error, n = 3)

Irrigated Soil (0.0–0.2 m Depth)	Helminth Eggs/100g	*Salmonella*
TW (n = 5)	46 ± 33b	5/5
MW (n = 5)	56 ± 40b	5/5
GW (n = 5)	11 ± 7a	5/5

Note: Different letters mean significant differences between treatments ($P < 0.05$) according to HSD test.

11.4 DISCUSSION

11.4.1 Nutrient Supply and Quality of the Irrigation Waters

The results from the present study show a beneficial effect of MW and TW as potential plants' nutrient supply sources. The N supply from TW and MW almost covered the lettuce N requirement. This might be mainly due to the high N content in these sources, consisting of (i) organic N from different points of market waste discharges or disposal (human excreta and solid waste), (ii) direct discharge of inorganic N, e.g., human urine into the market wastewater, and (iii) anaerobic decomposition of organic compounds present in the water itself in the presence of specific microorganisms. The total mineral N was higher in the TW than MW probably because of the mineralization of N during the water treatment through microbial processes. Indeed, the most important stage of the water treatment is the decantation-anaerobic digestion of the organic matter, which is mineralized by the microorganisms (Yadav et al., 2002, Rattan et al., 2005 and Toze 2006). Previous studies reported that wastewater nutrient supply depends on the chemical element involved (N, P, K, etc.), region, methods and levels of the wastewater treatment (Feigin et al., 1991; Bouhoum et al., 2002; da Fonseca et al., 2007; Akponikpè et al., 2011). The results from this study also indicate that N content in TW and MW exceeded the WHO acceptable limits, suggesting possible eutrophication of the surface waters, and pollution of the underground water. These effects could be reduced through moderate or no N, P and K fertilizer application during vegetable crop irrigations with reclaimed waters, thereby reducing the quantity and costs of inorganic fertilizer application.

The microbial pathogen load in TW and MW exceeded the acceptable limits as recommended by Blumenthal et al. (2000) and WHO (2006). This might be due to water contamination through the daily activities of the stakeholders of the market. The pathogen load was lower in TW compared to MW, and the lowest load occurred in GW, mainly because this source of the water (from underground water) was not exposed to human activities. In fact, wastewater contains a wide range and various amounts of microorganisms that are harmful to human beings (Toze 2006; Akpor and Muchie 2011; Chahal et al., 2016;). The prevalence of gastro-intestinal infections in the population will depend on the concentration of these microorganisms in the irrigation water, as well as the consumption of the affected vegetable products (Gerba and Rose, 2003). Therefore, the presence of pathogenic microorganisms in wastewater can pose a serious problem to agricultural use, and hence to human health. The adequate treatment of wastewater is highly recommended before any reuse, more specifically for fresh consumed vegetable crop irrigation.

11.4.2 Effects of Irrigation Waters on Growth, Yield and Quality of Lettuce

The growth and yield parameters of lettuce were affected by the irrigation water sources. The application of TW and MW irrigations significantly improved the crop growth parameters and increased the fresh yield compared to GW irrigation. The high quantities of nutrients present and available for plant uptake in TW and MW than GW, specifically the N content in TW and MW, could be the reason for the stimulation of the vegetative growth of lettuce to produce more leaves and biomass. Previous research similarly reported highest lettuce yield and growth parameters using reclaimed wastewater (Manas et al., 2009; Urbano et al., 2017). The use of inorganic fertilizer together with the reclaimed municipal wastewater significantly increased lettuce growth and yields, because of additional nutrients provided by the inorganic fertilizer. Obviously, the combination of organic and inorganic nutrient sources was more beneficial to crop production than the application of only one nutrient source. Previous authors found that the application of combined inorganic and organic fertilizers significantly improves jute mallow (*Corchorus olitorius* L.) and maize (*Zea mays*) growth and yield (Tavassoli et al., 2010 and Tovihoudji et al., 2015). However, an over-application of the fertilizers such as N, P, K and other micronutrients can lead to toxicity problems, and an excessive application along with reclaimed waters might inhibit crop growth and production (Gaye and Niang, 2002). Indeed, a frequent and long-term use of high nutrient content wastewaters, especially N, can lead to the closure of crop stomata due to osmotic phenomenon, thereby negatively affecting crops' growth and yield (Xanthoulis et al., 2002; da Fonseca et al., 2007; Bedbabis et al., 2010; Singh et al., 2012). This challenge can be overcome through microdosing fertilization technology (Tovihoudji et al, 2017) along with wastewater irrigation for sustainable vegetable crop production (Kpadonou et al., 2019b; Adjogboto et al., 2019).

Lettuce is a fresh consumed leafy vegetable for which the WHO (2006) recommended to avoid wastewater irrigation. Rather, wastewaters in acceptable limits of 1000 CFU/100 ml of total/fecal coliforms can be used to irrigate non-fresh consumed vegetable crops, since the cooking process might eliminate microorganisms (WHO 2006). The lettuce crops obtained using reclaimed waters (treated and untreated) from the wetland of Parakou city were all contaminated, because of the soil contamination and/or direct contact of the leaves with irrigation water through surface irrigation methods. Fardous and Jamjoum (1996) and Cissé (1997) also reported high number of coliforms on the leaves of lettuce and maize plants under reclaimed wastewater irrigations. Here, the leaves from MW and TW irrigations had higher contamination level than GW; therefore, the bacteria contamination of vegetable crops can highly depend on the irrigation water source. Several studies also reported that vegetable crops irrigated with highly contaminated effluents have higher number of pathogens than those irrigated with relatively low pathogens (Armon et al., 1994; Tiimub et al., 2012; Castro et al., 2013; Balkhair, 2016; Al-Quraan et al., 2020). Consequently, the use of high-quality reclaimed water significantly reduces the risk of fresh vegetable crop contamination (Bastos and Mara, 1995).

Bouhoun et al. (2002) and Yacouba et al. (2003) similarly reported that no helminth occurred in the leaves from all the three water sources. However, Akponikpè et al. (2011) reported a high amount of helminth eggs in vegetable crops irrigated with untreated wastewater, and the present study found *Salmonella* species in lettuce leaves under MW and TW irrigations. The contamination of vegetable crops by *Salmonella* highly depended on the quality (presence and concentrations) of wastewater. Unfortunately, GW which led to less contamination, was not sufficiently available on the production sites to cover the entire crops' water requirements. On the other hand, the quality of the wastewater depends on the source of the water, the treatment methods and systems, and canalization system for irrigation purposes, among others. In this regard, Jablasone et al. (2004) found no *Salmonella* in tomato fruits from plants grown in pots and irrigated with water containing 10^5 *Salmonella* per ml, using direct watering method (on the soil surface) every day for five weeks. Apparently, this value fits the minimum tolerable concentration level of irrigation water from which contamination can occur in the vegetable crops, compared to the present study wastewater sources.

The heavy metal accumulation in the lettuce leaves showed relatively low concentrations under the various irrigation schemes. The highest heavy metal concentration occurred under MW and TW irrigations. Heavy metals contained in such irrigation waters may be translocated directly or through the contaminated soil into plants, either by adsorption, ionic exchange, redox reaction, precipitation or dissolution as described by Smical et al. (2008). Several authors also found high heavy metal concentration in the edible parts of some vegetable crops after reclaimed municipal wastewater irrigation (Bedbabis et al., 2010; Singh et al., 2012; Boamponsem et al., 2012; Ahmed et al., 2013). Finally, the heavy metal content in the soil was under the WHO limits, likely because of the relatively short period of irrigation during our experiment.

11.4.3 Effects of Irrigation Waters on Soil Nutrients and Microbial Contamination

The soil chemical analysis reveals a significant change in the soil fertility status. Overall, the soils were more fertile under MW and TW irrigations than GW irrigation. Previous investigations, including long- and short-term studies, showed that soil fertility increases as a result of the application of wastewaters (Singh et al., 2012; Becerra-Castro et al., 2015; Sousa and Figueiredo, 2016; Börjesson and Kätterer, 2018). Thus, the valorization of wastewater as an important source of plant nutrients might contribute to minimize the use of inorganic fertilizers and overcome the decline of soil fertility. However, the continuous application of wastewaters over years, followed by the inappropriate application of inorganic fertilizers, may increase the concentration of salts and heavy metals in soils. In short, wastewater caused a slight translocation of heavy metals in soils probably due to the relatively low heavy metal concentrations in effluents. High levels of heavy metals in soils and their translocation in the various parts of the plant may lead to phytotoxic problems, thereby increasing their concentration in the edible parts of the crops that can threaten human health (Al-Lahham et al., 2007; Zakari et al., 2021).

Fecal coliform and streptococci in soils varied significantly among irrigation waters, with the highest concentration under MW and TW. In fact, some bacteria are adsorbed in soil after watering with wastewater, and their survival is better in alkaline soil than acidic soil (Malkawi and Mohammad, 2003).

11.5 CONCLUSIONS

The growth and yield of vegetable crops were significantly increased under reclaim water irrigations than groundwater irrigation. This resulted from the high amount of macronutrients in the wastewater. Similarly, the combination of inorganic fertilizer with reclaimed waters significantly improved vegetable crops' growth and yield. However, the use of wastewaters resulted in high contamination of the soils and the edible parts of lettuce crops, and the accumulation of heavy metals such as Cd, Cu and Pb in the soils and leaves did not exceed the WHO recommended limits. Frequent and longer period of irrigation using these sources of water may increase the concentration and translocation of the heavy metals in the plants, which can threaten human health. Hence, the quality of the wastewaters should be judiciously investigated (temporal dynamics evaluation) before their application for irrigation purpose. Therefore, based on the findings from this study, we recommended that the untreated municipal wastewater should not be directly used for irrigation in Parakou city. Vegetable famers are advised to use either treated municipal wastewater or underground water from wells. Moreover, reclaim waters must be treated and used under drip irrigation rather than surface irrigation methods (sprinkler, watering cans, etc.) to avoid the direct contact of water with crop leaves. Finally, the characteristics of the market waste materials must be carefully investigated to identify any severe sources of contamination of the wastewater, both from the stakeholder shops (wastewater production sites) and the channel system.

ACKNOWLEDGMENTS

The authors are grateful to the technical team of the "Laboratoire d'Analyse des Eaux de la Direction Départementale de l'Hydraulique de Parakou" and Laboratoire des Sciences de Sol, de l'Environnement et de l'Eau (INRAB/Cotonou) for their assistance in the physico-chemical and biological laboratory analyses. The valuable contribution by the anonymous reviewers is also gratefully acknowledged.

REFERENCES

Abebe, E., Gugsa, G., Ahmed M. (2020). Review on Major Food-Borne Zoonotic Bacterial Pathogens. *J. Trop. Med.* 2020, 2020, 4674235.

Adjogboto, A., Likpete, D.D., Akponikpe, P.I., Diallo, M.B., Kpadonou, G.E., Djenontin, A.J. and Fatondji, D. (2019). Technique de fertilisation microdose sur les légumes feuilles traditionnels au Bénin en Afrique de l'Ouest: performances et recommandations. *Ann. UP, Série Sci. Nat. Agron.* Hors-série n°3, Spécial Valorisation des Légumes Feuilles Traditionnels, Novembre 2019: 9–21.

AFNOR. (1990). Eaux: Méthodes d'essais, recueil de normes françaises. Association Française de Normalisation, AFNOR, Paris.

Agossou, J., Afouda, L., Adédémy, J.D., Noudamadjo, A., N'da Tido, C., Ahohoui, E. S. and Ayivi, B. (2014). Risques de fièvres typhoïdes et paratyphoïdes liés à l'utilisation des eaux usées en agriculture urbaine et périurbaine: cas du maraîchage dans la ville de Parakou (Bénin). *Environnement, Risques & Santé.* 13(5), 405–416.

Ahmed, H.R., Ahmed, H.H., Hashem, E.D.M. and Ahmed, S. (2013). Soil contamination with heavy metals and its effect on growth, yield and physiological responses of vegetable crop plants (turnip and lettuce). *Journal of Stress Physiology & Biochemistry.* 9(4), 145–162.

Akponikpè, P.B.I., Wima, K., Yacouba, H. and Mermoud, A. (2011). Reuse of domestic wastewater treated in macrophyte ponds to irrigate tomato and eggplant in semi-arid West-Africa: Benefits and risks. *Agricultural Water Management* 98, 834–840.

Akpor, O. B. and Muchie, M. (2011). Environmental and public health implications of wastewater quality: Reviewreview. *African Journal of Biotechnology* Vol. 10 (13), pp. 2379–2387, 28 March, 2011 Available online at http://www.academicjournals.org/AJB DOI: 10.5897/AJB10.1797 ISSN 1684–5315 © 2011 Academic Journals

Al-Lahham, O., El Assi, N.M. and Fayyad, M. (2007). Translocation of heavy metals to tomato (Solanum lycopersicom L.) fruit irrigated with treated wastewater. *Scientia Horticulturae.* 113(3), 250–254.

Al-Quraan, N.A., Abu-Rub, L. I. and Sallal, A.K. (2020). Evaluation of bacterial contamination and mutagenic potential of treated wastewater from Al-Samra wastewater treatment plant in Jordan. *Journal of Water and Health.* 18(6), 1124–1138.

Amissah-Reynolds, P.K., Yar, D.D., Aboagye, V., Monney, I., Nuamah, F. and Ndego, E.A. (2020). Parasitic contamination in ready-to-eat salads in the Accra metropolis, Ghana. *South Asian Journal of Parasitology.* 3(4), 1–11.

APHA (American Public Health Association), (1998). Standard Methods for the Examination of Water and Wastewater, 20th ed. American Public Health Association, Washington, DC.

Armon, R., Dosoretz, C.G., Azov, Y. and Shelef, G. (1994). Residual contamination of crops irrigated with effluent of different qualities: a field study. *J. Water Sci. Technol.* 30 (9), 239–248.

Ayers R.S. and Westcot D.W. (1985). Water quality for agriculture. *Irrigation and Drainage.* 29, 99–104. Rev.1: FAO

Bastos, R.K.X. and Mara, D.D. (1995). The bacterial quality of salad crops drip and furrow irrigated with waste stabilization pond effluent: an evaluation of the WHO guidelines. *J. Water Sci. Technol.* 31 (12), 425–430.

Bedbabis, S., Ferrara, G., Rouina, B.B., and Boukhris, M. (2010). Effects of irrigation with treated wastewater on olive tree growth, yield and leaf mineral elements at short term. *Scientia Horticulturae.* 126, 345–350.

Bennett, D.R. and Harms, T.E. (2011). Crop Yield and Water Requirement Relationships for Major Irrigated Crops in Southern Alberta. *Canadian Water Resources Journal.* 36, 159–170. https://doi.org/10.4296/cwrj3602853

Blumenthal, U., Mara, D., Peasey, A., Ruiz-Palacios, G. and Stott, R. (2000). Guidelines for the microbiological quality of treated wastewater used in agriculture: recommendations for revising WHO guidelines. *Bull. World Health Organ.* 78, 1104–1116.

Boamponsem, G. A., Kumi, M. and Debrah, I. (2012). Heavy metals accumulation in cabbage, lettuce and carrot irrigated with wastewater from Nagodi mining site in Ghana. *Int. J. Sci. Technol. Res.* 1(11), 124–129.

Börjesson, G. and Kätterer, T. (2018). Soil fertility effects of repeated application of sewage sludge in two 30-year-old field experiments. *Nutr. Cycl. Agroecosyst.* 112, 369–385. https://doi.org/10.1007/s10705-018-9952-4

Bouhoum, K., Amahmid, O. and Asmama, S. (2002). Wastewater reuse for agricultural purposes: Effects on population and irrigated crops. Proceeding of international symposium environmental pollution control and waste management. *EPCOWM.Tunis*, Part II, 582–586.

Bouhoum, K., Amahmid, O. and Asmama, S. 2002. Wastewater reuse for agricultural purposes: Effects on population and irrigated crops. Proceeding of international symposium environmental pollution control and waste management. EPCOWM, Tunis, Part II, 582–586.

Bouhoun, M. L., Blondeau, P., Louafi, Y. and Andrade, F. J. (2021). A paper-based potentiometric platform for determination of water hardness. *Chemosensors.* 9(5), 96.

Castro, E., Mañas, P. and De Las Heras, J. (2013). Effects of Wastewater Irrigation in Soil Properties and Horticultural Crop (*Lactuca sativa* lL.). *J.ournal of Plant Nutr.ition*, 36(11), 1659–1677. DOI: 10.1080/01904167.2013.805221.

Chahal, C., van den Akker, B., Young, F., Franco, C., Blackbeard, J., & and Monis, P. (2016). Pathogen and Particle Associations in Wastewater: Significance and Implications for Treatment and Disinfection Processes. *Adv.ances in aAppl.ied mMicrobiol.ogy*, 97, 63–119. https://doi.org/10.1016/bs.aambs.2016.08.001

Chau, H.L.Q., Thong, H.T., Chao, N.V., Son Hung, P.H., Hai, V.V., An, L.V., Fujieda, A., Ueru, T. and Akamatsu, M. (2014). Microbial and parasitic contamination on fresh vegetables sold in traditional markets in Hue City, Vietnam. *Journal of Food and Nutrition Research.* 2(12), 959–964. https://doi.org/10.12691/jfnr-2-12-16

Cissé, G. (1997). Impact sanitaire de l'utilisation d'eaux polluées en agriculture urbaine. Cas du maraîchage à Ouagadougou (Burkina Faso). Thèse, Ecole Polytechnique Fédérale de Lausanne, Suisse, 446 pp.

da Fonseca, A.F., Herpin, U., De Paula, A.M., Victoria, R.L. and Melfi, A.J. (2007). Agricultural use of treated sewage effluents: Agronomic agronomic and environmental implications and perspectives for Brazil. *Scientia. Agricola.* 64(2), 194–209.

Djaouga, N.T., Diogo, R.V.C., Baimey, H. and Godau, T. (2020). Développement du gboma (*Solanum macrocarpon* L.) sous l'influence du biochar, de la fréquence d'arrosage et des nématodes à galles en conditions de serre. *Annales de l'Université de Parakou-Série Sciences Naturelles et Agronomie.* 10(1), 33–40.

FAO, (2017). Water for Sustainable Food and Agriculture: A report produced for the G20 Presidency of Germany.

Fardous, A. and Jamjoum, K., (1996). Corn production and environment effects associated with the use of treated wastewater in irrigation of Khirbet Al-Samra Region. Annual Report, NCARTT, Amman, Jordan.

Feigin, A., Ravina, I. and Shalhevet, J. (1991). *Irrigation with Treated Sewage Effluent: Management for Environmental Protection.* Springer-Verlag, Berlin, p. 224.

Gaye, M. and Niang, S. (2002). Epuration extensive des eaux usées pour leur réutilisation dans l'agriculture urbaine: des technologies appropriées en zone sahélienne pour la lutte contre la pauvreté. Etudes et recherches 225-226-227, ENDA Dakar, p. 17-19-20-213-214-216-223.

Gee, G.W. and Or, D. (2002). Particle-size analysis. J.H. Dane, G.C. Topp (Eds.), Methods of Soil Analysis: Part 4 Physical Methods, Soil Science Society of America, Madison, WI, pp. 255–293.

Gerba, C.P. and, Rose, J.B., (2003). International guidelines for water recycling: microbiological considerations. *Water Sci. Technol. Water Suppl.* 3, 311–316.

Gerland, P., Raftery, A.E., Sevcíková, H., Li, N., Gu, D., Spoorenberg, T. and Bay, G. (2014). World Population Stabilization Unlikely This Century. *Science*, 346, 234–237. https://doi.org/10.1126/science.1257469

Hamilton A.J., Stagnitti F., Premier R., Boland A.-M., Hale G. (2006). Quantitative microbial risk assessment models for consumption of raw vegetables irrigated with reclaimed water. *Applied and Environmental Microbiology*. 72(5), 3284–3290.

Houba V., Van der Lee J., Novozamsky I. (1995). Soil Analysis Procedures; Other Procedures (Soil and Plant Analysis, Part 5B) Department of Soil Science and Plant Nutrition. Wageningen Agricultural University, Wageningen, Netherlands (1995), (217p).

Jablasone, J., Brovko, L.Y. and, Griffiths, M.W., (2004). A research note: the potential for transfer of *Salmonella* from irrigation water to tomatoes. *Journal of the Science of Food and Agriculture* 84 (3), 287–289.

Jalali M., H. Merikhpour, H., M.J. Kaledhonkar, M.J. and S.E.A.T.M. Van Der Zee, S.E.A.T.M. (2008). Effects of wastewater irrigation on soil sodicity and nutrient leaching in calcareous soils. *Agricultural Water Management*. 95(2), 143–153, ISSN 0378-3774.

Kalavrouziotis, I.K., Robolas, P., Koukoulakis, P.H. and Papadopoulos, A.H. (2008). Effects of municipal reclaimed wastewater on the macro- and micro-elements status of soil and of Brassica oleracea var. Italica, and B. oleracea var. Gemmifera. *Agric.ultural Water Manag.ement*, 95 (4), 419–426.

Kpadonou G.E., A. Adjogboto, A., D.D. Likpètè, D.D., Z.P. Dassigli, Z.P., P.B.I. Akponikpe, P.B.I. and A.J. Djenontin, A.J.. (2019a). Improving traditional leafy vegetables production through water use efficiency and fertilizer microdosing technology in Benin Republic. *Acta Hortic*. 1238, 55–64, DOI: 10.17660/ActaHortic.2019.1238.7

Kpadonou G. E., P.B. I. akponikpè, P.B. I., J Adanguidi, J., Zougmore, R. B., A. Adjogboto, A., Likpete, D. D., Sossa-Vihotogbe, C. N. A., Djenontin, A. J. and, Baco, M. N. (2019b). Quelles bonnes pratiques pour une Agriculture Intelligente face au Climat (AIC) en production maraîchère en Afrique de l'Ouest? *Annales de l'Université de Parakou* Série « Sciences Naturelles et Agronomie ».

Lautze J., Stander E., Drechsel P., Da Silva A.K., and Keraita B. (2014). Global experiences in water reuse. CGIAR Res. Progr. Water, L. Ecosyst. (WLE). *Int. Water Manag. Inst.* (IWMI), Colombo, Sri Lanka 31.

Malkawi, H.I. and Mohammad, M.J. (2003). Survival and accumulation of microoganismes in soils irrigated with secondary treated wastewater. *J. Basic Microbiol.*, 43 (1): 47–55.

Manas, P., Castro, E. and Heras, D.L.J. (2009). Irrigation with treated wastewater: effects on soil, lettuce (*Lactuca sativa* L.) crop and dynamics of microorganisms. *Journal of Environmental Science and Health*. 44, 1261–1273. https://doi.org/10.1080/10934520903140033

Matheyarasu R., Bolan N.S., and Naidu R. (2016). Abattoir Wastewater wastewater Irrigation irrigation Increases increases the Availability availability of Nutrients nutrients and Influences influences on Plant plant Growth growth and Developmentdevelopment. *Water Air Soil Pollut*. 2016; 227, 253. https://doi.org/10.1007/s11270-016-2947-3. Epub 2016 Jul 5. PMID: 27440946

Nasiru, M., Auta, T. and Bawa, J. (2015). Geohelminth contamination of fruits and vegetables cultivated on land irrigated with wastewater in Gusau Local Government Area, Zamfara State, Nigeria. *Zoo*. 13, 7–10.

Nyenje, P.M., Foppen, J.W., Uhlenbrook, S., Kulabako, R., & and Muwanga, A. (2010). Eutrophication and nutrient release in urban areas of sub-Saharan Africa–a review. *The Science of the tTotal eEnvironment*. 408, 447–55.

Page, A.L., Miller, R.H. and Keeny D.R. (1982). Methods of Soil Analysis. Part 2. Chemical and Microbiological Properties (second ed.), American Society of Agronomy, Soil Science Society of America, Madison, WI, USA.

Pescod, M.B. (1992). Wastewater treatment and use in agriculture. *Irrigation and Drainage.* Paper 47. Rome, Italy: FAO.

Rattan, R.K., Datta, S.P., Chhonkar, P.K., Suribabu, K. and Singh, A.K. (2005). Long-term impact of irrigation with sewage effluents on heavy metal content in soils, crops and groundwater—A case study. *Agric. Ecosyst. Environ.* 109(3), 310–322. https://doi.org/10.1016/j.agee.2005.02.025

Sare, E.B.N., Hounkpatin, A.S.Y., Adjahossou, V.N., Bagoudou, A.F. and Sourou, A.B. (2021). Assessment of bacterial contamination of irrigation water and market gardening products at Parakou (A city in northern Benin). *International Journal of Biological and Chemical Sciences*, 15(1), 241–250.

Sato, T., Qadir, M., Yamamoto, S., Endo, T., Zahoor, A. (2013). Global, regional, and country level need for data on wastewater generation, treatment, and use. *Agric. Water Manag.* 130, 1–13. https://doi.org/10.1016/j.agwat.2013.08.007

Singh, P.K., Deshbhratar, P.B. and Ramteke, D.S., (2012). Effects of sewage wastewater irrigation on soil properties, crop yield and environment. *Agricultural Water Management.* 103, 100–104, ISSN 0378-3774. https://doi.org/10.1016/j.agwat.2011.10.022

Smical A., H. Vasile, H., V. Oros, V., J. Jozsef, J. and P. Elena, P. (2008). Studies of Transfer transfer and Bioaccumulation bioaccumulation of Heavy heavy Metals metals from soil into lettuce. *Environ.mental Eng.ineering and Manag.ement Journal.* Vol.7, pp 609–615, 2008.

Sousa, A.A.T.C. and Figueiredo, C.C. (2016). Sewage sludge biochar: effects on soil fertility and growth of radish,. *Biol.ogical Agric.ulture & Hortic.ulture*, 32: (2), 127–138, DOI: 10.1080/01448765.2015.1093545

Tavassoli, A., Ghanbari, A., Amiri, E. and Paygozar, Y. (2010). Effect of municipal wastewater with manure and fertilizer on yield and quality characteristics of forage in corn. *African J.ournal of Biotechnol.ogy* Vol. 9 (17), pp. 2515–2520.

Tiimub, B. M., Kuffour, R. A. and Kwarteng, A. S. (2012). Bacterial Contamination contamination Levels levels of Lettuce lettuce Irrigated irrigated with Waste waste Water water in the Kumasi Metropolis. *Journal of Biology, Agriculture and Healthcare.* 2(10).

Tovihoudji, G.P., Djogbenou, C.P., Akponikpe, P.B.I., Kpadonou, E., Agbangba, C.E. and Dagbenonbakin, D.G. (2015). Response of Jute Mallow (Corchorus olitorius L.) to organic manure and inorganic fertilizer on a ferruginous soil in North-eastern Benin. *Journal of Applied Biosciences.* 92, 8610–8619.

Tovihoudji, P.G., Akponikpè, P.I., Agbossou, E.K., Bertin, P. and Bielders, C.L. (2017). Fertilizer microdosing enhances maize yields but may exacerbate nutrient mining in maize cropping systems in northern Benin. *Field Crops Research.* 213, 130–142. https://doi.org/doi.org/10.1016/j.fcr.2017.08.003

Toze, S. (2006). Reuse of effluent: benefits and risks. *Agric. Water Manage.* 80, 147–159.

Toze, S., (2006). Reuse of effluent water—benefits and risks. *Agric. Water Manage.* 80,147–159.

Ungureanu, N., Vlăduţ, V. and Voicu, G. (2020). Water scarcity and wastewater reuse in crop irrigation. *Sustainability.* 12(21), 9055. https://doi.org/10.3390/su12219055

Urbano, V.R., Mendonça, T.G., Bastos, R.G. and Souza, C.F. (2017). Effects of treated wastewater irrigation on soil properties and lettuce yield. *Agricultural Water Management.* 181, 108–115.

van Reeuwijk, L.P. (1993). Procedures for Soil Analysis. Technical Paper No 9, (fourth edition), International Soil Reference and Information Center (ISRIC), Netherlands (1993).

Walkley, A. and Black, I.A. (1934). An examination of the Degtjareff method for determining soil organic matter, and a proposed modification of the chromic acid titration method. *Soil Sci.* 37, 29–38.

Wang Z., Li, J. and Li, Y. (2017). Using reclaimed water for agricultural and landscape irrigation in China: a review. *Irrig. Drain.* 66, 672–686.

WHO, (2006). Guidelines for the safe use of wastewater, excreta and greywater. Volume 2, Wastewater uUse in aAgriculture., 213 pp.

Wu, W., Xu, C., Liu, H., Hao, Z., Ma, F. and Ma, Z. (2010). Effect of reclaimed water irrigation on yield and quality of fruity vegetables. *Trans. Chin. Soc. Agric. Eng.* 26, 36–40.

Xanthoulis, D., Rejeb, S., Chenini, E., Khelil, M.N. and Chaabouni, Z. (2002). Optimisation de la reutilisation des eaux usées traitées en irrigation. Faculté universitaire des sciences agronomiques de Gembloux, Rapport de synthèse, 60 pp.

Xanthoulis, D., Rejeb, S., Chenini, E., Khelil, M.N. and Chaabouni, Z. (2002). Optimisation de la reutilisation des eaux usées traitées en irrigation. Faculté universitaire des sciences agronomiques de Gembloux, Rapport de synthèse, 60 pp.

Xu, C., Wu, W., Liu, H., Shi, Y. and Yang, S. (2010). Effect of reclaimed water irrigation on the yield and quality of leaf vegetables. *Journal of Irrigation and Drainage.* 29(5): 23–26.

Yacouba, H., Akponikpè, P.B.I., Wima, K. and Mermoud, A. (2003). Impact de la réutilisation des eaux usées épurées par lagunage à macrophytes flottants sur la culture d'aubergine. *Annales de l'Université de Ouagadoudou – Série C.* 001, 86–105.

Yadav, R.K., Goyal, B., Sharma, R.K., Dubey, S.K. and Minhas, P.S. (2002). Post-irrigation impact of domestic sewage effluent on composition of soils, crops and ground water - – A a case study. *Environ.ment Int.ernational*, 28 (6): 481–486.

Youssouf, I. and Lawani, M. (2002). Les sols béninois: classification dans la base de référence mondiale. Quatorzième réunion du sous-comité ouest et centre africain de corrélation des sols pour la mise en valeur des terres, 9–13 octobre 2000 Abomey, Bénin. Rapport sur les Ressources en Sols du Monde n° 98, FAO, Rome (2002), pp. 29–50.

Zakari, S., Xiaojin J., Zhu, X., Liu, W., Allakonon, G., Singh, A., Chunfeng, C, Zou, X, Akponikpè, P.B., Dossa, G. and Yang, B. (2021). Influence of sulfur amendments on heavy metals phytoextraction from agricultural contaminated soils: A meta-analysis. *Environ. Pollution.* 288, 117820. 10.1016/j.envpol.2021.117820.

Zhang, Y.L., Dai, J.L., Wang, R.Q. and Zhang, J. (2008). Effects of long-term sewage irrigation on agricultural soil microbial structural and functional characterizations in Shandong, China. *Eur. J. Soil Biol.* 44, 84–91. https://doi.org/10.1016/j.ejsobi.2007.10.003

12 Artificial Intelligence and Machine Learning of Petroleum Wastewater Treatment by Nanofilteration Membranes

El-Sayed Gamal Zaki, Shyma Mohamed Elsaeed, and Hassan Hefni Hassan

12.1 INTRODUCTION

Scarcity is the main problem which affects most countries in the world, so we must find nonconventional sources of water. Treatment of wastewater and reuse in different fields such as agriculture, industries, and irrigation of trees is the main goal of our research. Micropollutants are those found in traces up to micrograms per liter in various water bodies. Pharmaceuticals, personal care products, steroid hormones, industrial chemicals, pesticides, polyaromatic hydrocarbons, and other recently discovered compounds make up the bulk of these emerging organic pollutants. This study does not focus on inorganic micropollutants such as heavy metals. Micropollutants only refer to organic micropollutants in this context. Contaminants have been linked to a variety of negative outcomes, including hormonal disruption, acute and chronic toxicity to various species, and the emergence of antibiotic resistance. Continuous release of micropollutants into the environment, even at very low levels, may elevate reproductive and developmental abnormalities in sensitive species. Treatment by nanofiltration (NF) membranes for wastewater is considered the most suitable for countries such as Egypt which has huge amount of wastewater that can be used to overcome water problem. The success of NF membranes so far has been owing to their unique ability to selectively separate the two species sought for separation. So, the future prospective are fouling mitigation and suitable modeling to optimization parameters in the preparation of membranes and its application; also, application of machine learning in the use of NF membranes.

DOI: 10.1201/9781003354475-12

12.2 VALORIZATION OF AGROWASTE FOR MEMBRANE PROCESS

Plantation crops, farming and mariculture, and forestry all fall under the agricultural sector, which generates a significant amount of waste. It's common to see a variety of agricultural waste, including animal, food, and crop waste, in our daily lives. The fact that agricultural and forestry waste and residues are abundant, renewable, and typically require little energy for conversion has long been established [1, 2]. Several industries, including water treatment, biopolymer production, energy generation, food production, and pharmaceuticals, have demonstrated the usefulness of agricultural waste [3]. Adsorbents made from agricultural waste, such as dyes, heavy metals and micropollutants, have been widely used to remove water-polluting contaminants [4–7]. From biofuel to a wide range of high-value pharmaceuticals, biomass has been transformed through biorefineries [8–11]. For example, the production of earth blocks and bricks can be aided by the use of biomass. Polymer and post-consumer plastics also make up a significant portion of the waste stream. Figures from the last 60 years show that 8.3 billion tonnes of plastic products have been produced worldwide. Out of the 6.3 billion metric tonnes of plastic waste produced, only 9 percent [12] of it was recycled. As a low-cost and easily collected post-consumer item, plastic waste can be recycled and converted into many value-added and futuristic products that can be potentially used in energy, fuel, and construction [13–15]. Plastics with a nano- or micrometer-sized particle size can be used to make media paints and cosmetics. Both agricultural waste and plastic waste could be attractive raw materials for polymeric membranes, which represent a high-value option for the upcycling of their chemical composition and properties [16].

As depicted in Figure 12.1a, there are many ways to reduce waste and make the economy more sustainable, such as recycling, reusing, and reducing the amount of waste that is produced, based on the European Union's Waste Management Directive's five-step waste hierarchy (Figure 12.1b).

FIGURE 12.1 The illustration of the concept of (a) circular economy and (b) waste hierarchy. (Europa.eu (official website of European Union) [16], with permission from mdpi.)

12.3 ARTIFICIAL INTELLIGENCE IN BIOPOLYMER MEMBRANE: OPTIMIZATION AND SUSTAINABILITY

Operating feed flow and pressure set points for the control system are generated using artificial neural network (ANN) models that attempt to keep the permeate recovery rate within a predetermined range. However, the membrane system was pushed beyond its acceptable operating limits by some instantaneous values of this rate (with a very low occurrence level). Not the ANN models, but the frequency converter and proportional-integral-derivative control algorithms, which regulate the pressure safety valve, are to blame for these permeate recovery rate outliers, (and hence the operating pressure). Permeate recovery rates exceeding acceptable operating limits necessitate a precise adjustment of the parameters of the controllers in order to prevent overheating of membrane systems. As a result, the researchers who wrote the paper ahead of time have already begun the necessary research and testing to meet this objective. Other machine learning techniques should also be examined to see if the error rates generated by the ANNs used can be decreased.

Neural networks (NNs) are mathematical models designed to mimic certain aspects of neurological functioning of the brain. A NN is a parallel structure consisting of nonlinear processing elements (neurons or nodes) interconnected by fixed or variable weights. The nodes are grouped into layers. A typical network consists of an input layer, at least one hidden layer and an output layer. The most widely employed networks have one hidden layer only. For a feedforward NN, the information propagates in one direction only—the forward direction. An example of a three-layer feedforward NN is shown in Figure 12.2. In this case, each node within a given layer

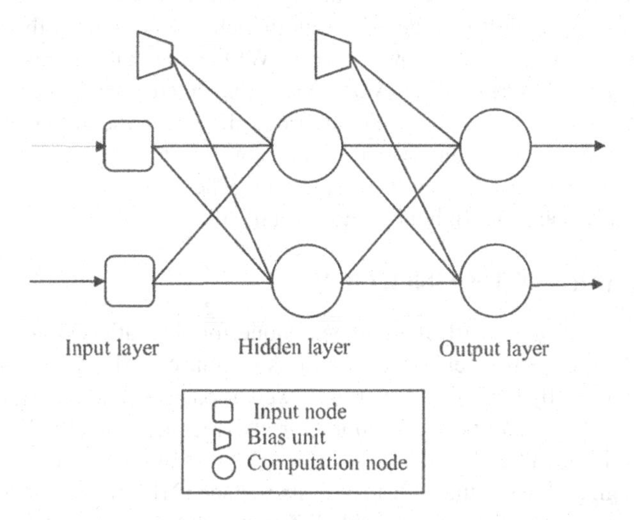

FIGURE 12.2 Structure of a feedforward network with two input nodes, one hidden layer with two nodes and two output nodes [17]. The feed flow and operating pressure setpoints of the membrane can be managed by ANN models. Over the whole spectrum of process variables, such as pressure, solute concentration, temperature, and surface flow velocity, NNs are often used to estimate membrane performance characteristics such as permeate flux and rejection.

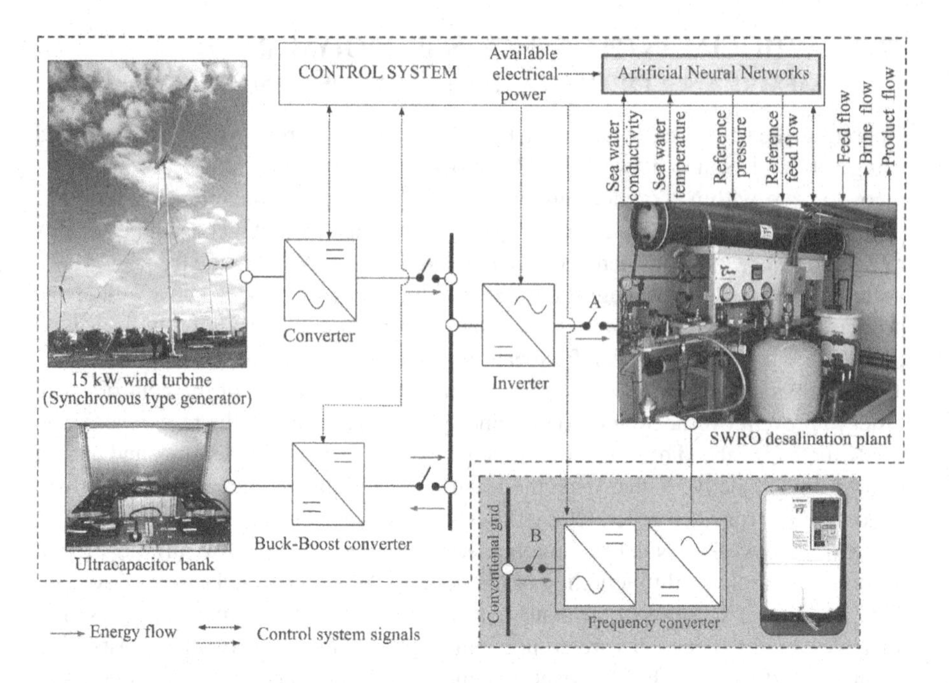

FIGURE 12.3 Seawater reverse osmosis desalination plant [18].

is connected to all the nodes of the previous layer. The node sums up the weighted inputs and a bias and passes the result through a linear or nonlinear function.

Future microgrids will include this plant, which can be seen in the dash line of Figure 12.3. When necessary, however, the SWRO desalination plant was disconnected from the microgrid (switch A open) and connected (switch B closed) through a frequency converter to the conventional grid (indicated in the following figure by a dash-dot line). The flow of energy and water, as well as the input and output signals of the control system, is depicted in this diagram. The control system's ANNs' input and output signals are highlighted in more detail [18].

12.4 MEMBRANE TECHNOLOGY

Membrane technology is still the best technique for separation systems and water treatments. Due to its high efficiency, easiness in control and applications, compact system, economically in time and energy, as well as one phase for separating, it is not needed to time for adsorption, desorption, or also regeneration [19–22]. Membrane phenomenon is considered as the physical barrier that allows the desired materials to pass through and reject the undesired at its surface [21]. The favorable membrane should contain functional materials with different structures to give high permeation rate, high selectivity for different species, long and reliable service life, resistance to fouling, and sufficient resistance to thermal, chemical, and mechanical influences under operating conditions [22]. Membrane technology could widely apply in different fields according to its composition, such as water treatments, petroleum industry, pharmaceuticals, gas purification and separation, food processing, environmental protection, and medical applications [23]. The classification of membranes, according to their composition, is organic (polymeric) membranes and inorganic membranes. The inorganic membranes are prepared from metals or ceramics materials to produce stable mechanical, thermal structural, and high selectivity membranes. However, the use of this membrane is limited due to its lower permeability, making it unfavorable in some applications [24] Polymeric membranes are generally made from organics (inexpensive materials) to form good film possessing favorable properties such as greater flexibility, mechanical strength, chemical stability, high perm selectivity, and desirable pore sizes for various filtration processes. The conventional polymers for the fabrication of polymeric membranes are polyethersulfone (PES), polyacrylonitrile (PAN), polyvinylidene fluoride (PVDF), polyvinyl alcohol (PVA), polypropylene (PP), polyvinyl chloride (PVC), polyimide (PI), polyamide (PA), polyethylene (PE), cellulose acetate (CA), and chitosan [25]. Polymeric membranes are also classified according to their pore size and application in pressure-driven processes as follows: Microfiltration (MF) membrane possesses pore size between 1 μm and 0.1 μm, with applied pressure driven from 0.1 to 2 bar for the separation of zooplankton, algae, opacities, bacteria, and suspension particles (its size more than 0.1 μm). Ultra-filtration (UF) membrane has a pore size between 0.1 μm and 0.01 μm, and the applied driven pressure from 0.1 to 5 bar for the separation of macromolecules, viruses, and colloidal). NF membrane is a dense membrane with pore size between 0.01 μm and 0.001 μm, and the applied pressure between 3 and 20 bar for the separation of organic compounds and divalent ions. Reverse osmosis (RO) membrane is a high dense membrane than NF, and its pore size is less than 0.001 μm while the applied driven pressure starts from 10 to 100 bar for the separation of monovalent ions [26–31].

Over the years, the membrane has become the necessary one for society. The beauty of membrane research is that it is a dynamic process that moves forward. The coming together of the chitosan and membranes is a moment bristling with possibilities and challenges.

12.5 TREATMENT OF PETROLEUM WASTEWATER WITH BIOCHAR MEMBRANES

Increased attention has been paid to biochar as an environmentally friendly and low-cost material made from organic waste such as agricultural or forestry residue or municipal wastes, which has led to an increase in environmental applications. Pyrolysis, hydrothermal carbonization (HTC), gasification, and torrefaction are all methods for converting organic waste into char. Gasification, torrefaction, and HTC chars do not generally meet the definition of biochar specified in the guidelines for the European Biochar Certificate, whereas pyrolysis is the traditional carbonization method (EBC). The enhanced properties of biochar and its activated derivatives, such as high carbon content, enhanced surface area, high cation/anion exchange capacity, and stable structure, have been reported as very efficient materials to remove various contaminants, including pathogenic organisms [32–36], inorganics like heavy metals [37, 38], and organic contaminants like dyes [39, 40]. Batch experiments have shown that biochar-based systems can remove pollutants from drinking water and wastewater, but there is a lack of information on the design and optimization of these systems. Adsorption capacity of biochar is directly linked to its physico-chemical characteristics such as elemental composition, surface area, distribution of pore size, surface functional groups, and cation/anion exchange capacity [41–44]. These physico-chemical properties vary with the nature of the feedstock and the preparation methods and conditions. The properties of biochar should be modulated to allow for better removal efficiency for some recalcitrant molecules, which are present at low concentrations. For biochar modification, the most common methods are divided into two categories: chemical and physical. Chemical methods include acid modification, alkalinity modification, and oxidizing agent modifications, while physical methods include gas purging. Despite the fact that biochar has demonstrated a wide range of potential applications in wastewater remediation, the potential negative effects of its use must also be considered. There are a variety of heavy metals and other contaminants that could be released into aqueous solutions while using biochar, and this varies depending on the feedstock and the conversion technique used to make it [45, 46]. Biochar's stability and its relationship with the experimental conditions used in biochar production need to be studied further. Recent research on biochar pyrolysis and other thermal processes for the removal of organic and inorganic pollutants in wastewater has been summarized in this review. These mechanisms, in addition to recent advances in the use of biochar as filter mediums and catalyst support as well as in wastewater treatment processes that utilize anaerobic digestion, will all be examined [47–53].

Predict reservoir response from injector wells, an intelligent model with the Extreme Gradient Boosting method was created Twenty-model ensemble water flood and a CO_2 flood in a heterogeneous reservoir with complex topography [54]. Figure 12.4 represents the artificial intelligence-assisted history matching workflow for reservoir properties tuning.

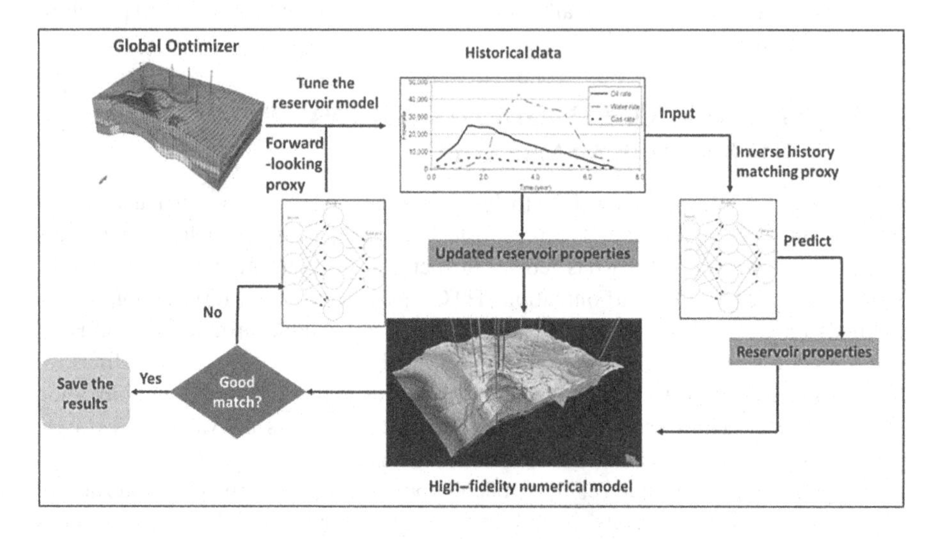

FIGURE 12.4 *Reservoir modeling* outline using *artificial neural network.* (With permission from Elsevier, [54].)

12.6 CONCLUSION

In this chapter, the problem of water scarcity was shown and also valorization of agrowaste in Egypt to save water. Several industries, including water treatment, biopolymer production, energy generation, food production, and pharmaceuticals, have demonstrated the usefulness of agricultural waste.

The use of artificial intelligence (AI) has been successful in areas such as biopolymer membrane for wastewater treatment monitoring, areas like the detection and removal of contaminated wastes, and in detection of levels necessary to start up a process.

The success of NF membranes so far has been owing to their unique ability to selectively separate the two species sought for separation. Based on the progress that has been made thus far, the following aspects are likely to be crucial in terms of NF membrane research and development: Modeling of NF membrane mechanisms for prediction: In essence, the extended Nernst–Planck equation can be used to simulate the transport of an infinite number of ions and dissolved substances over the NF membrane.

Preventing and mitigating fouling, fouling research has been done for a long time, but due to its complexity, it continues to be a concern for membrane application.

The use of biochar as filter mediums and catalyst support as well as in wastewater treatment membranes.

The future perspectives involve the use of artificial intelligence in *Orobanche aegyptiaca extract* for wastewater treatment.

REFERENCES

1. Debnath, B.; Haldar, D.; Purkait, M.K. A critical review on the techniques used for the synthesis and applications of crystallinecellulose derived from agricultural wastes and forest residues. Carbohydr Polym 2021, 273, 118537.
2. Bushra, R.; Mohamad, S.; Alias, Y.; Jin, Y.; Ahmad, M. Current approaches and methodologies to explore the perceptive adsorption mechanism of dyes on low-cost agricultural waste: A review. Microporous Mesoporous Mater 2021, 319, 111040.
3. Freitas, L.C.; Barbosa, J.R.; da Costa, A.L.C.; Bezerra, F.W.F.; Pinto, R.H.H.; de Carvalho Junior, R.N. From waste to sustainable industry: How can agro-industrial wastes help in the development of new products? Resour Conserv Recycl 2021, 169, 105466.
4. Kwikima, M.M.; Mateso, S.; Chebude, Y. Potentials of agricultural wastes as the ultimate alternative adsorbent for Cadmium removal from wastewater. A review Sci Afr 2021, 13, e00934.
5. Lewoyehu, M. Comprehensive review on synthesis and application of activated carbon from agricultural residues for the remediation of venomous pollutants in wastewater. J Anal Appl Pyrolysis 2021, 159, 105279.
6. Solangi, N.H.; Kumar, J.; Mazari, S.A.; Ahmed, S.; Fatima, N.; Mubarak, N.M. Development of fruit waste derived bio-adsorbents for wastewater treatment: A review. J Hazard Mater 2021, 416, 125848.
7. Kadhom, M.; Albayati, N.; Alalwan, H.; Al-Furaiji, M. Removal of dyes by agricultural waste. Sustain Chem Pharm 2020, 16, 100259.
8. Zhang, K.; Zhang, F.; Wu, Y.R. Emerging technologies for conversion of sustainable algal biomass into value-added products: A state-of-the-art review. Sci Total Environ 2021, 784, 147024.

9. Yan, G.; Chen, B.; Zeng, X.; Sun, Y.; Tang, X.; Lin, L. Recent advances on sustainable cellulosic materials for pharmaceutical carrier applications. Carbohydr Polym 2020, 244, 116492.
10. Abdulyekeen, K.A.; Umar, A.A.; Patah, M.F.A.; Daud, W.M.A.W. Torrefaction of biomass: Production of enhanced solid biofuel from municipal solid waste and other types of biomass. Renew Sustain Energy Rev 2021, 150, 111436.
11. Sharma, P.; Gaur, V.K.; Sirohi, R.; Varjani, S.; Hyoun Kim, S.; Wong, J.W.C. Sustainable processing of food waste for production of bio-based products for circular bioeconomy. Bioresour Technol 2021, 325, 124684.
12. Jannat, N.; Hussien, A.; Abdullah, B.; Cotgrave, A. Application of agro and non-agro waste materials for unfired earth blocks construction: A review. Constr Build Mater 2020, 254, 119346.
13. Al-Fakih, A.; Mohammed, B.S.; Liew, M.S.; Nikbakht, E. Incorporation of waste materials in the manufacture of masonry bricks: An update review. J Build Eng 2019, 21, 37–54.
14. Li, L.; Zuo, J.; Duan, X.; Wang, S.; Hu, K.; Chang, R. Impacts and mitigation measures of plastic waste: A critical review. Environ Impact Assess Rev 2021, 90, 106642.
15. Sharma, B.; Shekhar, S.; Sharma, S.; Jain, P. The paradigm in conversion of plastic waste into value added materials. Clean EngTechnol 2021, 4, 100254.
16. Goh, P.S.; Othman, M.H.D.; Matsuura, T. Waste Reutilization in Polymeric Membrane Fabrication: A New Direction in Membranes for Separation. Membranes 2021, 11, 782.
17. Abbas, A.; Al-Bastaki, N. Modeling of an RO water desalination unit using neural networks. Chem. Eng. J. 2005, 114, 139–143.
18. Cabreraa, P.; Cartaa, J.A.; Gonzálezb, J.; Meliánc, G. Artificial neural networks applied to manage the variable operation of a simple seawater reverse osmosis plant. Desalination 416 (2017) 140–156.
19. Baker, R., Future directions of membrane gas-separation technology. Membr. Technol. 2001, 2001(138), 5–10.
20. Xu, D.; Hein, S.; Wang, K., Chitosan membrane in separation applications. Mater. Sci. Technol. 2008, 24(9), 1076–1087.
21. Shi, W.; He, B.; Ding, J.; Li, J.; Yan, F.; Liang, X., Preparation and characterization of the organic–inorganic hybrid membrane for biodiesel production. Bioresour. technol. 2010, 101(5), 1501–1505.
22. Yeo, Z. Y.; Chew, T. L.; Zhu, P. W.; Mohamed, A. R.; Chai, S.-P., Conventional processes and membrane technology for carbon dioxide removal from natural gas: A review. J. Nat. Gas Chem. 2012, 21(3), 282–298.
23. Jafari Sanjari, A.; Asghari, M., A review on chitosan utilization in membrane synthesis. ChemBioEng Rev 2016, 3(3), 134–158.
24. Joshi, R.K.; Alwarappan, S.; Yoshimura, M.; Sahajwalla, V.; Nishina, Y., Graphene oxide: The new membrane material. Appl Mater Today 2015, 1(1), 1–12.
25. Ulbricht, M., Advanced functional polymer membranes. Polymer 2006, 47(7), 2217–2262.
26. Drioli, E.; Fontananova, E., Integrated membrane processes. Membr Ops: Innov Sep Transform 2009, 265–283.
27. Farrell, S.; Sirkar, K. K., Mathematical model of a hybrid dispersed network-membrane-based controlled release system. J Control Rel 2001, 70(1–2), 51–61.
28. Jhaveri, J. H.; Murthy, Z. V. P., A comprehensive review on anti-fouling nanocomposite membranes for pressure driven membrane separation processes. Desalination 2016, 379, 137–154.
29. Kochkodan, V., Johnson, D. J.; Hilal, N., Polymeric membranes: Surface modification for minimizing (bio) colloidal fouling. Adv Colloid Interface Sci 2014, 206, 116–140.

30. Rao, M. B.; Sircar, S., Performance and pore characterization of nanoporous carbon membranes for gas separation. J Membr Sci 1996, 110(1), 109–118.
31. Yampolskii, Y., Polymeric gas separation membranes. Macromolecules 2012, 45(8), 3298–3311.
32. Meyer, S.; Glaser, B.; Quicker, P., Technical, economical, and climate-related aspects of biochar production technologies: A literature review. Environ Sci Technol 2011, 45, 9473–9483.
33. Rizwan, M.; Ali, S.; Qayyum, M.F.; Ibrahim, M.; Ziaurrehman, M.; Abbas, T.; Ok, Y.S. Mechanisms of biochar-mediated alleviation of toxicity of trace elements in plants: A critical review. Environ Sci Pollut Res 2016, 23, 2230–2248.
34. Reddy, K.R.; Xie, T.; Dastgheibi, S. Evaluation of biochar as a potential filter media for the removal of mixed contaminants from urban storm water runoff. J Environ Eng 2014, 140, 04014043.
35. Molaei, R. Pathogen and Indicator Organisms Removal in Artificial Greywater Subjected to Aerobic Treatment. Master's Thesis, Department of Energy and Technology, the Swedish University of Agricultural Science in Uppsala, Uppsala, Sweden, February 2014.
36. Kaetzl, K.; Lübken, M.; Gehring, T.; Wichern, M. Efficient low-cost anaerobic treatment of wastewater using biochar and woodchip filters. Water 2018, 10, 818.
37. Kaetzl, K.; Lübken, M.; Nettmann, E.; Krimmler, S.; Wichern, M. Slow sand filtration of raw wastewater using biochar as an alternative filtration media. Sci Rep 2020, 10, 1229.
38. Yang, W.; Wang, Z.; Song, S.; Han, J.; Chen, H.; Wang, X.; Sun, R.; Cheng, J. Adsorption of copper(II) and lead(II) from seawater using hydrothermal biochar derived from *Enteromorpha*. Mar Pollut Bull 2019, 149, 110586.
39. Gwenzi, W.; Musarurwa, T.; Nyamugafata, P.; Chaukura, N.; Chaparadza, A.; Mbera, S. Adsorption of Zn^{2+} and Ni^{2+} in a binary aqueous solution by biosorbants derived from sawdust and water hyacinth (*Eichhornia crassipes*). Water Sci Technol 2014, 70, 1419–1427.
40. Chen, Y.; Lin, Y.-C.; Ho, S.-H.; Zhouc, Y.; Ren, N. Highly efficient adsorption of dyes by biochar derived from pigments-extracted macroalgae pyrolyzed at different temperature. Bioresour Technol 2018, 259, 104–110.
41. Park, J.-H.; Wang, J.J.; Meng, Y.; Wei, Z.; DeLaune, R.D.; Seo, D.-C. Adsorption/desorption behavior of cationic and anionic dyes by biochars prepared at normal and high pyrolysis temperatures. Colloids Surf APhysicochem Eng Asp 2019, 572, 274–282.
42. Mohan, D.; Sarswat, A.; Ok, Y.S.; Pittman, C.U., Jr. Organic and inorganic contaminants removal from water with biochar, a renewable, low-cost and sustainable adsorbent: A critical review. Bioresour Technol 2014, 160, 191–202.
43. Gai, X.; Wang, H.; Liu, J.; Zhai, L.; Liu, S.; Ren, T.; Liu, H. Effects of feedstock and pyrolysis temperature on biochar adsorption of ammonium and nitrate. PLoS ONE 2014, 9, e113888.
44. Joyce, S.C.; Suzanne, B.; Ted, M.K.; Joseph, M.; Cliff, T.J.; Brad, J. Initial biochar properties related to the removal of As, Se, Pb, Cd, Cu, Ni, and Zn from an acidic suspension. Chemosphere 2017, 170, 216–224.
45. Brassard, P.; Godbout, S.; Raghavan, V. Soil biochar amendment as a climate change mitigation tool: Key parameters and mechanisms involved. J Environ Manag 2016, 181, 484–497.
46. Kim, J.H.; Ok, Y.S.; Choi, G.H.; Park, B.J. Residual perfluorochemicals in the biochar from sewage sludge. Chemosphere 2015, 134, 435–437.
47. Hill, T.; Lewicki, P. Statistics. Methods and Applications, 1st edition, StatSoft, Oklahoma, 2006.
48. J.D. Gibbons, S. Chakraborti, Nonparametric Statistical Inference, 1st edition, CRC Press, London, 2011.

49. M. Hollander, D.A. Wolfe, E. Chicken, Nonparametric Statistical Methods, 3rd edition, John Wiley & Sons, Inc., New Jersey, 2014.
50. G.E.P. Box, D.A. Pierce. Distribution of residual autocorrelations in autoregressiveintegrated moving average time series models. J Am Stat Assoc 1970, 65, 1509–1526.
51. Ljung, G.; Box, G. On a measure of lack of fit in time series models. Biometrika 1978, 65, 297–303.
52. D'Agostino, R.; Stephens, M. Goodness of Fit Techniques, 1st edition, Dekker, New York, 1986.
53. Box, G.; Cox, D. An analysis of transformations. J R Stat Soc 1964, 26, 211–243.
54. Sircar, A.; Yadav, K.; Rayavarapu, K.; Bist, N.; Oza, H. Application of machine learning and artificial intelligence in oil and gas industry. Pet Res 2021, 6, (4), 379–391.

Index

Note: Locators in *italics* represent figures and **bold** indicate tables in the text

C

D

E

H

I

J

K

L

M

Q

R

S

X

Y

Z

Printed in the United States
by Baker & Taylor Publisher Services